T0183910

Lecture Notes in Computer Science 11706

More information about this series at http://www.springer.com/series/7409

Sven Hartmann · Josef Küng ·
Sharma Chakravarthy · Gabriele Anderst-Kotsis ·
A Min Tjoa · Ismail Khalil (Eds.)

Database and Expert Systems Applications

30th International Conference, DEXA 2019
Linz, Austria, August 26–29, 2019
Proceedings, Part I

Springer

Editors
Sven Hartmann
Clausthal University of Technology
Clausthal-Zellerfeld, Germany

Sharma Chakravarthy
The University of Texas at Arlington
Arlington, TX, USA

A Min Tjoa
Software Competence Center Hagenberg
Hagenberg im Mühlkreis, Austria

Josef Küng
Johannes Kepler University of Linz
Linz, Austria

Gabriele Anderst-Kotsis
Johannes Kepler University of Linz
Linz, Austria

Ismail Khalil
Johannes Kepler University of Linz
Linz, Austria

ISSN 0302-9743 ISSN 1611-3349 (electronic)
Lecture Notes in Computer Science
ISBN 978-3-030-27614-0 ISBN 978-3-030-27615-7 (eBook)
https://doi.org/10.1007/978-3-030-27615-7

LNCS Sublibrary: SL3 – Information Systems and Applications, incl. Internet/Web, and HCI

This Springer imprint is published by the registered company Springer Nature Switzerland AG
The registered company address is: Gewerbestrasse 11, 6330 Cham, Switzerland

Preface

This volume contains the papers presented at the 30th International Conference on Database and Expert Systems Applications (DEXA 2019), which was held in Linz, Austria, during August 26–29, 2019. On behalf of the Program Committee, we commend these papers to you and hope you find them useful.

Database, information, and knowledge systems have always been a core subject of computer science. The ever-increasing need to distribute, exchange, and integrate data, information, and knowledge has added further importance to this subject. Advances in the field will help facilitate new avenues of communication, to proliferate interdisciplinary discovery, and to drive innovation and commercial opportunity.

DEXA is an international conference series that showcases state-of-the-art research activities in database, information, and knowledge systems. The conference and its associated workshops provide a premier annual forum to present original research results and to examine advanced applications in the field. The goal is to bring together developers, scientists, and users to extensively discuss requirements, challenges, and solutions in database, information, and knowledge systems.

DEXA 2019 solicited original contributions dealing with any aspect of database, information, and knowledge systems. Suggested topics included, but were not limited to:

- Acquisition, Modeling, Management, and Processing of Knowledge
- Authenticity, Privacy, Security, and Trust
- Availability, Reliability, and Fault Tolerance
- Big Data Management and Analytics
- Consistency, Integrity, Quality of Data
- Constraint Modeling and Processing
- Cloud Computing and Database-as-a-Service
- Database Federation and Integration, Interoperability, Multi-Databases
- Data and Information Networks
- Data and Information Semantics
- Data Integration, Metadata Management, and Interoperability
- Data Structures and Data Management Algorithms
- Database and Information System Architecture and Performance
- Data Streams and Sensor Data
- Data Warehousing
- Decision Support Systems and Their Applications
- Dependability, Reliability, and Fault Tolerance
- Digital Libraries and Multimedia Databases
- Distributed, Parallel, P2P, Grid, and Cloud Databases
- Graph Databases
- Incomplete and Uncertain Data
- Information Retrieval

- Information and Database Systems and Their Applications
- Mobile, Pervasive, and Ubiquitous Data
- Modeling, Automation, and Optimization of Processes
- NoSQL and NewSQL Databases
- Object, Object-Relational, and Deductive Databases
- Provenance of Data and Information
- Semantic Web and Ontologies
- Social Networks, Social Web, Graph, and Personal Information Management
- Statistical and Scientific Databases
- Temporal, Spatial, and High-Dimensional Databases
- Query Processing and Transaction Management
- User Interfaces to Databases and Information Systems
- Visual Data Analytics, Data Mining, and Knowledge Discovery
- WWW and Databases, Web Services
- Workflow Management and Databases
- XML and Semi-Structured Data

Following the call for papers, which attracted 157 submissions, there was a rigorous review process that saw each submission refereed by 3 to 6 international experts. The 32 submissions judged best by the Program Committee were accepted as full research papers, yielding an acceptance rate of 20%. A further 34 submissions were accepted as special research papers.

As is the tradition of DEXA, all accepted papers are published by Springer. Authors of selected papers presented at the conference were invited to submit substantially extended versions of their conference papers for publication in special issues of international journals. The submitted extended versions underwent a further review process.

The success of DEXA 2019 was the result of collegial teamwork from many individuals. We wish to thank all authors who submitted papers and all conference participants for the fruitful discussions.

We are grateful to Dirk Draheim, (Technical University of Tallinn), Vladimir Marik (Technical University of Prague), Axel Polleres (Vienna Business School), and Stefanie Rinderle Ma (University of Vienna) for their keynote talks.

This edition of DEXA also featured four international workshops covering a variety of specialized topics:

- BIOKDD 2019: The 10th International Workshop on Biological Knowledge Discovery from Data
- IWCFS 2019: The Third International Workshop on Cyber-Security and Functional Safety in Cyber-Physical Systems
- MLKgraphs 2019: The First International Workshop on Machine Learning and Knowledge Graphs
- TIR 2019: The 16th International Workshop on Technologies for Information Retrieval

We would like to express our thanks to all institutions actively supporting this event, namely:

- Johannes Kepler University Linz (JKU)
- Software Competence Center Hagenberg (SCCH)
- International Organization for Information Integration and Web based applications and Services (@WAS)

Finally, we hope that all the participants of DEXA 2019 enjoyed the program that was put together.

August 2019

Sven Hartmann
Josef Küng
Sharma Chakravarthy

Organization

General Chair

A Min Tjoa	Technical University of Vienna, Austria

Program Committee Chairs

Sharma Chakravarthy	University of Texas at Arlington, USA
Sven Hartmann	Clausthal University of Technology, Germany
Josef Küng	Johannes Kepler University Linz, Austria

Steering Committee

Gabriele Anderst-Kotsis	Johannes Kepler University Linz, Austria
A Min Tjoa	Software Competence Center Hagenberg, Austria
Ismail Khalil	Johannes Kepler University Linz, Austria

Program Committee and Reviewers

Sonali Agarwal	Indian Institute of Information Technology Allahabad, India
Riccardo Albertoni	Institute of Applied Mathematics and Information Technologies, Italian National Council of Research, Italy
Idir Amine Amarouche	University Houari Boumediene, Algeria
Rachid Anane	Coventry University, UK
Mustafa Atay	Winston-Salem State University, USA
Faten Atigui	CNAM, France
Ladjel Bellatreche	ENSMA, France
Nadia Bennani	INSA Lyon, France
Karim Benouaret	Université Claude Bernard Lyon 1, France
Djamal Benslimane	Lyon 1 University, France
Morad Benyoucef	University of Ottawa, Canada
Mikael Berndtsson	University of Skovde, Sweden
Catherine Berrut	Grenoble University, France
Vasudha Bhatnagar	Delhi University, India
Athman Bouguettaya	University of Sydney, Australia
Omar Boussaid	University of Lyon/Lyon 2, France
Stephane Bressan	National University of Singapore, Singapore
Barbara Catania	DISI, University of Genoa, Italy
Sharma Chakravarthy	The University of Texas at Arlington, USA
Cindy Chen	University of Massachusetts Lowell, USA

Max Chevalier	IRIT - SIG, Université de Toulouse, France
Soon Ae Chun	City University of New York, USA
Alfredo Cuzzocrea	University of Trieste, Italy
Deborah Dahl	Conversational Technologies, USA
Jérôme Darmont	Université de Lyon (ERIC Lyon 2), France
Soumyava Das	Teradata, USA
Vincenzo Deufemia	Università degli Studi di Salerno, Italy
Juliette Dibie-Barthélemy	AgroParisTech, France
Dejing Dou	University of Oregon, USA
Cedric du Mouza	CNAM, France
Johann Eder	University of Klagenfurt, Austria
Suzanne Embury	The University of Manchester, UK
Markus Endres	University of Augsburg, Germany
Noura Faci	Lyon 1 University, France
Bettina Fazzinga	ICAR-CNR, Italy
Stefano Ferilli	University of Bari, Italy
Flavio Ferrarotti	Software Competence Center Hagenberg, Austria
Vladimir Fomichov	School of Business Informatics, National Research University Higher School of Economics, Russia
Flavius Frasincar	Erasmus University Rotterdam, The Netherlands
Bernhard Freudenthaler	Software Competence Center Hagenberg GmbH, Austria
Steven Furnell	Plymouth University, UK
Joy Garfield	University of Worcester, UK
Claudio Gennaro	ISTI-CNR, Italy
Manolis Gergatsoulis	Ionian University, Greece
Javad Ghofrani	HTW Dresden University of Applied Sciences, Germany
Vikram Goyal	IIIT Delhi, India
Carmine Gravino	University of Salerno, Italy
Sven Groppe	Lübeck University, Germany
William Grosky	University of Michigan, USA
Francesco Guerra	Università degli Studi Di Modena e Reggio Emilia, Italy
Giovanna Guerrini	University of Genova, Italy
Allel Hadjali	ENSMA, France
Abdelkader Hameurlain	Paul Sabatier University, France
Ibrahim Hamidah	Universiti Putra Malaysia, Malaysia
Takahiro Hara	Osaka University, Japan
Ionut Emil Iacob	Georgia Southern University, USA
Sergio Ilarri	University of Zaragoza, Spain
Abdessamad Imine	INRIA Grand Nancy, France
Yasunori Ishihara	Nanzan University, Japan
Peiquan Jin	University of Science and Technology of China, China
Anne Kao	Boeing, USA
Dimitris Karagiannis	University of Vienna, Austria

Stefan Katzenbeisser	University of Passau, Germany
Anne Kayem	Hasso-Plattner-Institute, Germany
Uday Kiran Rage	University of Tokyo, Japan
Carsten Kleiner	University of Applied Sciences and Arts Hannover, Germany
Henning Koehler	Massey University, New Zealand
Michal Krátký	Technical University of Ostrava, Czech Republic
Petr Kremen	Czech Technical University in Prague, Czech Republic
Anne Laurent	LIRMM, University of Montpellier 2, France
Lenka Lhotska	Czech Technical University, Czech Republic
Wenxin Liang	Dalian University of Technology, China
Chuan-Ming Liu	National Taipei University of Technology, Taiwan
Hong-Cheu Liu	University of South Australia, Australia
Jorge Lloret Gazo	University of Zaragoza, Spain
Hui Ma	Victoria University of Wellington, New Zealand
Qiang Ma	Kyoto University, Japan
Zakaria Maamar	Zayed University, UAE
Elio Masciari	ICAR-CNR, Università della Calabria, Italy
Brahim Medjahed	University of Michigan, USA
Jun Miyazaki	Tokyo Institute of Technology, Japan
Lars Moench	University of Hagen, Germany
Riad Mokadem	IRIT, Paul Sabatier University, France
Anirban Mondal	Ashoka University, India
Yang-Sae Moon	Kangwon National University, Republic of Korea
Franck Morvan	IRIT, Paul Sabatier University, France
Francesc Munoz-Escoi	Universitat Politècnica de València, Spain
Ismael Navas-Delgado	University of Málaga, Spain
Wilfred Ng	Hong Kong University of Science and Technology, SAR China
Javier Nieves Acedo	IK4-Azterlan, Spain
Marcin Paprzycki	Polish Academy of Sciences, Warsaw Management Academy, Poland
Oscar Pastor Lopez	Universitat Politècnica de València, Spain
Clara Pizzuti	Institute for High Performance Computing and Networking, National Research Council, Italy
Elaheh Pourabbas	National Research Council, Italy
Rodolfo Resende	Federal University of Minas Gerais, Brazil
Claudia Roncancio	Grenoble University, LIG, France
Viera Rozinajova	Slovak University of Technology in Bratislava, Slovakia
Massimo Ruffolo	ICAR-CNR, Italy
Marinette Savonnet	University of Burgundy, France
Florence Sedes	IRIT, Paul Sabatier University, France
Nazha Selmaoui	University of New Caledonia, New Caledonia
Michael Sheng	Macquarie University, Australia
Patrick Siarry	Université Paris 12, LiSSi, France

Hala Skaf-Molli	Nantes University, France
Srinivasa Srinath	IIIT Bangalore, India
Bala Srinivasan (Retried)	Monash University, Australia
Olivier Teste	IRIT, University of Toulouse, France
Stephanie Teufel	University of Fribourg, Switzerland
Jukka Teuhola	University of Turku, Finland
Jean-Marc Thevenin	University of Toulouse 1 Capitole, France
Vicenc Torra	Maynooth University, Ireland
Traian Marius Truta	Northern Kentucky University, USA
Lucia Vaira	University of Salento, Italy
Krishnamurthy Vidyasankar	Memorial University of Newfoundland, Canada
Marco Vieira	University of Coimbra, Portugal
Ming Hour Yang	Chung Yuan Christian University, Taiwan
Xiaochun Yang	Northeastern University, China
Haruo Yokota	Tokyo Institute of Technology, Japan
Qiang Zhu	The University of Michigan, USA
Yan Zhu	Southwest Jiaotong University, China
Marcin Zimniak	Leipzig University, Germany
Ester Zumpano	University of Calabria, Italy

Organizers

Abstracts of Keynote Talks

AI in Manufacturing

Vladimír Mařík

CIIRC, Czech Technical University,
Jugoslávských partyzánů 3, 160 00 Prague 6, Czech Republic
marik@cvut.cz

Abstract. The Industry 4.0 paradigm strongly influences industrial manufacturing area and highlights Artificial Intelligence as a driving force for increasing production efficiency and developing new business models. Intelligent networks of virtual twins of all physical units engaged in production enable smart system integration as well as permanent system optimization. The agent-based approaches influence both the system and the SW architectures (typically SOA) and lead to such novel concepts like production as a service, smart services of products and smart services provided to products during the whole life-cycle etc. AI has been deployed in many critical tasks like planning and scheduling, big data analytics, optimization by negotiations, learning from experience, predictive maintenance, and system resilience. Experimental testbeds do play the key role in developing new manufacturing solutions. System integration and human-machine interfaces are supported by virtual and augmented reality tools. Interesting industrial use cases will be presented and discussed.

Keywords. Artificial Intelligence · Agent-based solutions · Virtual twins · System integration · System optimization · Production planning and scheduling · Testbeds

Future Perspectives of Association Rule Mining Based on Partial Conditionalization

Dirk Draheim

Tallinn University of Technology, Akadeemia tee 15a, 12618 Tallinn, Estonia
dirk.draheim@taltech.ee

Abstract. In today's practice, major chunks of data analytics are still done rather interactively (OLAP, MOLAP, with tools such as Cognos or SAP-BW, and with related techniques such as conjoint analysis); i.e., they lack the exploitation of machine learning (AI) approaches. A middle position with respect to this gap between interactive and automatic data analytics is taken by association rule mining, which explains the current success of tools such as Rapidminer. Unfortunately, current association rule mining suffers two categories of shortcomings (both theoretically and practically, i.e., in the tool landscape). First, association rule mining works only for discrete-valued columns, i.e., numerical-valued columns cannot be handled, which is a painful shortcoming in practical scenarios. What is needed, is what we call a 'grand pivot report', i.e., a full combinatorial multi-factor analysis against common aggregate functions. Such grand reports needed to come with appropriate, generalized measures of interestingness. Second, and even more critical, today's analysts are not systematically supported in avoiding all kinds of data misinterpretations such as the Yule-Simpson effect. In this talk we discuss, how partial conditionalization can be exploited to generalize association rule mining from conditional probabilities to conditional expected values, including the necessary theoretical measures of interest that are (i) adequate to support multi-factor impact analysis and are (ii) explicitly robust against data misinterpretation. (The theoretical foundations of this endeavor are laid down in the book "Generalized Jeffrey Conditionalization" that can be downloaded for free from: https://www.ttu.ee/?id=170467).

Process Technology as Key Driver for Smart Manufacturing

Stefanie Rinderle-Ma

Faculty of Computer Science, University of Vienna
stefanie.rinderle-ma@univie.ac.at

Abstract. The keynote reports on the journey of process technology in smart manufacturing: from connecting and controlling machines in the EU project ADVENTURE, over developing a modular LEGO factory demonstrating Industry 4.0 concepts to centurio.work [1] – the manufacturing orchestration engine applied in real-world settings. On this journey we had to face several challenges that arise from the interplay of machines, data, systems, and humans. Examples comprise bridging the gap between software and the physical world, finding the right granularity for controlling the machines, connecting the machines along the process logic (horizontal integration), and integrating systems across the levels of the automation pyramid. Particularly horizontal and vertical integration bear several benefits, for example, the possibility to collect data not only in an isolated manner, e.g., per machine or per system, but in a connected way, following the process logic. This opens the door for advanced analysis of manufacturing data, employing a mix of techniques from cross-sectional and process mining [2].

References

1. Pauker, F., Mangler, J., Rinderle-Ma, S., Pollak, C: centurio.work - Modular secure manufacturing orchestration. In: BPM Industrial Track, pp. 164–171 (2018). http://ceur-ws.org/Vol-2196/BPM_2018_paper_33.pdf
2. Ehrendorfer, M., Fassmann, J.-A., Mangler, J., Rinderle-Ma, S.: Combining Conformance Checking and Classification of XES Log Data for the Manufacturing Domain. CoRR abs/1904.05883 (2019)

How Do Linked Data, Open Data, and Knowledge Graphs Interplay?

Axel Polleres

Institute for Information Business, WU Vienna
axel.polleres@wu.ac.at

Abstract. "A Knowledge Graphs is a graph of data with the intent to compose knowledge" ... this admit-tedly vague definition was the result of a recent Dagstuhl Seminar on the trending topic of Knowledge Graphs in industry and academia and its meaning for Knowledge Representation on the (Semantic) Web. In this talk I will tackle the question of what's actually new about Knowledge Graphs and how they can help us to get closer to a more natural understanding of data, particularly Open Data on the Web, from a personal perspective of +15 years of research on Semantic Web standards, Linked Data, and Open Data.

Contents – Part I

Big Data Management and Analytics

Data Structures and Data Management

Management and Processing of Knowledge

Contents – Part II

Distributed, Parallel, P2P, Grid and Cloud Databases

Information Retrieval

Temporal, Spatial, and High Dimensional Databases

Knowledge Discovery

Web Services

Big Data Management and Analytics

Optimization of Row Pattern Matching over Sequence Data in Spark SQL

Kosuke Nakabasami[1,3](✉), Hiroyuki Kitagawa[2](✉), and Yuya Nasu[3](✉)

[1] Railway Technical Research Institute,
Hikari-cho 2-8-38, Kokubunji-shi, Tokyo, Japan
nakabasami.kosuke.39@rtri.or.jp

[2] Center for Computational Sciences, University of Tsukuba,
Tennodai 1-1-1, Tsukuba-shi, Ibaraki, Japan
kitagawa@cs.tsukuba.ac.jp

[3] Graduate School of Systems and Information Engineering, University of Tsukuba,
Tennodai 1-1-1, Tsukuba-shi, Ibaraki, Japan
{nakabasami,yuuya.n}@kde.cs.tsukuba.ac.jp

Abstract. Due to the advance of information and communications technology and sensor technology, a large quantity of sequence data (time series data, log data, etc.) are generated and processed every day. Row pattern matching for the sequence data stored in relational databases was standardized as SQL/RPR in 2016. Today, in addition to relational databases, there are many frameworks for processing a large amount of data in parallel and distributed computing environments. They include MapReduce and Spark. Hive and Spark SQL enable us to code data analysis processes in SQL-like query languages. Row pattern matching is also beneficial in Hive and Spark SQL. However, computational cost of the row pattern matching process is large and it is needed to make this process efficient. In this paper, we propose two optimization methods to realize the reduction of computational cost for row pattern matching process. We focus on Spark and show design and implementation of the proposed methods for Spark SQL. We verify by the experiments that our optimization methods really contribute to the reduction of the processing time of Spark SQL queries including row pattern matching.

Keywords: Sequence data · Pattern matching · Row Pattern Recognition · MATCH_RECOGNIZE · Spark SQL · Optimization

1 Introduction

Due to the advance of information and communications technology and sensor technology, a large amount of various data is generated and transmitted over the internet. Then, a huge amount of data are stored and processed. They include sequence data such as stock market data, power consumption data, person trip data, and so on. In the analysis of such sequence data, row pattern matching is essential. Row pattern matching is a method of detecting pattern occurrences in

S. Hartmann et al. (Eds.): DEXA 2019, LNCS 11706, pp. 3–17, 2019.
https://doi.org/10.1007/978-3-030-27615-7_1

a sequence of rows. Typically, each row in the sequence data is a tuple composed of the sequence identifier (we call it partition attribute) like stock-id or person-id, an attribute representing the row order like time, and other attributes. By partitioning data by the partition attribute and sorting them by the ordering attribute, we get sequences of rows. When the user defines a pattern of the row sequence including attribute value conditions, row pattern matching extracts all occurrences of the given pattern, namely subsequences of rows. Row pattern matching is essentially important in sophisticated sequence data analysis.

The row pattern matching to the sequence data stored in relational databases was standardized in SQL2016 as SQL/RPR (Row Pattern Recognition) [1]. SQL/RPR is the extension of SQL for realizing row pattern matching. The user uses MATCH_RECOGNIZE clause, which defines the pattern to be extracted from the sequence.

Today, in addition to relational databases, there are many frameworks for processing a large amount of data in parallel and distributed computer environments. Hadoop [6] and Spark [14] are well known frameworks to support such big data analysis. Hive [10] and Spark SQL [3] enable us to code data analysis processes in SQL-like query languages. Row pattern matching is also very beneficial in Hive and Spark SQL in analyzing big sequence data. Therefore, the row pattern matching feature similar to SQL/RPR will be incorporated in Hive and Spark SQL as well. Actually, naive implementation of the row pattern matching takes much time in query processing as we demonstrate in Sect. 5, and we need to make this process more efficient. In this paper, we focus on Spark. Spark is a well known and standard framework for big data analysis. Therefore, realizing efficient row pattern matching in Spark SQL is an important technical issue.

We assume SQL/RPR is incorporated into Spark SQL, and implement this SQL/RPR extension by the front-end approach. We propose two optimization methods to efficiently execute the row pattern matching. More specifically, we provide the front-end module, which scans the Spark application codes including the extended Spark SQL calls using MATCH_RECOGNIZE clause and rewrite them into codes which include row pattern matching but runs on Spark. In addition, we propose two optimization methods for filtering input data using conditions included in MATCH_RECOGNIZE clause. One is *Sequence Filtering*, which filters out sequences whose rows do not satisfy the conditions in the MATCH_RECOGNIZE clause. Another is *Row Filtering*, which filters out rows which do not satisfy the conditions from the sequence.

The organization of the remaining part of this paper is as follows. We discuss related work in Sect. 2. Section 3 explains the MATCH_RECOGNIZE clause standardized in SQL/RPR. In Sect. 4, we explain the front-end approach to process the extended Spark SQL queries including the MATCH_RECOGNIZE clause and the processing flows. In Sect. 5, we show the processing cost of row pattern matching. Section 6 presents the proposed optimization methods: Sequence Filtering and Row Filtering. In Sect. 7, we show the experimental evaluation. Section 8 concludes this paper and discusses future works.

2 Related Work

Complex Event Processing (CEP) is a well-known pattern matching method to stream data. Examples are Cayuga, SASE, and ZStream. Cayuga [5] and SASE [11] do the pattern matching by using Non-deterministic Finite Automaton (NFA). Jagrati et al. proposed the processing method of the pattern matching using NFA and its optimization [2]. ZStream [9] adopts an original method using tree-based query plans to do the pattern matching. Bruno et al. suggest a row pattern matching method including preprocessing phase and pattern matching phase [4]. The method of [4] discusses data filtering in the context of event pattern matching. Their problem is different from ours in that we consider efficient row pattern matching over sequence data stored in Spark. In addition, they do not consider sequence filtering nor allow pattern variables without conditions.

Oracle implements MATCH_RECOGNIZE clause in their RDBMS and employs some optimization methods for efficient execution of row pattern matching [8]. They include selection of efficient sort algorithms, optimization of the query plan, and selection of DFA or NFA. Their optimization of the query plan is to push down selection conditions in WHERE clause which are relevant to the attribute designated in PARTITION BY in MATCH_RECOGNIZE clause. By doing this, some sequences are filtered out.

Our proposed optimization methods are different from the above optimization method in the following two points. The first point is the attribute used in filtering the sequences. In the above method, sequences are filtered only by the conditions on partition attributes. In the proposed methods, all the attributes can be used in filtering the sequences. The second point is row filtering. The above method filters out only sequences. The proposed methods filter out unnecessary rows in sequences which pass sequence filtering as well. Therefore, our methods can reduce more input data for row pattern matching and can reduce processing cost of pattern matching.

3 MATCH_RECOGNIZE in SQL/RPR

We explain the MATCH_RECOGNIZE clause, which was newly introduced in SQL/RPR. This clause enables the user to specify the row pattern matching over the sequence data and get a pattern occurrences table as the query output.

The syntax of MATCH_RECOGNIZE clause is shown in Figs. 1 and 2 represents an example query using MATCH_RECOGNIZE clause. The example query of Fig. 2 applies row pattern matching to the table of a person's trip data (moving_table), which has 3 attributes. The first attribute represents a person_id (identifying each person), the second one represents time, and the third one represents a location. MATCH_RECOGNIZE clause is specified within FROM clause. The <table name> before MATCH_RECOGNIZE is the target table storing the sequence data. In Fig. 2, moving_table is the target table. MATCH_RECOGNIZE clause specifies how to partition the whole sequence data into sequences in PARTITION BY clause. Based on the attributes specified

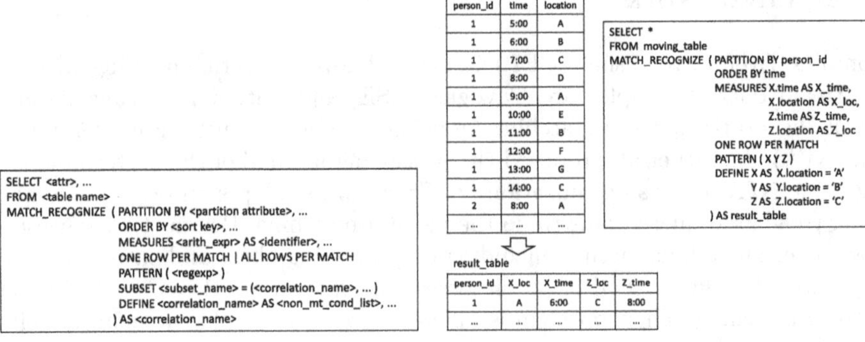

Fig. 1. Syntax of MATCH_RECOGNIZE clause

Fig. 2. Example query using MATCH_RECOGNIZE clause

in this clause, each sequence is formed. In Fig. 2, person_id is specified and a sequence is formed from trip log for each person. The order of rows within a sequence is decided by attributes in ORDER BY clause. In Fig. 2, time is specified and rows in each trip log sequence are ordered by time. The user defines a pattern that he/she wants to extract by using PATTERN clause. To define a pattern in PATTERN clause, the user uses *pattern variables.* Each pattern variable is defined as an arbitrary string and it corresponds to each row which he/she wants to match. The pattern variables and their regular expression are arranged for the desired row pattern. In Fig. 2, the sequential moving pattern of A -> B -> C is defined. In DEFINE clause, the user defines conditions for each pattern variables in PATTERN clause. Each attribute of a row is represented by "<pattern variable>.<attribute name>", and the user defines conditions for it. In Fig. 2, the pattern variables X, Y, Z are defied. DEFINE clause can include two kinds of conditions for the pattern variables. One is a condition comparing some attribute of a pattern variable with values (i.e., real number, string, and date). We call them *value conditions.* Another is a condition comparing attributes of two pattern variables. We call them *pattern variables conditions.* An example of a value condition is shown in Fig. 2. In DEFINE clause in Fig. 2, the location attribute is compared with a string of the location name such as "A". An example of the pattern variables condition is "Y AS Y.time < X.time + 3600". This condition says that the time attribute of Y must be less than the time attribute of X plus 3600 s. In MEASURES clause, the user defines the measures, which are attributes of the output *pattern occurrence table*, by using "<pattern variable>.<attribute name>" format, aggregation function, etc. In Fig. 2, the location and the time of the rows which match the pattern variable X and Z are specified as the measures.

The processing flow of MATCH_RECOGNIZE clause is as follows. First, the sequence of rows are formed by using the attributes defined in PARTITION BY clause and ORDER BY clause with respect to the input data defined in FROM clause. Then, the row pattern matching is executed by using DFA or NFA based

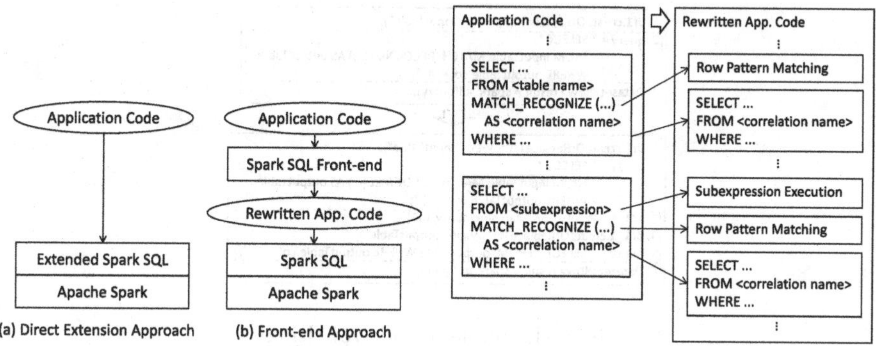

Fig. 3. Architecture for implementation of MATCH_RECOGNIZE clause

Fig. 4. Overview of query rewriting

on the information defined in PATTERN clause and DEFINE clause. Finally, the pattern occurrence table with the attributes defined in MEASURES clause is output.

4 Architecture for Row Pattern Matching on Spark

In this study, we assume SQL/RPR is incorporated into Spark SQL, and discuss design and implementation of the SQL/RPR extension. This section explains architecture to realize the SQL/RPR extension on Spark by using the front-end approach.

4.1 Overall System Architecture

Spark SQL is a module to code processes for structured data analysis using SQL-like queries on Spark. We consider system architecture to execute Spark SQL queries including MATCH_RECOGNIZE clause. There are two alternatives as shown in Fig. 3. One is to directly modify the current implementation of Spark SQL to accept the SQL/RPR extension. We call it Direct Extension Approach. The other is the Front-end Approach. We prepare the Front-end module, which analyzes Spark application codes which include row pattern matching specified in MATCH_RECOGNIZE clause and rewrites them into ones which run on the ordinary Spark.

In this paper, we take the Front-end approach. The reason is that it is easier to implement since we do not need to touch details of the current implementation of Spark SQL. However, the proposed optimization methods are useful even when we take the Direct Extension Approach.

4.2 Query Rewriting

We explain rewriting flow of Spark codes in the Spark SQL Front-end Module. Figure 4 gives overview of query rewriting. The Front-end module scans

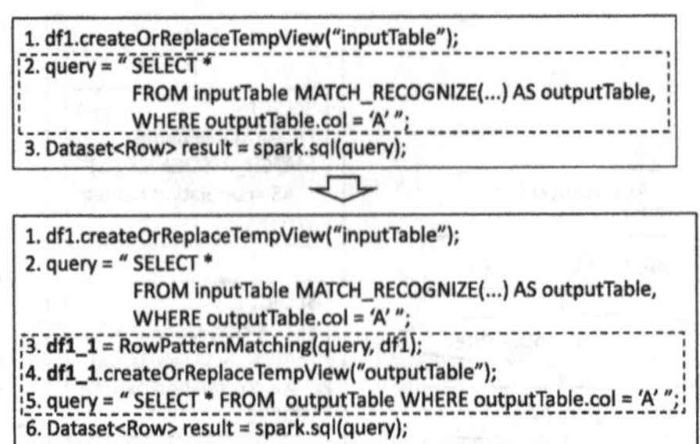

Fig. 5. Example of code rewriting

Algorithm 1. Pseudocode of Algorithm of Rewriting

Input: Application Code c
Output: Rewritten App. Code

1: **while** *scan* c **do**
2: $q \leftarrow searchQuery(c)$
3: $insert(c, makeCode(q))$
4: **end while**
5: **return** c

the application code including the extended Spark SQL calls. When the MATCH_RECOGNIZE clause is found in the query, the Front-end module inserts two components. The first one is the code executing the row pattern matching. The second one is rewritten Spark SQL query without the MATCH_RECOGNIZE clause. In its FROM clause, it designates the correlation name of the table obtained as the result of the row pattern matching. If MATCH_RECOGNIZE is appended to a subexpression to derive some tables, the inserting involves an additional component which corresponds to the subexpression and derives the target table for row pattern matching. It needs to be evaluated before the row pattern matching. Through these steps, the rewritten application code is prepared as shown in Fig. 4.

We explain the flow of the code rewriting in more detail. We use the query in Fig. 5 as an example. Algorithm 1 shows the pseudo-code of rewriting. From line 1 to 4, it scans the application code. In line 2, when it finds a query including MATCH_RECOGNIZE clause, it stores the query into q. In Fig. 5, the codes surrounded by the dotted line are stored into q. In line 3, the function makeCode() generates the executable code in Spark from the query stored in q and the function insert() inserts the code executable on the ordinary Spark after the original query including MATCH_RECOGNIZE clause.

Algorithm 2 shows the detail of the makeCode(). In line 2, the empty rewritten code *MRcodes* is prepared. In line 3, the specification corresponding to MATCH_RECOGNIZE clause in the query (we call it MATCH_RECOGNIZE specification) is extracted by the function extractMRspec(). MATCH_breakRECOGNIZE clause appears in FROM clause as follows: " <table

Algorithm 2. Pseudocode of makeCode()

```
1: function makeCode(q)
2:     MRcodes ← null
3:     mr ← extractMRspec(q)
4:     if subExp(mr) = true then
5:         MRcodes ← add(execSubExp(mr))
6:     end if
7:     MRcodes ← add(execRPR(mr))
8:     MRcodes ← add(rewrite(q))
9:     return MRcodes
10: end function
```

name> MATCH_RECOGNIZE(...) AS <correlation name>". Using this extracted MATCH_RECOGNIZE specification, the code executing row pattern matching is generated. In line 7, Nondeterministic Finite Automaton (NFA) is constructed and the code executing row pattern matching is inserted to *MRcodes*. In Fig. 5, this code corresponds to line 3 in the rewritten code. The name of the dataframe or the subquery shown in bold letters in Fig. 5 is assigned by the Front-end module automatically. In line 8, the query which eliminates MATCH_RECOGNIZE specification from the query stored in q is compiled and is inserted to *MRcodes*. Specifically, table name specified in FROM clause is replaced with correlation name in MATCH_RECOGNIZE clause. In Fig. 5, it corresponds to line 5 in the rewritten code. Since Spark needs the view of input table for query execution, the function createOrReplaceTempView() is also inserted. In Fig. 5, it corresponds to line 4 in the rewritten code.

If the query has a subexpression, in Algorithm 2, the code from line 4 to 6 is executed. The function execSubExp() generates the code corresponding to the subexpression and inserts to the top of *MRcodes*.

5 Processing Cost of Row Pattern Matching

This section shows preliminary experiments to evaluate the processing cost of row pattern matching for sequence data. We compare the processing cost of row pattern matching queries with selection queries and join queries. All queries are executed on Spark. The query including MATCH_RECOGNIZE clause is implemented following the architecture of Front-end approach in Fig. 3(b).

Now, we use the table (test_table) which has three attributes c1, c2, and c3. c1 starts from 1 and is incremented by 1 by every 10,000 rows. c2 starts from 1 and is incremented by 1 by each row. If c2 reaches 10,000, it restarts from 1. c3 is an attribute of the character, A, B, and C. We compare the processing time of three queries shown in Fig. 6. Selection Query selects rows whose c3 value is "A". Join Query joins two tables. test_table2 has 10,000 rows. MATCH_RECOGNIZE Query extracts the sequential pattern "A", "B", and "C" in c3. In this experiment, we use a cluster machine composed of six nodes

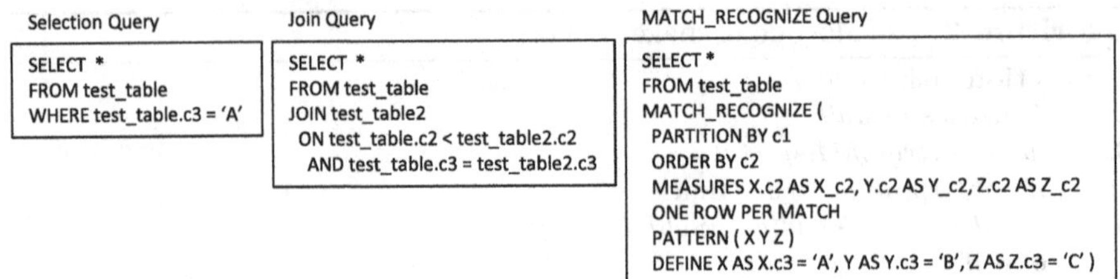
Selection Query

```
SELECT *
FROM test_table
WHERE test_table.c3 = 'A'
```

Join Query

```
SELECT *
FROM test_table
JOIN test_table2
  ON test_table.c2 < test_table2.c2
   AND test_table.c3 = test_table2.c3
```

MATCH_RECOGNIZE Query

```
SELECT *
FROM test_table
MATCH_RECOGNIZE (
 PARTITION BY c1
 ORDER BY c2
 MEASURES X.c2 AS X_c2, Y.c2 AS Y_c2, Z.c2 AS Z_c2
 ONE ROW PER MATCH
 PATTERN ( X Y Z )
 DEFINE X AS X.c3 = 'A', Y AS Y.c3 = 'B', Z AS Z.c3 = 'C' )
```

Fig. 6. Three queries

Table 1. Processing time of three queries

n (# of rows)	Selection query (s)		Join query (s)		MATCH_RECOGNIZE query (s)	
10,000	41.52	(7.55)	46.09	(12.34)	67.04	(33.20)
100,000	41.79	(8.40)	61.77	(28.26)	73.41	(39.57)
1,000,000	43.27	(9.76)	113.92	(80.41)	105.09	(71.54)
10,000,000	46.58	(13.29)	304.16	(270.49)	438.559	(405.13)

with AMD Opteron^{TM}Processor 2435 at 2.60 GHz and 8 GB RAM operating under Ubuntu 14.04 LTS.

In the cluster, one node works as a master node and the other five nodes serve as worker nodes. We use HDFS as a distributed file system and use YARN as a resource manager.

Table 1 shows the results of the overall processing time and the processing time of query processes for different numbers of rows in test_table. (The latter is shown in parenthesis.) These values are the average of five measurements. The processing time of Selection Query is almost stable. However, the more rows, the longer processing times Join Query and MATCH_RECOGNIZE Query take. This result suggests that row pattern matching is costly and making the row pattern matching efficient will result in much reduction of the total processing time.

6 Optimization of Pattern Matching over Sequence Data

6.1 Optimization for Efficient Pattern Matching

We propose optimization to make the row pattern matching more efficient. Basic idea of our proposal is to reduce the number of rows before the row pattern matching by filtering out non-relevant rows which do not contribute to the results. More concretely, we propose two optimization methods. The first one is reducing the number of the sequences (called *Sequence Filtering*) and the second one is reducing the number of the rows (called *Row Filtering*).

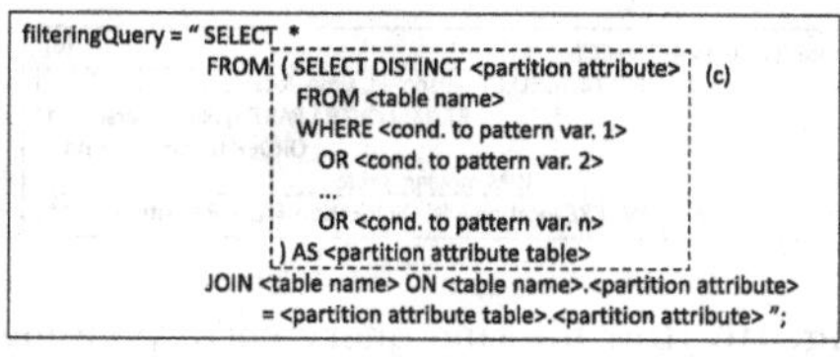

Fig. 7. Sequence filtering query

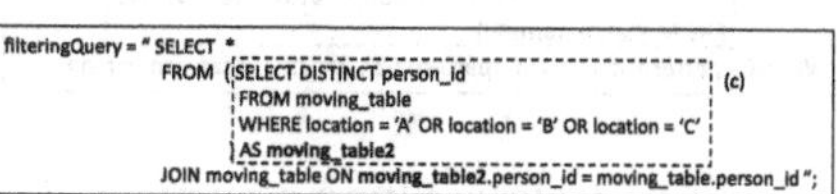

Fig. 8. Sequence filtering query corresponding to MATCH_RECOGNIZE specification in Fig. 2

Sequence Filtering. In Sequence Filtering, we focus on each sequence formed by the partition attribute. If there is no row meeting the value conditions defined in DEFINE clause in a given sequence, the whole sequence can be filtered out before the row pattern matching. By filtering out these sequences, the processing cost can be reduced. In this paper, we focus on only value conditions for simplicity since filtering out rows using pattern variable conditions is more sophisticated. We leave it as a future research issue. The condition to apply Sequence Filtering is that "there is at least one value condition for some pattern variables in DEFINE clause".

Sequence Filtering is done by a query named *Sequence Filtering query*. Let us consider the query in Fig. 2. It contains value conditions for pattern variables X, Y, Z, and we can apply Sequence Filtering. Figure 7 shows a template for the Sequence Filtering query. Part (c) constructs a table which contain partition attribute values of rows meeting the value conditions. After that, it is joined with the input table. Therefore, sequences which do not contain rows meeting the value conditions are eliminated. The row pattern matching is done after this filtering. Figure 8 shows Sequence Filtering query for the original query in Fig. 2. Here, only person_id values of rows which meet value conditions are selected in part (c).

Row Filtering. In Sequence Filtering, sequences which may contain row pattern occurences are retained, and the other sequences are eliminated. In other words, sequences are ftering units. Row filtering eliminate rows which cannot contribute to the row pattern matching results. Rows are filtering units in Row Filtering. The condition to apply Row Filtering is that "there is at least one value condition for all pattern variables in DEFINE clause".

Compared to the condition for Sequence filtering, it is more restircted. If Row Filtering is applicable, Sequence Filtering is also applicable but not vice versa.

In SQL/RPR, rows composing a pattern occurrence must be adjacent to each other in the input table. If Row Filtering filters out rows which do not meet value conditions in DEFINE clause, it becomes difficult to tell whether the remaining rows reside adjacent to each other in the original input table. To address this problem, we append an additional attribute which represents the sequential row number in the input table before Row Filtering. After Row Filtering, NFA is

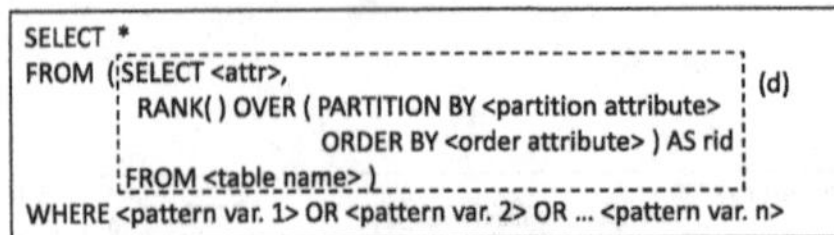

Fig. 9. Row filtering query

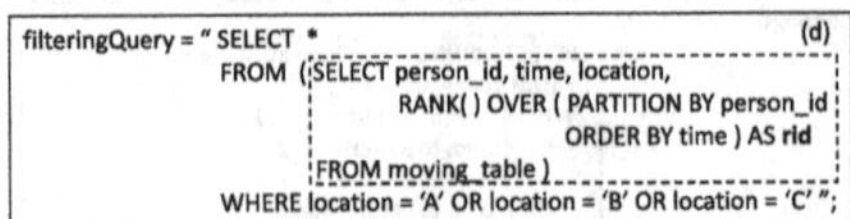

Fig. 10. Row filtering query corresponding to MATCH_RECOGNIZE specification in Fig. 2

used to do the row pattern matching. The NFA refers to the sequential row numbers to check whether the incoming rows reside adjacent to each other in the original input table. By doing this, we guarantee the correctness of the final query results.

Figure 9 shows a template for the Row Filtering query, which executes Row Filtering. Using RANK function, the sequential row number (rid (row id)) is appended to the input table as shown in part (d) in Fig. 9. Then, rows which do not satisfy the value conditions are filtered out. Figure 10 show the Row Filtering query for the query in Fig. 2.

Row Filtering can be applied alone, or applied after Sequence Filtering. We will compare performance of No Optimization, Sequence Filtering, Row Filtering, and Sequence Filtering + Row Filtering in Sect. 7.

6.2 Rewriting Flow of MATCH_RECOGNIZE Clause Including Optimization

We explain how to apply our optimization methods in the rewriting process of Sect. 4.2. Figure 11 illustrates the code rewriting including the optimization. The shaded boxes are extensions from the original rewriting process. The optimization code is inserted before the row pattern matching code. By doing this, the number of rows for the row pattern matching is reduced and the processing cost of the row pattern matching can be reduced.

The rewritten code of Fig. 5 changes to the one in Fig. 12 if we include the optimization. We also show the pseudocode of the function makeCode() including the optimization in Algorithm 3. The difference from Algorithm 2 is the function optCode() in line 7 which generates the optimization code before constructing NFA. In the rewritten code in Fig. 12, lines 3 and 4 are the optimization code generated by optCode().

7 Evaluation Experiment

7.1 Experiment Using Synthetic Data

Through the experiment, we evaluate effectiveness of the proposed optimizations by measuring their processing time. Here, we show the experimental result using synthetic data.

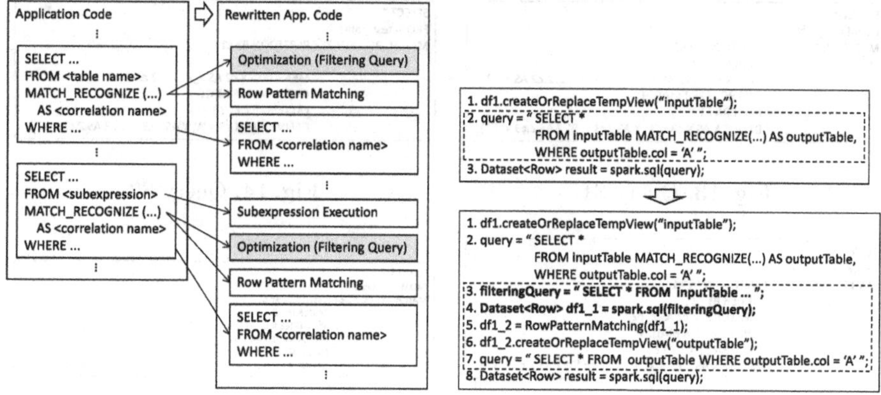

Fig. 11. Image of rewriting code including optimization

Fig. 12. Code rewriting including optimization

Algorithm 3. Pseudocode of makeCode() including Optimization

---

function makeCode(q)
  $MRcodes \leftarrow null$
  $mr \leftarrow extractMRspec(q)$
  if $subExp(mr) = true$ **then**
    $MRcodes \leftarrow add(execSubExp(mr))$
  end if
  $MRcodes \leftarrow add(optCode(mr))$
  $MRcodes \leftarrow add(execRPR(mr))$
  $MRcodes \leftarrow add(rewrite(q))$
  return $MRcodes$
end function

We use four queries (from Query S1 to Query S4) which have different patterns defined in DEFINE clause. Sequence Filtering and Row Filtering can be applied to all these queries. They are shown in Figs. 13, 14, 15 and 16.

The schema of the synthetic data used in this experiment is the same as test_table in Sect. 5. The number of the rows is 10,000,000 and the number of the sequences is 10,000. We set the two parameters for the input table. One is the ratio of the number of sequences which pass Sequence Filtering to the number of all the input sequences; α. The other is the ratio of the number of rows which pass Row Filtering to the number of rows in each sequence that passes Sequence Filtering; β. We evaluate the processing time for the following five configurations. Configuration 1: $\alpha = 0.0, \beta = 0.0$, Configuration 2: $\alpha = 0.2, \beta = 0.2$, Configuration 3: $\alpha = 0.2, \beta = 0.8$, Configuration 4: $\alpha = 0.8, \beta = 0.2$, and Configuration 5: $\alpha = 0.8, \beta = 0.8$. The experimental environment is the same as that shown in Sect. 5.

```
SELECT *
FROM test_table
MATCH_RECOGNIZE ( PARTITION BY c1
                  ORDER BY c2
                  MEASURES X.c2 AS X_c2, Y.c2 AS Y_c2, Z.c2 AS Z_c2
                  ONE ROW PER MATCH
                  PATTERN ( X Y Z )
                  DEFINE X AS X.c3 = 'A', Y AS Y.c3 = 'B', Z AS Z.c3 = 'C' )
```

Fig. 13. Query S1

```
SELECT *
FROM test_table
MATCH_RECOGNIZE ( PARTITION BY c1
                  ORDER BY c2
                  MEASURES X.c2 AS X_c2, Z.c2 AS Z_c2
                  ONE ROW PER MATCH
                  PATTERN ( X Y* Z )
                  DEFINE X AS X.c3 = 'A', Y AS Y.c3 = 'B', Z AS Z.c3 = 'C' )
```

Fig. 14. Query S2

```
SELECT *
FROM test_table
MATCH_RECOGNIZE ( PARTITION BY c1
                  ORDER BY c2
                  MEASURES Z.c2 AS Z_c2
                  ONE ROW PER MATCH
                  PATTERN ( ( X | Y ) Z )
                  DEFINE X AS X.c3 = 'A', Y AS Y.c3 = 'B', Z AS Z.c3 = 'C' )
```

Fig. 15. Query S3

```
SELECT *
FROM test_table
MATCH_RECOGNIZE ( PARTITION BY c1
                  ORDER BY c2
                  MEASURES X.c2 AS X_c2
                  ONE ROW PER MATCH
                  PATTERN ( X ( Y | Z W )* )
                  DEFINE X AS X.c3 = 'A', Y AS Y.c3 = 'B', Z AS Z.c3 = 'C', W AS W.c3 = 'D' )
```

Fig. 16. Query S4

Figure 17 shows the experimental results. It shows the execution time with/without optimization methods for each configuration. The vertical axis represents the average processing time of five trials. Each bar has an error bar representing the standard deviation of the processing time.

From this result, we can see that Sequence Filtering and Row Filtering are very effective except for Configuration 5. In Query S3 of Configuration 1, Sequence Filtering can reduce the processing time by 79.56% compared to that of No Optimization. This is the maximum reduction. When α is large and β is small as in the case of Configuration 4, Sequence Filtering does not effective. Note that Sequence Filtering + Row Filtering is not so advantageous. The reason is that Sequence Filtering + Row Filtering and Row filtering have almost the same filtering capability, but the former needs more cost for Sequence Filtering involving join.

7.2 Evaluation Using Real Data

We evaluate the proposed methods using real data. We use Foursquare Dataset. Foursquare [7] is a social networking service utilizing user location information. The users "check-in" at the specific place called "venue" by a portable device and can get the reward in its application service.

In this experiment, we use Global-scale Check-in Dataset [12,13] selected among the data sets disclosed as Foursquare Dataset. This dataset is a checking-in data collected for about 18 months in the world by Foursquare. We use five row pattern matching queries to evaluate effectiveness of our proposed optimization methods.

The data we use in the experiment are dataset_TIST2015_Checkins and dataset_TIST2015_POIs in the Global-scale Check-in Dataset. We use five queries (from Query F1 to Query F5) which have different patterns and different conditions as shown in Figs. 18, 19, 20, 21 and 22. Sequence Filtering and Row Filtering can be also applied to all these queries.

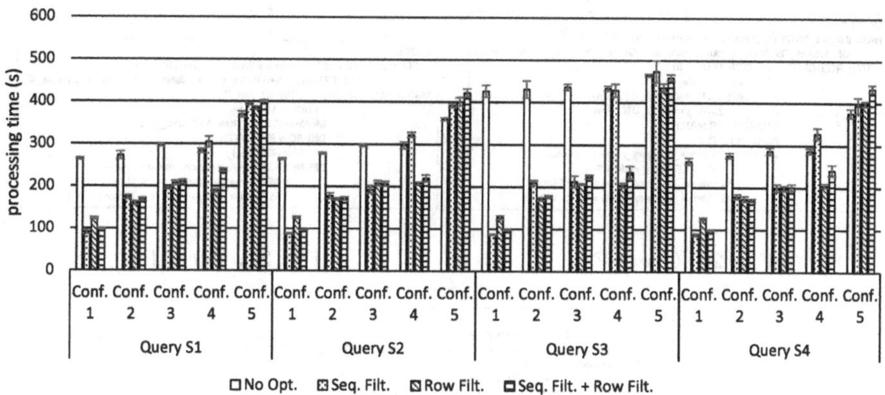

Fig. 17. Experimental results of synthetic data

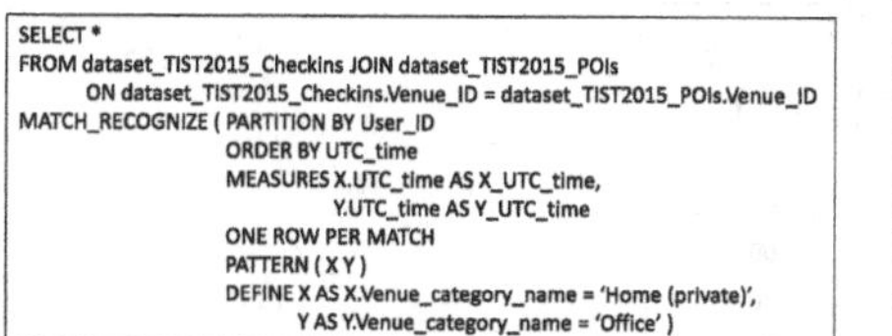

```
SELECT *
FROM dataset_TIST2015_Checkins JOIN dataset_TIST2015_POIs
    ON dataset_TIST2015_Checkins.Venue_ID = dataset_TIST2015_POIs.Venue_ID
MATCH_RECOGNIZE ( PARTITION BY User_ID
                  ORDER BY UTC_time
                  MEASURES X.UTC_time AS X_UTC_time,
                           Y.UTC_time AS Y_UTC_time
                  ONE ROW PER MATCH
                  PATTERN ( X Y )
                  DEFINE X AS X.Venue_category_name = 'Home (private)',
                         Y AS Y.Venue_category_name = 'Office' )
```

Fig. 18. Query F1

```
SELECT *
FROM dataset_TIST2015_Checkins JOIN dataset_TIST2015_POIs
    ON dataset_TIST2015_Checkins.Venue_ID = dataset_TIST2015_POIs.Venue_ID
MATCH_RECOGNIZE ( PARTITION BY User_ID
                  ORDER BY UTC_time
                  MEASURES X.UTC_time AS X_UTC_time,
                           Z.UTC_time AS Z_UTC_time
                  ONE ROW PER MATCH
                  PATTERN ( X Y+ Z )
                  DEFINE X AS X.Venue_category_name = 'Home (private)',
                         Y AS Y.Venue_category_name = 'Office',
                         Z AS Z.Venue_category_name = 'Mall' )
```

Fig. 19. Query F2

The experimental result using single worker node is shown in Fig. 23 and 5 worker nodes is shown in Fig. 24. The vertical axis represents the average processing time of five trials. Figure 23 shows that Sequence Filtering and Row Filtering are effective except for Sequence Filtering of Queries F1 and F2. Especially, Row Filtering can reduce the processing time by 54.90% compared to No Optimization for Query F3. Noted that there is no result in No Optimization in Query F4 because the execution was not terminate within 1 h. Figure 24 shows that Sequence Filtering and Row Filtering are effective except for Sequence Filtering of Query F2. Especially, Row Filtering can reduce the processing time by 73.38% compared to No Optimization for Query F4. This is the maximum reduction. Therefore, the execution time can be reduced a lot by these optimization methods. The processing time of Sequence Filtering is longer than that of No Optimization for Query F2. This reason is that the number of the input rows can be cut only 3.15% by Sequence Filtering. Focus on the difference of the number of nodes, 5 worker nodes is about from 40% to 50% faster than single worker node.

```
SELECT *
FROM dataset_TIST2015_Checkins JOIN dataset_TIST2015_POIs
    ON dataset_TIST2015_Checkins.Venue_ID = dataset_TIST2015_POIs.Venue_ID
MATCH_RECOGNIZE ( PARTITION BY User_ID
                  ORDER BY UTC_time
                  MEASURES X.UTC_time AS X_UTC_time,
                           Z.UTC_time AS Z_UTC_time
                  ONE ROW PER MATCH
                  PATTERN ( X Y* Z )
                  DEFINE X AS X.Venue_category_name = 'Hospital'
                              AND X.Country_code = 'FR',
                         Y AS Y.Venue_category_name = 'Post Office',
                         Z AS Z.Venue_category_name = 'Restaurant' )
```

Fig. 20. Query F3

```
SELECT *
FROM dataset_TIST2015_Checkins JOIN dataset_TIST2015_POIs
    ON dataset_TIST2015_Checkins.Venue_ID = dataset_TIST2015_POIs.Venue_ID
MATCH_RECOGNIZE ( PARTITION BY User_ID
                  ORDER BY UTC_time
                  MEASURES X.UTC_time AS X_UTC_time
                  ONE ROW PER MATCH
                  PATTERN ( X ( Y | Z ) )
                  DEFINE X AS X.Venue_category_name = 'Market',
                         Y AS Y.Venue_category_name = 'Shoe Store'
                         Z AS Z.Venue_category_name = 'Liquor Store' )
```

Fig. 21. Query F4

```
SELECT *
FROM dataset_TIST2015_Checkins JOIN dataset_TIST2015_POIs
    ON dataset_TIST2015_Checkins.Venue_ID = dataset_TIST2015_POIs.Venue_ID
MATCH_RECOGNIZE ( PARTITION BY User_ID
                  ORDER BY UTC_time
                  MEASURES X.UTC_time AS X_UTC_time
                  ONE ROW PER MATCH
                  PATTERN ( X Y * ( Z | W ) )
                  DEFINE X AS X.Venue_category_name = 'Airport',
                         Y AS Y.Country_code = 'DE',
                         Z AS Z.Country_code = 'CH',
                         W AS W.Country_code = 'FR' )
```

Fig. 22. Query F5

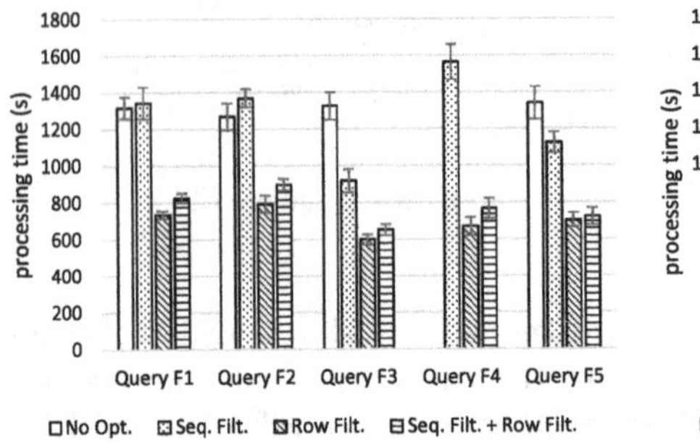

Fig. 23. Experimental result of foursquare data using single worker node

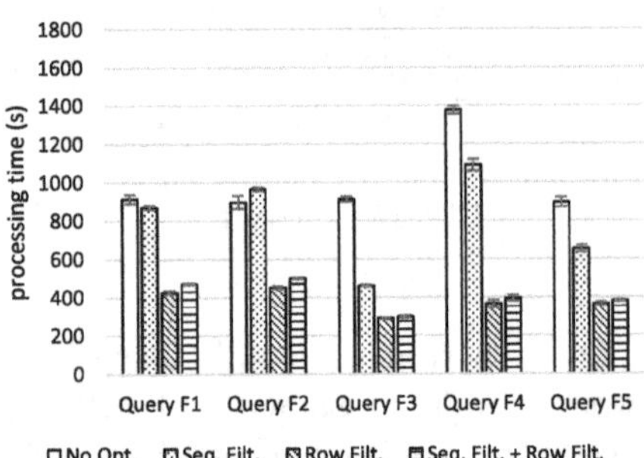

Fig. 24. Experimental result of foursquare data using 5 worker nodes

8 Conclusion and Future Work

In this paper, we have proposed two optimization methods: Sequence Filtering and Row Filtering to execute the row pattern matching efficiently. We have focused on Spark and implemented MATCH_RECOGNIZE clause in Spark SQL by the Front-end Approach. Through the evaluation experiments, we have proved the effectiveness of our proposed optimization.

Future research issues include cost models of the query including the row pattern matching and selection of appropriate optimization methods. We will examine the cost of sort and filtering. We will also examine the optimization method considering the pattern variables conditions and pattern variables which are not associated with any conditions. Treatment of aggregation functions in MEASURES clause is another interesting research issue.

Acknowledgement. This work was partly supported by Grant-in-Aid for Scientific Research (B) (#19H04114) from JSPS.

References

1. 19075-5:2016(E), I.T.: Information technology - database languages - sql technical reports - part 5: row pattern recognition in sql. technical report. Technical report, ISO copyright office (2016)
2. Agrawal, J., Diao, Y., Gyllstrom, D., Immerman, N.: Efficient pattern matching over event streams. In: Proceedings of the 2008 ACM SIGMOD International Conference on Management of Data, pp. 147–160 (2008)
3. Armbrust, M., et al.: Spark SQL: relational data processing in spark. In: Proceedings of the 2015 ACM SIGMOD International Conference on Management of Data, pp. 1383–1394 (2015)
4. Cadonna, B., Gamper, J., Böhlen, M.H.: Efficient event pattern matching with match windows. In: Proceedings of the 18th ACM SIGKDD International Conference on Knowledge Discovery and Data Mining (KDD2012), pp. 471–479 (2012)
5. Demers, A., Gehrke, J., Panda, B., Riedewald, M., Sharma, V., White, W.: Cayuga: a general purpose event monitoring system. In: CIDR 2007, pp. 412–422 (2007)
6. Foundation, T.A.S.: Hadoop (2018). http://hadoop.apache.org/
7. Foursquare: Foursquare (2018). https://foursquare.com
8. Laker, K.: A technical deep dive into pattern matching using match_recognize (2016). http://www.oracle.com/technetwork/database/bi-datawarehousing/mr-deep-dive-3769287.pdf
9. Mei, Y., Madden, S.: ZStream: a cost-based query processor for adaptively detecting composite events. In: Proceedings of the 2009 ACM SIGMOD International Conference on Management of Data, pp. 193–206 (2009)
10. Thusoo, A., et al.: Hive - a petabyte scale data warehouse using Hadoop. In: Proceedings of the 26th International Conference on Data Engineering (ICDE2010) (2010)
11. Wu, E., Diao, Y., Rizvi, S.: High-performance complex event processing over streams. In: SIGMOD 2006, pp. 407–418 (2006)
12. Yang, D., Zhang, D., Chen, L., Qu, B.: NationTelescope: monitoring and visualizing large-scale collective behavior in LBSNs. J. Netw. Comput. Appl. (JNCA) **55**, 170–180 (2015)
13. Yang, D., Zhang, D., Qu, B.: Participatory cultural mapping based on collective behavior data in location based social networks. In: ACM Trans. on Intelligent Systems and Technology (TIST) (2015)
14. Zaharia, M., Chowdhury, M., Franklin, M.J., Shenker, S., Stonica, I.: Spark: cluster computing with working sets. In: Proceedings of the 2nd USENIX Conference on Hot Topics in Cloud Computing (HotCloud2010), vol. 55, p. 10 (2010)

Rainfall Estimation from Traffic Cameras

Remmy Zen[1(✉)], Dewa Made Sri Arsa[2], Ruixi Zhang[1], Ngurah Agus Sanjaya ER[2], and Stéphane Bressan[1]

[1] National University of Singapore, Singapore, Singapore
{remmy,zhangruixi}@u.nus.edu, steph@nus.edu.sg
[2] Universitas Udayana, Bali, Indonesia
{dewamsa,agus_sanjaya}@unud.ac.id

Abstract. We propose and evaluate a method for the estimation of rainfall from images from a network of traffic cameras and rain gauges. The method trains a neural network for each camera under the supervision of the rain gauges and interpolates the results to estimate rainfall at any location. We study and evaluate variants of the method that exploit feature extraction and various interpolation methods. We empirically and comparatively demonstrate the superiority of a hybrid approach and of the inverse distance weighting interpolation for an existing comprehensive network of publicly accessible weather stations and traffic cameras.

Keywords: Rainfall estimation · Convolutional neural network · CCTV camera

1 Introduction

The seemingly increasing recurrence of extreme meteorological events [43] with their dramatic consequences and the heated debates on climate change with their grim prospect emphasise the necessity of scientific and engineering methods and tools for the modelling, hindcasting and forecasting of weather and climate. The significance of this concern is reflected in the mission statement of the American National Weather Service: "The National Weather Service provides weather, water and climate data, forecasts and warnings for the protection of life and property and enhancement of the national economy".

Rainfall is of particular importance among other atmospheric phenomena as it concerns all aspect of human activities [31] from the most mundane for the functioning of our societies such as communication and transportation to the most essential for our survival such as the availability of drinking water, irrigation for agriculture and power production. Meteorologists and hydrologists use rain gauges and weather radar to measure and estimate rainfall amounts. The Meteorological Service Singapore (www.weather.gov.sg) operates a network of over sixty rain gauges. The National Environment Agency makes rainfall readings available on the Singapore government's portal (data.gov.sg).

The convergence of the commoditisation of cheap connected sensing devices with the development of efficient and effective machine learning solutions for the

S. Hartmann et al. (Eds.): DEXA 2019, LNCS 11706, pp. 18–32, 2019.
https://doi.org/10.1007/978-3-030-27615-7_2

analysis of big data allows the designers of smart cities and smart countries to revisit the design of monitoring infrastructures. For instance, the total number of closed-circuit television cameras installed by private and corporate owners and by government agencies in Singapore can probably be estimated to be in the hundreds of thousands. Several government agencies such as the Public Utilities Board, the Land Transport Authority and the Singapore Police Force make some data sets related to these closed-circuit television cameras available on Singapore government's portal (data.gov.sg).

Leveraging such smart city and smart nation open infrastructure, we propose a method for the estimation of rainfall value from traffic camera images. The method is twofold. Firstly, the method estimates the current rainfall value at the location observed by each camera with a neural network analysing the current image. We compare two architectures for the neural network. The first architecture consists of a supervised regression network. The second architecture is hybrid. It augments the regression network with a feature extraction autoencoder that extracts rain streaks from the image. Secondly, the method estimates the current rainfall value at any given location from the rainfall values estimated at the location of every camera using one of three spatial interpolation methods: nearest neighbour, inverse distance weighting and Kriging.

We use a data set covering most of Singapore consisting of actual readings from rain gauges of the Meteorological Service Singapore and the corresponding images of forty traffic cameras of the Land Transport Authority of Singapore. We empirically and comparatively demonstrate the viability of the proposed approach and the superiority of the hybrid approach with the inverse distance weighting interpolation.

The remainder of this paper is structured as follows. Section 2 synthesises the main related works on spatial rainfall distribution, rainfall estimation and rain streak removal. Section 3 presents and discusses the methods devised and applied. Section 4 describes the data set and the performance evaluation for the experiments. Section 5 presents and analyses the result of an empirical and comparative performance evaluation of the methods proposed. Section 6 concludes the paper and highlights future directions for this work.

This work constitutes not only an explained practical solution to the improvement of rain estimation tools leveraging an existing infrastructure at almost no additional cost but also, more generally, provides a blueprint for the modelling, hindcasting and forecasting of other environmental phenomena leveraging the now available plethora of sensors.

2 Related Work

Rainfall is measured with rain gauges. A standardised rain gauge measures the rainfall amount in millimetre where one millimetre corresponds to one litre of rain falling on one square metre. Rainfall intensity is measured in millimetre per hour.

Direct measurements are only available where rain gauges are located and the continuous spatial distribution of rainfall is estimated by interpolation. The two state-of-the-art rainfall interpolation methods are the *inverse distance weighting method* [38] and *Kriging* [42]. Shepard, in [38], proposed to compute the rainfall at a target location as the sum of the rainfall at all source locations where it is known. The sum is weighted by the inverse distance between the source locations and the target location, respectively. We refer to this method as the inverse distance weighting method. The method has been used for locations in Austria [2], Hawaii [32] and Taiwan [9]. Kriging [42] models the rainfall distribution as a Gaussian process. Kriging has been used for locations in India [7], Taiwan [10] and Tunisia [6].

We refer to the now-casting or hindcasting problem of finding current or past rainfall as *rainfall estimation* and to the corresponding forecasting problem as *rainfall prediction*. This work is concerned with rainfall estimation.

Hernandez et al. [16] use meteorological data such as temperature, humidity and wind speed to predict the next-day rainfall. The authors use a neural network architecture combining autoencoder and multilayer perceptron and frame the problem as a regression problem. Kashiwao et al. [20] use meteorological data such as temperature, pressure and wind speed to predict the rainfall and frame the problem as a binary classification problem. They use a multilayer perceptron and a radial basis function network to classify rainfall. Several works used historical rainfall data and frame the problem as a time-series forecasting problem. Different work used artificial neural network [15] combined with wavelet [35], genetic algorithms [46], convolutional neural network [14] or recurrent neural network [5].

Kusiak et al. [26] compare different machine learning algorithms to estimate rainfall from radar reflectivity data. Similarly, Meyer et al. [33] use optical satellite data for classification and regression. The authors of [3,21,47] use a combination of convolutional neural networks and recurrent neural networks to estimate rainfall from satellite images.

The authors of [4,8,13,25] use image processing techniques to estimate rainfall from ground-level images. Roser and Moosman [37] classify rainfall with support vector machines from images of a camera mounted on a vehicle. Lee et al. [28,29] extract weather information from closed-circuit television cameras. Using image processing techniques, they extract features from the video and use them to categorise rainfall. Sirirattanapol et al. [40] extract several road environment information from closed-circuit television camera images in the streets of Bangkok. They frame the problem as a multi-label classification problem. They use a convolutional neural network to label the images with such labels as "rain" or "not rain". The ground truth labels are assigned manually. The reader may also refer to [12,36] for a survey of machine learning methods for rainfall estimation and prediction.

Finally, raindrop detection and removal from video and camera images are performed with image processing techniques [17,18] and neural network architectures [30,34,48,49]. Readers should refer to [45] for a survey of raindrop removal from videos with image processing techniques. We use a density-aware multi-stream densely connected convolutional neural network-based algorithm proposed by Zhang et al. [49] to extract rain streaks from images.

3 Methodology

3.1 Problem Definition and Outline of the Solution

The general problem that we are considering is that of the estimation of an environmental parameter from a network of primary and secondary geo-located sensors provide time-stamped data. The primary sensors provide golden standard measurements of the environmental parameter. The secondary sensors are existing sensors possibly deployed for other purposes that may be seamlessly diverted from their original application to provide information for a new and different task. The secondary sensors may have many advantages such as being available in larger quantities, being available in locations or at times where or when it is not possible or difficult to place or access the primary sensors, being cheaper and cheaper to operate. They may, however, not provide direct measurements of the environmental parameter. Inferring a measurement from the data provided by the secondary sensor possibly involves processing by a statistical machine learning algorithm. The concept of primary and secondary sensors has been used in the smart sensors research, for instance, in [44].

A primary sensor reading is a tuple $p_i = (x_p, l_p, t_p)$ where x_p is the measurement provided by the sensor at location l_p and time-stamp t_p. We call $\mathcal{P}$ the set of all such tuples. Similarly, a secondary sensor reading is a tuple $s = (x_s, l_s, t_s)$. We call $\mathcal{S}$ the set of all such tuples. x_s is not a measurement; it is the raw data provided by the secondary sensor. Therefore, we want to infer a measurement v_s for each secondary sensor s. Note that the characteristics of different secondary sensors may differ. For each secondary sensor, we need to learn a local function $f(x_s|\mathcal{S} \cup \mathcal{P}) = v_s$ and infer a spatiotemporal function that estimates the parameter value at any location and time-stamp. The method that we propose is therefore twofold. For each secondary sensor, we train a supervised machine learning model to learn the local function. We then create a function $f(l, t) = v$ that estimates the parameter value at any given location and time-stamp from the estimated values for each secondary sensor available and the values for each primary sensor available.

In the problem at hand, the environmental parameter that we study is rainfall. The primary sensors are rain gauges and the secondary sensors are traffic cameras.

The method estimates the current rainfall amount at the location of the camera for each camera using the current measurements of the rain gauges. To do so, we train a supervised neural network based regression model for each camera. We use three spatial interpolation techniques discussed in Sect. 2, namely nearest neighbour, inverse distance weighting and Kriging, to interpolate the value for training the neural network from the current measurements of the rain gauges only[1]. The details of this first step are presented in Subsect. 3.2.

The trained models for each camera are used to estimate rainfall at any location. Again, we do so by using one of the three spatial interpolation methods. The details of this first step are presented in Subsect. 3.3.

We evaluate the performance of the proposed method using the absolute ground truth of the rainfall measured by the rain gauges at their location.

3.2 Learning the Local Rainfall Function

We train a supervised regression model for each camera to estimate the current rainfall at the location of the camera given the current image of the camera. We interpolate the value used for training from the rainfall values measured at the rain gauges. We comparatively used the three methods described in Subsect. 3.3 to compute this training value. Note that this value is an interpolated estimate. It is hopefully sufficient for training purposes but it may not be correct. We, therefore, do not use it for the evaluation of the performance of our model as a reference ground truth.

We use a convolutional neural network model [24]. Convolutional neural networks use convolutional and pooling operations to extract features from an image. Convolutional neural networks have been proven to be the state-of-the-art model for many applications, such as image classification and object detection [27]. We compare two convolutional neural network-based architectures. The first architecture is a standard convolutional neural network model that extracts features from the image and use a fully connected neural network for regression. We refer to this architecture as the *standard model*. The second architecture is a hybrid model. The model combines the regression network with another model that extract rain streaks from the image. The model fuses the feature of the image with an image of the rain streaks extracted from the image. Figure 1 shows a diagram of the standard and hybrid model. As discussed earlier several algorithms have been developed to remove rain streaks from an image. We use DID-MDN algorithm [49]. We chose DID-MDN for its availability and ease of use. The algorithm consists of two networks. The first network classifies the rain density of the image. The second network extracts the features of the image. The networks need to be trained on many images with different rain density. Unfortunately, this cannot easily be done otherwise than by using synthetic rain images. The density label from the first network is fused with the feature from the second network. The fused features are up-sampled to get the rain streak from the image making it similar to an autoencoder architecture.

We do not wish to remove but rather to emphasise the rain streaks. We create a rain streak image by subtracting the original image with the output image of the algorithm.

[1] For this work, we neither consider the temporal sequences of measurements nor the sequence of images.

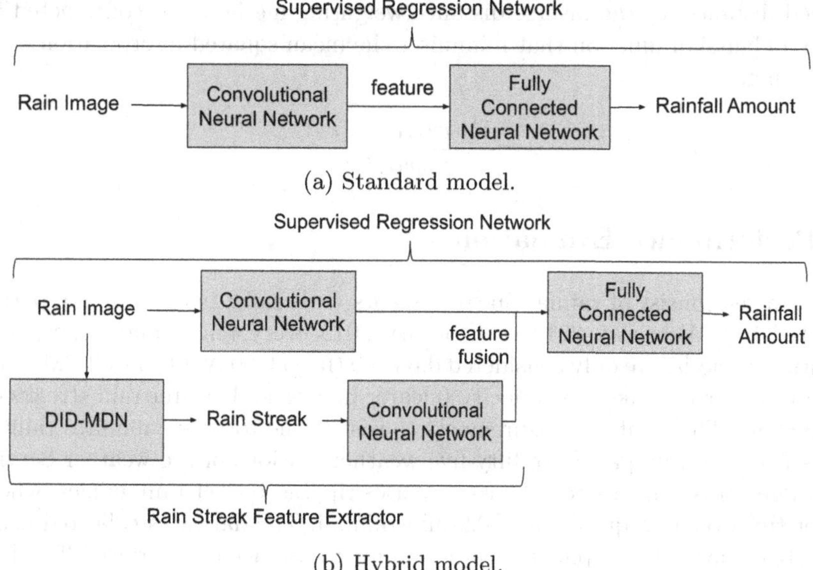

Fig. 1. Comparison of the standard and hybrid model for learning the local rainfall function.

3.3 Interpolating the Global Rainfall Function

We estimate the current rainfall at any location by interpolating learned rainfall amount for the cameras and the rainfall measurements by the rain gauges. We comparatively use the three spatial interpolations: nearest neighbour, inverse distance weighting and Kriging.

The nearest neighbour method serves as a baseline. It uses the nearest available measurement. Formally, the rainfall value v at a location l is:

$$v = v_{\text{argmin}_x(\text{distance}(l, l_x))}$$

where x ranges over the rain gauges for learning the local rainfall function and over all the available cameras and rain gauges for interpolating the global rainfall function. The distance is the geodesic distance, namely the shortest distance on the surface of an ellipsoidal model of the earth [19].

The inverse distance weighting assigns rainfall by normalizing the sum of the available estimates and measurements weighted by the inverse distance between their location and the location of the calculated estimate.

Kriging uses the spatial arrangement of the available measurements and estimates with a Gaussian function for the weighting.

Formally, the rainfall value v at a location l is defined in Eq. 1, where x ranges over the rain gauges for learning the local rainfall function and over all the available cameras and rain gauges for interpolating the global rainfall function and w is a weighting function. The weighting function w is defined as the inverse squared distance for the inverse distance weighting method and constructed by fitting a Gaussian function that minimises the mean squared interpolation error for Kriging.

$$v = \frac{\Sigma_x (w(l, l_x) \times v_x)}{\Sigma_x w(l, l_x)} \tag{1}$$

4 Performance Evaluation

The data set consist of rainfall measurements and traffic images for the period collected from February 2018 to February 2019 every ten minutes except for downtime periods. We only considered daytime (from 6:00 AM to 19:00 PM) data and images for the sake of simplicity (clearer images and clearer rain streaks on the images). The rainfall measurements consist of the total five minutes rainfall values from the rain gauges of fifty-five weather stations of the weather Service Singapore across Singapore. The agency uses tipping bucket rain gauges, whose bucket tips over in a quantum of 0.2-mm rainfall. The data was collected using the publicly available application programming interface of a service offered by the Meteorological Service Singapore[2]. The traffic images consist of images from eighty-five traffic cameras operated by the Land Transport Authority of Singapore overlooking expressways across Singapore. The cameras have one of two resolutions. We refer to cameras and images with resolution 640 × 480 as high-resolution cameras and images and with resolution 320 × 240 as low-resolution cameras and images, respectively. Figure 2a and b shows an example image for the high-resolution and low-resolution camera, respectively. We selected a set of forty cameras with twenty high-resolution and twenty low-resolution cameras. Figure 3 shows the location of the selected cameras and weather stations.

(a) Example image of high-resolution camera with 640 × 480 resolutions.

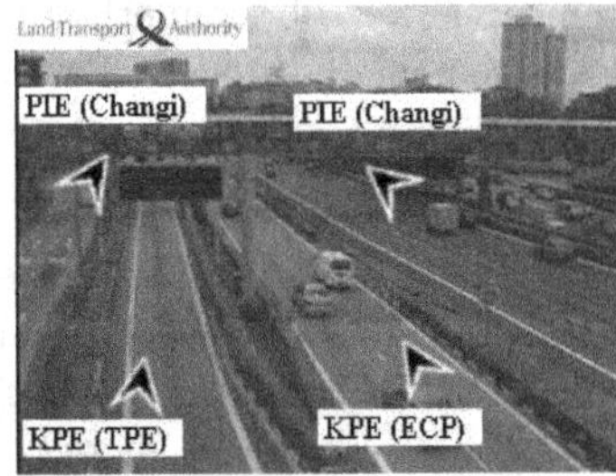

(b) Example image of low-resolution camera with 320 × 240 resolutions.

Fig. 2. Example images from high-resolution and low-resolution cameras.

[2] https://data.gov.sg/dataset/realtime-weather-readings?resource_id=8bd37e06-cdd7-4ca4-9ad8-5754eb70a33d.

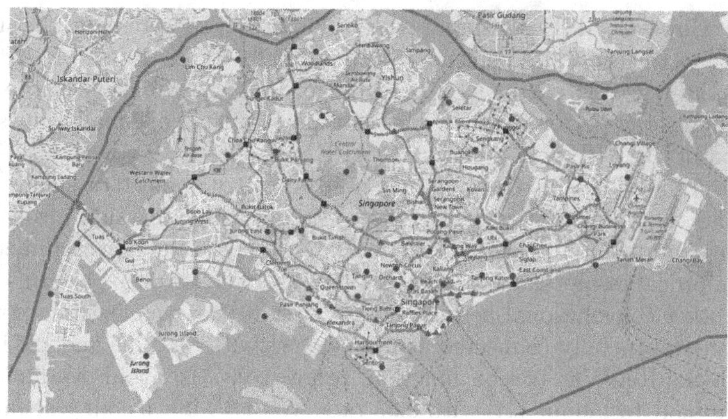

Fig. 3. Locations of the high-resolution cameras, low-resolution cameras and weather station denoted by square, triangle and circle marker, respectively.

All programs run on a machine with a 3 Ghz AMD Ryzen™ 7-1700 eight-core processor. The machine is equipped with a 64 GB DDR4 memory and an NVIDIA GTX 1080Ti GPU card with 3584 CUDA cores and 11 GB memory. The operating system is Ubuntu 18.04 OS. We implement the method with Keras [11] on top of the Tensorflow™ library [1] using Python™ 2.7.15 as the scripting language.

For learning the local rainfall function for the cameras, we train two models: the standard model and the hybrid model. After evaluation, we only present the result for the most performant interpolation method, namely inverse distance weighting, to compute the reference rainfall value. We divide our experiments into three phases: training, validation and testing. We use data from February 2018 to October 2018 for the training and validation phase and the rest for the testing phase. We split the training and validation ratio to 70% and 30%, respectively.

We faced the problem of a seriously imbalanced data set. There are more dry than rainy data and images. In average, we found that there are only 3.31% of rain images at each camera. Therefore, we use the random oversampling method where we pick rain images at random with replacement to construct the data set during the training and validation phase to balance it. The training phase trains the model, the validation phase tests the model against a balanced data set and the test phase tests the model against the original imbalanced data.

We use the mean square error (MSE) and mean arctangent absolute percentage error (MAAPE) [22] as the evaluation metrics for the validation and testing phase. The MSE calculates the squared residual for every data point. The calculation of MSE is as follows MSE$= \frac{1}{N}\sum_{i=1}^{N}(y_i - \hat{y}_i)^2$ where N is the number of data, y is the actual rainfall amount and $\hat{y}$ is the estimated rainfall amount. The smaller the MSE the better. MSE penalises higher for a mistake made in the outlier data. In our case, this is important since the rainy data and image is an outlier. However, MSE is generally hard to interpret so we use another metric that measures the error as a percentage and normalised to 0–100%. MAAPE [22] measures the regression error as a percentage. It is an extension of the mean absolute percentage error since it does not work with 0 or small values. The calculation of MAAPE is as follows $\text{MAAPE} = \frac{100}{N*0.5*\pi}\sum_{i=1}^{N}(\arctan(|\frac{y_i-\hat{y}_i}{\hat{y}_i}|))$. Since arctan function is used, we do not need to worry about 0 or small values. The smaller the MAAPE the better. Since there is randomization involved, we repeat the experiment 10 times and report the average value for each metric.

Our empirical results show that VGG [39] with 19 layers deep performs better than the other convolutional neural network architectures as feature extractor. We use three layers fully-connected neural network with 512, 256 and 128 hidden units with the ReLU activation function, respectively. To avoid overfitting, we use the dropout technique [41] with 0.25 rate. We use MSE as our lose function and use a gradient descent algorithm with ADAM [23] optimiser to train the fully-connected neural network. We use the early stopping technique to stop the training after the validation loss stopped decreasing.

We test our method by comparing the value of the global rainfall function for the locations of the rain gauges with actual corresponding measurement. This provides us an absolute ground truth. For this reason and in the interest of fairness, we excluded those available measurements in the definition of the global rainfall function that we evaluate. The global rainfall function that we evaluate estimates the rainfall by interpolation of the estimates obtained from the cameras only. This restriction can be removed in an operational system that can freely combine measurements and estimates from secondary and primary sensors, respectively. The overall performance can only be improved.

We use MSE and MAAPE between the estimated amount and the real amount as the evaluation metric. We only test model trained with similar spatial interpolation methods. This means that we do not test a model trained with inverse distance weighting for the ground truth and determine the rainfall value at the rain gauges with Kriging for testing. We use the data at the testing phase to test the model.

5 Results

5.1 Training and Validation of the Local Rainfall Function with the Cameras

We comparatively evaluate the supervised regression model for each camera with the standard and hybrid model. We report the mean squared error (MSE) and mean arctangent percentage error (MAAPE) during the validation and test phase. We use inverse distance weighting as the ground truth for training the cameras. We use inverse distance weighting since it is the best spatial interpolation method to test with the rain gauges shown in Subsect. 5.2. Figure 4 shows the overall performance of the standard and hybrid model. Figure 4a and b show the whisker plot of the MSE and the MAAPE metric, respectively, on the validation and test phase. The lower the MSE and MAAPE metric the better.

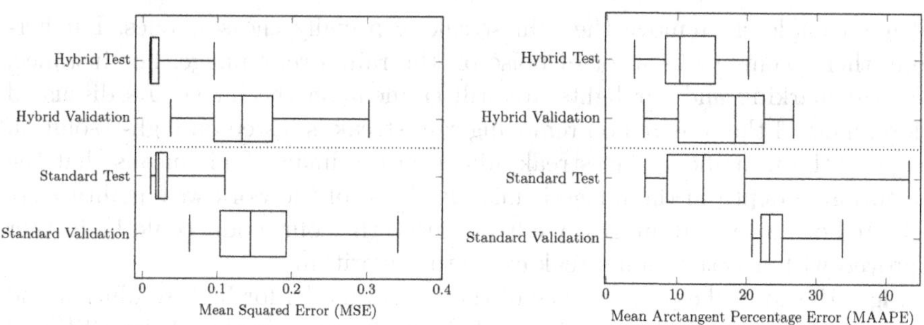

(a) Mean squared error of the validation and test phase.

(b) Mean arctangent percentage error of the validation and test phase.

Fig. 4. Local performance of the standard and hybrid models during the validation and test phase with inverse distance weighting as the ground truth rainfall amount value. For both metrics, the lower the value the better.

We see that the hybrid method is statistically better than the standard model for both metrics and both phases. The median of the MAAPE metric for the hybrid model improved the performance of the regression model by about 3% and 4% than the standard model for the validation and test phase, respectively. We also see that the performance of the model is statistically better during the test phase than in the validation phase. This means that the model performs well in an imbalanced data set which is prominent in the real-world scenario.

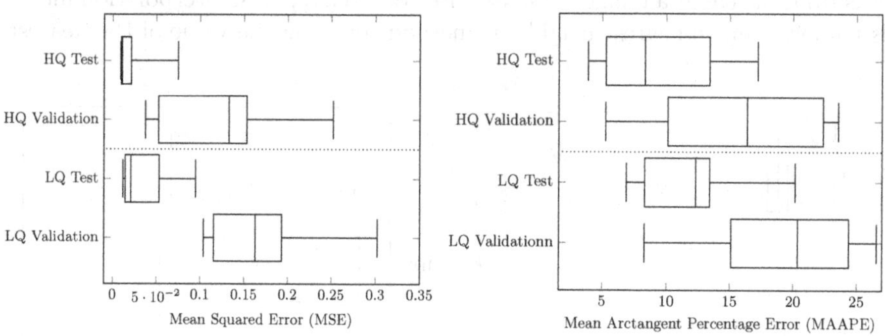

(a) Mean squared error of the validation and test phase.

(b) Mean arctangent percentage error of the validation and test phase.

Fig. 5. Local performance of the hybrid model for training the during the validation and test phase with high-resolution (HQ) and a low-resolution (LQ) camera, respectively. For both metrics, the lower the value the better.

We qualitatively evaluate the DID-MDN algorithm [49] used as the rain streak feature extractor for our hybrid model. We observe that the algorithm did not completely remove the rain streaks especially the soft ones. Furthermore, there seems to be a lot of noise on the rain streak image. For instance, the road marking and car lights are still eminent on the image. As discussed before, most of the research on removing rain streak is tested on high-resolution images with an artificial rain streak added on the image. This means that the perception of depth in the image is missing. Most of the work was qualitatively evaluated on real-world images. We hypothesise that our model could be further improved with a better rain streak extractor algorithm.

Figure 5 shows the performance of the hybrid model for high-resolution and low-resolution cameras. Figures 5a and b show the whisker plot of the MSE and the MAAPE metric, respectively, on the validation and test phase. It is not surprising to observe that the performance of the model on the high-resolution cameras is statistically better than that on the low-resolution cameras in both metrics and phases. As seen in Fig. 2b, the quality of the image in the low-resolution camera is lower and a lot of noise obstruct the image. Therefore, the rain streak might not be clearly visible for the model to learn. The median of the MAAPE metric for high-resolution cameras is better by about 4% than that of the low-resolution cameras in both phases. This suggests that the quality of the sensors plays a role in the model performance and should be taken into account when deploying the system.

5.2 Testing the Global Rainfall Function with the Weather Stations

We evaluate the global performance of the hybrid model by comparing the rainfall measurements at the rain gauges with a value estimated by the neural networks from the camera images. We use three different spatial interpolation methods namely a naïve nearest neighbour method selecting the value of the nearest

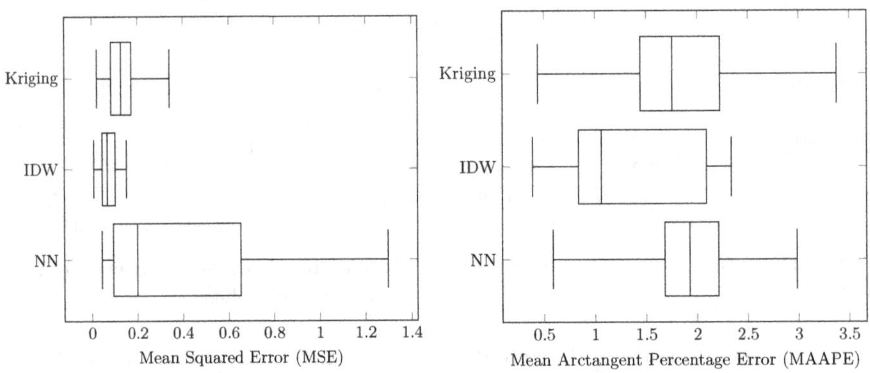

(a) Mean squared error of testing with the rain gauges.

(b) Mean arctangent percentage error of testing with the rain gauges.

Fig. 6. Global performance of the hybrid model with the three different spatial interpolation methods nearest neighbour (NN), inverse distance weighting (IDW) and Kriging.

camera and the two state-of-the-art methods inverse distance weighting and Kriging. Figure 6 shows the performance with the three different spatial interpolation methods. Figures 6a and b show the whisker plot of the mean squared error (MSE) and the mean arctangent percentage error (MAAPE) metric, respectively. Unexpectedly, we observe that inverse distance weighting is statistically better than Kriging. This may be due to the climate and elevation of Singapore. As expected, the nearest neighbour yields the worst performance.

6 Conclusion

We have proposed a method for the estimation of rainfall from closed-circuit television camera images. The method consists of two parts. The first part trains a regression neural network, with or without the adjunction of an autoencoder for preliminary feature extraction, to estimates, a local rainfall function, i.e. the current rainfall value at the location of an individual camera. The second part estimates a global rainfall function, i.e. the current rainfall value at any given location from the rainfall values estimated at the location of every available camera. The second part uses one of three interpolation methods: nearest neighbour, inverse distance weighting and Kriging. The resulting system is ready to be put into operation with weather stations and images from traffic cameras in Singapore.

We trained and evaluated the system with one year of data and learned the local and global rainfall functions. We empirically and comparatively demonstrated the viability of the proposed approach and the superior effectiveness of the hybrid approach with the inverse distance weighting interpolation. The average estimation error of the best variant of our method is as low as nearly 1%.

We are currently exploring several natural extensions to these already practical results. In addition to considering variants of the task including classification and estimation of the surface water, we would like to further understand the role and significance of feature extraction in a hybrid system from the point of view of robustness and transfer learning for different geographic regions. This may include considering sets and sequences of images instead of individual images. We are also considering leveraging other sensors, in particular those available on mobile devices such as dashboard and mobile phones cameras.

Acknowledgment. This work is supported by the National University of Singapore Institute for Data Science project WATCHA: WATer CHallenges Analytics and by Singapore Ministry of Education project Janus.

References

1. Abadi, M., et al.: Tensorflow: a system for large-scale machine learning. In: 12th {USENIX} Symposium on Operating Systems Design and Implementation, {OSDI} 2016, pp. 265–283 (2016)
2. Ahrens, B.: Distance in spatial interpolation of daily rain gauge data. Hydrol. Earth Syst. Sci. Discuss. **2**(5), 1893–1922 (2005)

3. Akbari Asanjan, A., Yang, T., Hsu, K., Sorooshian, S., Lin, J., Peng, Q.: Short-term precipitation forecast based on the PERSIANN system and LSTM recurrent neural networks. J. Geophys. Res. Atmos. **123**(22), 12–543 (2018)
4. Allamano, P., Croci, A., Laio, F.: Toward the camera rain gauge. Water Resour. Res. **51**(3), 1744–1757 (2015)
5. Aswin, S., Geetha, P., Vinayakumar, R.: Deep learning models for the prediction of rainfall. In: 2018 International Conference on Communication and Signal Processing (ICCSP), pp. 0657–0661. IEEE (2018)
6. Bargaoui, Z.K., Chebbi, A.: Comparison of two kriging interpolation methods applied to spatiotemporal rainfall. J. Hydrol. **365**(1–2), 56–73 (2009)
7. Basistha, A., Arya, D., Goel, N.: Spatial distribution of rainfall in indian himalayas-a case study of Uttarakhand region. Water Res. Manag. **22**(10), 1325–1346 (2008)
8. Cerqueira, R.F., Mantripragada, K.: Estimating rainfall precipitation amounts by applying computer vision in cameras, uS Patent App. 14/748,125, 26 May 2016
9. Chen, F.W., Liu, C.W.: Estimation of the spatial rainfall distribution using inverse distance weighting (IDW) in the middle of Taiwan. Paddy Water Environ, **10**(3), 209–222 (2012)
10. Chen, Y.C., Wei, C., Yeh, H.C.: Rainfall network design using kriging and entropy. Hydrol. Process. Int. J. **22**(3), 340–346 (2008)
11. Chollet, F., et al.: Keras (2015). https://keras.io
12. Cramer, S., Kampouridis, M., Freitas, A.A., Alexandridis, A.K.: An extensive evaluation of seven machine learning methods for rainfall prediction in weather derivatives. Expert Syst. Appl. **85**, 169–181 (2017)
13. Garg, K., Nayar, S.K.: Vision and rain. Int. J. Comput. Vis. **75**(1), 3–27 (2007)
14. Haidar, A., Verma, B.: Monthly rainfall forecasting using one-dimensional deep convolutional neural network. IEEE Access **6**, 69053–69063 (2018)
15. Hardwinarto, S., Aipassa, M., et al.: Rainfall monthly prediction based on artificial neural network: a case study in Tenggarong station, East Kalimantan-indonesia. Proc. Comput. Sci. **59**, 142–151 (2015)
16. Hernández, E., Sanchez-Anguix, V., Julian, V., Palanca, J., Duque, N.: Rainfall prediction: a deep learning approach. In: Martínez-Álvarez, F., Troncoso, A., Quintián, H., Corchado, E. (eds.) HAIS 2016. LNCS (LNAI), vol. 9648, pp. 151–162. Springer, Cham (2016). https://doi.org/10.1007/978-3-319-32034-2_13
17. Jiang, T.X., Huang, T.Z., Zhao, X.L., Deng, L.J., Wang, Y.: Fastderain: a novel video rain streak removal method using directional gradient priors. IEEE Transact. Image Process. **28**(4), 2089–2102 (2019)
18. Kang, L.W., Lin, C.W., Fu, Y.H.: Automatic single-image-based rain streaks removal via image decomposition. IEEE Transact. Image Process. **21**(4), 1742–1755 (2012)
19. Karney, C.F.: Algorithms for geodesics. J. Geodesy **87**(1), 43–55 (2013)
20. Kashiwao, T., Nakayama, K., Ando, S., Ikeda, K., Lee, M., Bahadori, A.: A neural network-based local rainfall prediction system using meteorological data on the internet: a case study using data from the Japan meteorological agency. Appl. Soft Comput. **56**, 317–330 (2017)
21. Kim, S., Hong, S., Joh, M., Song, S.k.: Deeprain: ConvLSTM network for precipitation prediction using multichannel radar data. arXiv preprint arXiv:1711.02316 (2017)
22. Kim, S., Kim, H.: A new metric of absolute percentage error for intermittent demand forecasts. Int. J. Forecast. **32**(3), 669–679 (2016)
23. Kingma, D.P., Ba, J.: Adam: a method for stochastic optimization. arXiv preprint arXiv:1412.6980 (2014)

24. Krizhevsky, A., Sutskever, I., Hinton, G.E.: ImageNet classification with deep convolutional neural networks. In: Advances in neural information processing systems, pp. 1097–1105 (2012)
25. Kurihata, H., et al.: Rainy weather recognition from in-vehicle camera images for driver assistance. In: IEEE Proceedings of Intelligent Vehicles Symposium, 2005, pp. 205–210. IEEE (2005)
26. Kusiak, A., Wei, X., Verma, A.P., Roz, E.: Modeling and prediction of rainfall using radar reflectivity data: a data-mining approach. IEEE Transact. Geosci. Remote Sens. **51**(4), 2337–2342 (2013)
27. LeCun, Y., Bengio, Y., Hinton, G.: Deep learning. Nature **521**(7553), 436 (2015)
28. Lee, I.J.: Big data processing framework of learning weather information and road traffic collision using distributed CEP from CCTV video: cognitive image processing. In: 2017 IEEE 16th International Conference on Cognitive Informatics and Cognitive Computing (ICCI* CC), pp. 400–406. IEEE (2017)
29. Lee, J., Hong, B., Shin, Y., Jang, Y.J.: Extraction of weather information on road using CCTV video. In: 2016 International Conference on Big Data and Smart Computing (BigComp), pp. 529–531. IEEE (2016)
30. Liu, J., Yang, W., Yang, S., Guo, Z.: D3r-Net: dynamic routing residue recurrent network for video rain removal. IEEE Transact. Image Process. **28**(2), 699–712 (2019)
31. Loucks, D.P., van Beek, E.: Water resources planning and management: an overview. In: Water Resource Systems Planning and Management, pp. 1–49. Springer, Cham (2017). https://doi.org/10.1007/978-3-319-44234-1_1
32. Mair, A., Fares, A.: Comparison of rainfall interpolation methods in a mountainous region of a tropical island. J. Hydrol. Eng. **16**(4), 371–383 (2010)
33. Meyer, H., Kühnlein, M., Appelhans, T., Nauss, T.: Comparison of four machine learning algorithms for their applicability in satellite-based optical rainfall retrievals. Atmos. Res. **169**, 424–433 (2016)
34. Mu, P., Chen, J., Liu, R., Fan, X., Luo, Z.: Learning bilevel layer priors for single image rain streaks removal. IEEE Sig. Process. Lett. **26**(2), 307–311 (2019)
35. Ramana, R.V., Krishna, B., Kumar, S., Pandey, N.: Monthly rainfall prediction using wavelet neural network analysis. Water Resour. Manage **27**(10), 3697–3711 (2013)
36. Rani, B.K., Govardhan, A.: Rainfall prediction using data mining techniques-a survey. Comput. Sci. Inform. Technol. **3**, 23–30 (2013)
37. Roser, M., Moosmann, F.: Classification of weather situations on single color images. In: 2008 IEEE Intelligent Vehicles Symposium, pp. 798–803. IEEE (2008)
38. Shepard, D.: A two-dimensional interpolation function for irregularly-spaced data. In: Proceedings of the 1968 23rd ACM national conference, pp. 517–524. ACM (1968)
39. Simonyan, K., Zisserman, A.: Very deep convolutional networks for large-scale image recognition. arXiv preprint arXiv:1409.1556 (2014)
40. Sirirattanapol, C., Nagai, M., Witayangkurn, A., Pravinvongvuth, S., Ekpanyapong, M.: Bangkok CCTV image through a road environment extraction system using multi-label convolutional neural network classification. ISPRS Int. J. Geo-Inf. **8**(3), 128 (2019)
41. Srivastava, N., Hinton, G., Krizhevsky, A., Sutskever, I., Salakhutdinov, R.: Dropout: a simple way to prevent neural networks from overfitting. J. Mach. Learn. Res. **15**(1), 1929–1958 (2014)
42. Stein, M.L.: Interpolation of Spatial Data: Some Theory for Kriging. Springer, Heidelberg (2012). https://doi.org/10.1007/978-1-4612-1494-6

43. Stocker, T.F., et al.: Climate change 2013: The physical science basis (2013)
44. Toledo-Moreo, R., et al.: Positioning and digital maps. In: Intelligent Vehicles, pp. 141–174. Elsevier (2018)
45. Tripathi, A.K., Mukhopadhyay, S.: Removal of rain from videos: a review. Sig. Image Video Process. **8**(8), 1421–1430 (2014)
46. Wu, J., Long, J., Liu, M.: Evolving RBF neural networks for rainfall prediction using hybrid particle swarm optimization and genetic algorithm. Neurocomputing **148**, 136–142 (2015)
47. Xingjian, S., Chen, Z., Wang, H., Yeung, D.Y., Wong, W.K., Woo, W.c.: Convolutional LSTM network: a machine learning approach for precipitation nowcasting. In: Advances in neural information processing systems, pp. 802–810 (2015)
48. Yang, W., Tan, R.T., Feng, J., Liu, J., Yan, S., Guo, Z.: Joint rain detection and removal from a single image with contextualized deep networks. In: IEEE Transactions on Pattern Analysis and Machine Intelligence (2019)
49. Zhang, H., Patel, V.M.: Density-aware single image de-raining using a multi-stream dense network. In: Proceedings of the IEEE Conference on Computer Vision and Pattern Recognition, pp. 695–704 (2018)

Towards Identifying De-anonymisation Risks in Distributed Health Data Silos

Nikolai J. Podlesny(✉), Anne V. D. M. Kayem(✉), and Christoph Meinel(✉)

Hasso Plattner Institute, University of Potsdam, Potsdam, Germany
{Nikolai.Podlesny,Anne.Kayem,Christoph.Meinel}@hpi.de

Abstract. Accessing distributed and isolated data repositories such as medical research and treatment data in a privacy-preserving manner is a challenging problem. Furthermore, in the context of high-dimensional datasets, adhering to strict privacy legislation can be projected to a W[2]-complete problem whereby all privacy violating attribute combinations must be identified. While traditional anonymisation algorithms incur high levels of information loss when applied to high-dimensional data, they often do not guarantee privacy, which defeats the purpose of anonymisation. In this paper, we extend our previous work and address these issues by using Bayesian networks to handle data transformation for anonymisation [29]. By computing conditional probabilities linking attribute pairs for all attribute pair combinations the privacy exposure risk can be assessed. Attribute pairs differing by a high conditional probability indicate a high risk of de-anonymisation, similar to quasi-identifiers in syntactic anonymisation schemes, and can be separated instead of deleted. Attribute compartmentation removes the risk of privacy exposure, and deletion avoidance results in a significant reduction in information loss. In other words, assimilating the conditional probability of outliers directly in the adjacency matrix in a greedy fashion is efficient and privacy-preserving. Further, we offer deeper evaluation insights for optimising Bayesian networks with multigrid solver for aggregating state space explosion.

1 Introduction

In the recent past, there has been an increasing news footprint of de-anonymisation incidents in industries like telecommunication [11], transportation [31] and financial transactions [40] resulting in serious consequences [26] such as a class action lawsuit [3]. Simultaneously, data collections grow in size both in describing attributes and data records themselves. Those increasing data repositories can be leveraged for better diagnosis and predictions in digital health to enable personalised treatments [2,32]. Especially in the pharmaceutical industry, or for medical practitioners a large comparison population is crucial to assess assimilable courses of diseases under treatment. Accessing such data repositories is a huge struggle and restricted by privacy legislation requiring that all personally identifiable information (PII) must be removed or explicit consent must

S. Hartmann et al. (Eds.): DEXA 2019, LNCS 11706, pp. 33–43, 2019.
https://doi.org/10.1007/978-3-030-27615-7_3

be obtained for each use case. The challenge of data anonymisation is within the W[2]-completeness of properly handling multi-attribute high-dimensional datasets to find quasi-identifiers [5,27], adding the complexity of distributed and siloed health data repositories [9,46].

Problem Statement: In the digital health sector, recent developments highlight the absence of efficient data exposure identification solutions, particularly for distributed and high-dimensional data repositories [2,16,22]. Increasing dimensional complexity with often several hundred describing attributes such as in genome sequencing serve as an important knowledge-base for medical research and drug triage [33]. Additionally, strong guarantees of anonymity must be provided to abide to recent privacy regulation like GDPR [22]. Existing work focuses on certain use case specific scenarios without considering the W[2]-completeness of high-dimensional datasets or their nature of separated health repositories.

Contributions: We extend previous work [29] of using Bayesian networks for a consolidated model across several health data repositories optimised by multigrid solvers. The model is generated by learning probabilistic inferences between attribute values, risks of data exposure cannot only be identified but also be circumvented. The rest of the paper is structured in the following manner: We discuss related work in Sect. 2. In Sect. 3, we describe the approach of *Bayesian networks* and their optimisation with multigrid solver. Section 3.1 will provide an in-depth assessment of our proposed approach based on a real-world use case and dataset. Finally, Sect. 4 summarises and concludes our contributions.

2 Related Work

Recent developments highlight the absence of efficient data exposure identification solutions, particularly for distributed and high-dimensional data repositories [2,16,22].

There is some work that addresses the challenge of anonymising high-dimensional data sets. In the medical field, Kohlmayer et al. [18] present a flexible approach on top of k-anonymity, l-diversity and t-closeness as well as heuristic optimisation to anonymize distributed and separated data silos using a newly introduced Secure Multi-party Computing (SMC) protocol. Major challenges for distributed anonymisation in the healthcare industry are outlined by Mohammed et al. [25] who additionally propose *LKC-privacy* to achieve privacy in both centralized and distributed scenarios promising scalability for anonymising large datasets. *LKC-privacy* primarily leverages the thought that acquiring background knowledge is nontrivial and therefore limiting the length of quasi-identifier tuples to a predefined size L. In spite of the fact that this intention is fully comprehensible, it violates recent privacy regulations[1] since anonymity cannot be guaranteed. Other work uses a MapReduce technique based on the

[1] https://ec.europa.eu/info/law/law-topic/data-protection/reform/what-does-general-data-protection-regulation-gdpr-govern_en.

Hadoop to boost computation performance [45], this, however, does not resolve the issue of anonymisation of distributed health data silos.

Zhang et al. propose a data release technique *PrivBayes* which leverages Bayesian networks [44] and demonstrate the accuracy of Bayesian theorem for anonymisation purposes. Yet, due to the NP-hardness of *PrivBayes*, there is the need to improve its scalability. Additionally, Zhang et al. outline future work for multi-database settings also in a non-distributed environment [44]. Using the advantages of Bayes' theorem is however still seldom in the privacy community especially when it comes to conquer high-dimensional datasets siloed in multiple distributed repositories. An insider privacy intrusion detection system based on a Bayesian network technique for database systems has been proposed by An et al. [1] while Takbiri et al. leverage Markov chains to identify users' movement pattern for location-based privacy [37]. Meng et al. presented a random, projection-based method of using inferences in Bayesian networks for a set of linear equations given private dataset distributed across several parties [23]. A privacy-preserving protocol based on the K2 algorithm is introduced by Wright et al. [42] for learning the Bayesian network structure of heterogeneous, distributed data among different parties covering the underlying data. Zhang et al. investigated feature selection methods for high-dimensional micro-data with binary (categorical) features from a theoretical perspective [43].

Established generalisation algorithms like k-anonymity, l-diversity, t-closeness, and extensions [4,36] are of NP-hard nature [24] and offer vulnerabilities of de-anonymisation based on exploiting the semantics of the data. In the past, the privacy community pivoted from syntactic data anonymisation towards semantic approaches like *differential privacy*. Yet, differential privacy and its statistical methods are highly optimised for predefined use cases and bulk data processing [12,21], which knowledge might not be always available in advance. Postponing the anonymisation of data towards the runtime of a query leaves working surface for data leakage. Vulnerabilities of colluding users are also well known [17]. Leoni introduces "non-interactive" differential privacy by applying its statistical treatments a priori to a user query [19]. But almost none support strong and GDPR required guarantees whereas under any circumstance not a single data record may be linked back to its individual owner through an arbitrary family of attribute sets which uniquely identifies at least one original entity or row given arbitrary efforts [14].

This work offers a solution on how to use the Bayesian theorem to identify data exposure in high-dimensional data silos. We now describe our proposed solution.

3 Leveraging Conditional Probabilities for Data Exposure Identification

In probability theory, Bayesian networks are probabilistic directed acyclic graphical models [13] where conditional probabilities may be delineated in adjacency matrices or lists. The underlying Bayes theorem can also be described as the inter-dependency of two different events A and X, where A occurs knowingly that X already occurred. This corresponds to the conditional probability or dependency that A occurs when X has already taken place: $P(A|X)$. The Bayesian network requires the storage of $O(n^2)$ edges and doubles implementing the assigned probabilities, where n corresponds to the number of nodes.

In previous work we explained how the concept of conditional probabilities can be used to represent inter-attribute dependencies within an adjacency matrix which can be used to identify data exposure risks [29]. In detail, the cycle inference as representative of the conditional probability is summed over a certain row or column within the matrix. As an additional indicator, the mean value is calculated which combined act as metric. Any deviation can be considered as risk candidates, while an increasing deviation correlates to a higher risk in data exposure for *summed cycle inferences* [29].

3.1 Aggregating State Space Explosion Through Multigrid Solver

It is widely known that composing and sampling from a Bayesian network comes with the cost of inconvenient time complexity. As nodes in Bayesian networks traditionally represent one single state, and a high-dimensional data set with many attributes holding many different states result in a huge growth of nodes, the NP-hardness of the exact [8] or even approximate [10] inference may be a bottleneck.

Given the context of high-dimensional data sets, we pick up the measures of reducing the given complexity. Multigrid solver, which expresses the concept of coarsening the state space, originates from numerical analysis and forms a family of efficient algorithms to approximate solutions of equation systems derived from the discretization of partial differential equations [6,29,35]. One of the main differences is, that an adjacency matrix could be symmetric using multigrid (unidirectional arcs instead of directed arcs). The multigrid solver method in an exact way aggregates nodes of the Bayesian network representation which is defined by the conditional probability on each edge or correspondingly, the inferences in the adjacency matrix. The aggregation is purely based on the conditional probability along the edges emerge across all attribute values and only reflect node of disjunctive attribute subsets. As a result of such aggregation, a new node is created and replaces the former existing ones with new edges.

This *exact approach* when being in all probability may be weakened to an *approximated one* for accelerate computation time but involves higher information loss. In the latter case, nodes are aggregated given some threshold for high likelihoods of collective appearances. First experiments support a threshold for a specific dataset as >95% conditional probability. The determination of such threshold should be realised in a greedy way to ensure anonymity overall. Leveraging such methodology dramatically increases the nodes under aggregation but might result in losing 5% information as well especially the rare outliers. Respectively, by altering any threshold this balance between computation time and information loss is balanced. However, in this work we focus on leveraging the exact multigrid solver approach.

This way, the exploding complexity in large and high-dimensional datasets can be capture. By reducing the number of nodes successively in an exponential space, the runtime for sampling and analysing decreases significant since fewer node combinations and therefore edges must be evaluated. First evaluations show promising results. To better understand the effectiveness of a multigrid techniques, we refer to the security research field. Here, multigrid approaches proved to solves the Poisson-Equation for elliptic curves in $O(n)$ instead of $O(n^{2.5})$ [15,39]. In addition to multigrid, manifold learning could be used to aggregate those Bayesian network nodes as well. Manifolds as a topological space originates in mathematical physics where it has been used to reduce complicated geometric structures into simpler topological properties by resembling Euclidean space near the targeted point of observation. By formulating the dimensionality reduction problem as a classical problem in Riemannian geometry [7,20], the number of nodes can be efficiently aggregated to reduce its processing complexity (see Fig. 2). Various optimisation like the adaptive manifold learning proposed by Wang et al. [41] are used in different geometric research fields and could serve as basis for further state aggregation in Bayesian networks. Luke Olson and Jacob Schroder developed an algebraic multigrid (AMG) solver library [28] which implements a "multilevel technique for solving large-scale linear systems with optimal or near-optimal efficiency". Unlike geometric multigrid, AMG requires little or no geometric information about the underlying problem and develops a sequence of coarser grids directly from the input matrix. This feature is especially important for problems discretised on unstructured meshes and irregular grids.

The algebraic solvers serve as basis of combining and subsuming nodes as well as corresponding edges within the network to reduce the complexity of the network particularly during sampling.

4 Assessment

For deeper insights in the advantages and disadvantages of using Bayesian networks to identify data exposure across high-dimensional health data silos, a solid and in-depth evaluation is required. Therefore, the setup and metrics at hand will be defined, and a real-world use case will be assessed further to determine time complexity and data exposure. To guarantee anonymity, we will compare inferential results with unique attribute combinations (mpmUCCs) and check if individuals can be linked through any patterns. Since the novel method is fundamentally different to pre-existing work, a comparison with differential privacy is lopsided and therefore an uneven comparison.

4.1 Setup and Metrics

To run the assessment, we have organised a machine equipped with an Intel Xeon Gold 6140 CPU (12x cores) and 36 GB RAM. As underlying dataset, real world data has been gathered from multiple sources and enriched with fake profiles in close adjustment with real world data distributions. This way a fair evaluation set shall be ensured which can also be published in full for reasons of confirmability and traceability without endangering single entities through their personally identifiable information (PII). In total, the enriched data set consists of 105 attributes and a bit more than 1M rows. Composition sources for the real-world data include but may not limited to study appendices, publications, as well as official government websites. A complete list of all data sources and their mixture is available on github.com [30] whereat the dataset itself contains disease details and disease-disease relations, blood type distribution, drug as well as SNP and genome data and relations.

For the leakage measurement, we use similar metrics as Smith introduced [34], yet some complexity of the operationally-significant measures have been dropped. Since any kind of leakage is unbearable under recent privacy legislation, no unique column combination may remain serving also as main metric for leakage.

4.2 Experiment

To gain closer understanding of the Bayesian network impact in regard to anonymisation, the metric of time complexity, data informative value and precision especially in the high-dimensional context will be addressed. Figure 1b delineates the increase of the number of edges and consumed processing time over the describing attribute amount. Consequently, the time complexity for creating the exact adjacency matrix growths proportional to the number of edges and permutations. The increase, however, stays relatively reasonable as shown in Fig. 1a.

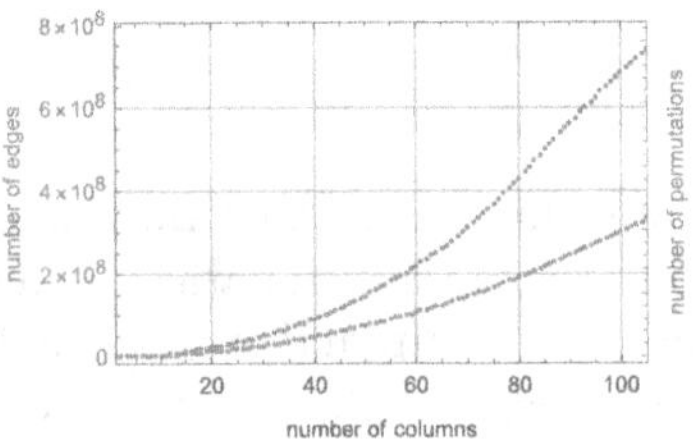

(a) Edge against permutation growth

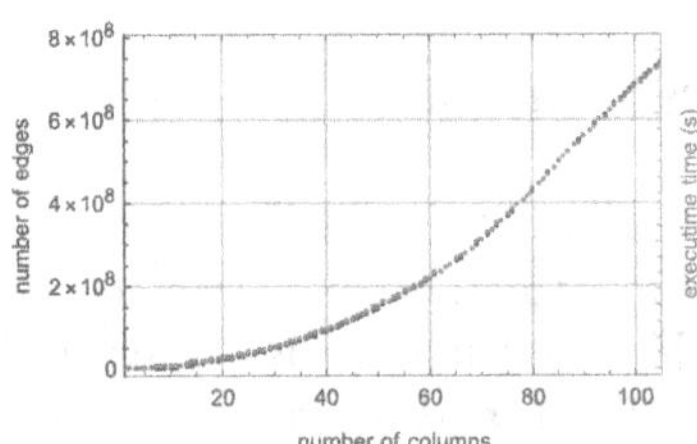

(b) Time complexity

Fig. 1. Time complexity over permutation growth

When projecting the processing time on a logarithmic scale, the growth seems to decelerate with increasing column amount. Same applies to the edges. On first sight, this might seem surprising, however, it can be explained through more available combination options within the complexity reduction procedure.

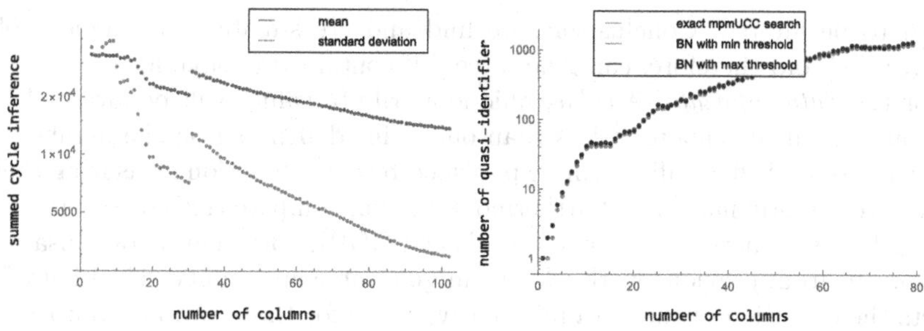

(a) Mean summed cycle inferences and its standard deviation

(b) Precision of data exposure identification through Bayesian inferences

Fig. 2. Metric visualisation over the number of columns (Color figure online)

To gain deeper discretion in the trustworthiness of the data exposure identification, the approach of summed cycle inferences can be compared to the actual unique column combinations (mpmUCCs). MpmUCCs are attribute combinations identifying at least one single row and therefore, potentially at least one individual uniquely [27]. Figure 2a illustrates the evolution of the mean summed cycle inference and its deviation. In this context, the mean summed cycle inference describes simply all inference for one record summed up, and the mean over all sums. Further, with an increasing number of describing attributes, a decreasing mean summed cycle inference as well as a decreasing standard deviation can be observed. That corresponds to our expectations, since having more options the inferences gravitate towards their lower boundary. The observed spike can be explained by the addition of a high cardinality containing attribute. Statistically speaking, the outliers to the mean summed cycle inference still exist, however, it becomes more difficult to find them. Those outliers represent data records (rows) with a higher risk of compromising anonymity. As mentioned above, trade-offs between altering the given conditional probability to counter de-anonymisation attacks and potentially loosing outliers need to be made knowing that information becomes acquainted. As long as no unique tuple (mpmUCCs) remain, which can possibly be utilised to unite original data records, we can ensure anonymity [27,38]. Cutting off the ramp of the inferential distribution on both sides supports us in the preliminary work for achieving anonymisation. The trade-off is defined by the threshold, first testing results indicate a trustworthy basis starting at 10% of the outliers. In order to validate such thresholds, we compared each identified column tuple record through the summed cycle inference with the result set of all unique column combinations (mpmUCCs) as a source of truth in Fig. 2b. The green evolution shows the increase of quasi-identifiers over the increasing number of attributes. The other lines represent the quasi-identifier as inferences found given different thresholds in our approach. The larger the delta between the approaches in Fig. 2b, the higher also the effort in finding the quasi-identifiers and the higher the potential information loss as more inferences needed to be altered. Conclusively, we find and cross-match all unique tuple (mpmUCCs) and therefore, can guarantee self-contained anonymity.

For the *data informative value*, unique attribute values will be awarded by one point, duplicates guerdon less than one point depending on the number of duplicates and falsified values will be punished by some function measuring their distance to the original value. Furthermore, we will compare this novel approach of using Bayesian networks against established GDPR compliant anonymisation mechanisms from previous work which can guarantee anonymity [27]. Figure 3b depicts the evolution of the data informative value for those various approaches while Fig. 3b only delineates the top performing one. It is essential to point out, that the original dataset is represented as grey line, the Bayesian network is applied across a distributed dataset and its model is combined for each data repository while all other approaches consume the entire dataset as bulk. Accordingly, the performance of applying Bayesian network in a decentralized setting performs significantly better than established privacy compliant anonymisation techniques like attribute compartmentation, suppression and generalisation. The previously introduced results indicate significant performance boost through the complexity reduction in the state-place-explosion. In fact, the time complexity directly corresponds to the number of edges rather than being an exponential power. Implementing such greedy approach for data exposure identification, we can guarantee to detect all quasi-identifiers as unique attribute combinations (mpmUCCs) which is required by recent privacy regulations. Following their identification, we showed that smoothing the corresponding inferences achieves to counter de-anonymisation attacks. Considering the fundamental difference of this method to pre-existing work, any comparison may be lopsided and is therefore avoided for now.

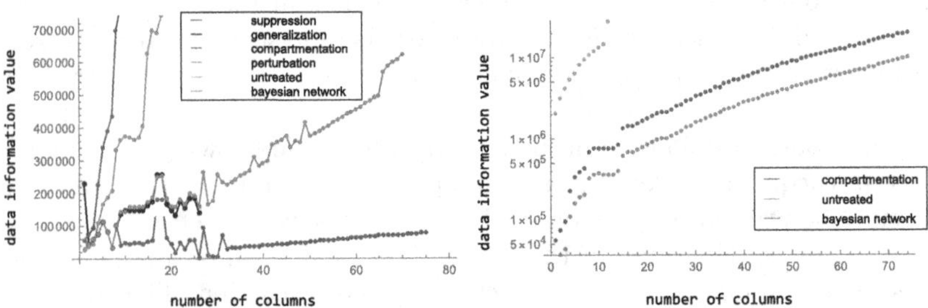

(a) Various GDPR compliant anonymisation mechanisms

(b) Top GDPR compliant anonymisation mechanisms

Fig. 3. Data informative value evolution comparing distributed processing

5 Summary and Conclusions

This paper contributes an in-depth evaluation of using Bayesian networks for anticipate private data exposure and consequently avert such in high-dimensional and distributed data repositories. By transforming relational data into an adjacency matrix representation, several benefits can be leveraged including the privacy compliant coalescence of high-dimensional data in distributed health data repositories. We showed how multigrid solver support complexity reduction in Bayesian networks by aggregating the state-space-explosion. Further, we assessed how data exposure is anticipated and circumvented based on conditional probability specific metrics within any adjacency matrix. We published additional documentation and exemplary source code publicly [30].

References

1. An, X., Jutla, D., Cercone, N.: A Bayesian network approach to detecting privacy intrusion. In: Proceedings of the 2006 IEEE/WIC/ACM International Conference on Web Intelligence and Intelligent Agent Technology, pp. 73–76. IEEE Computer Society (2006)
2. Aue, G., Biesdorf, S., Henke, N.: ehealth 2.0: how health systems can gain a leadership role in digital health. McKinsey & Company, December 2015
3. Barbaro, M., Zeller, T.: A face is exposed for AOL searcher no. 4417749, August 2006. http://www.nytimes.com/2006/08/09/technology/09aol.html
4. Bayardo, R.J., Agrawal, R.: Data privacy through optimal k-anonymization. In: 2005 Proceedings of 21st International Conference on Data Engineering, ICDE 2005, pp. 217–228. IEEE (2005)
5. Bläsius, T., Friedrich, T., Schirneck, M.: The parameterized complexity of dependency detection in relational databases. In: Guo, J., Hermelin, D. (eds.) 11th International Symposium on Parameterized and Exact Computation (IPEC 2016). Leibniz International Proceedings in Informatics (LIPIcs), vol. 63, pp. 6:1–6:13. Schloss Dagstuhl-Leibniz-Zentrum fuer Informatik, Dagstuhl, Germany (2017). http://drops.dagstuhl.de/opus/volltexte/2017/6920
6. Briggs, W.L., Henson, V.E., McCormick, S.F.: A Multigrid Tutorial. SIAM, Philadelphia (2000)
7. Carr, J.: Applications of Centre Manifold Theory, vol. 35. Springer, New York (2012)
8. Chickering, D.M., Geiger, D., Heckerman, D., et al.: Learning Bayesian networks is NP-hard. Technical Report, MSR-TR-94-17, Microsoft Research (1994)
9. Crossfield, S.S., Clamp, S.: Electronic health records research in a health sector environment with multiple provider types. In: HEALTHINF, pp. 104–111 (2013)
10. Dagum, P., Luby, M.: Approximating probabilistic inference in Bayesian belief networks is NP-hard. Artif. Intell. **60**(1), 141–153 (1993)
11. De Montjoye, Y.A., Hidalgo, C.A., Verleysen, M., Blondel, V.D.: Unique in the crowd: the privacy bounds of human mobility. Sci. Rep. **3**, 1376 (2013)
12. Dwork, C.: Differential privacy. In: van Tilborg, H.C.A., Jajodia, S. (eds.) Encyclopedia of Cryptography and Security, pp. 338–340. Springer, Boston (2011). https://doi.org/10.1007/978-1-4419-5906-5_752
13. Efron, B.: Bayes' theorem in the 21st century. Science **340**(6137), 1177–1178 (2013)
14. European Commission: opinion 05/2014 on anonymisation techniques, April 2014. https://www.pdpjournals.com/docs/88197.pdf
15. Fulton, S.R., Ciesielski, P.E., Schubert, W.H.: Multigrid methods for elliptic problems: a review. Mon. Weather Rev. **114**(5), 943–959 (1986)

16. Kayyali, B., Knott, D., Van Kuiken, S.: The big-data revolution in us health care: accelerating value and innovation, April 2013
17. Kifer, D., Machanavajjhala, A.: No free lunch in data privacy. In: Proceedings of the 2011 ACM SIGMOD International Conference on Management of Data, SIGMOD 2011, pp. 193–204. ACM, New York (2011). https://doi.org/10.1145/1989323.1989345
18. Kohlmayer, F., Prasser, F., Eckert, C., Kuhn, K.A.: A flexible approach to distributed data anonymization. J. Biomed. Inform. **50**, 62–76 (2014)
19. Leoni, D.: Non-interactive differential privacy: a survey. In: Proceedings of the First International Workshop on Open Data, pp. 40–52. ACM (2012)
20. Lin, T., Zha, H.: Riemannian manifold learning. IEEE Trans. Pattern Anal. Mach. Intell. **30**(5), 796–809 (2008)
21. Liu, F.: Generalized Gaussian mechanism for differential privacy. arXiv preprint arXiv:1602.06028 (2016)
22. Massey, R.: How the GDPR will impact life sciences and health care, February 2017
23. Meng, D., Sivakumar, K., Kargupta, H.: Privacy-sensitive Bayesian network parameter learning. In: 2004 Fourth IEEE International Conference on Data Mining, ICDM 2004, pp. 487–490. IEEE (2004)
24. Meyerson, A., Williams, R.: On the complexity of optimal k-anonymity. In: Proceedings of the Twenty-Third ACM SIGMOD-SIGACT-SIGART Symposium on Principles of Database Systems, pp. 223–228. ACM (2004)
25. Mohammed, N., Fung, B., Hung, P.C., Lee, C.K.: Centralized and distributed anonymization for high-dimensional healthcare data. ACM Trans. Knowl. Discov. Data (TKDD) **4**(4), 18 (2010)
26. Narayanan, A., Shmatikov, V.: How to break anonymity of the netflix prize dataset. CoRR abs/cs/0610105 (2006). http://arxiv.org/abs/cs/0610105
27. Podlesny, N.J., Kayem, A.V.D.M., von Schorlemer, S., Uflacker, M.: Minimising information loss on anonymised high dimensional data with greedy in-memory processing. In: Hartmann, S., Ma, H., Hameurlain, A., Pernul, G., Wagner, R.R. (eds.) DEXA 2018. LNCS, vol. 11029, pp. 85–100. Springer, Cham (2018). https://doi.org/10.1007/978-3-319-98809-2_6
28. Olson, L.N., Schroder, J.B.: PyAMG: algebraic multigrid solvers in Python v4.0 (2018). release 4.0, https://github.com/pyamg/pyamg
29. Podlesny, N., Kayem, A.V., Meinel, C.: Identifying data exposure across high-dimensional health data silos through Bayesian networks optimised by multigrid and manifold. In: 2019 IEEE 17th International Conference on Dependable, Autonomic and Secure Computing (DASC). IEEE (2019)
30. Podlesny, N.J.: Enriched health dataset (2017). https://github.com/jaSunny/MA-enriched-Health-Data
31. Rubinstein, I.S., Hartzog, W.: Anonymization and risk. 91 Washington Law Review, p. 703 (2016)
32. Sajda, P.: Machine learning for detection and diagnosis of disease. Annu. Rev. Biomed. Eng. **8**, 537–565 (2006)
33. Schadt, E., Chilukuri, S.: The role of big data in medicine, November 2015
34. Smith, G.: Recent developments in quantitative information flow (invited tutorial). In: Proceedings of the 2015 30th Annual ACM/IEEE Symposium on Logic in Computer Science (LICS), pp. 23–31. IEEE Computer Society (2015)
35. Stüben, K.: An introduction to algebraic multigrid. Multigrid, pp. 413–532 (2001)

36. Sweeney, L.: Achieving k-anonymity privacy protection using generalization and suppression. Int. J. Uncertain. Fuzziness Knowl.-Based Syst. **10**(05), 571–588 (2002)
37. Takbiri, N., Houmansadr, A., Goeckel, D.L., Pishro-Nik, H.: Fundamental limits of location privacy using anonymization. In: 2017 51st Annual Conference on Information Sciences and Systems (CISS), pp. 1–6. IEEE (2017)
38. Terrovitis, M., Mamoulis, N., Kalnis, P.: Privacy-preserving anonymization of set-valued data. Proc. VLDB Endow. **1**(1), 115–125 (2008)
39. Vaněk, P., Mandel, J., Brezina, M.: Algebraic multigrid by smoothed aggregation for second and fourth order elliptic problems. Computing **56**(3), 179–196 (1996)
40. Vessenes, P., Seidensticker, R.: System and method for analyzing transactions in a distributed ledger, US Patent 9,298,806, 29 March 2016. https://www.google.com/patents/US9298806
41. Wang, J., Zhang, Z., Zha, H.: Adaptive manifold learning. In: Advances in Neural Information Processing Systems, pp. 1473–1480 (2005)
42. Wright, R., Yang, Z.: Privacy-preserving Bayesian network structure computation on distributed heterogeneous data. In: Proceedings of the Tenth ACM SIGKDD International Conference on Knowledge Discovery and Data Mining, pp. 713–718. ACM (2004)
43. Zhang, B., Dave, V., Mohammed, N., Hasan, M.A.: Feature selection for classification under anonymity constraint. arXiv preprint arXiv:1512.07158 (2015)
44. Zhang, J., Cormode, G., Procopiuc, C.M., Srivastava, D., Xiao, X.: Privbayes: private data release via bayesian networks. ACM Trans. Database Syst. (TODS) **42**(4), 25 (2017)
45. Zhang, X., Yang, L.T., Liu, C., Chen, J.: A scalable two-phase top-down specialization approach for data anonymization using mapreduce on cloud. IEEE Trans. Parallel Distrib. Syst. **25**(2), 363–373 (2014)
46. Zillner, S., Neururer, S.: Big data in the health sector. In: Cavanillas, J.M., Curry, E., Wahlster, W. (eds.) New Horizons for a Data-Driven Economy, pp. 179–194. Springer, Cham (2016). https://doi.org/10.1007/978-3-319-21569-3_10

An Attribute-Based Fine-Grained Access Control Mechanism for HBase

Liangqiang Huang, Yan Zhu(✉), Xin Wang, and Faisal Khurshid

Southwest Jiaotong University, Chengdu 611756, China
hwanglq@163.com, {yzhu,xinwang}@swjtu.edu.cn,
faisalnit@gmail.com

Abstract. In the current age of big data, the access control mechanism of HBase, a kind of NoSQL big data management system, needs to be improved, because there are some limitations of Role-Based Access Control (RBAC) in HBase. The coarse-grained access permissions produce little effect in many cases, and the elements used for authorization are not comprehensive enough. Attribute-Based Access Control (ABAC) is suitable for the authorization of NoSQL data storages due to its flexibility. However, it has not been investigated in HBase deeply. The objective of this paper is to study the data access control in HBase and to develop an ABAC-based mechanism for the security of HBase data. In light of the wide column feature of HBase, an Attribute-Based Fine-Grained Access Control mechanism (AGAC) is proposed, which covers two aspects, users' atomic operations and five granularity levels. When a user needs to access data in HBase storage, the AGAC will give the permission or deny by verifying user's atomic operations and by analyzing user's attributes according to the access control policies related to the data granularity level. This access control mechanism is verified on publically available email dataset and is proven to be effective to improve the access control capability of HBase.

Keywords: Column-oriented big data management system-HBase · Database access control · Attribute-Based · Fine-grained · NoSQL (Not Only SQL)

1 Introduction

Data is the treasure of any organizations and data security is absolutely vital. HBase is one kind of NoSQL (Not Only SQL) databases and a wide column big data storage based on Hadoop. It has outstanding capabilities for managing large data and is adopted by many enterprises and organizations, such as *Twitter*, *Facebook*, *Alibaba* etc. However, data access control in HBase is a big challenge for researchers. The *Kerberos* technology is used to verify the identity of user in HBase [1]. RBAC currently provides five coarse-grained permissions in HBase, i.e.; *Admin*, *Create*, *Write*, *Read*, and *Execute* [1]. For example, if Smith is authenticated via *Kerberos* and he has a right to *Write* data, then he can perform any one of *Write* operations such as *put*, *delete*, *append*, etc., because all sub-operations under *Write* are permitted to use. Data in HBase may be maliciously deleted or modified on an account that may cause some

S. Hartmann et al. (Eds.): DEXA 2019, LNCS 11706, pp. 44–59, 2019.
https://doi.org/10.1007/978-3-030-27615-7_4

serious consequences and will be very dangerous for an organization. In addition, RBAC is not applied to the scenarios where some attributes such as time or places that are also important authorization elements. Therefore, a more effective fine-grained access control mechanism should be developed for HBase in order to overcome the current problem.

In recent years, Attribute-Based Access Control (ABAC) [2] attracts the attention of many researchers due to highly customized ability on data protection [3]. Entities (subjects and objects) can be effectively distinguished by their characteristics, which are attributes of entities. In our study, the subject represents a user interacting with HBase, and an object denotes data at a specified granularity level of HBase. A mechanism based on ABAC is developed, which takes the attributes as the basic authorization elements, including entities, operations, and the environment data relevant to an access. Authorization rules based on these elements are defined in the access control policy. Multiple aspects of any entity are described by attributes, e.g., time, IP address, role, so that access control policies can be specified according to the actual situation, and an access to the object will be permitted or not by evaluating access control policies. Compared with RBAC, this ABAC-based mechanism can flexibly use attributes to verify whether an access operation should be authorized or not and overcomes the main limitations of RBAC. For example, an access control policy can be defined to only permit Smith's role to perform the fine-grained operation *put* over a specific time slot instead of a coarse one *write*.

In brief, this paper proposes an Attribute-Based Fine-Grained Access Control mechanism (AGAC) for HBase. The main contributions are as follows.

1. The proposed access control mechanism covers two fine-grained aspects, which are atomic operations and five granularity levels of HBase data.
2. A set of access control policies is defined and managed in HBase. Policies should be correlated with relevant objects in each permission verification by a binding method.
3. An access controller containing parser and verifier is developed for authorization verification, which is the key module of AGAC.

The structure of this paper is as follows. The related work is briefly introduced in Sect. 2. Section 3 discusses two aspects of access control of HBase. The principles of designing and parsing of access control policy are given in Sect. 4. Section 5 addresses the enforcement of our access control mechanism. The experimental verifications are discussed in Sect. 6. Finally, we conclude the AGAC mechanism and give future research directions in Sect. 7.

2 Related Work

Many recent studies have been carried out for the fine-grained access control mechanisms in NoSQL databases. In recent survey paper [4, 5], the authors discusses the differences of access control between big data platforms and traditional data management systems, and summarizes the research status of NoSQL databases access control. Fine grained access control within NoSQL databases is still in the very early stage [5].

In [3] Colombo et al. proposed a unified ABAC mechanism for NoSQL databases, but this access control mechanism is oriented to document storage and does not consider features of HBase comprehensively. The key point in [3] is about database query rewriting based on SQL++ technique, but HBase does not support SQL++ [6]. In [7], Kulkarni proposed a fine grained access control model for wide-column NoSQL databases, but it only consider a specified scenario. Longstaff et al. discussed an ABAC-based mechanism [8], which is supposed to be applicable to several NoSQL databases, but it may work only when these databases support a high-level SQL-like language.

In [9], Lai et al. proposed an access control mechanism with fine-grained authorizations to resolve the coarse-grained permissions problem in HBase. Their work improve the access control capabilities to some extent. However, there are still some shortcomings according to their experimental results. For example, if Mr. Smith has *SCAN* permission, he can not only perform *scan* operation but also execute *get* operation in HBase, which may lead to security risk. The studies in [10–12] only focused on document storage of NoSQL, which is different from column-oriented databases, such as HBase.

This paper develops an effective fine-grained mechanism for data access control in HBase, which integrates ABAC techniques with HBase features for improving security of big data management.

3 Aspects of Access Control in HBase

The access control mechanism of HBase must cover two aspects, one is data that will be accessed, and another is the operations which are used to manipulate data resources.

For example, Mr. Smith submits a query *q* to obtain data at column *cdata* of column family *cf* in table *t* under the namespace *default*. An example of HBase data is shown in Fig. 1.

Fig. 1. An example of data in table *t* under the namespace *default*.

Mr. Smith uses *GET* (see Table 1) operation to fetch the data value *15.78* from *default.t.cf.cdata*. When Smith has submitted *q*, HBase should verify whether Smith has right to execute *GET* and whether he is permitted to access the data located by *default.t.cf.cdata*. To this end, the fine grained access control is needed.

3.1 Design of Granularity Levels of Access Control

Relational databases (e.g., Microsoft SQL Server) can authorize users to access data at the levels of database, table, and column, etc. MongoDB, a document storage of NoSQL, grants users to access data at the level of database, collection, document, and field. HBase is a column-oriented database, which is designed to accommodate semi-structured big data that could vary in file size, data types and columns. The data model in HBase consists of several logical components such as namespaces, tables, rows, column families, columns, and cells. Leaned from the access control mechanisms of RDBMS and MongoDB, a fine-grained access control mechanism for HBase should consider five levels of granularity at global database, namespace, table, column family, and column. These five granularity levels form a tree structure to demonstrate the path for addressing a data value, where the global database is the root node, columns are leaf nodes, and each node in the tree can inherit all access control policies from its parent node. For example in Fig. 2, the table *t* can inherit all access control policies from global database and namespace *default*.

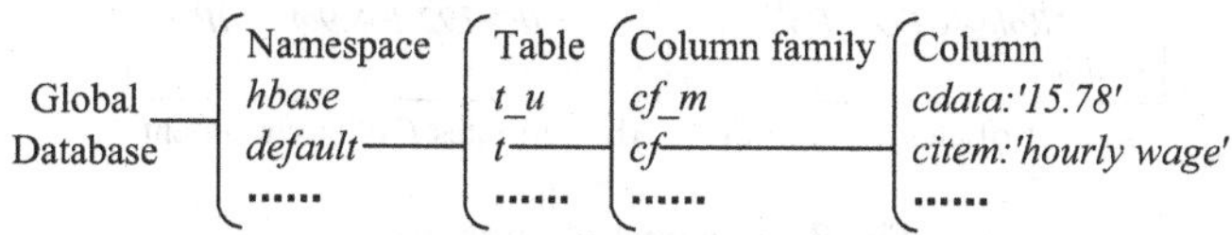

Fig. 2. The policy inheritance relationship between five granularity levels.

3.2 Design of Access Operation Control

RDBMS can authorize users data access operations by SQL, for example, *GRANT SELECT ON TABLE_1 TO SMITH.* HBase does not support SQL, but provides the basic way of Java API to manipulate data. By calling these basic Java APIs, the functionalities can be implemented similarly to those in terms of SQL. The Java APIs are classified in terms of data manipulation functions, which are grouped as 8 types and 46 kinds operations [13]. Some are shown in Table 1. The letter in a parenthesis represents the finest grained access level of operation. G, N, T, CF and C stand for global database, namespace, table, column family, and column respectively.

Table 1. Examples of operation types and operation sets.

Type	Set
DDL	*ALTER(CF), CREATE(N), DESCRIBE(T), LIST(T), DROP(T)*
DML	*APPEND(C), DELETE(C), GET(C), TRUNCATE(T), SCAN(C), PUT(C)*
Snapshot	*SNAPSHOT(T), RESTORE_SNAPSHOT(T), DELETE_SNAPSHOT(G)*

These operations are strong tools to manage and query big data, but may be misused. So there is a security risk posed to HBase. To this end, operations provided by HBase will be studied in our fine-grained access control mechanism in order to develop policies for managing them reasonable.

4 Access Control Policy and Its Parser

4.1 Definition of Policy

The key elements of AGAC are attributes of subjects, objects, operations and the environment relevant to a request. These elements constitute an access control policy. An access control policy is a XML file. We define the key concepts of policy as follows and several of them drew inspiration from [3].

Definition 1. (Subject): Subjects define users interacting with HBase. A subject S is a set of attributes which characterizes a user. A subject attribute s_i is a pair *<aid, val>*, where *aid* is the attribute identifier, and *val* represents the current value of s_i.

For example, if Mr. Smith's role is CEO, the subject of Smith is defined in Fig. 3(a).

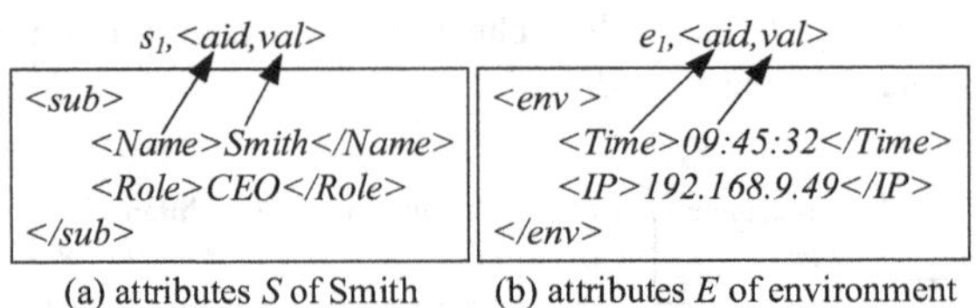

Fig. 3. Example of attributes.

Definition 2. (Object): An object O denotes data at a specified granularity level. For example, data at the specified column family *cf*, which path is *default.t.cf.*

Definition 3. (Environment): An environment E is a set of attributes which characterize the access request context, which can be used to influence authorization. An environment attribute e_i is a pair *<aid, val>*, where *aid* is the attribute identifier, and *val* specifies the value of e_i.

For example, an environment E is defined by the time and IP address, then E can be defined in Fig. 3(b).

Definition 4. (Access Control Policy): An access control policy is a triple *<operations, sub, env>*, where *operations* is a set of permitted operations, *sub* specifies attributes of subject that must meet the conditions of the authorization, *env* specifies which attributes of environment must meet the conditions of the authorization

Two policy instances are given in Fig. 4.

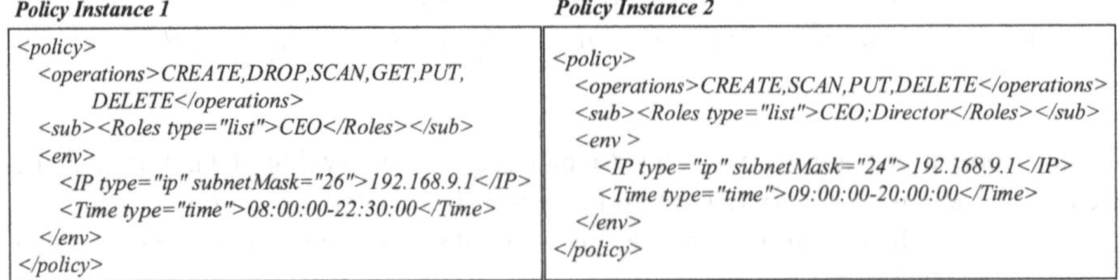

Fig. 4. Example of policy instances.

For example, if *Policy Instance 1* is bound with the object *default.t.cf.cdata*, that means all operations on object *default.t.cf.cdata* are controlled by *Policy Instance 1*.

If Mr. Smith's role is CEO, he uses *GET* to fetch a data from *default.t.cf.cdata* at time *14:20:23*, and the IP address of Smith's device is *192.168.9.49*, his data access will be permitted. This is because CEO is in the authorized list of *Policy Instance 1*, IP address belongs to the authorized IP address segment (*192.168.9.1-192.168.9.62*) which is calculated by subnet mask and IP start address, and the access time is in the authorized time slot. Therefore, Smith is permitted to access data located by *default.t.cf.cdata*.

On the contrary, if Smith's role is CTO and/or the access time is *07:58:00*, then Smith is rejected to access data located by *default.t.cf.cdata*.

4.2 Binding Policies with Objects

A policy is bound with an object to determine to which object the policy applies. A policy can be bound with multiple objects and multiple policies can be bound with one object. An object is identified by the row key, and a policy instance is specified by the policy serial number. Connecting the row key and the policy serial number will bind them together. Such a binding information will be managed in a HBase table, AGACL (Attribute-Based Fine-Grained Access Control List).

AGACL is created under the namespace *hbase* for managing policies and policies binding information. The detailed binding method is described as follows.

Row Key of AGACL. The row key of AGACL is designed by using the feature of row keys in HBase, which means the HBase data is stored uniquely in a table according to the sort order of dictionary of row keys. They can be designed as:

$$<namespace\ name\ [:table\ name\ [:column\ family\ name\ [:column\ name]]]>$$

The value of row key denotes the path of the objects to which the policies are applied. For example, the value of row key *"default:t:cf:cdata"* means the policies stored in this row of AGACL table, which will be used to control the access of the data located at the cell addressed with *default.t.cf.cdata*. The special value *'hbase:global'* specifies the policies are applied on the global database.

Column Family of AGACL. As the column can only be attached to the column family, the column family *'abcf'* (Attribute-Based Column Family) is created in AGACL.

Column of AGACL. Multiple policies can be applied to an object, so that a serial number is designed to identify each policy. The column can be designed as:

$$<policy : serial\ number>$$

Suppose we want to bind *Policy Instance 1* and *Policy Instance 2* with column *cdata* in column family *cf* in table *t* under the namespace *default*, the format of policy in AGACL is shown in Fig. 5. The parts marked with black box should be replaced with actual content of the XML files (see Fig. 4), which is simplified here to reduce the space.

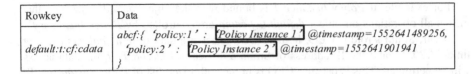

Rowkey	Data
default:t:cf:cdata	*abcf:{ 'policy:1' : 'Policy Instance 1' @timestamp=1552641489256,* *'policy:2' : 'Policy Instance 2' @timestamp=1552641901941* *}*

Fig. 5. Format of policy bound in AGACL.

4.3 Policy Processing

From the operational point of view, binding or unbinding means to write a policy into or delete a policy from AGACL according to the path of object. In our access control mechanism, we use PUT operation to write a policy into AGACL, and DELETE to take the policy away from AGACL. The *Superuser* in HBase has all permissions and can initialize the access control mechanism of HBase.

We developed a parser using *dom4j* [14] (a flexible XML framework for Java). The parser is used to analyze a policy in XML to produce an authorization condition set. The parsing procedure is as follows.

Step 1: Load the XML file to the parser.
Step 2: Get the list of operations according to the label *operations*.
Step 3: Get subject attributes, including the list of roles.
Step 4: Get environment attributes, including the IP address segment calculation by subnet mask and IP start address, and the time slot.

5 Enforcement of AGAC

5.1 The Framework of AGAC

HBase provides a framework-*Coprocessor* [1] that allows developers to write their own program in the server to implement fine-grained access control. In this study, we develop an Attribute-Based Fine-Grained Access Controller (AGACer) based on *Coprocessor*, which is integrated into HBase. AGACer can perform policy parsing and access authorization verification based on users' attributes and policies, when users submit a data access request. The framework of our AGAC mechanism is shown in Fig. 6.

The workflow of our framework consists of seven main steps. ① A user submit a data access request. ② Send the request and the attributes to verifier. ③ Search and find the policies based on the path of the required object. ④ Send the relevant policies to the parser. ⑤ Send the key attributes of the parsed policies to verifier. ⑥ If the verification is successful, the request is permitted to execute, else the permission is denied, exception is returned. ⑦ Return the result of the request under the successful permission.

AGACer contains parser and verifier. The parser is used to analyze policies in XML to produce the authorization conditions (ref. Sect. 4.3). The verifier evaluates whether the operation in the request and user's attributes meet the authorization conditions.

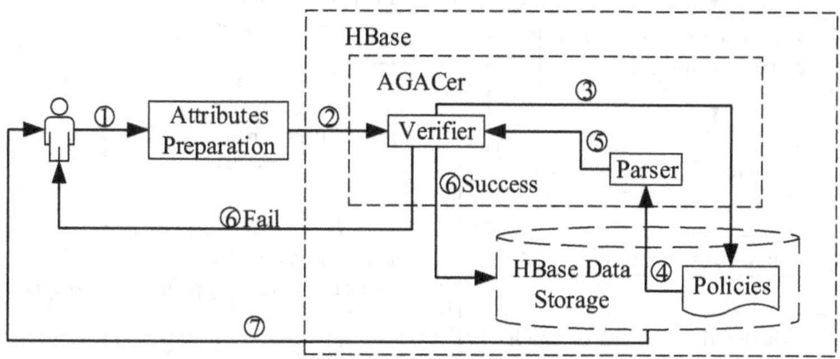

Fig. 6. The framework of AGAC.

5.2 Preparation of Users' Attribute Set

The subject is users' static attributes, such as CEO, Project Leader, and Department Manager. In our work, subject is prepared and managed through information transformation from HR department or from the e-forms filled by users when they register in HBase system.

The environment parameters are users' dynamic attributes, such as IP address, request time, and place. Environment attributes are prepared when a user submits an access request to HBase data, which is accomplished automatically by our mechanism.

5.3 Process of Access Control

We use an example to show the procedure of access control. Let's assume that *Policy Instance 1* (Fig. 4) is already bound with the Namespace *default* by *Superuser*. Mr. Smith submitted a database request *Q* which uses *CREATE* operation to create a table *t* at namespace *default*, the process of authorization verification is addressed in Fig. 7.

Attributes Preparation. An example of the attributes of Mr. Smith and the environment attributes relevant to *Q* are shown in Fig. 8.

Policy Acquiring. The access object of *Q* is the namespace *default*. Because of the policy inheritance, all policies from its parents, global database, should be firstly applied to control the object access. However, there are no policies on the global database, therefore we can use all policies from the namespace *default*, where the data object locates.

Policy Evaluation. Based on the policies acquired at the previous step, the attribute values of User Smith and *CREATE* are evaluated as whether they meet the authorization conditions. For example, the IP address of Smith should belong to the authorized IP address segment. If the attributes and operation together conform to one

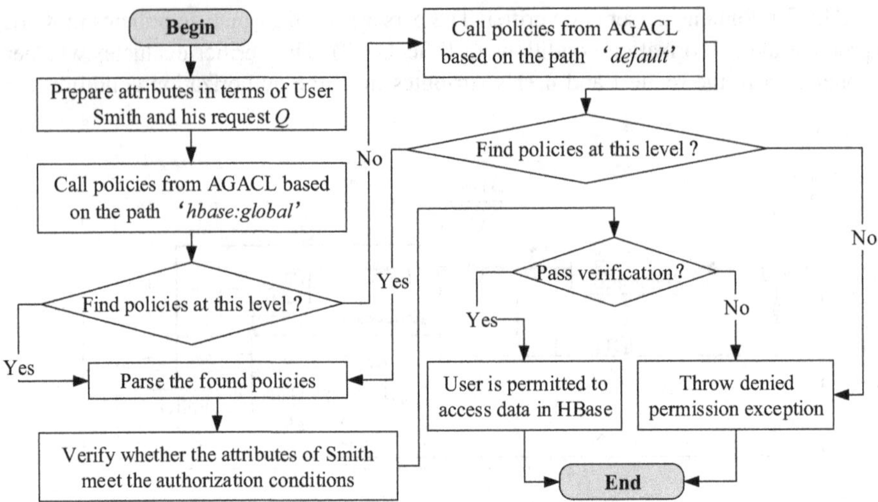

Fig. 7. The verification process of authorization.

Attributes of Smith:

<attributes> *_{<Role>CEO</Role>}* *<env><IP >192.168.9.49</IP><Time>10:30:21</Time></env>* *</attributes>*

Fig. 8. Attributes of Mr. Smith

Table 2. Policy evaluation algorithm.

Input: *policies* obtained from previous step; an operation *op*; *S* denotes user attributes; *E* denotes environment attributes.
Output: *true* denotes the access is successfully permitted, *false* denotes the fail case.

Begin
For each *policy* in *policies* do
If *policy* does not contain *op*, then go to next 'For' loop.
If *S* does not meet the authorization requirements of *policy*, then go to next 'For' loop.
Else if *E* meet the authorization conditions of *policy*, then return *true*.
End For
Return *false*;
End

of the policies, Mr. Smith is permitted to manipulate access object using *CREATE*. The policy evaluation algorithm is described in Table 2.

In the example, Mr. Smith passed the access authorization verification and can execute *Q*, because both attributes and access operation conform to the authorization conditions of *Policy Instance 1*.

6 Experimental Verification and Analysis

We have designed two experiments to verify the effectiveness and policy inheritance. The access control mechanism is effective, if AGAC can control the access operations over HBase data correctly. In policy inheritance evaluation, parents' policies should be inherited by children for managing data access. However, policies cannot be inherited between peers or from children to parents.

To the best of our knowledge, the ABAC-based mechanism in HBase has not been applied. It is the first study to suggest this solution, which is verified by applying different test cases to confirm effectiveness of our proposed AGAC mechanism.

The experiments are carried out on a cluster built by three PCs. The cluster is based on the master-slave architecture with one master node and two slave nodes. The environment configuration is as follows, *Ubuntu 16.04 LTS, HBase 1.2.6, Hadoop 2.7.6, Zookeeper 3.10.4, JDK 1.8*.

6.1 HBase Data Storage Construction and Experiment Preparation

The dataset used in this study is *Enron Corpus* [15] publically available, which contains emails and related information from a company employees. The dataset contains the subject, recipients, body and sending date of email, etc. The dataset is preprocessed and imported into HBase. Accessing data will be managed with 5 granularity levels from global database, namespace, table, and column family to column. The schema is shown in Table 3.

Table 3. Schema of enron dataset in HBase.

Table	Rowkey	Column family
enronEmail	*sender:messageId*	*message :{body, subject, recipientList, sendDate ...}*
enronUser	*emailId*	*info :{firstName, lastName, status ...}*

It is supposed that two users (U_1 and U_2) want to manipulate the Enron email data in HBase using a set of operations. Those operations with high usage frequency are selected in the assessment, which are listed in Table 4.

Table 4. Requests for testing.

Request	Content
Req1	Create a specified table *testTable1* under the namespace *default* using *CREATE* operation
Req2	Use *DROP* to drop the table *testTable2* under the namespace *default*
Req3	Use *GET* to query a specified email record sent by *lisa.gang@enron.com*
Req4	Use *SCAN* to query from email address *marie.heard@enron.com* the historic email records
Req5	Add the body of an email record using *PUT*
Req6	Delete the body of an email record using *DELETE*

The attributes of U_1 and U_2 are prepared (shown in Fig. 9). The test time is from 1 p.m. to 3 p.m (obtained from server in real time). Two access control policy instances from Sect. 4.1 are used in the experiments.

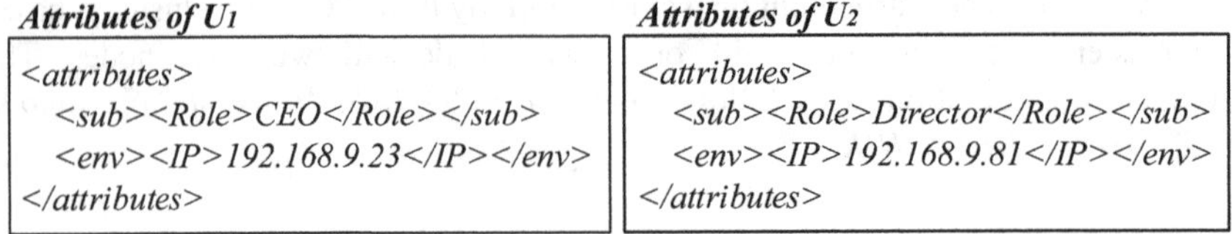
Attributes of U1
<attributes>
_{<Role>CEO</Role>}
<env><IP>192.168.9.23</IP></env>
</attributes>
Attributes of U2
<attributes>
_{<Role>Director</Role>}
<env><IP>192.168.9.81</IP></env>
</attributes>

Fig. 9. The attributes of U_1 and U_2.

6.2 Experiments and Analysis

The access control of five granularity levels will be evaluated on the hierarchy from top downwards.

Case 1. Binding *Policy Instance 1* with the global database. U_1 and U_2 execute requests in Table 4, respectively. If users' requests are permitted to execute, HBase return the execute results. For example, Fig. 10 shows the result of *Req3* executed by U_1.

```
body: According to the scheduling sheet, deal#380414 is 400mw total.  This is a 50 mw deal.  25mw was booked
recipientList: TO:jennifer.thome@enron.com
reference: From:          Thome, Jennifer  Sent:  Monday, June 24, 2002 2:05 PMTo:          Gang, LisaSubject:
sendDate: 2002/6/24 15:10:37
subject: RE: BPA - Dec. 2001
```

Fig. 10. The result of *Req3* executed by U_1.

If users' requests are rejected, HBase throw the permission denied exceptions. Figure 11 shows the rejection information of *Req1* executed by U_2.

```
org.apache.hadoop.hbase.security.AccessDeniedException: Insufficient permissions:[operate=CREATE;namespace:default]
        at org.apache.hbase.security.AGAC.AGACer.preCreateTable(AGACer.java:143)
        at org.apache.hadoop.hbase.master.MasterCoprocessorHost$11.call(MasterCoprocessorHost.java:216)
        at org.apache.hadoop.hbase.master.MasterCoprocessorHost.execOperation(MasterCoprocessorHost.java:1146)
        at org.apache.hadoop.hbase.master.MasterCoprocessorHost.preCreateTable(MasterCoprocessorHost.java:212)
```

Fig. 11. Rejection information of *Req1* executed by U_2.

All access control results of Case 1 are shown in Table 5, where *Pass* means access is permitted, and *Reject* means access is not allowed.

Table 5. Results of Case 1.

	Req1	*Req2*	*Req3*	*Req4*	*Req5*	*Req6*
Results of U_1	*Pass*	*Pass*	*Pass*	*Pass*	*Pass*	*Pass*
Results of U_2	*Reject*	*Reject*	*Reject*	*Reject*	*Reject*	*Reject*

The results show that the proposed access control mechanism is effective and can make correct authorization verification. Since *Director* is not in the authorized list of *Policy Instance 1*, the IP address of U_2 does not belong to the authorized IP address segment, all requests of U_2 are rejected.

The result of *Req1* executed by U_1 shows that the policy inheritance is successful, because *Policy Instance 1* is bound with the global database, the access permission of namespace *default* is inherited from that of root level (global database), although *Req1* is to access data in the namespace *default.*

Case 2. Unbinding *Policy Instance 1* from global database firstly, then binding *Policy Instance 2* with the namespace *default.* U_1 and U_2 execute requests in Table 4, respectively. The access control results are shown in Table 6. Since neither *DROP* nor *GET* is an operation permitted by *Policy Instance 2*, *Req2* and *Req3* issued by both users are rejected. It shows that the proposed access control mechanism can control fine-grained operations.

Table 6. Results of Case 2.

	Req1	*Req2*	*Req3*	*Req4*	*Req5*	*Req6*
Results of U_1	*Pass*	*Reject*	*Reject*	*Pass*	*Pass*	*Pass*
Results of U_2	*Pass*	*Reject*	*Reject*	*Pass*	*Pass*	*Pass*

Case 3. Unbinding *Policy Instance 2* from namespace *default*, then binding *Policy Instance 1* and *Policy Instance 2* with table *enronEmail.* U_1 and U_2 execute each request in Table 4, respectively.

The access control results are shown in Table 7. Because the execution level of *CREATE* is the namespace level, *Req1* is rejected, although *CREATE* is permitted on the table level. This proves that high levels cannot inherit low-level policies. The execution level of *DROP* is at the table level, but there are no policies bound with *testTable2* and policies cannot be inherited between peers, so that *Req2* is rejected, because *Req2* should be permitted or not to execute could not be judged.

Table 7. Results of Case 3.

	Req1	*Req2*	*Req3*	*Req4*	*Req5*	*Req6*
Results of U_1	*Reject*	*Reject*	*Pass*	*Pass*	*Pass*	*Pass*
Results of U_2	*Reject*	*Reject*	*Reject*	*Pass*	*Pass*	*Pass*

Case 4. Unbinding *Policy Instance 1* and *Policy Instance 2* from table *enronEmail.* Binding *Policy Instance 1* with column family *message.* U_1 and U_2 execute each request of Table 4, respectively. The authorization verification is as the same as the cases above. The access control results are shown in Table 8.

Table 8. Results of Case 4.

	Req1	*Req2*	*Req3*	*Req4*	*Req5*	*Req6*
Results of U_1	*Reject*	*Reject*	*Pass*	*Pass*	*Pass*	*Pass*
Results of U_2	*Reject*	*Reject*	*Reject*	*Reject*	*Reject*	*Reject*

Case 5. Unbinding *Policy Instance 1* from column family *message.* Binding *Policy Instance 1* with column *body.* U_1 and U_2 execute the requests given in Table 4, respectively. The authorization verification is as the same as the cases above. The access control results are shown in Table 9.

Table 9. Results of Case 5.

	Req1	*Req2*	*Req3*	*Req4*	*Req5*	*Req6*
Results of U_1	*Reject*	*Reject*	*Reject*	*Reject*	*Pass*	*Pass*
Results of U_2	*Reject*	*Reject*	*Reject*	*Reject*	*Reject*	*Reject*

AGAC mechanism proposed in this paper is proven to be effective by the experiments. It can safeguard HBase data and system to a very large extent by verifying who can use which operations to manage what kind of data.

6.3 Time Overhead Analysis

We use *executeTime* in milliseconds to record the process time of AGACer.

$$execute\,Time = endTime - startTime \tag{1}$$

Time overhead for verifying authorization is affected by many factors; such as network bandwidth of the cluster and machine performance. The time cost of our access control mechanism has two parts. One is the time for accessing AGACL to acquire the policy set for authorization. Another is the time for verifying authorization.

For example, the time overhead of Case 1 is shown in Fig. 12. The time overhead for U_1's successful authorization verification is significantly less than the time overhead for rejecting U_2. This is because U_1 passes authorization verification at the global level, which is the coarsest grained level and did not need much time to obtain the policies, while verifying authorization for U_2 needs more time to search and obtain the polices along the fine-grained path.

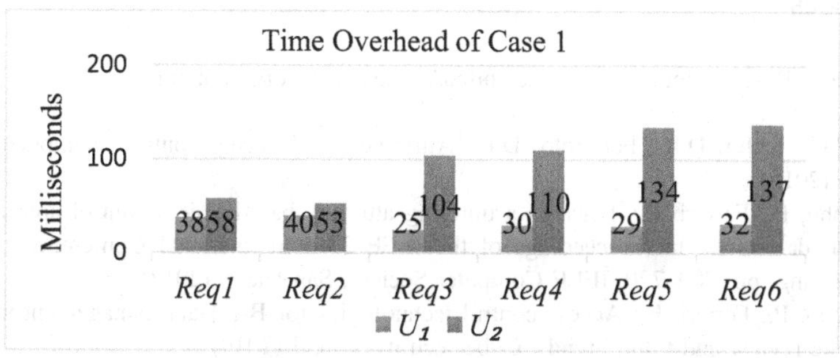

Fig. 12. The time overhead of Case 1.

In addition, the number of policies used for each request is also an important factor affecting time overhead, which cannot be ignored.

7 Conclusions and Future Work

Based on the in-depth study of HBase, we proposed a novel access control mechanism AGAC for improving access control capability of HBase.

AGAC mechanism consists of three key parts, i.e., defining access control policies and its parser, storing and managing the policies, and developing an access controller. AGAC covers two fine-grained aspects, which are atomic operations and five data granularity levels, therefore HBase data access is controlled carefully and fine-grained. Our experimental results prove that AGAC improves the access control capability and supports flexibility authorization of HBase. Besides, the proposed AGAC mechanism can be implemented quickly in HBase.

In the future, we plan to improve and extend AGAC mechanism based on following aspects.

1. Policy conflict checking and a default policy will be investigated.
2. The technique for extracting the complete entity attributes and the approach for automatically correlating attributes and permissions should be studied in depth.
3. A better model than XML should be investigated for policy description and quickly parsing, in order to improve the performance.
4. NoSQL databases are in varied forms [16], such as key-value storage, graph databases. To improve the access control ability of these NoSQL databases is still a big challenge. Modifying and integrating AGAC mechanism in graph storages of NoSQL will be our next focus.

Acknowledgement. This work is supported by the Sichuan Science and Technology Program (No 2019YFSY0032).

References

1. Apache HBase™ Reference Guide. http://hbase.apache.org/book.html. Accessed 20 Feb 2019
2. Hu, V.C., Kuhn, D.R., Ferraiolo, D.F.: Attribute-based access control. Computer **48**(2), 85–88 (2015)
3. Colombo, P., Ferrari, E.: Towards a unifying attribute based access control approach for NoSQL datastores. In: Proceedings of the IEEE 33rd International Conference on Data Engineering, pp. 709–720. IEEE Computer Society, San Diego (2017)
4. Colombo, P., Ferrari, E.: Access control technologies for Big Data management systems: literature review and future trends. Cybersecurity **2**(1), 3 (2019)
5. Colombo, P., Ferrari, E.: Access control in the era of big data: state of the art and research directions. In: Proceedings of the 23rd ACM Symposium on Access Control Models and Technologies, pp. 185–192. ACM, Indianapolis (2018)
6. Ong, K.W., Papakonstantinou, Y., Vernoux, R.: The SQL++ unifying semi-structured query language, and an expressiveness benchmark of SQL-on-Hadoop, NoSQL and NewSQL databases. Comput. Sci. CoRR, abs/1405.3631 (2014)
7. Kulkarni, D.: A fine-grained access control model for key-value systems. In: Proceedings of the 3rd ACM Conference on Data and Application Security and Privacy, pp. 161–163. ACM, San Antonio (2013)
8. Longstaff, J., Noble, J.: Attribute based access control for big data applications by query modification. In: Proceedings of the IEEE Second International Conference on Big Data Computing Service and Applications, pp. 58–65, IEEE, Oxford (2016)
9. Lai, Y.Y., Qian, Q.: HBase fine grained access control with extended permissions and inheritable roles. In: Proceedings of the 2015 IEEE/ACIS 16th International Conference on Software Engineering, Artificial Intelligence, Networking and Parallel/Distributed Computing, pp. 181–185, IEEE, Takamatsu (2015)
10. Colombo, P., Ferrari, E.: Towards virtual private NoSQL datastores. In: Proceedings of the 32nd IEEE International Conference on Data Engineering, pp. 193–204, IEEE, Helsinki (2016)

11. Colombo, P., Ferrari, E.: Enhancing MongoDB with purpose based access control. IEEE Trans. Dependable Secure Comput. **14**(6), 591–604 (2017)
12. Colombo, P., Ferrari, E.: Fine-grained access control within NoSQL document-oriented datastores. Data Sci. Eng. **1**(3), 127–138 (2016)
13. Huang, L.Q., Zhu, Y., Tao, X.: Research on fine-grained access control method based on HBase. Appl. Res. Comput. (2019). https://doi.org/10.19734/j.issn.1001-3695.2018.08.0648. (In Chinese)
14. Flexible XML framework for Java. https://dom4j.github.io/. Accessed 20 Feb 2019
15. Klimt, B., Yang, Y.: The enron corpus: a new dataset for email classification research. In: Boulicaut, J.-F., Esposito, F., Giannotti, F., Pedreschi, D. (eds.) ECML 2004. LNCS (LNAI), vol. 3201, pp. 217–226. Springer, Heidelberg (2004). https://doi.org/10.1007/978-3-540-30115-8_22
16. DB-Engines Ranking. https://db-engines.com/en/ranking. Accessed 20 Feb 2019

Data Structures and Data Management

Lopper: An Efficient Method for Online Log Pattern Mining Based on Hybrid Clustering Tree

Jiawei Liu[1], Zhirong Hou[1], and Ying Li[2(✉)]

[1] School of Software and Microelectronics, Peking University, Beijing, China
{colordown,hou.zhirong}@pku.edu.cn
[2] National Engineering Center of Software Engineering, Peking University, Beijing, China
li.ying@pku.edu.cn

Abstract. Large-scale distributed system suffers from the problem that system manager can't discover, locate and fix system anomaly in time when system malfunctions. People often use system logs for anomaly detection. However, manually inspecting system logs to detect anomaly is unfeasible due to the increasing scale and complexity of distributed systems. As a result, various methods of automatically mining log patterns for anomaly detection have been developed. Existing methods for log pattern mining have drawbacks of either time-consuming or low-accuracy. In order to address these problems, we propose Lopper, a hybrid clustering tree for online log pattern mining. Our method accelerates the mining process by clustering raw log data in one-pass manner and ensures the accuracy by merging and combing similar patterns with different kernel functions in each step. We evaluate our method on massive sets of log data generated in different industrial applications. The experimental results show that Lopper achieves the accuracy with 92.26% on average which is much better than comparative methods and remains high efficiency at the same time. We also conduct experiments on system anomaly detection task using the log patterns generated by Lopper, the results show an average F-Measure performance of 91.97%, which further proves the effectiveness of Lopper.

Keywords: Log analysis · Log pattern mining · System anomaly detection · Data structure

1 Introduction

Due to the complex composition and operation logic of large-scale distributed system, it is often difficult for system managers to discover, locate, and diagnose system anomaly in the event of a system failure. How to quickly and accurately identify, detect and even predict the failure of large-scale distributed system has become an important research issue.

At present, the mainstream technology of anomaly diagnosis for large-scale distributed systems is based on system logs [1, 2, 4, 6, 15]. System logs based anomaly detection has following advantages: 1. System logs can track program execution logic

S. Hartmann et al. (Eds.): DEXA 2019, LNCS 11706, pp. 63–78, 2019.
https://doi.org/10.1007/978-3-030-27615-7_5

and capture exceptions across components and services. 2. System logs describe the system state more fine-grained which can locate specific error log, event information and even program code.

System logs are used to record the system runtime information. In general, system logs are consisted of constants and variables. The constant is used to describe the role of current code or the reason for triggering log print statement and the variable reflects the dynamic system runtime data, such as IP address and parameter information [4, 6].

Researchers often focus on the constant part when using system logs for anomaly detection [3, 4, 6]. To better describe the work, we come up with the concept of log pattern: **A log pattern** [3, 13] is composed of the constant part of a log record and wildcards which are used for replacing variables. Moreover, a log pattern is used to identify a specific type of system log and essentially corresponds to a log print statement in program code. **Log pattern mining** [8, 9, 13] refers to the process of mining log patterns from massive raw log data. The following example shows a set of original system log records and their corresponding log patterns (see Fig. 1).

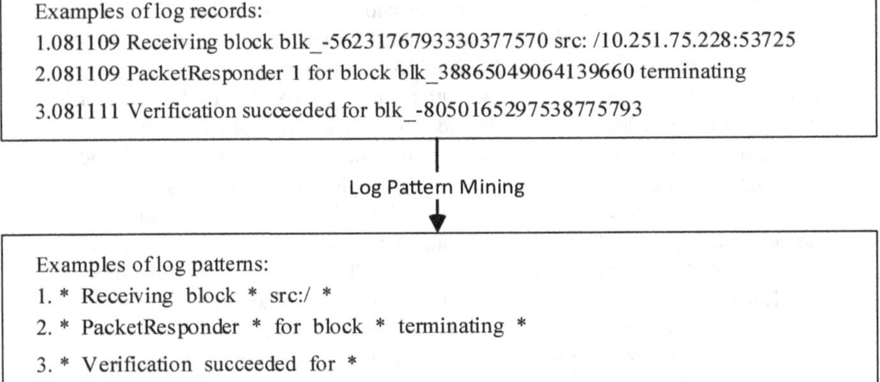

Fig. 1. Examples of log records and log patterns

The research of log pattern mining has been a hotspot for many years, and there have been many research results (related works will be further discussed in detail in Sect. 2). Vaarandi [12] design a data clustering algorithm which learns the idea from the classic Apriori algorithm for mining frequent item sets. However, this method couldn't perform well on large data set due to its high time complexity. Xu et al. [6] propose a method using abstract syntax tree for mining log pattern based on source code, which could be useful but only apply to the condition where source code is accessible. Tang et al. [10] come up with a model named LogSig for clustering log data according to their longest common sequence (LCS), which is not efficient because the process of calculating the LCS of two log records is time-consuming. Makanju et al. [8] design IPLoM which clusters system logs using iterative partitioning, but their method has high space complexity. All the early approaches for log pattern mining act in offline mode and we observe that they are not accurate enough, with the increasing demand for real-time analysis of online log data, online log pattern mining with high accuracy [3, 7, 9, 14] has become a trend and necessity.

Existing methods have drawbacks of either time-consuming or low-accuracy [2, 3]. In this paper, we propose an online log pattern mining method named Lopper to improve the performance of log pattern mining task. Lopper is a hybrid clustering tree with three layers, the first of which is used for clustering raw log data, the second is for merging similar clusters and the last is for combining candidate patterns to get final patterns. Lopper accelerates the mining process by clustering raw log data in one-pass manner and ensures the accuracy by merging and combing similar patterns with different kernel functions in each step. We evaluate our method on massive sets of log data generated in different industrial applications, the experimental results show that Lopper achieves the accuracy with 92.26% on average which is much better than comparative methods and remains high efficiency at the same time. We also conduct experiments on system anomaly detection task using the log patterns generated by Lopper, the results show an average F-Measure performance of 91.97%, which further proves the effectiveness of Lopper.

The rest of the paper is organized as follows: Sect. 2 introduces related work on log pattern mining, Sect. 3 presents our online log pattern mining method, Lopper. In Sect. 4, we evaluate the performance of Lopper on massive log data sets and Sect. 5 concludes the paper.

2 Related Work

In this section, we will give a detailed survey on the research of log pattern mining.

2.1 Offline Log Pattern Mining

As we discussed before, Log pattern mining technology refers to the process of mining log patterns from massive raw log data. According to specific technical approach, log pattern mining technology can be classified as following: clustering based methods, frequent item sets mining based methods, static code analysis based methods and other methods.

Clustering based log pattern mining technology [4, 7, 10, 13] introduces clustering algorithm into log pattern mining. The core idea is using the similarity between log records to calculate the distance for clustering. The clustering results are extracted to form the final log patterns. Clustering based method is the mainstream method in log pattern mining technology. This method usually goes as following: 1. Data preprocessing: Using regular expressions based on domain knowledge to replace commonly-used variables in log records with wildcards, such as IP address and timestamp. Data preprocessing can reduce the complexity of following work effectively. 2. Clustering log records: Clustering log data using clustering algorithms based on the distance between log records, such as edit distance [16]. 3. Log pattern extraction: Clusters are formed after clustering and different clusters correspond to different log patterns. Pattern extraction is the process of extracting one certain log pattern from a cluster.

Frequent item sets mining based method [12] is first used in the field of log pattern mining, this method can discover association rules between data and give frequent item sets. Log data has the feature that constants come up more frequent than variables [2, 3], hence by mining the frequent item sets in log records we can get the log patterns. Taking the classic Apriori algorithm as an example, each log record is treated as a transaction and each token (constant or variable) in the log record is treated as a commodity. The frequent item sets mined are often the combination of log constants, which can be reorganized to form log patterns.

Log pattern mining based on static code analysis [6] is aimed at the system whose source code is available. This approach differs significantly from the previous methods because it analyzes the source code instead of log data. The core idea of this method is directly mining log print statement from the source code and generating log patterns.

Other research includes that Zhu et al. [5] treat logs as natural language from the perspective of Natural Language Processing (NLP) and propose an incremental learning model to study the schemes of log data. Cheng et al. [11] design a deep convolutional neural network for automated classification of anomalous events detected from the distributed system logs. Lu et al. [1] come up with a CNN-based model which can automatically learn event relationships in system logs and detect system anomaly.

2.2 Online Log Pattern Mining

All the early approaches for log pattern mining act in offline mode, but with the increasing demand for real-time analysis of online log data, online log pattern mining has become a trend and necessity. Mizutani [9] design the first online log pattern mining model named SHISO, which is a tree with predefined rule for clustering log data in online way. SHISO has to update each time when a new log record comes in, so the model is way too cumbersome. Du et al. [14] propose an online streaming method Spell, which is based on longest common subsequence to parse system logs. This approach is not efficient because the time complexity of this algorithm is O (n^2). He et al. [3] design Drain, an online log parsing approach with fixed depth tree. Drain is more efficient compared to SHISO [9], but Drain firmly relies on the preprocessing of log data and the rules for constructing Drain are arbitrary so that Drain is not widely applicable. Hamooni et al. [7] propose LogMine, which adopts a one-pass algorithm for clustering with high efficiency. But the merging step in LogMine is based on Smith–Waterman Algorithm [17] which is time-consuming.

3 Methodology

In this section, we will introduce our online log pattern mining method, Lopper. Lopper is named on the basis of **<u>lo</u>g <u>p</u>attern min<u>er</u>**, and the word "lopper" itself has the meaning of a worker who prunes redundant trees, so we expect Lopper to be a superior "lopper" who can finish the job of pruning log data with high accuracy and efficiency. We will give a brief overview of our method first and then explain in detail each step of the mining process.

3.1 Definition

We will first define some terms for the purpose of clearer presentation. System logs are used to record the system runtime information. In general, system logs are consisted of constant and variable. The **Constant** is used to describe the role of the current code or the reason for triggering log print statement and the **Variable** reflects the dynamic system runtime data, such as IP address and parameter information. Either constant or variable is described as a **Token**, so one log record is actually a series of tokens. **A log pattern** is composed of the constant part of a log record and wildcards which are used for replacing variables [2–4, 6, 13].

To be clear, we define T_n^C which means the n-th token in a log record (or log pattern) is a constant and T_n^V which means the n-th token in a log record (or log pattern) is a variable. The **Length** of a log record (or log pattern) is the number of tokens that form the log record (or log pattern).

For example, the log record {Receive 120 KB from 10.0.2.17} has length of 5 and can be formally expressed as $\{T_1^{C1}\, T_2^{V1}\, T_3^{C2}\, T_4^{C3}\, T_5^{V2}\}$, and the corresponding log pattern {Receive * KB from *} also has length of 5 and can be formally expressed as $\{T_1^{C1}\, T_2^{V}\, T_3^{C2}\, T_4^{C3}\, T_5^{V}\}$. As we can see from the difference between two expressions, all variables in log pattern are replaced by wildcards because we give equal treatment to variables in log pattern.

3.2 Model Overview

The structure of Lopper is illustrated in Fig. 2. Log records will be first preprocessed using regular expressions based on domain knowledge and commonly-used variables in log records such as IP address and timestamp should be replaced by wildcards in this step. All preprocessed log data will be then sorted into different groups according to their length, and each separate group is an aggregation of log records with identical length.

One-pass clustering will take place in each individual group and generate clusters with different ID. In order to further accelerate the speed of clustering process, we adopt a relatively simple similarity measure function for one-pass clustering, so the preliminary clustering results are not accurate enough. To solve this problem, we will merge similar clusters employing more precise similarity measure function right after the one-pass clustering, hence similar clusters in the same group are merged to form the candidate log patterns. It's possible that candidate log patterns from different groups actually belong to same log pattern, therefore we need to combine similar candidate log patterns across groups to generate the final log patterns. To better demonstrate the log pattern mining process intuitively, we give a running example in Fig. 3.

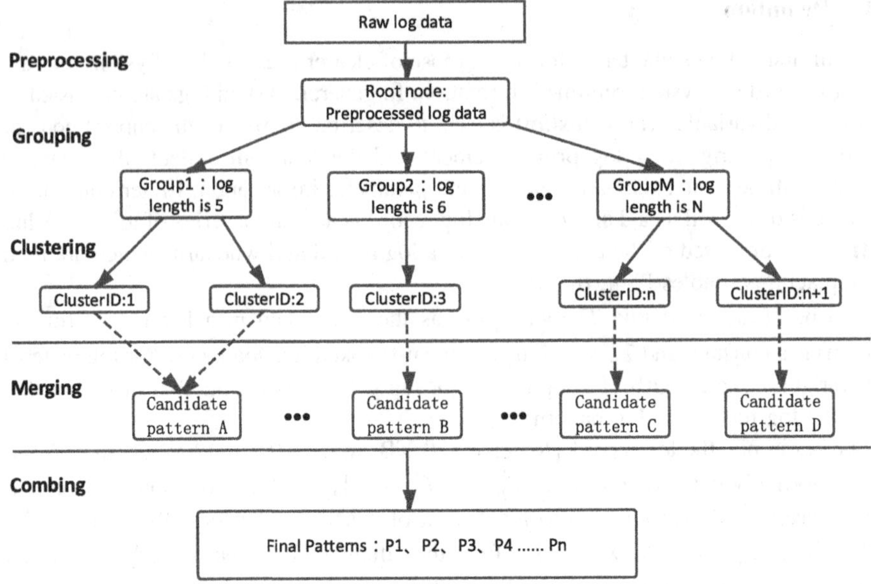

Fig. 2. Structure of Lopper

3.3 Preprocessing and Grouping Log Data by Length

Log pattern mining refers to the process of mining log patterns from massive raw log data, in which variables are distinguished and replaced while constants are extracted to form the final patterns. In fact, preprocessing plays the same role. In this step we use simple regular expressions to parse some commonly-used variables (e.g., IP address, timestamp and number). As shown in Fig. 3, numbers and IP addresses in log records are replaced by wildcards. Preprocessing reduces the complexity of following work to a certain extent. The more pre-defined rules are set up in this step, the more convenient later work will be. But there is an obvious problem that we can't always know the exact content and structure of the log data we are going to parse, hence just relying on preprocessing is unrealistic. In order to make our model widely adaptable, we only adopt limited pre-defined rules in Lopper.

After preprocessing, we then sort all log records into different groups according to their length. We do this for the following reasons: 1. Based on previous research work [2, 15] and our statistical analysis of log data, we have the assumption that log records belong to the same patterns always share identical log length; 2. It will be more convenient and more precise to design similarity measure function if log records in one group are of identical log length; 3. Previous work [7, 8, 13] processes all log data with unequal length at one time and then tries to align log records in order to extract log patterns. By classifying log records at the very start, there is no need for alignment operation which will greatly raise the mining speed.

```
Log record1: Verification succeeded for blk_-8050
Log record2: Receiving block blk_-5623 src: /10.251.75.228:53725
Log record3: Receiving block blk_-5009 src: /10.251.199.19:52622
Log record4: PacketResponder 1 for block terminating
Log record5: PacketResponder 0 for block terminating
Log record6: BLOCK* ask 10.251.126.5:50010 to delete blk_-9016
Log record7: BLOCK* ask 10.251.126.5:50010 to delete blk_1671 blk_2654 blk_5202
```

Preprocessing and grouping log data by length

```
Group1: Verification succeeded for blk_-8050
------------------------------------------------
Group2: Receiving block blk_-5623 src: /*
Group2: Receiving block blk_-5009 src: /*
------------------------------------------------
Group3: BLOCK* ask * to delete blk_-9016
Group3: PacketResponder * for block terminating
Group3: PacketResponder * for block terminating
------------------------------------------------
Group4: BLOCK* ask * to delete blk_1671 blk_2654 blk_5202
```

One-pass clustering of log data

```
Cluster1: Verification succeeded for blk_-8050
------------------------------------------------
Cluster2: Receiving block blk_-5623 src: /*

Cluster3: Receiving block blk_-5009 src: /*
------------------------------------------------
Clusler4: BLOCK* ask * to delete blk_-9016

Cluster5: PacketResponder * for block terminating
Cluster5: PacketResponder * for block terminating
------------------------------------------------
Cluster6: BLOCK* ask * to delete blk_1671 blk_2654 blk_5202
```

Merging clusters and identifying variables

```
Candidate pattern1: Verification succeeded for *
------------------------------------------------
Candidate pattern2: Receiving block * src: /*
------------------------------------------------
Candidate pattern3: BLOCK* ask * to delete *

Candidate pattern4: PacketResponder * for block terminating
------------------------------------------------
Candidate pattern5: BLOCK* ask * to delete * * *
```

Combining candidates

```
Pattern1: Verification succeeded for *
------------------------------------------------
Pattern2: Receiving block * src: /*
------------------------------------------------
Pattern3: BLOCK* ask * to delete *
------------------------------------------------
Pattern4: PacketResponder * for block terminating
```

Fig. 3. Running Example of log pattern mining by Lopper

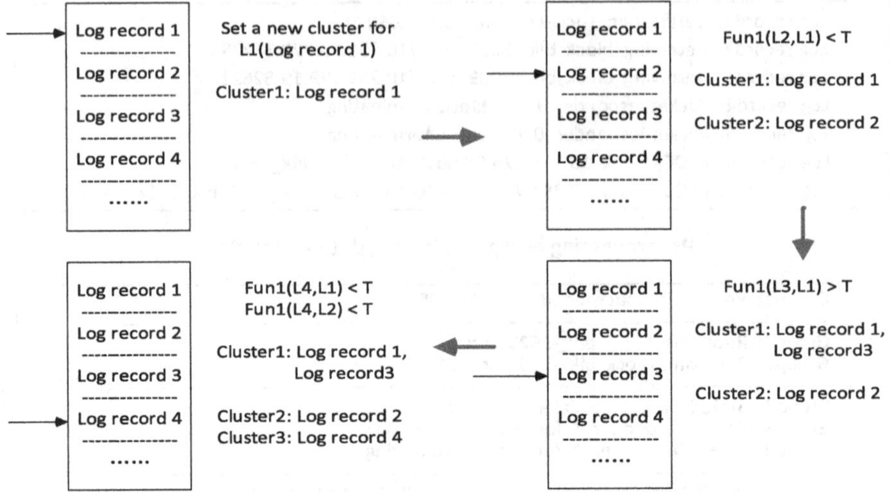

Fig. 4. Running example of one-pass clustering algorithm

3.4 One-Pass Clustering of Log Data

In order to cluster log data with high efficiency, we adopt one-pass clustering algorithm which cluster log data in streaming way. Before describing the algorithm in detail, we first give our similarity measure function. The similarity measure function ***Fun1*** between two log records is defined as following:

$$Fun1(log1, log2) = \frac{\sum_{i=1}^{n} S1(log1(i), log2(i))}{\text{n}} \tag{1}$$

Where *log1* and *log2* stand for two separate log records; $log1(i)$ or $log2(i)$ represents the i-th token in this log record; n is the log length of *log1* (the length of *log1* and *log2* is the same); Function $\boldsymbol{S}1(\boldsymbol{x},\boldsymbol{y})$ is used to determine whether two tokens are the same which is defined as following:

$$S1(x,y) = \begin{cases} 1, \ if\, x = y \\ 0, \ otherwise \end{cases} \tag{2}$$

Where x and y are two tokens in log1 and log2 respectively. $\boldsymbol{Fun}1$ reflects the ratio of identical tokens of two log records. We will set a similarity threshold T, if $\boldsymbol{Fun}1(log1, log2) >$ T, we consider the two log records as similar records and put them into the same cluster otherwise they will be sorted into different clusters.

One-pass clustering algorithm starts from the first log record and goes through all log records in streaming way. When the very first log record L1 arrives, we create a new cluster for L1 and L1 is the only log record in this cluster. And then comes the second log record L2, we will calculate the similarity of L1 and L2 using ***Fun1***, if $\boldsymbol{Fun}1(L1, L2) >$ T, we send L2 into the same cluster with L1, otherwise we will create a new cluster for L2 and L2 will be the only log record in the new cluster. When a new log record Ln arrives, we first calculate the similarity of L1 and Ln, if they are similar log records, Ln will be put into the same cluster with L1 and we go on processing next log record. If Ln and L1 are not similar, we will continue to compare the similarity between Ln and log records in other clusters until we find a suitable cluster for Ln. If Ln is not similar to any log record in all existing clusters, we will establish another new cluster for Ln. We use a running example to demonstrate the whole process in Fig. 4.

There is one key point to stress: if there're more than one log record in a cluster, we only have to choose a representative in this cluster because log records in same cluster are already similar. Thus, the complexity of the algorithm is relevant to the number of clusters which is O (n*m), in which n stands for the number of log records and m stands for the number of clusters. Taking log data set HDFS as an example (more examples can be seen in Sect. 4.1), there are 2000 log records but only 25 clusters, it's obvious that the number of log records is far more than the number of clusters ($n \gg m$), so the actual complexity of one-pass clustering algorithm is approximately equal to O(n) which is really efficient.

3.5 Merging Clusters and Identifying Variables

The results generated by one-pass clustering are not accurate enough, because the similarity measure function ***Fun1*** which calculates the ratio of identical tokens is too simple and therefore many special cases are not taken into account. As shown in Fig. 3, according to ***Fun1***, log records *blk_-5623* and *blk_-5009* will be put into different clusters because the third token is different. But it's obvious that the third token is a variable and these two log records actually belong to the same cluster, so we propose a more precise similarity measure function ***Fun2*** which reflects the ratio of similar tokens in two log records to check whether clusters need to be further merged. ***Fun2*** is defined as following:

$$Fun2(log1, log2) = \frac{\sum_{i=1}^{n} S2(log1(i), log2(i))}{\text{n}} \tag{3}$$

Where *log1* and *log2* stand for two separate log records in different clusters; $log1(i)$ or $log2(i)$ represents the i-th token in this log record; n is the log length of *log1* (the length of *log1* and *log2* is the same); The function $\boldsymbol{S2(x,y)}$ is used to determine whether two tokens are alike which is defined as following:

$$S2(x,y) = \begin{cases} 1, if\, x\, is\, similar\, to\, y \\ 0, otherwise \end{cases} \tag{4}$$

Where x and y are two tokens in log1 and log2 respectively. How to measure whether two tokens are similar is easy, we can use edit distance [16] or Cosine distance. We set another similarity threshold T* and choose a random log record from each cluster as representative. If $\boldsymbol{Fun2}(L1, L2) > \mathrm{T}^*$, we will merge the two clusters which L1 and L2 stand for. We go through all clusters in pairs and merge similar clusters, although the time complexity of merging is $\mathrm{O}(m^2)$, this step takes up very little time because m is a small number (m stands for the number of clusters).The following task after merging is identifying variable which is illustrated in Fig. 5. The rules for identifying is simple: *If all tokens in the same column are the same, they are constants, otherwise, they are variables.* The goal of identifying variable is telling which tokens in a log record are variables and then generating a candidate pattern for each cluster.

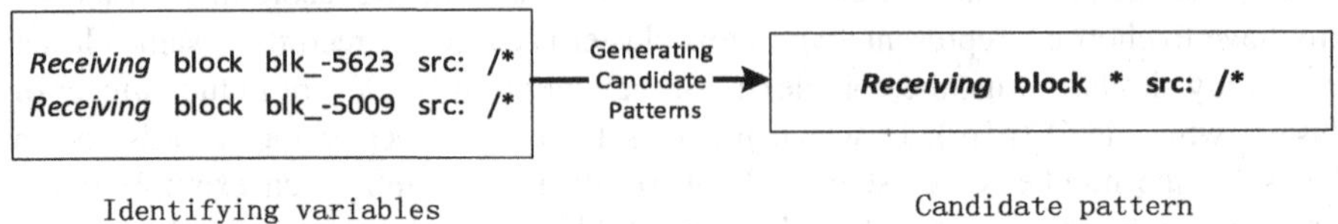

Fig. 5. Identifying variables and generating candidate patterns

3.6 Combining Candidates

We have the assumption that log records belong to the same log patterns will be very likely to share identical log length, so we group log data by length at first step. It is a smart approach because we avoid a lot of trouble caused by dealing with log records with unequal length and hence accelerate the process of clustering and merging. But nothing is perfect because log records of different log length may also belong to same log pattern, that's why we have the step of combination!

The goal of combination is to combine candidate patterns which actually come from the same log pattern. We define the instruction of combination as following: *Continuous variables will be integrated into one variable and if constant parts of two candidate log patterns are the same after they are segmented by variable, we combine them.* In order to better elaborate, we give some examples in Fig. 6.

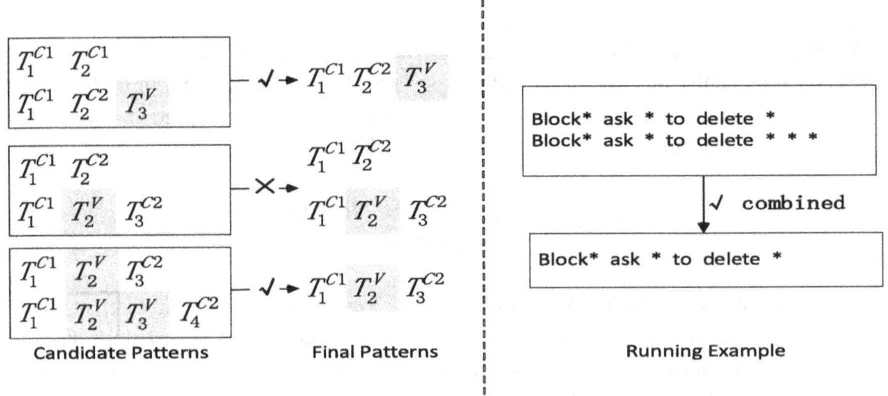

Fig. 6. Rules for combination operation

Let's explain why pattern $\{T_1^{C1}T_2^{C2}\}$ and pattern $\{T_1^{C1}T_2^{V}T_3^{C2}\}$ are not combined: There is no variable in pattern $\{T_1^{C1}T_2^{C2}\}$ so it remains as usual, but pattern $\{T_1^{C1}T_2^{V}T_3^{C2}\}$ is segmented by variable T_2^{V} and becomes $\{T_1^{C1}\},\{*\},\{T_3^{C2}\}$. It's obvious these two candidate patterns are different after segmentation so we don't combine them.

Same as the merging step, the time complexity of combination is O(c^2), in which c stands for the number of candidate patterns. So time complexity of the whole algorithm is: O(n) + O(m^2) + O(c^2). However, the number of log records (n) is far greater than the number of clusters (m) and the number of candidate log patterns (c), taking log data set HDFS as an example (more examples can be seen in Sect. 4.1), there are 2000 log records but only 25 clusters and 16 candidate patterns, which means n $\gg$ m $> c$. So actual time complexity of the whole algorithm is O(n).

4 Evaluation

4.1 Experimental Settings

Log Data Sets and Comparative Algorithms. The log data sets [2, 3] we used to evaluate our work are summarized in Table 1, We collect eight log data sets generated from real systems which includes distributed system logs (HDFS and Zookeeper), supercomputer logs (BGL and HPC), software logs (Proxifier) and system running logs from an anonymous Chinese financial institution (Dataset A B C).

Table 1. Summary of collected log datasets

System	Description	#Log Records	#Log Length	#Log Pattern
HDFS	Hadoop File System	11,175,629	8 to 29	116
BGL	BlueGene/L Supercomputer	4,747,963	10 to 104	376
Proxifier	Proxy Client	10,108	10 to 27	8
Zookeeper	Distributed System Coordinator	74,380	8 to 27	80
HPC	High Performance Cluster	433,490	6 to 104	105
Dataset A	System log data	3,000	5 to 17	7
Dataset B	System log data	50,314	8 to 35	78
Dataset C	System log data	126,397	8 to 33	36

We choose 4 representational existing log patterns mining methods to compare with Lopper and we test their performance based on some commonly adopted evaluation metrics. The comparative algorithms are briefly introduced as following:

- **Drain** [3]: Drain is an online log parsing tree with pre-defined rules for classifying log data and generating log patterns.
- **LogMine** [7]: LogMine is an online log pattern mining model which adopts one-pass clustering algorithm for fast log pattern mining.

- **LogSig** [10]: LogSig is an offline log pattern mining model which uses LCS (Longest Common Subsequence) for classifying log data.
- **IPLoM** [8]: IPLoM adopts a three-layer hierarchical partitioning model for clustering log records and generates log patterns from each cluster.

Table 2. Experimental environment

OS	CPU	Bits	Memory
Windows 7	Inter(R) Core(TM) i5-3230M	64	8G
Windows 7	Inter(R) Core(TM) i5-6200U	64	8G

Evaluation Metric and Experimental Setups: We use F-Measure [18] to evaluate the accuracy of Lopper, which are defined as the following:

$$F - measure = \frac{2 * Precision * Recall}{Precision + Recall} \tag{5}$$

$$Precision = \frac{TP}{TP + FP} \tag{6}$$

$$Recall = \frac{TP}{TP + FN} \tag{7}$$

Where TP stands for True Positive Rate which means true positive sample is inferred as positive. FP stands for False Positive Rate which means true negative sample is inferred as positive. FN stands for False Negative Rate which means true positive sample is inferred as negative. Experimental environment is summarized in Table 2 and we conduct each experiment several times on these two machines to reduce bias errors.

4.2 Accuracy of Lopper

Accuracy is very important because high quality log patterns are the guarantees of subsequent log mining task. We evaluate the accuracy of Lopper and other four comparative algorithms based on the log data sets described in Table 1. Because the time complexity of LogSig is O (n^2), we can't run big data set on it, therefore we randomly divide some original data sets into several log data sets with 2000 log records.

The metric "Recall" reflects the ability to mine log patterns as completely as possible. As we can see from Table 3, Lopper beats other four algorithms on 6 log data sets while IPLoM and Drain achieve the best the performance on Zookeeper (97.52%) and HPC (93.85%) respectively. LogMine has a relatively mediocre performance compared to others and LogSig is not doing very well. The metric "Precision" reflects the quality of the mined patterns which is listed at Table 4. It's obvious that Lopper still wins this game in most cases and Drain together with IPLoM follow closely.

Table 3. Recall of log pattern mining models

Log	Lopper	Drain	LogMine	IPLoM	LogSig
HDFS(2k)	**100.0%**	92.86%	87.11%	85.71%	71.43%
BGL(2K)	**96.43%**	89.29%	94.64%	95.54%	86.61%
Proxifier(2k)	**100.0%**	99.17%	98.89%	**100.0%**	87.52%
Zookeeper(2k)	96.67%	95.24%	95.24%	**97.62%**	90.48%
HPC(2k)	92.31%	**93.85%**	90.77%	95.38%	87.12%
Dataset A(3k)	**100.0%**	84.72%	80.01%	**100.0%**	71.43%
Dataset B(50k)	**91.84%**	79.15%	76.26%	86.42%	52.15%
Dataset C(126k)	**94.74%**	85.24%	84.55%	92.40%	76.25%

Table 4. Precision of log pattern mining models

Log	Lopper	Drain	LogMine	IPLoM	LogSig
HDFS(2k)	96.11%	**98.19%**	95.14%	96.25%	94.15%
BGL(2K)	83.17%	83.12%	80.72%	**84.12%**	80.14%
Proxifier(2k)	**87.17%**	86.26%	78.15%	85.20%	82.15%
Zookeeper(2k)	**93.12%**	92.15%	92.14%	90.84%	88.45%
HPC(2k)	**91.24%**	85.23%	88.96%	90.22%	60.23%
Dataset A(3k)	**85.71%**	79.43%	82.16%	83.72%	66.75%
Dataset B(50k)	**82.46%**	76.33%	79.25%	76.92%	70.21%
Dataset C(126k)	**88.89%**	77.78%	80.84%	82.08%	78.24%

Table 5. F-Measure of log pattern mining models

Log	Lopper	Drain	LogMine	IPLoM	LogSig
HDFS(2k)	**98.02%**	95.45%	90.95%	90.67%	81.23%
BGL(2K)	**89.31%**	86.09%	87.13%	89.47%	83.25%
Proxifier(2k)	**93.15%**	92.27%	87.31%	92.01%	84.75%
Zookeeper(2k)	**94.86%**	93.67%	93.66%	94.11%	89.45%
HPC(2k)	91.77%	89.33%	89.86%	**92.73%**	71.22%
Dataset A(3k)	**92.31%**	81.99%	81.07%	91.14%	69.01%
Dataset B(50k)	**86.90%**	77.71%	77.73%	81.39%	59.85%
Dataset C(126k)	**91.72%**	81.34%	82.65%	86.93%	77.23%

In order to balance "Recall" and "Precision", we use a more scientific metric "F-Measure" to evaluate the accuracy, the results are shown in Table 5. Lopper has the best performance with 92.26% on average. IPLoM is second best due to its iterative clustering structure. Drain is doing well on first five data sets but its performance has fallen sharply on Dataset A, B and C, which means Drain is not so widely applicable because Drain depends heavily on predefined rules and manual parameter adjustment.

4.3 Efficiency of Lopper

We measure the running time of Lopper and four other models to evaluate the efficiency. The experimental results are demonstrated in Table 6:

Table 6. Running time of log pattern mining models

Log(size)	Lopper	Drain	LogMine	IPLoM	LogSig
HDFS(2k)	**0.17** s	0.18 s	0.21 s	0.24 s	5.01 s
BGL(2K)	0.34 s	**0.31 s**	0.46 s	0.32 s	4.98 s
Proxifier(2k)	0.28 s	**0.26 s**	0.27 s	**0.26 s**	5.27 s
Zookeeper(2k)	**0.19 s**	0.24 s	0.21 s	0.25 s	4.90 s
HPC(2k)	**0.28 s**	0.29 s	**0.28 s**	0.31 s	4.78 s
Dataset A(3k)	0.50 s	**0.46 s**	0.68 s	0.56 s	6.53 s
Dataset B(50k)	5.17 s	**4.52 s**	8.72 s	5.40 s	137.20 s
Dataset C(126k)	**9.76** s	9.80 s	16.55 s	12.59 s	325.82 s

The time complexity of Lopper is approximately equal to O(n) which is discussed in Sect. 3.6. Drain and IPLoM share the same time complexity which is also O(n) because both of them process each log record only once. In Drain, a log record has to traverse through a tree to match the right log pattern which takes extra time. And in IPLoM, the hierarchical structure requires a lot of memory space which also impacts on execution efficiency. LogMine adopts one-pass clustering algorithm which is efficient at first, but LogMine uses Smith–Waterman Algorithm [17] which aligns two sequences of length l1 and l2 in $O(l1 * l2)$ time for merging two log records, leading to a reduction in efficiency. LogSig is the first algorithm using longest common sequence (LCS) for parsing log data, the time complexity of LCS algorithm is $O(n^2)$, so LogSig is not efficient compared to other methods.

4.4 System Anomaly Detection Task Based on Lopper

As we discussed before, log pattern mining is a preliminary work for system anomaly detection, fault diagnosis and fault root cause analysis [1, 2, 4, 6, 11]. So subsequent work after log pattern mining is the key to validating the quality of log patterns. In this section, we choose system anomaly detection task as a case study to further prove the effectiveness of Lopper.

The summarized workflow of system anomaly detection model goes like the following: Each log record corresponds to a certain log pattern, so the sequence of log records will be transferred to a sequence of log patterns. We number each log pattern so we get a sequence of numbers which stands for the original log record flow. We then set up a sliding window of fixed size to intercept log sequence and count the frequency of each log pattern in this window which will be used as feature vector. Finally, we label each window either "normal" or "abnormal" and the labels together with feature vectors are the inputs for training bi-classification models.

In this case study, we choose two commonly-used machine learning algorithms Random Forest and Decision Tree for building anomaly detection model, while Lopper and IPLoM are used for mining log patterns. The reason we choose IPLoM is that IPLoM has the best F-Measure on average among the four comparative algorithms (Drain, LogMine, IPLoM, LogSig). The results are shown in Table 7, and we can see from the experimental results that Lopper based anomaly detection model achieves an average F-Measure performance of 91.97%, which further proves the effectiveness of Lopper.

Table 7. System anomaly detection task based on Lopper and IPLoM

Lopper based anomaly detection model	Precision	Recall	F-Measure
Random Forest	100.0%	85.00%	91.89%
Decision Tree	96.15%	88.29%	92.05%
IPLoM based anomaly detection model	Precision	Recall	F-Measure
Random Forest	100.0%	81.43%	89.76%
Decision Tree	100.0%	85.50%	92.18%

5 Conclusion

System log based analysis is crucial to the management of distributed platform. To better understand system logs and depict the inner structure of log data, log pattern plays an important role. In this paper, we propose Lopper, a hybrid clustering tree for online log pattern mining. Our method accelerates the mining process by clustering raw log data in one-pass manner and ensures the accuracy by merging and combing similar patterns with different kernel functions in each step. To sum up, Lopper works in **online** mode with high **accuracy** and **efficiency**. We conduct experiments on massive sets of log data to prove the effectiveness of Lopper, the experimental results show that Lopper outperforms existing log pattern mining methods in terms of regular evaluation metrics. We also conduct aided experiments on system anomaly detection task, the results prove that log patterns generated by Lopper are quite useful for anomaly detection.

Acknowledgments. This work is supported by UINNOVA Joint Innovation project.

References

1. Lu, S., et al.: Detecting anomaly in big data system logs using convolutional neural network. In: 2018 IEEE 16th International Conference on Dependable, Autonomic and Secure Computing, 16th International Conference on Pervasive Intelligence and Computing, 4th International Conference on Big Data Intelligence and Computing and Cyber Science and Technology Congress (DASC/PiCom/DataCom/CyberSciTech). IEEE (2018)

2. He, P., et al.: An evaluation study on log parsing and its use in log mining. In: 2016 46th Annual IEEE/IFIP International Conference on Dependable Systems and Networks (DSN). IEEE (2016)
3. He, P., et al.: Drain: an online log parsing approach with fixed depth tree. In: 2017 IEEE International Conference on Web Services (ICWS). IEEE (2017)
4. Fu, Q., et al.: Execution anomaly detection in distributed systems through unstructured log analysis. In: 2009 Ninth IEEE International Conference on Data Mining. IEEE (2009)
5. Zhu, K.Q., Fisher, K., Walker, D.: Incremental learning of system log formats. ACM SIGOPS Operating Syst. Rev. **44**(1), 85–90 (2010)
6. Xu, W., et al.: Detecting large-scale system problems by mining console logs. In: Proceedings of the ACM SIGOPS 22nd Symposium on Operating Systems Principles. ACM (2009)
7. Hamooni, H., et al.: LogMine: fast pattern recognition for log analytics. In: Proceedings of the 25th ACM International on Conference on Information and Knowledge Management. ACM (2016)
8. Makanju, A.A.O., Nur Zincir-Heywood, A., Milios, E.E.: Clustering event logs using iterative partitioning. In: Proceedings of the 15th ACM SIGKDD International Conference on Knowledge Discovery and Data Mining. ACM (2009)
9. Mizutani, M.: Incremental mining of system log format. In: 2013 IEEE International Conference on Services Computing. IEEE (2013)
10. Tang, L., Tao, L., Perng, C.-S.: LogSig: generating system events from raw textual logs. In: Proceedings of the 20th ACM International Conference on Information and Knowledge Management. ACM (2011)
11. Cheng, J., et al.: Deep convolutional neural networks for anomaly event classification on distributed systems. arXiv preprint arXiv:1710.09052 (2017)
12. Vaarandi, R.: A breadth-first algorithm for mining frequent patterns from event logs. In: Aagesen, F.A., Anutariya, C., Wuwongse, V. (eds.) INTELLCOMM 2004. LNCS, vol. 3283, pp. 293–308. Springer, Heidelberg (2004). https://doi.org/10.1007/978-3-540-30179-0_27
13. Vaarandi, R.: A data clustering algorithm for mining patterns from event logs. In: Proceedings of the 3rd IEEE Workshop on IP Operations & Management (IPOM 2003) (IEEE Cat. No. 03EX764). IEEE (2003)
14. Du, M., Li, F.: Spell: streaming parsing of system event logs. In: 2016 IEEE 16th International Conference on Data Mining (ICDM). IEEE (2016)
15. Stearley, J.: Towards informatic analysis of syslogs. In: 2004 IEEE International Conference on Cluster Computing (IEEE Cat. No. 04EX935). IEEE (2004)
16. Edit distance. https://en.wikipedia.org/wiki/Edit_distance
17. Smith–Waterman_algorithm. https://en.wikipedia.org/wiki/Smith-Waterman_algorithm
18. Manning, C., Raghavan, P., Schütze, H.: Introduction to information retrieval. Nat. Lang. Eng. **16**(1), 100–103 (2010)

Discord Monitoring for Streaming Time-Series

Shinya Kato(✉), Daichi Amagata, Shunya Nishio, and Takahiro Hara

Osaka University, Osaka, Japan
{kato.shinya,amagata.daichi,nishio.syunya,hara}@ist.osaka-u.ac.jp

Abstract. Many applications generate time-series and analyze it. One of the most important time-series analysis tools is anomaly detection, and discord discovery aims at finding an anomaly subsequence in a time-series. Time-series is essentially dynamic, so monitoring the discord of a streaming time-series is an important problem. This paper addresses this problem and proposes SDM (Streaming Discord Monitoring), an algorithm that efficiently updates the discord of a streaming time-series over a sliding window. We show that SDM is approximation-friendly, i.e., the computational efficiency is accelerated by monitoring an approximate discord with theoretical bound. Our experiments on real datasets demonstrate the efficiency of SDM and its approximate version.

Keywords: Time-series · Discord · Sliding window

1 Introduction

Motivation. Many real-world applications generate time-series and want to utilize it for obtaining useful knowledge [6]. Anomaly or outlier detection supports this, because it can find unusual observations and helps data cleaning, thereby time-series anomaly (or outlier) detection has been extensively studied [19,23]. One of the most effective time-series anomaly detection is discord discovery [10,11,21]. Given a time-series, the discord of the time-series is the subsequence with the largest distance to its nearest neighbor among all subsequences (the formal definition is introduced in Sect. 2). Figure 1 illustrates an example, and the red subsequence is the discord of an ECG time-series.

It has been shown that discord discovery can be employed in industry [3], medical care [20], and Web [21]. Note that time-series generated in the above applications is essentially dynamic [12]. Therefore, the discord of a time-series is updated over time. This fact renders an important problem of discord monitoring of a streaming time-series. We address this problem with a count-based sliding window setting in this paper.

Technical Challenge. When the window slides, we have a new subsequence and the oldest subsequence expires. Due to them, the nearest neighbor of each subsequence may change, i.e., the discord may also change. Applications, which

S. Hartmann et al. (Eds.): DEXA 2019, LNCS 11706, pp. 79–94, 2019.
https://doi.org/10.1007/978-3-030-27615-7_6

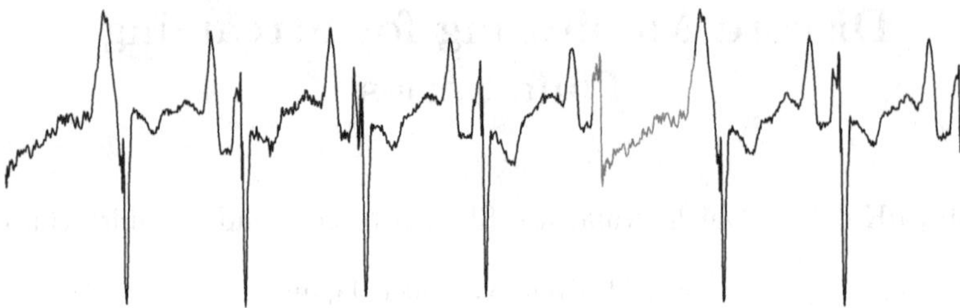

Fig. 1. A streaming ECG time-series and its discord (the red subsequence) (Color figure online)

employ discord monitoring, require real-time update of the discord, hence we need to evaluate whether the discord is updated or not.

A straightforward approach to achieve this is to re-evaluate the nearest neighbor of each subsequence whenever the window slides. However, this approach obviously incurs an expensive cost. Although an existing discord detection algorithm for *static* time-series [10] can also be utilized for discord monitoring, it cannot incrementally update the discord. Therefore, when the discord changes, it has to re-evaluate the discord from scratch, which is not suitable for real-time monitoring, as shown in our experiments.

From the above discussion, we see that an efficient solution needs to have the following properties: it can (i) evaluate whether the discord changes or not quickly and (ii) discover the discord when it is updated by the window slide as soon as possible.

Overview of Our Solution and Contribution. We devise SDM (Streaming Discord Monitoring) to achieve such a solution. SDM exploits a nearest neighbor search based on *sequential scan*, which is fast for high-dimensional data [18]. This approach is efficient for identifying the subsequences whose nearest neighbors change, thus satisfies the property (i). Besides, SDM maintains two kinds of nearest neighbor for each subsequence, to prune unnecessary computation when the discord changes, which enables result update without computing the discord from scratch, i.e., SDM satisfies the property (ii). SDM furthermore reduces the worst update time by using an approximation with bound guarantee.

To summarize, our contributions are as follows:

- We address the problem of time-series discord monitoring over a sliding window. To the best of our knowledge, this paper is the first to address this problem.
- We propose SDM to efficiently solve the problem.
- We propose an approximate version of SDM, namely A-SDM, which reduces the update time of the worst case and provides a theoretical guarantee w.r.t. the monitoring result.
- We empirically evaluate SDM and A-SDM on real datasets, and the results show that they quickly update the discord and the worst update time is significantly faster than those of competitors.

Road-Map. Section 2 defines the problem of this paper and Sect. 3 introduces related works. In Sect. 4, we present our algorithm SDM, and in Sect. 5, we show our experimental results. Finally, in Sect. 6, we conclude this paper.

2 Problem Definition

A streaming time-series t is an unbound sequence of real values and represented as $t = (t[1], t[2], ...)$. First of all we define subsequence of t.

Definition 1 (Subsequence). *Given a time-series t and a subsequence size l, the subsequence s_p, which starts at $t[p]$, is defined as follows.*

$$s_p = (t[p], t[p+1], ..., t[p+l-1])$$

Let $s_p[x]$ be the x-th value of s_p, then $s_p = (s_p[1], s_p[2], ..., s_p[l])$. Next, we use z-normalized Euclidean distance between two subsequences. Note that z-normalized Euclidean distance is often utilized to measure the similarity between time-series [15,22].

Definition 2 (Z-normalized Euclidean distance). *Given two subsequences s_p and s_q with length l, their z-normalized Euclidean distance, $dist(s_p, s_q)$, is.*

$$dist(s_p, s_q) = \sqrt{\sum_{i=1}^{l}(\frac{s_p[i] - \mu(s_p)}{\sigma(s_p)} - \frac{s_q[i] - \mu(s_q)}{\sigma(s_q)})^2},$$

where $\mu(s)$ and $\sigma(s)$ are respectively the mean and standard deviation of $\{s[1], ..., s[l]\}$.

It is obvious that $dist(s_p, s_{p+1})$ is small but this observation is not interesting and interrupts obtaining a meaningful result [2,14]. We therefore ignore trivial matched subsequences, which are defined below.

Definition 3 (Trivial match) [5]. *The set S_p of subsequences that have trivial match relationships with s_p is*

$$S_p = \{s_q \,|\, p - l + 1 \leq q \leq p + l - 1\}.$$

Here, the applications introduced in Sect. 1 are interesting only in recent values of a streaming time-series [13]. We hence employ a count-based sliding window to consider only the most recent w values of the time-series t, as with existing works [9,12]. That is, we can represent t on the sliding window as $t = (t[i], ..., t[i+w-1])$, and $t[i+w-1]$ is the latest value of t. When the window slides, a new subsequence, which has $t[i+w-1]$, is generated and the subsequence, which has $t[i-1]$, expires. Note that, when we refer to a subsequence s of t, s is on the window hereafter. Note furthermore that all subsequences on the window are memory-resident.

Our problem is discord monitoring, and the discord is obtained from the nearest neighbor of each subsequence. We therefore define the nearest neighbor of a given subsequence s.

Definition 4 (Nearest neighbor). *Given a subsequence* s_p *and a set* S *of all the subsequences of a streaming time-series* t*, the nearest neighbor of* s_p *is the subsequence that satisfies* $argmin_{s_q \in S \setminus s_p} dist(s_p, s_q)$.

Let $s_p.dist_{NN}$ be the distance between s_p and its the nearest neighbor, then our problem is formally defined as follows.

PROBLEM DEFINITION. *Given a streaming time-series* t*, a windows size* w*, and a subsequence size* l*, we monitor the discord* s^* *of* t *that satisfies*

$$s^* = \underset{s_p \in S}{\text{argmax}}\, s_p.dist_{NN}.$$

Applications that employ discord monitoring requires real-time update of the discord. We therefore aim at minimizing computation time to update s^*.

3 Related Work

Although there exist many works that mine useful information from time-series [6] and temporal data [1,7], we here focus on existing studies that have addressed discord detection/monitoring.

Discord Discovery of a Static Time-Series. Several works have proposed efficient discord discovery algorithms for a *static* time-series. Literature [10] has proposed HOT SAX to find the discord of a time-series that residents in-memory. HOT SAX basically computes the nearest neighbor for each subsequence s_p, but terminates the computation when we know that an upper-bound of $s_p.dist_{NN}$ is less than $s^*.dist_{NN}$. To enable this early termination efficiently, HOT SAX transforms subsequences to symbols (or strings), and uses an idea that the distance between two subsequences with similar symbols tends to be small. Hence, HOT SAX evaluates the distance of subsequences with similar symbols in an early iteration. We can employ HOT SAX for our environment but incurs significant cost when s^* expires, i.e., the update time in the worst case is too long. We show this in Sect. 5.

Literature [21] assumes that a given time-series is disk-resident and has proposed an algorithm based on linear scan, which deals with a totally different setting to ours. Literature [8] has proposed a parallel algorithm for discord detection. We note that parallel computation is beyond the scope of this paper.

Discord Monitoring of a Dynamic Time-Series. Literatures [17,22] have proposed discord monitoring algorithms for a streaming time-series. Literature [17] utilizes an R-tree to quickly evaluate whether a new subsequence can become the discord or not. On the other hand, literature [22] employs a data structure Matrix Profile, which has been originally proposed for discord discovery of a static time-series. Unfortunately, they consider append-only case, and dealing with deletions of subsequences is not trivial for them.

4 SDM: Streaming Discord Monitoring

Recall that, when the window slides, a new subsequence s_n is generated and the oldest subsequence s_e expires. They may change the discord, because

- s_n may become the discord,
- the subsequences, whose nearest neighbors were s_e, may become the discord,
- if s_e is the discord, s^* certainly changes, and
- the nearest neighbors of some subsequences may become s_n, which also may derive the discord update.

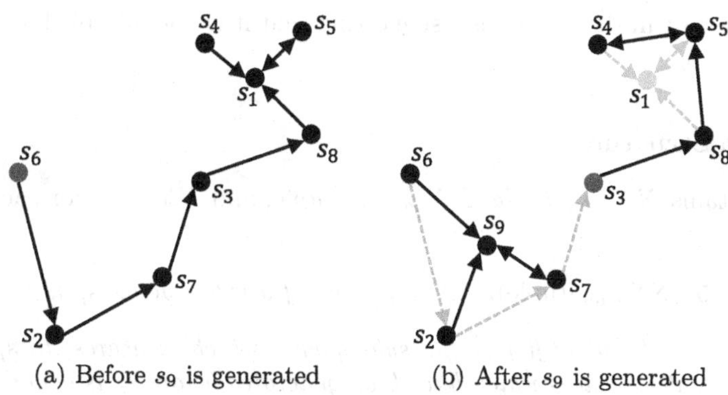

(a) Before s_9 is generated (b) After s_9 is generated

Fig. 2. A streaming time-series on a sliding window with $w = 8$. The arrows indicate the nearest subsequences, and an arrow length represents the distance to the nearest subsequence. The red subsequence is the discord. (Color figure online)

Example 1. *Figure 2 illustrates a streaming time-series t on a sliding window with $w = 8$. Assume $l = 2$, thus each subsequence s of t is represented by a two-dimensional point. Note that the arrows indicate the nearest subsequences (for example, the nearest subsequence of s_1 is s_5), and arrow length shows the distance to the nearest subsequence. In Fig. 2(a), the discord is s_6, and after the window slides, s_1 expires and s_9 is generated as shown in Fig. 2(b). We see that the nearest neighbors of s_4, s_5, s_6, s_7, and s_8 change, due to the expiration of s_1 and the generation of s_9. This results in the discord update (from s_6 to s_3).*

4.1 Main Idea

Our algorithm SDM achieves an efficient discord update by using simple data structures and by exploiting sequential scan. More specifically, we solve the problem with the following ideas:

1. If we maintain the nearest neighbor of each subsequence and the distance, we can easily identify the discord.

2. By scanning all subsequences, we can compute the nearest neighbor of s_n, update the nearest neighbors of the other subsequences incrementally, and identify the subsequences whose nearest neighbors were s_e before the window slides.

The scan-based approach (i.e., the second idea) seems simple, but it is effective for efficient discord monitoring. One may consider an index-based approach, e.g., HOT SAX, to compute the nearest neighbor of s_n fast by pruning some subsequences. Such an approach unfortunately loses chances of updating the nearest neighbors of pruned subsequences, although they may need to be updated. This consequently incurs an expensive cost to verify the nearest neighbors of many subsequences. In Sect. 5, we show that such an approach takes a significant update cost in the worst case, suggesting that it is not suitable for real-time monitoring.

4.2 Data Structure

SDM maintains NN_{older}*-tuple*, $NN_{younger}$*-tuple*, and *NN-tuple* for each subsequence.

Definition 5 (NN$_{older}$-tuple). *NN$_{older}$-tuple of a subsequence s_p is a pair of*

- $s_p.id_{NN_{older}}$*: the identifier of the subsequence which is nearest to s_p among the set of subsequences which have been generated before s_p is generated, and*
- $s_p.dist_{NN_{older}}$*: the distance to the above nearest subsequence.*

Definition 6 (NN$_{younger}$-tuple). *NN$_{younger}$-tuple of a subsequence s_p is a pair of*

- $s_p.id_{NN_{younger}}$*: the identifier of the subsequence which is nearest to s_p among the set of subsequences which have been generated after s_p is generated, and*
- $s_p.dist_{NN_{younger}}$*: the distance to the above nearest subsequence.*

Definition 7 (NN-tuple). *NN-tuple of a subsequence s_p is a pair of the identifier of the nearest neighbor of s_p and $s_p.dist_{NN}$.*

Note that, if $s_p.dist_{NN_{older}} < s_p.dist_{NN_{younger}}$, NN-tuple of s_p is its NN$_{older}$-tuple. Otherwise, NN-tuple of s_p is its NN$_{younger}$-tuple. Furthermore, if NN-tuple of s_p is its NN$_{younger}$-tuple, i.e., $s_p.id_{NN_{older}} \geq s_p.id_{NN_{younger}}$, we do not need to consider its NN$_{older}$-tuple, because its NN$_{older}$-tuple never becomes its NN-tuple. This is an important observation to avoid unnecessary distance computation.

Example 2. *Figure 3 summarizes NN$_{older}$-tuple, NN$_{younger}$-tuple, and NN-tuple of each subsequence in Fig. 2. In Fig. 3(a), for example, NN$_{older}$-tuple of s_2 is empty, because $s_2.dist_{NN_{older}} > s_2.dist_{NN_{younger}}$. In Fig. 3(b), highlighted parts are updated parts from Fig. 3(a), which are derived from the expiration of s_1 and the generation of s_9.*

We can see that the discord can be obtained from NN-tuple of each subsequence, e.g., s_6 and s_3 in Figs. 3(a) and (b), respectively. That is, if we can efficiently update NN-tuples, we can efficiently monitor the discord. We use sequential scan to do this, as discussed in Sect. 4.1.

Subsequence	NN_{older}-tuple	$NN_{younger}$-tuple	NN-tuple
s_1	-	⟨5,1.0⟩	⟨5,1.0⟩
s_2	-	⟨7,2.7⟩	⟨7,2.7⟩
s_3	-	⟨8,2.5⟩	⟨8,2.5⟩
s_4	⟨1,0.8⟩	⟨5,1.5⟩	⟨1,0.8⟩
s_5	⟨1,1.0⟩	⟨8,1.8⟩	⟨1,1.0⟩
s_6	⟨2,3.0⟩	⟨7,3.1⟩	⟨2,3.0⟩
s_7	⟨3,2.2⟩	⟨8,3.2⟩	⟨3,2.2⟩
s_8	⟨1,1.4⟩	-	⟨1,1.4⟩

(a) Before the window slides

Subsequence	NN_{older}-tuple	$NN_{younger}$-tuple	NN-tuple
s_2	-	⟨9,1.5⟩	⟨9,1.5⟩
s_3	-	⟨8,2.5⟩	⟨8,2.5⟩
s_4	-	⟨5,1.5⟩	⟨5,1.5⟩
s_5	⟨4,1.3⟩	⟨8,1.8⟩	⟨4,1.3⟩
s_6	-	⟨9,2.4⟩	⟨9,2.4⟩
s_7	-	⟨9,1.2⟩	⟨9,1.2⟩
s_8	⟨5,1.6⟩	⟨9,4.2⟩	⟨5,1.6⟩
s_9	⟨7,1.2⟩	-	⟨7,1.2⟩

(b) After the window slided

Fig. 3. NN_{older}-tuple, $NN_{younger}$-tuple, and NN-tuple of each subsequence in Fig. 2

Why We Use Older and Younger Nearest Neighbors. Let s_p be the older nearest neighbor of s_q, i.e., $s_q.id_{NN_{older}} = p$. Also, s_r be the younger nearest neighbor of s_q. It is trivial that s_p expires before s_q does. When s_p expires, if $dist(s_p, s_q) < dist(s_q, s_r)$, we do not know the nearest neighbor of s_q. (If $dist(s_p, s_q) \geq dist(s_q, s_r)$, s_r is the nearest neighbor of s_q, so we do nothing in this case.) However, if $s_q.dist_{NN_{younger}} = dist(s_q, s_r)$ is smaller than $s^*.dist_{NN}$, we can guarantee that s_q is not the discord, because $s_q.dist_{NN} \leq s_q.dist_{NN_{younger}}$. That is, we do not have to search its nearest neighbor, which reduces the update time. Furthermore, even if $dist(s_q, s_r)$ is not smaller than $s^*.dist_{NN}$, we just need to search the *older* nearest neighbor of s_q. The search space is only a set of subsequences s_p such that $p < q$, which is much smaller than the set of all subsequences.

We can see that if we employ only NN-tuple, we need a large update cost when the nearest neighbor of s_q expires. This is because we cannot prune its nearest neighbor search and its search space is large.

4.3 Algorithm Description

Rationale of Utilizing Sequential Scan. When the window slides, we need to confirm the discord update. This is to check the nearest neighbors of the other subsequences s_p. Recall that the nearest neighbor of s_p may become the new subsequence s_n. We see that for all s_p, s_n is younger, thereby the check (normally) corresponds to update the $NN_{younger}$-tuple of each subsequence. For fixed l, this check needs $O(w)$ time, because we need $O(1)$ time to check $\min\{s_p.dist_{NN_{younger}}, dist(s_p, s_n)\}$ for each s_p. This is exactly the same cost of sequential scan of all subsequences on the window.

Algorithm. SDM is designed based on the above analysis and the motivation of our data structure (Sect. 4.2), and is described in Algorithm 1. We first maintain a temporal discord s^*_{temp}, which is the previous discord (i.e., the one before the window slided) at initialization. Recall that S is the set of all subsequences. We second remove the expired subsequence s_e from S and insert the new subsequence s_n into S. Next, we initialize NN$_{older}$-tuple, NN$_{younger}$-tuple, and NN-tuple of s_n. We then execute Nearest-Neighbor-Search. In a nutshell, this function computes the nearest neighbor of s_n while updating NN$_{younger}$-tuples of the other subsequences.

Algorithm 1. SDM

Input: s_e: the expired subsequence, s_n: the new subsequence
Output: s^*: the discord

$s^*_{temp} \leftarrow$ the discord before the window slided
$S \leftarrow S \backslash \{s_e\}$, $S \leftarrow S \cup \{s_n\}$ ▷ S is the set of all subsequences on the window
$s_n.\langle\cdot,\cdot\rangle_{NN} \leftarrow \varnothing$, $s_n.\langle\cdot,\cdot\rangle_{NN_{older}} \leftarrow \varnothing$, $s_n.\langle\cdot,\cdot\rangle_{NN_{younger}} \leftarrow \varnothing$
$s^* \leftarrow$ Nearest-Neighbor-Search

Algorithm 2. Nearest-Neighbor-Search

Input: S: the set of all subsequences, s^*_{temp}: a temporal discord
Output: s^*: the discord

for $\forall s_p \in S \backslash S_n$ **do**
  $d \leftarrow dist(s_p, s_n)$
  if $s_n.dist_{NN_{older}} > d$ **then**
    $s_n.\langle\cdot,\cdot\rangle_{NN_{older}} \leftarrow \langle p, d\rangle$
  if $s_p.dist_{NN_{younger}} > d$ **then**
    $s_p.\langle\cdot,\cdot\rangle_{NN_{younger}} \leftarrow \langle n, d\rangle$
    if $s_p.\langle\cdot,\cdot\rangle \neq \varnothing \wedge s_p.dist_{NN} > s_p.dist_{NN_{younger}}$ **then**
      $s_p.\langle\cdot,\cdot\rangle_{NN} \leftarrow s_p.\langle\cdot,\cdot\rangle_{NN_{younger}}$
      if $s_p = s^*_{temp}$ **then**
        $s^*_{temp} \leftarrow$ Discord-Update
  if $s_p.id_{NN} = e$ **then**
    $s_p.\langle\cdot,\cdot\rangle_{NN_{older}} \leftarrow \varnothing$, $s_p.\langle\cdot,\cdot\rangle_{NN} \leftarrow \varnothing$
    if $s_p = s^*_{temp}$ **then**
      $s^*_{temp} \leftarrow$ Discord-Update
    if $s_p.dist_{NN_{younger}} > s^*_{temp}.dist_{NN}$ **then**
      $s^*_{temp} \leftarrow$ Older-Nearest-Neighbor-Search(s_p)
$s_n.\langle\cdot,\cdot\rangle_{NN} \leftarrow s_n.\langle\cdot,\cdot\rangle_{NN_{older}}$
$s^* \leftarrow s^*_{temp}$

Nearest-Neighbor-Search. Algorithm 2 details this function. Given a subsequence $s_p \in S \backslash \{s_n\}$, we do the following. First, we compute $dist(s_p, s_n)$. Then we update NN_{older}-tuple of s_n (lines 3–4) and $\mathrm{NN}_{younger}$-tuple of s_p (lines 5–7) if necessary. Note that if $\mathrm{NN}_{younger}$-tuple of s_p is updated, we update NN-tuple of s_p if necessary (lines 8–10). Besides, if NN-tuple of s_p is updated by the above operation and s_p is s^*_{temp}, the distance to the nearest neighbor of s^*_{temp} becomes smaller. This means that the discord may change, so we execute Discord-Update, which is introduced later, to obtain a temporal discord.

Next, we check whether the nearest neighbor of s_p expires. If so, we initialize NN_{older}-tuple and NN-tuple of s_p (line 12), and if s_p is s^*_{temp}, we execute Discord-Update. In addition, if $s_p.dist_{NN_{younger}} \leq s^*_{temp}.dist_{NN}$, we see that s_p is not s^*, thus we prune its nearest neighbor search, as mentioned in Sect. 4.2. Otherwise, we execute Older-Nearest-Neighbor-Search(s_p), which computes NN_{older}-tuple of s_p by sequential scan and updates s^*_{temp} if necessary.

After we do the above operations for all $s_p \in S \backslash \{s_n\}$, we obtain NN-tuple of s_n (line 17) and the correct s^* (line 18).

Discord-Update. We describe this function in Algorithm 3. When we need to update a temporal discord s^*_{temp}, we first initialize its NN-tuple (line 1). Then, by scanning a set of subsequence s_p such that their NN-tuple is not empty, we obtain s^*_{temp} (lines 2–4). In addition, we scan a set of subsequences s_p such that their NN-tuple is empty and satisfies $s_p.dist_{NN_{younger}} < s^*_{temp}.dist_{NN}$, to maintain them in a heap H (lines 5–7). Note that, the subsequences in H are sorted in descending order of $s_p.dist_{NN_{younger}}$.

Next, if H is not empty, we pop the front subsequence s_p of H and check whether s_p satisfies $s_p.dist_{NN_{younger}} < s^*_{temp}.dist_{NN}$ (lines 8–9). If so, we ter-

Algorithm 3. Discord-Update

Output: s^*_{temp}: a temporal discord

1 $s^*_{temp}.\langle\cdot,\cdot\rangle_{NN} \leftarrow \langle -1, 0\rangle$
2 **for** $\forall s_p \in S$ *such that* $s_p.\langle\cdot,\cdot\rangle_{NN} \neq \varnothing$ **do**
3   **if** $s^*_{temp}.dist_{NN} < s_p.dist_{NN}$ **then**
4         $s^*_{temp} \leftarrow s_p$
5 **for** $\forall s_p \in S$ *such that* $s_p.\langle\cdot,\cdot\rangle_{NN} = \varnothing$ **do**
6   **if** $s_p.dist_{NN_{younger}} > s^*_{temp}.dist_{NN}$ **then**
7   Up-Max-Heap(H, s_p)   ▷ s_p is inserted into a heap H
8 **while** $H \neq \varnothing$ **do**
9     $s_p \leftarrow$ Down-Max-Heap(H)   ▷ s_p is poped from H
10   **if** $s_p.dist_{NN_{younger}} < s^*_{temp}.dist_{NN}$ **then**
11   **break**
12   **else**
13         $s^*_{temp} \leftarrow$ Older-Nearest-Neighbor-Search(s_p)
14 **return** s^*_{temp}

minate Discord-Update (line 11). Otherwise, we execute Older-Nearest-Neighbor-Search(s_p) (line 13). It is important to note that, thanks to H and $\mathrm{NN}_{younger}$-tuple, we can avoid unnecessary executions of Older-Nearest-Neighbor-Search($\cdot$).

We here discuss the space and time complicities of SDM.

Theorem 1 (Space complexity). *The space complexity of SDM is $O(w)$.*

Proof. We maintain NN_{older}-tuple, NN_{older}-tuple, and NN-tuple for each subsequence, which requires $O(1)$ space. Because the number of subsequences on the window is $O(w)$, SDM requires only $O(w)$ space. □

Theorem 2 (Time complexity). *The time complexity of SDM is $O((1+c)wl + c'(w + h\log h))$, where c is the number of executions of* Older-Nearest-Neighbor-Search($\cdot$), *c' is the number of executions of* Discord-Update, *and h is the heap size at line 8 of Algorithm 3.*

Proof. The main cost of SDM is incurred by Nearest-Neighbor-Search. Because Nearest-Neighbor-Search scans S and computes distances between s_p and s_n, which requires at least $O(wl)$. In addition, during the scan, it may execute (some) Older-Nearest-Neighbor-Search($\cdot$), which requires $O(wl)$, and Discord-Update, which requires $O(w+h\log h)$. Therefore Nearest-Neighbor-Search requires $O(wl + cwl + c'(w + h\log h))$, which concludes the proof. □

4.4 Approximation of SDM

From Theorem 2, we can see that the cost of SDM is dependent on the execution numbers of Older-Nearest-Neighbor-Search($\cdot$) and Discord-Update. In the experiments, we observe that we usually have $c = c' = 0$ for each window slide in practice. However, if the nearest neighbors of many subsequences expire at a time, SDM incurs a large cost to update the discord, since c and c' become large. If applications do not require the correct result but are tolerant of an approximate (but highly accurate) result, we can alleviate the update time by decreasing c and c'.

We propose A-SDM (Approximate-SDM), which monitors an approximate discord with a theoretical guarantee. Surprisingly, A-SDM needs very slight extension to SDM. Given an approximation factor ϵ (> 1), which is specified by an application, we replace line 15 of Algorithm 2 and line 10 of Algorithm 3 with

$$s_p.dist_{NN_{younger}} > \epsilon \cdot s^*_{temp}.dist_{NN},$$

and replace line 6 of Algorithm 3 with

$$s_p.dist_{NN_{younger}} < \epsilon \cdot s^*_{temp}.dist_{NN}.$$

Then we have the following theorem.

Theorem 3 (Accuracy guarantee). *Let s_{out} be the subsequence monitored by A-SDM. We have*

$$s_{out}.dist_{NN} \geq \frac{s^*.dist_{NN}}{\epsilon}. \tag{1}$$

Proof. We first demonstrate that Discord-Update, i.e., Algorithm 3, holds Inequality (1). Recall that, if $s_p.dist_{NN_{younger}} \leq \epsilon \cdot s^*_{temp}.dist_{NN}$, Discord-Update does not insert s_p into H. Even if s_p is s^*, Inequality (1) holds. This is because $s_p.dist_{NN} \leq s_p.dist_{NN_{younger}} \leq \epsilon \cdot s^*_{temp}.dist_{NN}$. The same discussion is applied to line 10.

We turn our attention to line 15 of Algorithm 2. It is important to notice that at line 15, s^* is s_p or s^*_{temp}. If $s^*_{temp} = s^*$, A-SDM monitors the correct answer, so Inequality (1) holds. On the other hand, if $s_p = s^*$ and $s_p.dist_{NN_{younger}} \leq \epsilon \cdot s^*_{temp}.dist_{NN}$, we monitor $s^*_{temp} = s_{out}$. This does not violate Inequality (1) by using the above discussion at Discord-Update. We therefore conclude that Theorem 3 is true. □

5 Experiment

All experiments were conducted on a PC with Intel Xeon Gold 6154 (3.0 GHz) and 512 GB RAM.

5.1 Setting

Datasets. We used the following three real datasets in this paper.

- Bitcoin[1]: This is a streaming time-series of bitcoin transactions. Its length is 100,000.
- ECG [4]: This is a streaming time-series of electrocardiogram. Its length is 100,000.
- Google-cpu [16]: This is a streaming time-series of CPU usage rate generated by Google data center. Its length is 133,902.

Algorithms. We evaluated the following algorithms, and all of them were implemented in C++.

- HOT SAX [10]: This is a discord detection algorithm for a static time-series. We extended the original algorithm to deal with the discord update based on window sliding. This is a competitor of SDM.
- N-SDM: This is a variant of SDM, and employs only NN-tuple but does not employ NN_{older}-tuple and $\text{NN}_{younger}$-tuple. This algorithm is employed to investigate how efficiently NN_{older}-tuple and $\text{NN}_{younger}$-tuple function.
- SDM: This is the proposed algorithm in this paper.
- A-SDM: The approximation version of SDM.

We do not consider the other existing discord detection/monitoring algorithms, because all of them cannot deal with subsequence deletions and discord expirations. The original HOT SAX also cannot do it, but is a state-of-the-art algorithm that computes the discord from scratch.

Criteria. We measured the average update time per window sliding, the worst update time, and the practical approximation rate ($= s^*.dist_{NN}/s_{out}.dist_{NN}$).

[1] http://api.bitcoincharts.com/v1/csv/.

5.2 Results

We here show our experimental results. Note that the default values of l, w, and ϵ are 100, 10000, and 1.2. When we investigate the impact of a given parameter, the other parameters are fixed.

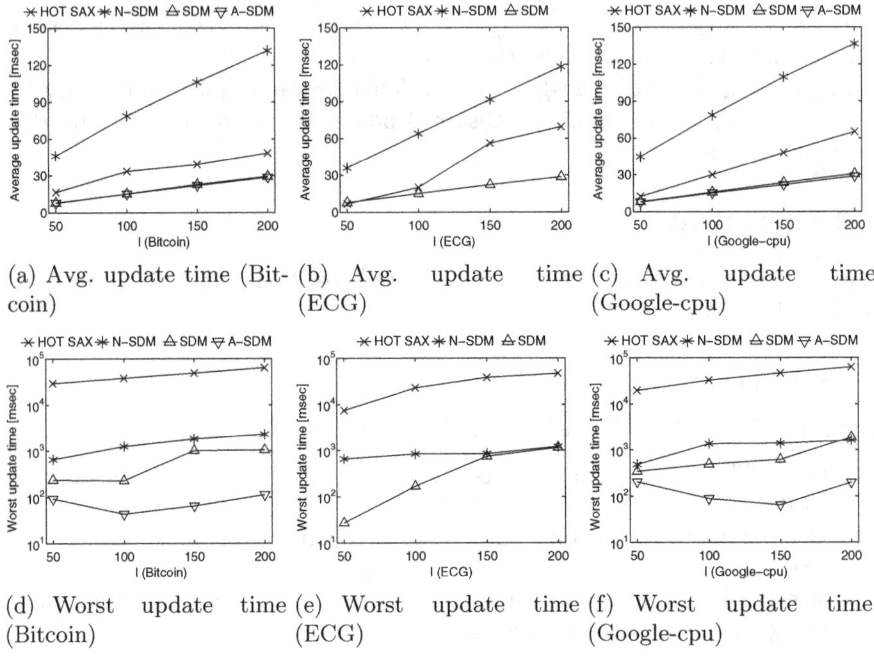

(a) Avg. update time (Bitcoin) (b) Avg. update time (ECG) (c) Avg. update time (Google-cpu)

(d) Worst update time (Bitcoin) (e) Worst update time (ECG) (f) Worst update time (Google-cpu)

Fig. 4. Impact of l (subsequence size)

Impact of l. We investigate how subsequence size affects the performance of each algorithm, and Fig. 4 illustrates the result. We here focus on average update time, which is shown in Figs. 4(a)–(c). The first observation is that SDM has a linear scalability w.r.t. l. This is reasonable, as Theorem 2 verifies. Second, SDM (and A-SDM) is (are) faster than the competitors. For each window slide, the average numbers of executions of Older-Nearest-Neighbor-Search(·) and Discord-Update are small, so SDM and A-SDM show similar update time. Also, this result suggests the effectiveness of the sequential scan based approach, as HOT SAX, which is an index-based approach, incurs more update time. Furthermore, compared with N-SDM, SDM is much faster. This is due to NN_{older}-tuple and $NN_{younger}$-tuple, as discussed in Sect. 4.2.

Next, we focus on Figs. 4(d)–(f), which depict the worst update time. It can be seen that SDM has a competitive (or better) performance with (than) N-SDM, which always holds the exact nearest neighbor for each subsequence, and is significantly faster than HOT SAX. (We omit the result of A-SDM on

ECG, because it shows a similar performance to SDM.) This result demonstrates that approaches for static time-series are not suitable for streaming time-series.

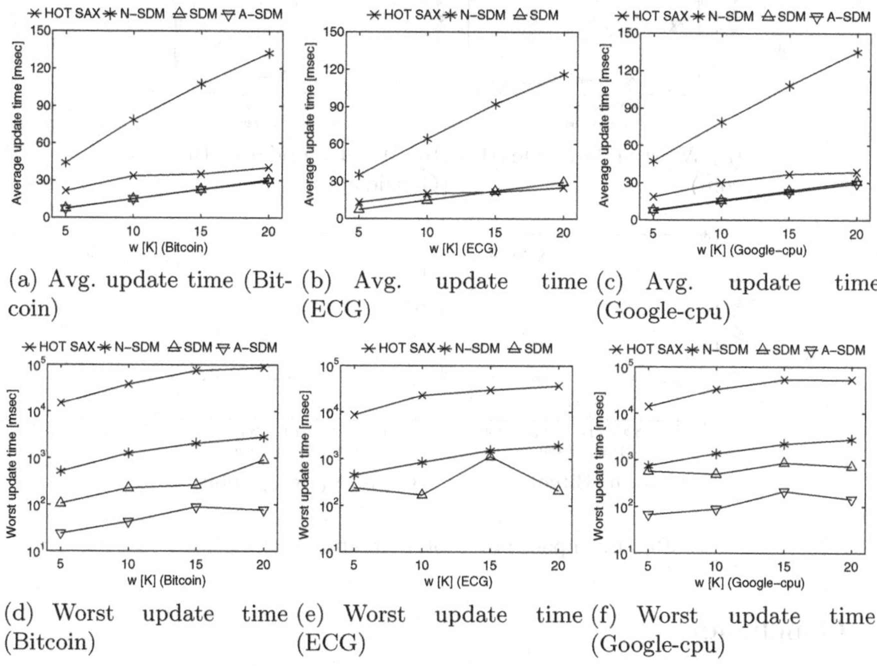

(a) Avg. update time (Bitcoin) (b) Avg. update time (ECG) (c) Avg. update time (Google-cpu)

(d) Worst update time (Bitcoin) (e) Worst update time (ECG) (f) Worst update time (Google-cpu)

Fig. 5. Impact of w (window size)

Impact of w. We next test the scalability of each algorithm w.r.t. the window size. Focus on average update time, and we can see that SDM and A-SDM are basically faster than the other algorithms, as shown in Figs. 5(a)–(c). Besides, SDM and A-SDM scale linearly w.r.t. the window size, which is also validated by Theorem 2.

Figures 5(d)–(f) illustrate that the worst update time of SDM is always faster than those of HOT SAX and N-SDM. This result is derived from the fact that SDM reduces unnecessary distance computation (i.e., nearest neighbor search) by its data structure. A-SDM furthermore reduces the worst update time, and its worst update time is much faster than those of HOT SAX and N-SDM.

Impact of ϵ. Finally, we study the impact of the approximation factor ϵ of A-SDM. Theoretically, a large ϵ provides small update time but inaccurate result. Figure 6 illustrates the practical relationship. From Figs. 6(a)–(b), as ϵ becomes large, the worst update time becomes shorter. This is a quite intuitive result. (We confirmed that the average update time is not affected by ϵ, so the result is omitted.) On the other hand, Figs. 6(c)–(d) show that the practical approximation rate is almost 1 (less than 1.03 actually), even if ϵ becomes large. That is, A-SDM monitors a highly accurate result continuously.

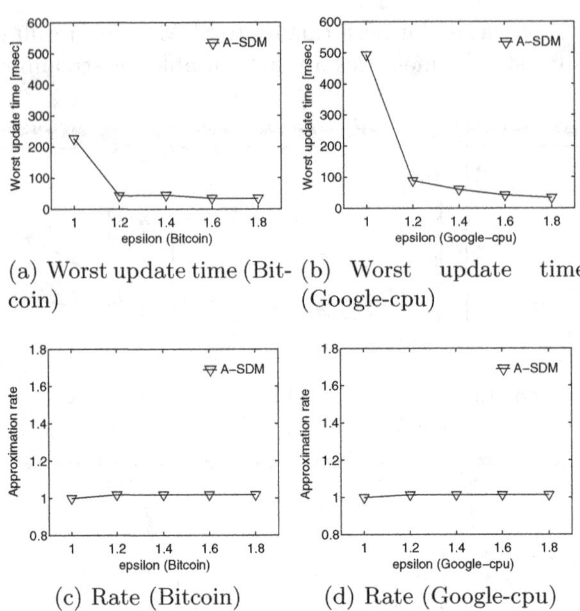

(a) Worst update time (Bitcoin) (b) Worst update time (Google-cpu)

(c) Rate (Bitcoin) (d) Rate (Google-cpu)

Fig. 6. Impact of ϵ (approximation factor)

6 Conclusion

Recent applications have been generating streaming time-series, and monitoring outlier from the time-series is an important operator for anomaly detection and data cleaning. Motivated by this observation, in this paper, we addressed a novel problem of monitoring time-series discord over a sliding window.

As an efficient solution for this problem, we proposed SDM (Streaming Discord Monitoring). SDM exploits nearest neighbor search based on sequential scan, to obtain the nearest neighbor of a new subsequence and identify the subsequences which need to update their nearest neighbor. We showed that SDM is simple, efficient, and easy to approximate the answer for further accelerating its efficiency. Our experiments using real datasets demonstrate that SDM can monitor the discord efficiently and A-SDM reduces the worst time update while keeping high accuracy.

Acknowledgment. This research is partially supported by JSPS Grant-in-Aid for Scientific Research (A) Grant Number 18H04095, JSPS Grant-in-Aid for Scientific Research (B) Grant Number JP17KT0082, and JSPS Grant-in-Aid for Young Scientists (B) Grant Number JP16K16056.

References

1. Amagata, D., Hara, T.: Mining top-k co-occurrence patterns across multiple streams. TKDE **29**(10), 2249–2262 (2017)

2. Begum, N., Keogh, E.: Rare time series motif discovery from unbounded streams. PVLDB **8**(2), 149–160 (2014)
3. Bu, Y., Leung, T.W., Fu, A.W.C., Keogh, E., Pei, J., Meshkin, S.: WAT: finding top-k discords in time series database. In: SDM, pp. 449–454 (2007)
4. Chen, Y., et al.: The UCR time series classification archive (2015). www.cs.ucr.edu/~eamonn/time_series_data/
5. Chiu, B., Keogh, E., Lonardi, S.: Probabilistic discovery of time series motifs. In: KDD, pp. 493–498 (2003)
6. Esling, P., Agon, C.: Time-series data mining. ACM Comput. Surv. **45**(1), 12 (2012)
7. Gupta, M., Gao, J., Aggarwal, C.C., Han, J.: Outlier detection for temporal data: a survey. TKDE **26**(9), 2250–2267 (2014)
8. Huang, T., et al.: Parallel discord discovery. In: Bailey, J., Khan, L., Washio, T., Dobbie, G., Huang, J.Z., Wang, R. (eds.) PAKDD 2016. LNCS (LNAI), vol. 9652, pp. 233–244. Springer, Cham (2016). https://doi.org/10.1007/978-3-319-31750-2_19
9. Kato, S., Amagata, D., Nishio, S., Hara, T.: Monitoring range motif on streaming time-series. In: Hartmann, S., Ma, H., Hameurlain, A., Pernul, G., Wagner, R.R. (eds.) DEXA 2018. LNCS, vol. 11029, pp. 251–266. Springer, Cham (2018). https://doi.org/10.1007/978-3-319-98809-2_16
10. Keogh, E., Lin, J., Fu, A.: Hot sax: efficiently finding the most unusual time series subsequence. In: ICDM, pp. 226–233 (2005)
11. Keogh, E., Lin, J., Lee, S.H., Van Herle, H.: Finding the most unusual time series subsequence: algorithms and applications. Knowl. Inf. Syst. **11**(1), 1–27 (2007)
12. Lam, H.T., Pham, N.D., Calders, T.: Online discovery of top-k similar motifs in time series data. In: SDM, pp. 1004–1015 (2011)
13. Li, Y., Zou, L., Zhang, H., Zhao, D.: Computing longest increasing subsequences over sequential data streams. PVLDB **10**(3), 181–192 (2016)
14. Li, Y., Yiu, M.L., Gong, Z., et al.: Quick-motif: an efficient and scalable framework for exact motif discovery. In: ICDE, pp. 579–590 (2015)
15. Linardi, M., Zhu, Y., Palpanas, T., Keogh, E.: Matrix profile X: VALMOD-scalable discovery of variable-length motifs in data series. In: SIGMOD, pp. 1053–1066 (2018)
16. Reiss, C., Wilkes, J., Hellerstein, J.L.: Google cluster-usage traces: format+ schema. Google Inc., White Paper, pp. 1–14 (2011)
17. Sanchez, H., Bustos, B.: Anomaly detection in streaming time series based on bounding boxes. In: Traina, A.J.M., Traina, C., Cordeiro, R.L.F. (eds.) SISAP 2014. LNCS, vol. 8821, pp. 201–213. Springer, Cham (2014). https://doi.org/10.1007/978-3-319-11988-5_19
18. Teflioudi, C., Gemulla, R., Mykytiuk, O.: LEMP: fast retrieval of large entries in a matrix product. In: SIGMOD, pp. 107–122 (2015)
19. Wang, X., Lin, J., Patel, N., Braun, M.: A self-learning and online algorithm for time series anomaly detection, with application in CPU manufacturing. In: CIKM, pp. 1823–1832 (2016)
20. Wei, L., Keogh, E., Xi, X.: Saxually explicit images: finding unusual shapes. In: ICDM, pp. 711–720 (2006)
21. Yankov, D., Keogh, E., Rebbapragada, U.: Disk aware discord discovery: finding unusual time series in terabyte sized datasets. Knowl. Inf. Syst. **17**(2), 241–262 (2008)

22. Yeh, C.C.M., et al.: Matrix profile i: all pairs similarity joins for time series: a unifying view that includes motifs, discords and shapelets. In: ICDM, pp. 1317–1322 (2016)
23. Zhang, A., Song, S., Wang, J., Yu, P.S.: Time series data cleaning: from anomaly detection to anomaly repairing. PVLDB **10**(10), 1046–1057 (2017)

Partially Indexing on Flash Memory

Wojciech Macyna(✉) and Michal Kukowski

Department of Computer Science, Faculty of Fundamental Problems of Technology, Wrocław University of Technology, Wrocław, Poland
wojciech.macyna@pwr.edu.pl, michalkukowski10@gmail.com

Abstract. Query indexing is a mature technique in relational databases. Organizing as tree-like structures, the indexes facilitate data access and speed up query processing. Nevertheless, the construction and modification of the indexes is very expensive and can slow down the database performance. Traditional approaches cover all records equally, even if some records are queried often and some never. To avoid this problem, partially indexing has been introduced. The core idea is to create indexes adaptively and incrementally as a side-product of query processing. In this way, only such records are indexed which take part in the queries. After emerging modern data storage technologies like: flash memory or phase change memory, the new index types appeared. They have been invented to overcome the limitations of such technologies. In this paper, we deal with partially indexing on flash memory. We propose a method which reduces the number of write and erase operations on flash memory during index creation. Due to employing optimization techniques specific for flash memory, the query response time is decreased twice in comparison to the traditional methods. As far as we know, it is the first approach which considers partially indexing on the physical data storage level. Thus, the paper may be the initiation of a new research direction.

Keywords: Partially indexing · Flash memory · Adaptive merging

1 Introduction

Traditional databases enable to create or remove indexes on tables and views depending on the query workload. Although the use of the index accelerates query processing, its creation and maintaining can slow down the performance of the database. In modern applications, the query workload is usually not predictable what makes the index creation more difficult. Moreover, traditional approaches cover all records equally, even if some records are queried often and some never. Let's assume that the database table *Sales* holds the history of selling transactions. If most of the queries concern records from the last year, it is not worth to index the entire table by the column *Date*. In such a case, it is more profitable to index just the data from the last year, since the data from other ranges are queried very rarely. To cope with the problem, the partially indexing was introduced. There are two main approaches: database cracking and adaptive merging. In the database

S. Hartmann et al. (Eds.): DEXA 2019, LNCS 11706, pp. 95–105, 2019.
https://doi.org/10.1007/978-3-030-27615-7_7

cracking (see [1–4]), the index is incrementally adapted in response to the actual workload. Its creation and optimization is a side effect of query execution, i.e. only such tables, columns and key ranges inside the table are indexed which are really queried. When a column is queried by a predicate for the first time, a new cracker index is initialized. Keys in a cracker index are partitioned into disjoint key ranges, but left unsorted within each partition. In most cases, each partitioning step creates two new sub-partitions. Figure 1 shows an example of the database cracking. First, the data for indexing are loaded into the unsorted array. Then, as a side-effect of query processing on the range 4 and 6, the initial partition is split into three others: (1) keys less than 4; (2) keys that fall between 4 and 6; (3) keys more than 6. After that, a new query for range 8 to 9 is executed. The partitions 1 and 2 are ignored, since their values are outside the query range. Partition 3 is split into three partitions. Subsequent queries may continue to crack the index until it will be optimal for the given workload.

An alternative approach to the index cracking is adaptive merging [5]. The main difference is that the data are sorted inside each partition. It means that before the indexing process the column is split to n sorted partitions. Each merge step only affects those key ranges that are relevant to actual queries, leaving records in all other key ranges in their initial places. This merge logic takes place as a side effect of query execution.

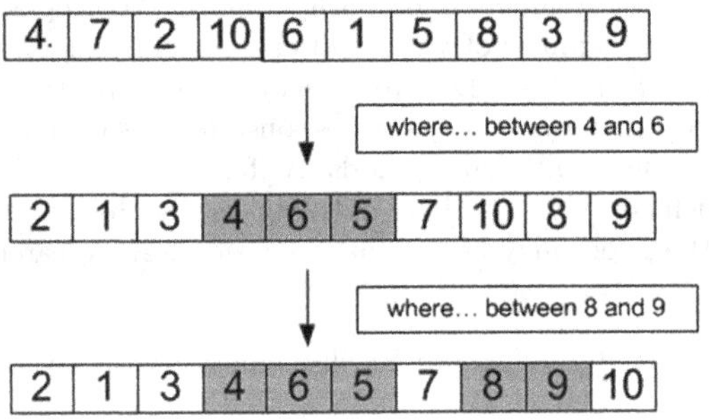

Fig. 1. Partially index creation

In this paper, we focus on adaptive merging for flash memory. Such storage type has completely different characteristics than a hard disk. It does not contain any moving parts, so the rotational delays are not dominant costs in reading or writing data. The flash memory chip consists of blocks and each block contains the fixed number of pages (32 or 64). One flash memory page has typically 2048–4096 bytes. The read and write operations are performed at the page granularity, but the way of data updating is completely different. The page cannot be overwritten. If the data modification occurs, a new version must be written to a free page. Simultaneously, the old page is checked as "dirty" and will be reclaimed later. To remove data physically, the whole block must be erased. It is carried out by the garbage collector. The serious drawback of flash memory is the limited number of erase data per block (about 100000 times). After that, the flash memory chip can be burn out and destroyed [6,7]. A distinguishing feature of flash memory is the asymmetry of the read and write speed. The read operation is significantly faster than the write one.

Such architecture has an impact on the data storage. Several types of flash-aware indexes have been proposed. Many of them, like BFTL [8] are write-optimized, but suffer poor search performance. Other index structures like: FD-tree [9] and LA-tree [10] are read and write optimized. All flash optimized index types have several common features. They reduce the number of writes and erases as much as possible. Moreover, they prefer to facilitate the sequential write over the random one, because the former is significantly faster.

The **contribution** of this paper can be summarized as follows. We propose the implementation of lazy adaptive indexing for flash memory. Our method overcomes the limitations of this memory type by reducing the number of write and erase operations. In this way, the query response time decreases twice in comparison to the traditional approach. We perform several experiments to prove the efficiency of the method. The method can work with all index types mentioned above. As far as we know, this is the first approach of implementing partially indexing for modern storage devices.

2 Lazy Adaptive Merging

In this section we propose the lazy adaptive merging (LAM) tailored for flash memory.

The main goal of LAM is to minimize the number of write and erase operations. In our method, we utilize the lazy delete technique. The data are not deleted immediately but only in the case when a lot of memory pages are not valid. As adaptive merging is applied to the secondary index, we define the index element as an entry: $<key, ptr>$, where *key* denotes a key value and *ptr* is a pointer to the record in the database table.

Our method works as follows. In the beginning, the entries for indexing are written to flash memory partitions. When a query arrives, all entries taking part in this query are copied from the partitions and to the flash index which is stored in the separate structures of flash memory. The entries are not immediately deleted from the partitions, but are removed in bulks during the reorganization process. Additionally, we put the query range into the special structure called *Usage Array*. It makes the already indexed entries easy to identify in the subsequent queries. The method works until all entries from the partitions are stored in the index.

2.1 System Architecture

The system has following components:

- *Sorting buffer.* The RAM resident buffer for sorting the entries.
- *Partition.* The structure for storing sorted entries. Its size is equal to the size of the sorting buffer. One partition can contain many memory blocks.

The entries within each partition are sorted by the key. So, it is easy to find an entry with a requested key using the binary search algorithm. Each partition holds in RAM the minimal (*min*) and maximal (*max*) key value. In this way, we can easily determine which partitions should be accessed after the particular query request.

- *Flash Index.* Any index optimized for flash memory (for example: FD-tree [9] or LA-tree [10]). The entires from the partitions are copied to the index to carry out the adaptive merging.
- *Usage Array.* The array which holds a number of entries copied to the index from the particular block of the partition. Its size is equal to the total number of blocks in all partitions. For example, if the block 5 holds 160 entries and 10 of them are already indexed, then $usage[5] = 10$.
- *Journal.* It holds a list key ranges already copied to the index. For example, after two queries: (b, g) and (a, c), the journal contains $(b, g), (a, c)$. In this way, it is easy to determine whether the key is already indexed or should be searched in the partitions.

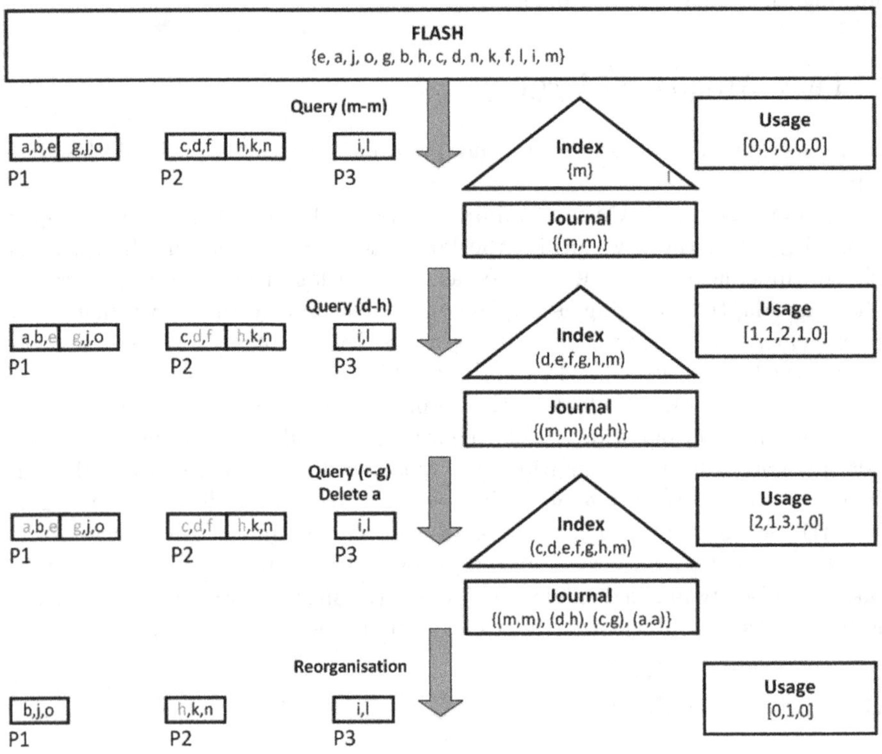

Fig. 2. LAM Process example

Figure 2 shows an example of our method. Let's consider the table containing 15 records with the following key values: $(e, a, j, o, g, b, h, c, d, n, k, f, l, i, m)$. We assume that one flash block can hold 3 entries with one entry per page. The sorting buffer may hold only 6 records, so one partition consists of 2 blocks. Our method uses a merge threshold (MT). In this example, we set $MT = 2$. It means that the reorganization process would take place if it is possible to reduce the number of used memory blocks by at least 2.

At the beginning, the entries are stored arbitrarily. Let's assume that we receive a query (m, m). The entries are sorted in the buffer, and then stored in the partitions. Thus, we write m into the index and the actual range (m, m) into the journal. The second query is of the range (d, h). First, we scan the journal to find whether one of the existing ranges overlaps the range of the actual query. Unfortunately, (m, m) is disjunctive with (d, h). In this case, we must take the requested entries from the partitions. We know that the partition $P3$ does not contain these values. So, there is no need to access the pages of $P3$. We take $\{e, g\}$ from $P1$ and $\{d, f, h\}$ from $P2$. Then, we write the entries into the index and update the journal. To avoid unnecessary writes, the values are not removed from the partitions. We use a shadowed font to mark the values which have already been written to the index. In our method, we remember the block usage. It denotes the number of already copied entries from the particular block to the index. For example, after two queries, we have usage[1] = 1. It is because one entry from this block is already in the index. In the last step of the algorithm, we check the possibility of partition reorganization. The process may be invoked when a number of memory blocks can be reduced by 2 ($MT = 2$). As we can see, the condition is not fulfilled so far. The situation changes in the third query (c, g). A range (d, g) can be found in the journal but c should be copied from partitions $P2$. So, the next range (c, g) will appear in the journal. We delete also the value a. Instead of physically removing it from the partition, we depict it as obsolete and write it into the index. Thus, the new entry (a, a) appears in the journal.

From the usage array, we can conclude that already seven entries have been copied from the partitions to the index. If one block may hold three entries, it is easy to determine that two blocks can be recovered ($MT = 2$) in the reorganization process: the blocks from $P1$ are merged and the first block is removed from $P2$. Clearly, the usage array must be also updated. It contains three elements now, since there are three memory blocks in the partitions. The example shows the advantage of our method over the standard adaptive merge. LAM carried out 3 erase and 3 write operations. At the same time, the standard adaptive merge without any optimization would require 6 erases and 8 writes. It is worth to see that a number of reads would stay the same. So, our approach has no impact on search performance, but drastically decreases the number of writes and erases.

2.2 LAM Algorithms

In this section, we present the algorithms of LAM.

Insertion. Insertion adds a new entry to the index. As the index is a tree optimized for flash memory (e.g. FD-tree, LA-tree), the operation is executed in

Algorithm 1.1. Search(input: Key *keys*[ ])

```
// Let E_k denote the entry of the key k
Entry result[ ] := ∅
Entry toInsert[ ] := ∅
Entry result[ ] := ∅
Key keysFromIndex[ ] := ∅
Key keysFromPartition[ ] := ∅
// It estimates on the basis of the journal which entries should be
   copied from the partitions
foreach Key k ∈ keys do
    if journalContains(k) = False then
        keysFromPartition := keysFromPartition ∪ k
    else
        keysFromIndex := keysFromIndex ∪ k
result := result ∪ findEntriesInIndex(keysFromIndex)
// Find partition to insert on the basis of their min, max
   attributes
Partition toLoad[ ] := ∅
foreach Key k ∈ keysFromPartitions do
    foreach Partition p do
        if k ≥ p.min AND k ≤ p.max then
            toLoad := toLoad ∪ p
foreach Partition p ∈ toLoad do
    foreach Key key ∈ keysFromPartition do
        if k ≥ p.min AND k ≤ p.max then
            Page page := findPage(p, E_k)
            Let b be a block containing page
            usage[b] := usage[b] + 1
// Insert the new data to the partial index
insertToIndex(toInsert)
result := result ∪ toInsert
Pair p := (min(keysFromPartition), max(keysFromPartition))
addToJournal(p)
if blockNmb - blockNmbReorganization ≥ MT then
    reorganization()
return result
```

the specific way for the underlying index (see [9,10]). Additionally, the keys of inserted entries must be added to the journal.

Deletion. The entry to remove may be stored in the index or one of the partitions. In the first case, the operation is executed in the specific way for the underlying index. In the second case, the partition holding the requested entry is determined. After that, the key is searched in the partition, the entry is marked as deleted and the usage array is updated. The data are not removed from the partitions immediately. The reorganization process may be invoked only in the case when it is possible to reduce the number of memory blocks by MT.

Update. Update operation is performed as a deletion of the old data version and insertion of the new one.

Search. Search is the most important operation in LAM (see Algorithm 1.1). The algorithm takes as an argument a sorted array of keys and returns the requested entries as a result. At first, the method checks in the journal which parts of the range can be obtained directly from the index and which from the partitions (lines 5–9). A subset of the keys already stored in the index can be obtained directly by the search index method. The implementation of this method is specific for the index (line 10). Lines 12–15 are responsible for deriving partitions with the entries of the requested keys. It is easy to determine the partition since each partition has the minimum and maximum key value. In lines 16–21 the pages with the requested entries are selected from the partitions and the entries are inserted into the index (line 22). Additionally, the journal must be updated (lines 24–25) and the keys are added to the array $toInsert$. As in the deletion algorithm, the reorganization process may be invoked (lines 26–27).

Reorganization. LAM uses lazy deletion. It means that the data are not immediately removed from the disk. Lazy deletion is necessary to avoid expensive write and erase operations in the flash memory. The reorganization process is invoked only in the case when many blocks in the partitions hold already indexed entries. It may be estimated on the basis of the usage array and the journal. We prefer to erase the blocks with the high value of the usage array (the blocks with many marked entries).

3 Experiments

In this section, we present some experiments which confirm the effectiveness of LAM. The experiments were carried out on Intel Core i7-6700k 4.0 Ghz (8CPUs) equipped with 32 GB RAM. The implementation was written in C compiled with gcc 8.1. For the experiments, we chose the disks with parameters presented in Table 1. It contains write and read latency and the block size of the flash disks.

In the experiments, we used the table *Orders* from TPC-H [11] with 10 million records. The size of one record is about 140 bytes, what makes 1,4 GB total table size. For indexing, a column of the integer type was selected.

Table 1. Disk parameters

Operation	Intel DC P4511	Toshiba VX500	Samsung 840
Sequential read	2000 MB/s	550 MB/s	540 MB/s
Sequential write	1430 MB/s	515 MB/s	520 MB/s
Random read	295 KIOPS	92 KIOPS	95 KIOPS
Random write	36 KIOPS	65 KIOPS	44 KIOPS
Block size	64 * 4096	64 * 4096	64 * 8192

The indexed column stores the values from 0 to 10 million with even distribution. To compare different partial indexing algorithms, we utilize a framework proposed in [12]. The authors determine four stages of adaptive indexing over the lifespan of a workload phase: from initialization till the full index creation.

1. *Planting.* This first stage is characterized by the fact that per-query costs of adaptive indexing exceed those of scan-based query evaluation. The initial step in adaptive indexing is always the extraction of future index entries from the original data collection. In this stage, the query response time is always higher than the table scan.
2. *Nursing.* With more queries being executed, the index becomes more detailed and complete, yet query execution benefits from the efforts of prior queries. Hence, the expenses for further refining the index decrease while at the same time the benefits of using the improving index increase. Consequently, the per-query costs of adaptive indexing decrease. In this stage, the query response time is always better than the time of table scan but slower than the full index time.
3. *Growing.* As index refinement proceeds, the cumulative benefits of the nursing stage eventually outweigh the cumulative investments during the planting stage. In this stage, there are the points where the query response time is similar to the time of using full index.
4. *Harvesting.* Finally, the index structure is fully optimized and query execution no longer includes side effects. Per-query execution costs reach a minimum.

In the experiments, we use the following patterns:

1. Random pattern - we draw a minimal key (min) with a uniform distribution. Then, we select the records starting with this key so that the fixed selectivity is reached. By selectivity, we denote which fraction of records will be returned by the query. Thus, we obtain the range $[min, min + rowsToRead]$ where $rowsToRead = totalRowsInTable * selectivity$.
2. Sequential pattern - we create the sequential access pattern as follows (see [13]): starting from the beginning of the value domain, the queried range is moved for each query by half of its size towards the end of the domain to guarantee overlapping query predicates. When the end is reached, the query range restarts from the beginning. The position to begin is randomly set in

the first 0.01% of the domain to avoid repetition of the same sequence in subsequent rounds.

3. New keys pattern - each query asks for non-index data. Theoretically, this is the worst case for our algorithm.

In the experiments, we compare the query response time using our method (LAM), the traditional adaptive merging (AM) without flash memory optimization, the scan access (without indexing) and the full index. The figures presented below show the number of queries (axis X) and the execution times of the particular queries (axis Y). Please recall that the assumption of partially indexing is to create index gradually. In the case of the full index, the query response time is very small, but the overhead at the beginning (which is not pointed out on the chart) is very high.

In the experiments, we used the sorting buffer with the size 14 MB (0.01% of the database size) and the merge threshold (MT) set as 42 MB/blocksize. Each query selects 5% records from the table. The experiment was carried out until the LAM index was build completely. So, it ended with starting of the harvesting stage. We present the results only for Toshiba (the results for Intel and Samsung are similar).

Figures 3, 4 and 5 present results for different patterns. For the random pattern (Fig. 3), we observe that AM reaches the nursing stage very fast, but does not approach the growing stage. On the contrary, LAM performs much better. Clearly, at the times of reorganization process, LAM is slower than the scan, but it happens very rarely. The growing stage is reached very fast.

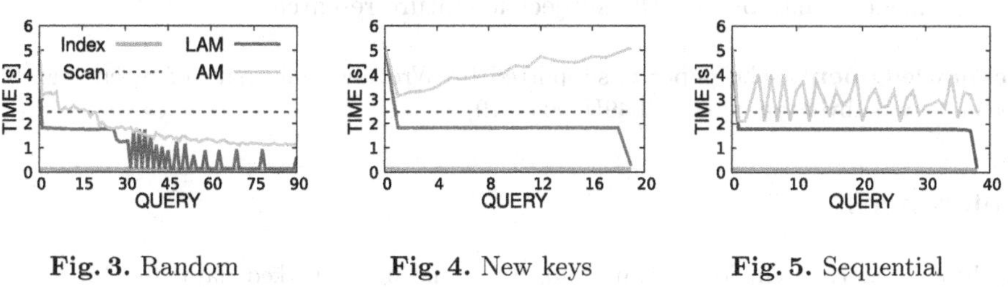

Fig. 3. Random **Fig. 4.** New keys **Fig. 5.** Sequential

Figure 4 contains the results of the same experiment, but for the new key pattern. We can observe that AM does not reach the nursing stage and is always worse than the scan. The query processing time grows because the data is read from random locations in each partition. As a consequence, the entries from many blocks are copied to the index and the larger number of blocks must be erased. Such a situation has no impact on LAM. Instead of removing the blocks immediately, it waits for the appropriate moment to trigger the reorganization process.

Figure 5 contains the results of the same experiments for the sequential pattern. Both algorithms achieve the nursing stage already in the second query. What is interesting, AM does not reach the growing stage, because the erase cost is too high. Let's note, however, that LAM ideally amortizes the cost of index merging. Its time is constant almost over the whole period of time. It is different in the case of AM. We see huge variations in the query processing time. Sometimes we load the entire blocks, sometimes the fraction of blocks. The second case happens, when we do not start from the beginning of the block. This causes the fluctuations visible on the charts. Clearly, it does not apply to LAM because it erases many blocks at once.

From the experiments presented above, we draw the following conclusions. The LAM index reaches the nursing stage very fast, but stays in it quite long. On the contrary, AM needs more time to achieve even the nursing stage before LAM reaches the growing stage. In all cases, the total processing time of LAM is reduced at least twice in comparison to AM. Additionally, due to MT parameter, we may determine the reorganisation frequency which can slow down query processing at the expense of memory overhead.

4 Conclusions

In this paper, we proposed the new method of partially indexing for devices based on flash memory technology. Our approach decreases the number of flash memory writes and erases and, as a consequence, it outperforms the traditional approach. We carried out several experiments which confirmed the efficiency of our approach. The future work may be very diverse. It would be interesting to deal with indexing merging for column oriented databases on flash memory. The application of partially indexing for other memory types like, for example, phase change memory may be also the subject for future research.

Acknowledgment. The paper is supported by Wroclaw University of Science and Technology (subvention number 049U/0044/19).

References

1. Idreos, S., Kersten, M.L., Manegold, S.: Updating a cracked database. In: Proceedings of the 2007 ACM SIGMOD International Conference on Management of Data, SIGMOD 2007, pp. 413–424. ACM, New York (2007)
2. Idreos, S., Kersten, M.L., Manegold, S.: Database cracking. In: CIDR (2007)
3. Idreos, S., Kersten, M.L., Manegold, S.: Self-organizing tuple reconstruction in column-stores. In: Proceedings of the 2009 ACM SIGMOD International Conference on Management of Data, SIGMOD 2009, pp. 297–308. ACM, New York (2009)
4. Kersten, M.L., Manegold, S., Kersten, M., Manegold, S.: Cracking the database store. In: CIDR (2005)
5. Graefe, G., Kuno, H.: Self-selecting, self-tuning, incrementally optimized indexes. In: Proceedings of the 13th International Conference on Extending Database Technology, EDBT 2010, pp. 371–381. ACM, New York (2010)
6. Park, D., Debnath, B.K., Du, D.H.C.: A dynamic switching flash translation layer based on page-level mapping. IEICE Trans. **99-D**(6), 1502–1511 (2016)

7. Wang, Y., et al.: A real-time flash translation layer for NAND flash memory storage systems. IEEE Trans. Multi-Scale Comput. Syst. **2**(1), 17–29 (2016)
8. Wu, C.H., Kuo, T.W., Chang, L.P.: An efficient B-tree layer implementation for flash-memory storage systems. ACM Trans. Embed. Comput. Syst. **6**(3) (2007)
9. Li, Y., He, B., Yang, R.J., Luo, Q., Yi, K.: Tree indexing on solid state drives. Proc. VLDB Endow. **3**(1–2), 1195–1206 (2010)
10. Agrawal, D., Ganesan, D., Sitaraman, R., Diao, Y., Singh, S.: Lazy-adaptive tree: an optimized index structure for flash devices. Proc. VLDB Endow. **2**(1), 361–372 (2009)
11. Barata, M., Bernardino, J., Furtado, P.: An overview of decision support benchmarks: TPC-DS, TPC-H and SSB. In: Rocha, A., Correia, A.M., Costanzo, S., Reis, L.P. (eds.) New Contributions in Information Systems and Technologies. AISC, vol. 353, pp. 619–628. Springer, Cham (2015). https://doi.org/10.1007/978-3-319-16486-1_61
12. Graefe, G., Idreos, S., Kuno, H., Manegold, S.: Benchmarking adaptive indexing. In: Nambiar, R., Poess, M. (eds.) TPCTC 2010. LNCS, vol. 6417, pp. 169–184. Springer, Heidelberg (2011). https://doi.org/10.1007/978-3-642-18206-8_13
13. Schuhknecht, F.M., Jindal, A., Dittrich, J.: The uncracked pieces in database cracking. Proc. VLDB Endow. **7**(2), 97–108 (2013)

HGraph: A Connected-Partition Approach to Proximity Graphs for Similarity Search

Larissa Capobianco Shimomura[1,2](✉) and Daniel S. Kaster[2]

[1] Eindhoven University of Technology, Eindhoven, The Netherlands
l.capobianco.shimomura@tue.nl
[2] University of Londrina, Londrina, PR, Brazil
dskaster@uel.br

Abstract. Similarity search is a common approach to support new applications that deal with complex data (e.g., images, videos, georeferenced data, etc.). As a consequence, appropriate indexing structures to support this task have been proposed in the literature. Recently, graph-based methods have shown to be very efficient for approximate similarity search. However, some of the main types of graphs used still suffer from two main drawbacks: (i) slow construction, and (ii) inaccurate retrieval. To reduce these drawbacks, in this paper, we propose the *HGraph* method. *HGraph* is a divide-and-conquer method for building graphs for similarity search that recursively partitions the input dataset and connect vertices across partitions at different levels. The method can be used with different types of graphs proposed in the literature to speed up the graph construction time as well as to increase the approximate search results quality through long-range edges connecting pivots of different partitions. We present experimental results using real datasets that show that *HGraph* applied to the *k-NNG* graph was able to decrease the construction time while increasing the approximate search recall when compared to the *k-NNG*. Regarding the application of *HGraph* to the *NSW* graph, the query recall also increased, however with a higher computational cost. An analysis of different combinations of the tested methods demonstrated *HGraph* query times given a recall rate were always among the top results regarding different setups.

Keywords: Proximity Graphs · Similarity search · Metric spaces

1 Introduction

A wide range of modern applications require similarity retrieval of complex data, such as images, videos, and georeferenced data. A complex datum is commonly represented through a feature vector describing to some extent its intrinsic characteristics. The similarity between two complex data is usually measured by

This work has been supported by the Brazilian agencies CAPES and CNPq.

S. Hartmann et al. (Eds.): DEXA 2019, LNCS 11706, pp. 106–121, 2019.
https://doi.org/10.1007/978-3-030-27615-7_8

applying a distance function to their respective feature vectors [2]. The combination between feature space and distance function is called similarity space, which can be modeled as a metric space [5].

However, similarity searches can have a high computational cost due to the high complexity of calculating similarity functions and the distribution of data in the search space [5,12]. A large number of access methods have been proposed in the literature to speed up this task. Recently, graph-based methods have emerged as a very efficient option for similarity searching [7,9].

A common approach to use graphs for similarity searching is to use vertices to represent complex data, and edges to connect two vertices as the similarity relationship between the pair of complex data objects [9,10,12]. The most common category of graphs used for complex data similarity retrieval is the Proximity Graphs. A proximity graph is a graph $G(V, E)$ (V is the set of vertices and E the set of edges) where each pair of vertices $u, v \in V$ is connected by an edge $e = (u, v)$, $e \in E$, if and only if u and v fulfill a defined property P. Property P is called the neighborhood criterion, and it defines the type of graph [10,12].

Some of the main types of graph used for similarity searches are the *k*-Nearest Neighbor Graph (*k-NNG*) [7,12] and the Relative Neighborhood Graph (*RNG*) [10]. However, one of the drawbacks of these types of graphs is the slow construction time, especially for exact construction. Notice that, even regarding exact construction, currently the most effective similarity search algorithms in these graphs are based on the so-called spatial approximation property and return approximate results. Additionally, to return high-quality results, these graphs structure must provide means so the search algorithm can approximate to the query answer in wherever graph region the correct answer is. The search must not concentrate efforts in only one region because it will probably result in a local optimal answer [14]. A recent proposal is the Navigable Small World Graph (*NSW*) [9], which has achieved in general a high recall. One of the main properties that lead *NSW* to reach accurate results is that it connects different graph regions by using the long-range edges [9]. However, *NSW*'s construction approach is incremental, which makes it dependent on the insertion order.

In this paper, we propose the *HGraph* method, which aims at reducing the drawbacks and increasing the approximate search quality of graph-based methods. *HGraph* is a divide-and-conquer method for building graphs for similarity search based on two pillars: (i) recursive partitioning of the input search space, and (ii) proper connection between vertices across partitions in different levels. The method can be used with different types of graphs. While the partitioning allows *HGraph* to reduce the cost of the base graph construction as the construction happens in smaller parts of the dataset, which can indeed be executed in parallel, the partitions should be connected in the resulting graph. *HGraph* employs two approaches for connecting partitions: replication of elements in an overlapping region in the border between two neighbor partitions, and connection of pivots of different partitions using long-range edges. As pivots become gradually closer to each other with partition steps, the length of the long-range edges decreases compared to the whole graph. As a result, the graph has long-range edges of varied sizes, which can be useful as shortcuts during the search.

We evaluated *HGraph* using several datasets. The results show the impact of the parameters in the method and how it behaves when applied to *k-NNG* and *NSW*. The application of *HGraph* to *k-NNG* enabled to decrease the (brute-force) construction time, increase the recall rate, and reduce query time in several situations. On the other hand, when applied to *NSW*, the method was only able to increase the recall rate; however, with a higher query execution time. When comparing the query time of different configurations for a given recall rate, *HGraph* settings were always among the top positions, including cases restricted to exact answers.

This paper is organized as follows: Sect. 2 describes the *HGraph* method to build a graph for similarity searching; Sect. 3 presents an analysis of the *HGraph* main parameters, and experiments comparing the method to the corresponding base graphs; and Sect. 4 concludes the paper.

2 The HGraph Method

The *HGraph* main objectives are to accelerate the graph construction time through partitioning and increase the approximate search results quality by adding long-range edges connecting pivots across different partitions. Long-range edges are edges that do not follow the proximity property of the graph and link vertices that are not close to each other according to the proximity property implied to build the graph. The *HGraph* foundation is twofold.

1. Divide-and-conquer strategy to build different types of graph. This strategy not only speeds up the graph construction but also can run in parallel. The *HGraph* method divides the dataset into smaller overlapping subsets according to elements selected as pivots. The division process is done recursively until the cardinality of a subset is smaller than a fixed value (m).
2. Addition of long-range edges connecting pivots in every partition. The method connects the partition pivots following the neighborhood criterion of a graph type defined as a parameter to the method. The added long-range edges are responsible for connecting different graph regions, reducing the local answer problem.

Algorithm 1 summarizes the *HGraph* method. The parameters are: number of pivots (n_p); pivot selection strategy; minimum number of elements per partition (m); overlap rate (o); type of the base graph (g_1, along with the corresponding construction parameters); and a second type of graph (g_2), which is employed for a final refinement procedure applied only to the pivots. The main procedure of the method is the recursiveConstruction, which is responsible for partitioning the dataset recursively in n_p partitions until every partition reaches at most m elements, adding long-range edges to the chosen pivots, and building the graph type provided as a parameter (g_1) in each subset. The refineHGraph method is a final refining method to connect pivots from different graph regions. This method relies on a global set (P) that stores all pivots. The next subsections explain each step of the *HGraph* method in more details.

Algorithm 1. *HGraph* Method

Input Dataset S, distance function δ, number of pivots n_p, pivot selection technique p_{sel}, minimum partition size m, overlap rate o, base graph type g_1, and refinement graph type g_2
Output *HGraph* G

function HGRAPH(S, x, m, o, g_1, g_2)
  $G, P \leftarrow \{\}$ ▷ G: HGraph; P: global set of pivots
  $G, P \leftarrow$ recursiveConstruction($S, \delta, G, P, n_p, p_{sel}, l = 0, o, m, g_1, g_2$)
  $G \leftarrow$ refineHGraph(G, P, g_2)
end function

2.1 Dataset Partitioning

The recursive partitioning technique employed by *HGraph* initially selects from the dataset n_p elements that will act as pivots according to a given pivot selection strategy (p_{sel}). Then the dataset is divided into n_p subsets using the generalized hyperplane partition approach [16]. The generalized hyperplane partition adds each element to the subset of its closest pivot. Therefore, it requires to calculate the distance of each element of the dataset to each one of the pivots.

Figure 1 illustrates this process. The dataset was divided into disjoint partitions according to the pivots (R, Q and E). Notice that using only the generalized hyperplane partition approach the final graph built by the *HGraph* method would not connect two elements from different subsets even if they are really close in the similarity space. Disconnected regions or cases in which two elements that are very similar but are not connected can restrain spatial approximation based search algorithms to find exact or high recall approximate similarity search answers. This problem remains even when adding the long-range edges to the pivots. For this reason, our proposal considers an overlap region between the subsets during the dataset partition (shown as dash lines in the figure), in which elements are duplicated to the corresponding neighbor partition to allow them to be connected to elements from the other partition that are close to the partition's boundary. Next, we present the approach adopted to calculate the overlap.

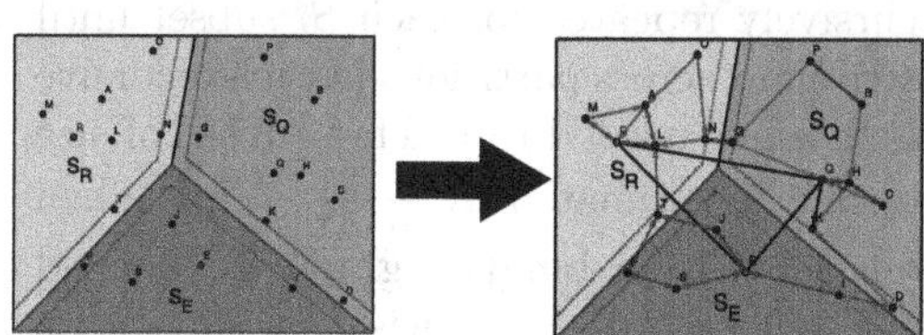

Fig. 1. *HGraph* method overview for a 2-NNG. (Color figure online)

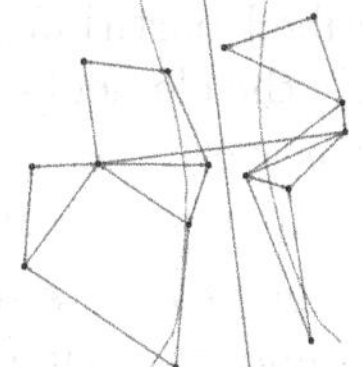

Fig. 2. Hyperplane curvature

2.2 Overlap Region Definition

An approach to select the overlap region elements is to use the covering radius from the pivots. However, a fixed distance threshold does not guarantee that there will be elements in the overlap region because the existence of elements at this distance from the pivots depends on the dataset distribution and shape of the used distance function. To assure a given number of elements in the overlap and reasonable distribution of them across the partition boundary, we propose to use the distance to the hyperplane that divides the neighbor partitions and define the size of the region according to the number of elements in the subset.

Precisely, our approach for overlap consists of estimating the number of elements in the overlap region using the input parameter o. The overlap region size between two subsets S_1 and S_2 is calculated as the proportion o of the whole set that is being partitioned: $|S_1 \cap S_2| = \lceil |o * S| \rceil$. However, since we are dealing with metric spaces, the hyperplane may not exist. As a result, instead of calculating the distance of an element/point to the hyperplane, we approximate this distance by using the difference between the distances from element to the two pivots of the neighbor partitions. In order to determine if an element $s_i \in S_1$ is in the overlap region regarding S_2, we consider the distance difference between the element and the pivots p_1 of S_1 and p_2 of S_2 to simulate the distance to the hyperplane: i.e., $\delta(s_i, p_1) - \delta(s_i, p_2)$. The elements in the overlap region are the top-$\lceil |o * S| \rceil$ elements with the smallest distance difference to the pivots, which are then duplicated and added to the neighbor's partition subset. This approximation produces a region whose boundaries are not parallel to the hyperplane when the hyperplane do exist. Figure 2 shows the curvature of the overlap when the elements have the same distance difference to the pivots. Moreover, the areas of two neighbor overlap regions may be different because the area of an overlap is defined by the farthest element in the region to the hyperplane. Finally, since the distance δ is calculated for all pivots during the generalized hyperplane partition, the overlap computation does not require additional distance calculations.

2.3 Recursive Construction

Algorithm 2 shows the recursive construction process in *HGraph*. We call in this text each division step of *HGraph* as level l, incremented every call (line 2). The described partition process is recursively repeated for each S_i subset until $|S_i| \leq m$. If the elements are evenly distributed across partitions in each recursive partition process, the total number of divisions needed to reach a subset of size less or equal to m, represented by L, can be estimated by $\frac{|S|}{(n_p)^L} \leq m$. When the partition process stops, the *HGraph* algorithm builds a graph g_1 (defined in the parameter settings) in each subset (lines 3–6). Otherwise, the algorithm chooses n_p pivots from S according to the pivot selection strategy p_{sel}, divide S into partitions $S_1, \ldots, S_{n_p}$ respecting the overlap o (lines 8–9). After that, the algorithm creates long-range edges connecting the chosen pivots, according to the neighborhood property of the refining graph type g_2, given as a parameter. The new vertices and edges are added to graph G (lines 10–12). All long-range edges added to the pivots (at any partition level) are undirected even when g_2 is a directed graph so that they can be traversed in both directions. The pivots are added to the global set P to be used in the final refinement (line 13).

Algorithm 2. HGraph Recursive Construction Algorithm

1: **function** RECURSIVECONSTRUCTION($S, \delta, G(V, E), P, n'_p, p_{sel}, l, o, m, g_1, g_2$)
2: $l \leftarrow l + 1$
3: **if** $|S| \leq m$ **then**
4: $G'(V', E') \leftarrow$createGraph(S, δ, g_1) ▷ Builds graph of type and parameters g_1 from set S
5: $V \leftarrow V \cup V'$
6: $E \leftarrow E \cup E'$
7: **else**
8: $P_S \leftarrow$ choosePivots(S, n'_p, n_{sel}) ▷ Selects n_p pivots from S
9: $S_1, S_2 \ldots S_{n'_p} \leftarrow$ partitionSet(S, P_S, o)
10: $G_p(V_p, E_p) \leftarrow$ connectPivots(P_S, g_2) ▷ Connect pivots according to g_2 (undirected)
11: $V \leftarrow V \cup V_p$
12: $E \leftarrow E \cup E_p$
13: $P \leftarrow P \cup P_S$
14: $s \leftarrow \frac{|S|}{(n_p)^l}$
15: **for** $i \leftarrow 1, \ldots, |P_S|$ **do**
16: $n'_p \leftarrow \frac{|S_i|}{s}$
17: recursiveConstruction($S_i, \delta, G, P, n'_p, p_s el, l, o, m, g_1, g_2$)
18: **end for**
19: **end if**
20: **return** G, P ▷ Graph and set of Pivots
21: **end function**

Depending on the dataset distribution, after partitioning, the subsets may not be balanced. Unbalanced subsets might degenerate the *HGraph* partitioning to an unnecessarily large number of recursive steps. To balance the sizes of the subsets in each level, we calculate the proportion of the actual subset size according to the expected size of the subset in that level and address the number of pivots according to this proportion. In the first division step, at level $l = 1$, the dataset is divided into parameter defined n_p subsets. From the second level ($l = 2$), the expected total number of pivots is $(n_p)^l$ and the expected size of each subset is $s = \frac{|S|}{(n_p)^l}$. According to the expected size of the subsets, the number of pivots addressed to each subset is calculated by $\frac{|S_i|}{s}$, which is the parameter passed to the subsequent recursive calls (lines 14–17 of Algorithm 2). The pivots chosen in the previous level are selected to be pivots in the next level as well. This way, the pivots will become hubs in the graph that connects different subsets as the dataset partitioning recursively continues. Using the pivots as the vertices with long-range edges has two advantages: (1) no additional computations are needed to select vertices to add long-range edges, and (2) the pivots act as "connectors" to other regions as there will be at least one connection from one subset to another.

2.4 Final Refinement

At the end of the recursiveConstruction function, the *HGraph* will have few long-range edges that connect graph regions that are far from each other. This is a result of connecting pivots only locally in each level subset. To better interconnect all partitions, *HGraph* calls the RefineHGraph procedure to add undirected edges between all pivots, following the neighborhood criterion of g_2.

Table 1. Datasets used for the experiments.

Dataset	Size	Dimensions	Source
USCities	25,374	2	Geographic coordinates from American cities[a]
Color Moments	68,040	9	Feature vector from *Corel* [11]
Texture	68,040	16	Co-ocurrence texture descriptor from *Corel* [11]
Color Histogram	68,040	32	Color histogram descriptor from *Corel* [11]
MNIST	70,000	784	Images of handwritten digits [8][b]

[a]http://www.census.gov/main/www/cen2000.html
[b]http://yann.lecun.com/exdb/mnist

3 Experiments and Results

In this section, we discuss how the *HGraph* parameters interfere in construction and search performance and show experimental results regarding varying parameter settings. We compare the performance of applying *HGraph* to the graph-based methods k-Nearest Neighbors Graph (k-*NNG*) and Navigable Small World Graph (*NSW*) to the performance of these graphs not using *HGraph*. We also compare *HGraph* to the tree-based method Spatial Approximation Tree (*SAT*).

We implemented *HGraph* in C++ as an extension of the Non-Metric Space Library [3]. For the graph-based methods, we used the *GS* and *GNNS* approximate search algorithms. Both algorithms rely on a greedy strategy to always approximate spatially to the query answer, based on the spatial approximation principle [14]. In results, the *NN* denotes the number of neighbors used to build the graph g_1, and the Restarts is the parameter R in the *GNNS* algorithm [7]. Table 1 shows the five datasets employed in our analysis and their respective sizes and dimensionalities. From each dataset, we removed 100 random elements to serve as query elements and used the remaining ones to build the graphs. We employed the Euclidean distance (L_2) as the similarity measure for all methods. To avoid too many settings, we fixed the parameter g_2 to a k-*NNG* with $2 * NN$ that is being used in g_1 for all experiments. We used average query time, number of distance computations, and recall (fraction of correct query answers retrieved) to evaluate search performance. We also evaluated the structure construction time of each tested method. The experiments were carried out on an Intel Core i7 (32 GB RAM) with a single thread for all methods on an Ubuntu GNU/Linux 18.04.1 64 bits.

3.1 Impact of Long-Range Edges in *HGraph*

In this subsection, we show how the long-range edges can increase the average recall rate for similarity searches. To evaluate only the effect of long-range edges in *HGraph*, we set the overlap proportion parameter o to 1. By setting $o = 1$,

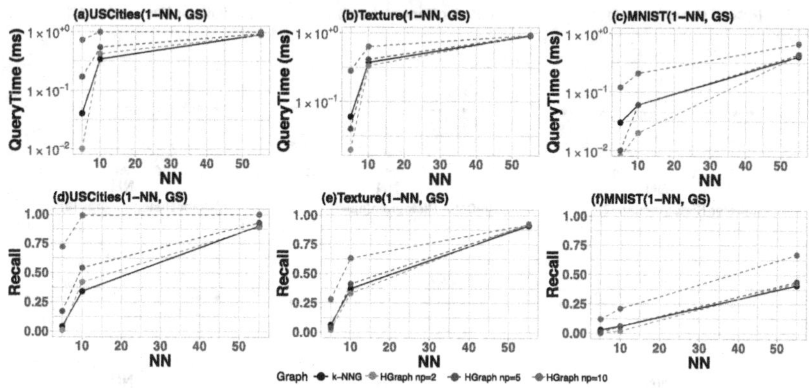

Fig. 3. Long-range edges for 1-NN *GS* algorithm.

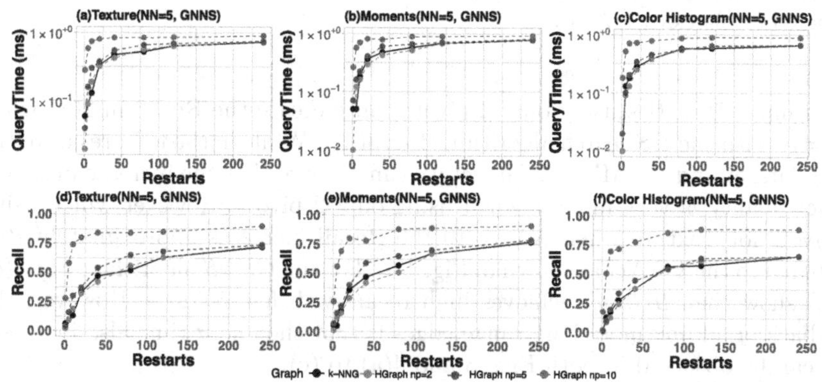

Fig. 4. Long-range edges for 1-NN and $NN = 5$ according to the number of restarts.

we can build an exact base graph g_1 but with the addition of long-range edges. Considering that $o = 1$, regardless of how many times we partition the dataset the size of the set will not reduce, in consequence, the stop condition of the *HGraph* was altered to run the experiments in this subsection.

In the experiments of this subsection, we tested $n_p = \{2, 5, 10\}$ and $g_1 = k$-*NNG* with parameter $NN = \{5, 10, 55\}$ for 1-Nearest Neighbor queries. Figure 3 shows the performance of the search regarding query time and recall according to the graph parameter *NN*, and Fig. 4 shows the performance of the search regarding query time and recall according to the number of Restarts for the *GNNS* search algorithm. The query time plots are in log_{10} scale.

As expected, as the number of pivots increases, the more impact it has on the query time and the recall. As a pattern, the Figs. 3 and 4 show that increasing the number of long-range edges improves the recall; however, the query time also increases. Thus, we can also clearly observe the trade-off between the query time and recall for *HGraph* when adding long-range edges. For smaller datasets,

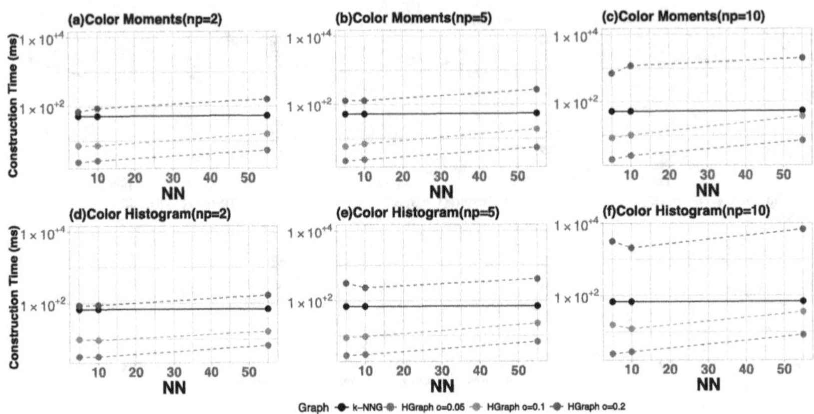

Fig. 5. Construction time of *HGraph* for Color Moments and Color Histogram datasets compared to the exact *k-NNG*.

for example, USCities using only 2 pivots can increase the Recall in 20% when $NN = 10$ with the *GS* search algorithm (Fig. 3(d)). While in the features from the Corel dataset and the MNIST dataset we can observe that the increase in search answer quality is very modest when using only 2 pivots. However, when using 10 pivots for Texture when $NN = 10$ the Recall increases approximately 70% compared to the *k-NNG* as shown in Fig. 3(e). For the *GNNS* algorithm, Fig. 4(d) to (e) show that *HGraph* achieves high recall with few restarts. However, the drawback of adding more long-range edges is that the query time also increases as it can be observed in both Figs. 3 and 4(a) to (c).

3.2 Analysis of the Overlap Size

In this subsection, we evaluate the parameter o of the *HGraph* construction. In order to isolate the parameter o, we did not add any long-range edges to the graphs built in this experiment. The o values tested were 0.05, 0.1 and 0.2, which means that in each partition step 5%, 10% and 20%, respectively, of the set to be partitioned in each recursive step are used as overlap elements.

Figure 5 show the construction time for *HGraph* in comparison to the brute force *k-NNG* for the Color Histogram and Moments features of the Corel dataset. From this figure, we can observe that when increasing the overlap rate (o), the construction time is affected and even takes more time to construct the graph than the brute force *k-NNG*. The higher the overlap rate o, the larger is each resulting subset of the partition process. As a consequence, the number of recursive calls in the partitioning step increases to reach the minimum value of m elements to build the graph. This happens when overlap rate $o = 0.2$ while when $o = 0.05$ or 0.1 the construction difference can be of at most 1 order magnitude faster compared to the *k-NNG* brute force construction. Another point we can observe in Fig. 5 is that the construction time for *HGraph* with any value of

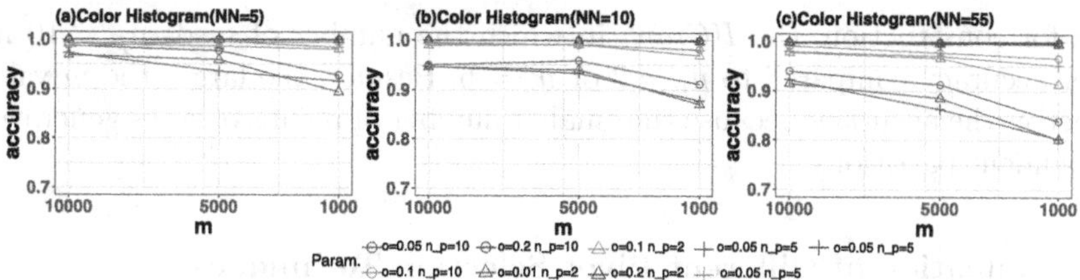

Fig. 6. Accuracy of the HGraph for Color Histogram datasets compared to the exact *k-NNG*.

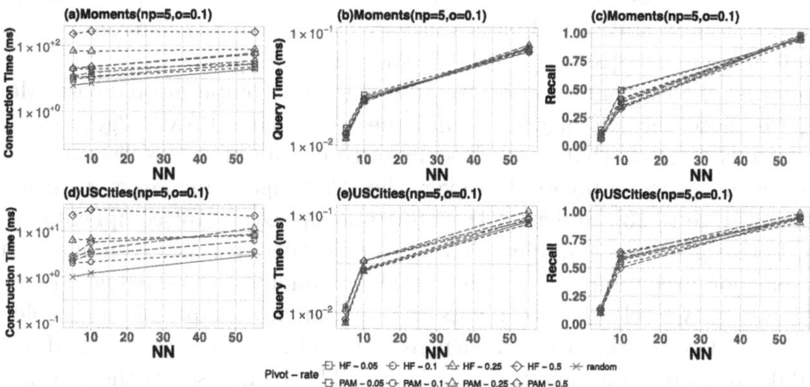

Fig. 7. Construction time (log_{10} scale), query time (log_{10} scale) and recall according to *NN* for Moments and USCities datasets, considering different pivot selection techniques and sample rates.

o grows along with the increase of the *NN* parameter, while for *k-NNG* the construction remains almost stable.

A second analysis is how much the proposed *HGraph* can approximate to the actual base graph g_1 and how the overlap rate affects it. To measure how much the HGraph-*k-NNG* approximates to the brute force *k-NNG* construction, we used the accuracy defined by [6]. The accuracy of a graph G1 compared to a graph G2 is the ratio between the number of edges that are in both G1 and G2 to the number of edges in G2. Figure 6 show the calculated accuracy of *k-NNG* using *HGraph* compared to the exact *k-NNG* for the Color Histogram dataset.

We can observe from the results that the higher the overlap rate value, the more accurate the resulting graph. For instance, the use of $o = 0.2$ was enough to give accuracy above 0.99 regardless of the number of pivots (n_p) used as a construction parameter. However, as discussed previously, $o \geq 0.2$ causes the construction of the *HGraph* to be slower than the brute-force construction, considering non-parallel execution. The overlap rate $o = 0.1$ achieved accuracy above 0.9 in all parameters settings, while $o = 0.05$ had the worst performance among the tested values. The results also show is that the more pivots, the better the accuracy in both datasets. Comparing the results that use the same overlap rate value for construction, the *HGraph* in which the number of seeds $n_p = 10$ has better accuracy compared to $n_p = 2$ or $n_p = 5$. Besides, the larger the m value, the better the accuracy because the smaller number of recursive steps generating independent partitions.

3.3 Evaluation of Different Pivot Selection Techniques

In this subsection, we present how distinct pivot selection techniques perform in *HGraph*. Survey articles, by Amato et al. [1] and Bustos et al. [4], introduced pivot selection techniques for similarity search in metric spaces. We evaluated three representative techniques to work on *HGraph*: (1) Random Selection, (2) Hull of Foci (HF) [15] and (3) k-Medoids (PAM) [1].

We chose the random selection because of its low computational cost. Meanwhile, we chose the HF algorithm and the k-Medoids (PAM) algorithm taking into consideration that both these algorithms minimize the distance of the dataset elements to its pivots. However, the PAM algorithm works inefficiently for large datasets [13], and the HF algorithm has $O(n)$ [15] complexity. As a result, both algorithms can affect *HGraph*'s construction time. To minimize the pivot selection execution time for the HF algorithm and the k-Medoids, in every partition step, we selected the pivots from a random sample of the whole set in that step. Considering that in each step the set size changes, we used a rate from 0.05 to 0.5 to determine the sample size. The sample size is the rate values times the set size in that step.

Figure 7 show the results for the Color Moments and USCities datasets for the construction parameters of HGraph: $m = 1000, n_p = 5, o = 0.1$, and $g_1 = k$-*NNG*. The figures show the construction time, query time, and recall for 1-NN searches using the *GS* algorithm. As expected, the use of HF and PAM algorithms to select pivots affects the graph construction time. We can observe from Figs. 7(a) and (d) that in both datasets the PAM algorithm resulted in the slowest HGraph construction because of its computational complexity. In the USCities dataset, the HF algorithm took longer to execute queries when compared to the PAM and random pivot selection for all rate values used in the experiment. However, in the Color Moments dataset, the query times for all pivot selection methods were similar for $NN = 5$ and 10. Unexpectedly, the recall in the Moments dataset using random pivot selection was better or very

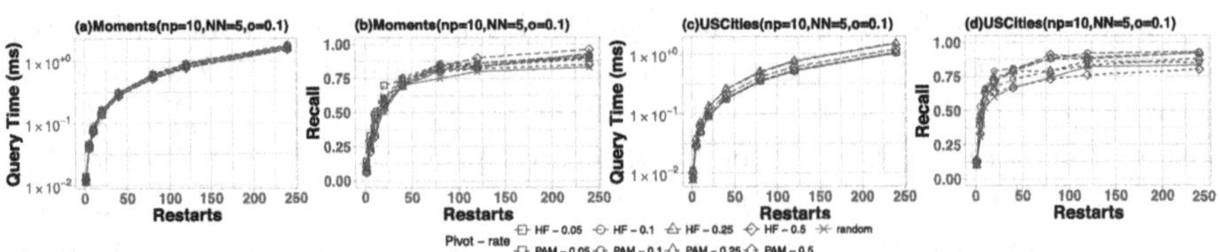

Fig. 8. Query time and recall according to number of restarts using *GNNS* algorithm, for Moments and USCities datasets.

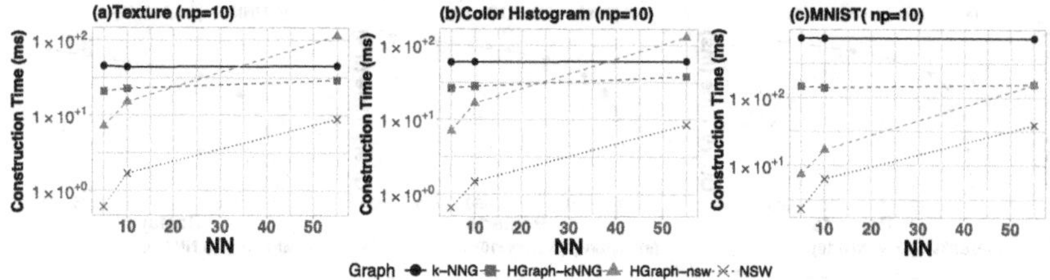

Fig. 9. Construction time on log_{10} scale for Texture, Color Histogram and MNIST datasets, for $n_p = 10$ and $m = 5000$.

close to the other pivot selection algorithms, while in the USCities datasets the random pivot selection recall stands in between PAM and HF algorithms.

Figure 8 shows the results for pivot selection techniques according to the number of Restarts in the *GNNS* algorithm. In both datasets, the HF algorithm for pivot selection performed the best using 0.5 sample rate according to the increase of restarts. On the other hand, unexpectedly, the PAM algorithm resulted in worst to close recall to the random pivot selection. In the Fig. 8 we can clearly see the tradeoff of query time and recall for Moments and USCities datasets. The *HGraph* constructions that took the longest to execute the search algorithm had the best resulting recall while the fastest ones had the worst recall. In both cases, the random pivot selection had intermediate results; however, it is the fastest algorithm to use in *HGraph* construction as already shown.

According to this experiment, there are no essential advantages in using the PAM algorithm for pivot selection as it takes a longer time to construct the graph and the search recall was close or worst when compared to the HF algorithm and the random pivot selection. The HF algorithm was the best choice among the tested methods for search recall. However, the algorithm affects the graph construction time and also takes a longer time to execute a search algorithm in the graph. On the other hand, the random pivot selection promotes the fastest graph construction and average performance regarding query time and recall. Therefore, for the experiments in the next sections, we employed both the HF algorithm with 0.1 sample rate and the random selection of pivots.

3.4 Effectiveness of *HGraph* for Different Types of Graph

Another critical parameter is the type of graph that *HGraph* will build, the parameter g_1. We have implemented mainly two graph types for g_1: *k-NNG* and *NSW*. We chose these two types of graphs because *NSW* is the current state of the art, and because *k-NNG* is one of the most used types of graphs in the literature along with its variations.

Figure 9 shows the Construction Time in log_{10} scale of the *HGraph* when $g_1 = k\text{-}NNG$ and *NSW* compared to the corresponding base graph. From the figure, we can observe that the *HGraph* drops the construction time for the *k-NNG*; however, it takes more time to construct when using $g_1 = NSW$ when

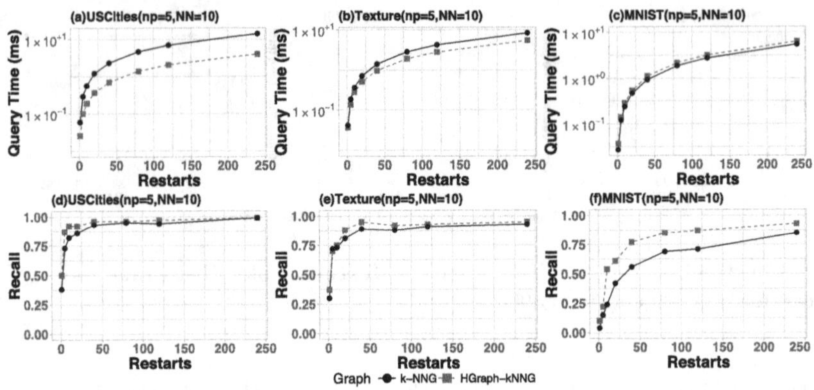

Fig. 10. Query time (log_{10} scale) and recall comparing *HGraph-k-NNG* to *k-NNG* for $x = 5$, $NN = 10$ and $m = 1000$, for USCities, Texture and MNIST datasets.

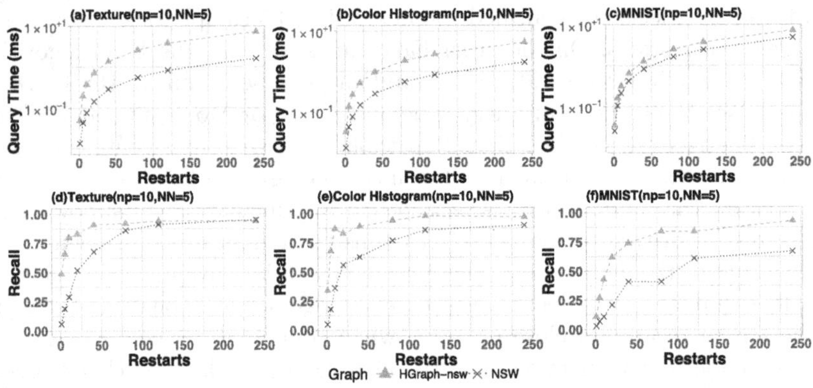

Fig. 11. Query time (log_{10} scale) and recall comparing *HGraph-NSW* to *NSW* for $x = 10$, $NN = 5$ and $m = 1000$, for Color Histogram, MNIST and Texture datasets.

compared to the *NSW*. As the *NN* parameter increases, we can observe from Fig. 9(a) and (b) that the *HGraph-NSW* can take more time to build than the brute-force *k-NNG*. This happens because it takes more time to partition the dataset into n_p subsets using the generalized hyperplane partition strategy than the algorithm used by the *NSW* to insert the vertices.

Next, we compared the search performance of *HGraph* compared to each of its base graph g_1. Figure 10 shows the query time and recall of the *HGraph-k-NNG* compared to the *k-NNG* according to the number of Restarts of the *GNNS* algorithm. We can observe from these results that the *HGraph* was able to increase the recall in all datasets. In the MNIST dataset the results showed a trade-off between query time and recall as the query time also increased for *HGraph-k-NNG*. However, in the datasets with lower dimensionality, such as USCities and Texture, *HGraph* was able to improve the query time as well.

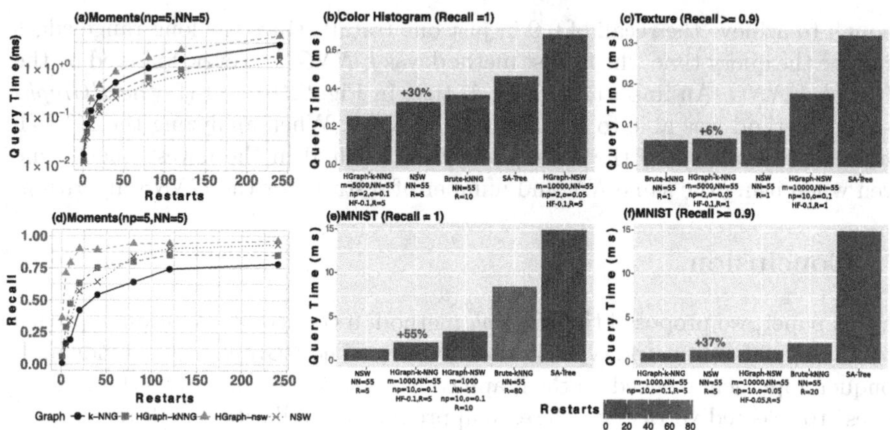

Fig. 12. Query time and recall according to the number of restarts for Moments dataset, and best configuration of each method for exact and approximate answers.

Figure 11 shows the query time and recall of the *HGraph-NSW* compared to *NSW*. In this case, we can observe from the results a clear trade-off between query time and recall. In the MNIST dataset, *HGraph-NSW* had the closest query time to the *NSW* and the most substantial recall increase. Unexpectedly, for larger *NN* values ($NN \geq 10$), *HGraph-NSW* did not increase the recall or decrease the query time compared to the *NSW*.

Next, we compared all the tested combinations regarding query time and recall according to the number of restarts for the *GNNS* algorithm for 1-*NN* queries. We can observe in this test (Fig. 12(a) and (d)) also the trade-off between query time and recall according to the number of restarts for *HGraph*.

Finally, in order to compare the different parameter settings for the graphs, we selected the best parameter setting of each method in query time that resulted in exact answers (recall = 1) and approximate answers (recall $\geq$ 0.9). We then analyzed the query time for 1-*NN* queries using *GNNS* algorithm (Fig. 12(b), (c), (e) and (f)). We also compared the search results to the *SAT*, which always produce exact answers. In these experiments, the *NN* parameter had the most impact on the results. We can also observe the best configuration of the graph-based methods executed the query in less time than the *SAT* for exact answers. To achieve the exact answer, the best configurations were with $NN = 55$ and a small number of restarts. For the Color Histogram dataset, *HGraph-k-NNG* had the best performance in both query time and distance computations followed by the *NSW* and the *k-NNG*. This reinforces what has been shown in previous results: the *HGraph* method can improve *k-NNG* in construction and query time.

However, the *NSW* can also outperform the *HGraph* in some cases. For example, in Fig. 12(e) the *NSW* showed the best performance, followed by the *HGraph-k-NNG* and the *HGraph-NSW*. In the Texture dataset, since the long-range edges increase the query time, the fastest method was *k-NNG*, however, followed by the *HGraph-k-NNG*. In Texture dataset, the high value for *NN* was enough to achieve the recall of 0.9 in just one restart, since the long-range edges increase the query time, the fastest method was *k-NNG*, however, followed by the *HGraph-k-NNG*. An important point is that in Fig. 12(b) and (f) the *HGraph-k-NNG* method was able to outperform the *NSW*. When analyzing the MNIST dataset for recall ≥ 0.9 the *HGraph-k-NNG* resulted in the lowest query time even when using the same *NN* and number of restarts for the *GNNS* algorithm.

4 Conclusion

In this paper, we proposed the *HGraph* method, a connected partition approach to building different graph types in the literature. The *HGraph* uses a divide-and-conquer approach to speed up the graph construction and adds edges (long-range edges) to selected vertices to increase approximate similarity search recall.

We presented an experimental evaluation of the *HGraph* main parameters. Through this evaluation we concluded: (1) long-range edges increased the answer quality compared to the base graph type; (2) the *HGraph-k-NNG* algorithm can build a graph that is approximate to the *k-NNG* with a small overlap rate of 0.1; (3) the random pivot selection algorithm and the HF algorithm had the best performances while the k-Medoids algorithm did not show advantages over the other two methods. Moreover, we compared the use of *HGraph* for *k-NNG* and *NSW* to the baseline *k-NNG* and *NSW* graphs, and also to the exact method *SAT*. *HGraph* was able to improve the baseline *k-NNG* in both construction time, query time and recall. Finally, we showed cases in which *HGraph-k-NNG* and the *HGraph-NSW* were able to outperform *NSW* and *SAT* when recall = 1.

As future works, we plan to run experiments the *HGraph* method using other base graph types and test a faster dataset partition strategy. We intend to investigate theoretically how each parameter affects the *HGraph* performance. Also, we plan to implement the *HGraph* method in an architecture that supports running the *HGraph* construction in parallel.

References

1. Amato, G., Esuli, A., Falchi, F.: A comparison of pivot selection techniques for permutation-based indexing. Inf. Syst. **52**(C), 176–188 (2015)
2. Barioni, M.C.N., Kaster, D.D.S., Razente, H.L., Traina, A.J., Júnior, C.T.: Advanced Database Query Systems. IGI Global (2011)
3. Boytsov, L., Naidan, B.: Engineering efficient and effective non-metric space library. In: Brisaboa, N., Pedreira, O., Zezula, P. (eds.) SISAP 2013. LNCS, vol. 8199, pp. 280–293. Springer, Heidelberg (2013). https://doi.org/10.1007/978-3-642-41062-8_28
4. Bustos, B., Navarro, G., Chavez, E.: Pivot selection techniques for proximity searching in metric spaces. In: SCCC, pp. 33–40, November 2001
5. Chávez, E., Navarro, G., Baeza-Yates, R., Marroquín, J.L.: Searching in metric spaces. ACM Comput. Surv. **33**(3), 273–321 (2001)
6. Chen, J., Fangand, H.R., Saad, Y.: Fast approximate KNN graph construction for high dimensional data via recursive Lanczos bisection. J. Mach. Learn. Res. **10**, 1989–2012 (2009)

7. Hajebi, K., Abbasi-Yadkori, Y., Shahbazi, H., Zhang, H.: Fast approximate nearest-neighbor search with k-nearest neighbor graph. In: IJCAI, pp. 1312–1317 (2011)
8. Lecun, Y., Bottou, L., Bengio, Y., Haffner, P.: Gradient-based learning applied to document recognition. Proc. IEEE **86**(11), 2278–2324 (1998)
9. Malkov, Y., et al.: Approximate nearest neighbor algorithm based on navigable small world graphs. Inf. Syst. **45**, 61–68 (2014)
10. Ocsa, A., Bedregal, C., Cuadros-Vargas, E.: A new approach for similarity queries using neighborhood graphs. In: Brazilian Symposium on Databases, pp. 131–142 (2007)
11. Ortega, M., Rui, Y., Chakrabarti, K., Porkaew, K., Mehrotra, S., Huang, T.S.: Supporting ranked boolean similarity queries in MARS. TKDE **10**(6), 905–925 (1998)
12. Paredes, R., Chávez, E.: Using the k-nearest neighbor graph for proximity searching in metric spaces. In: Consens, M., Navarro, G. (eds.) SPIRE 2005. LNCS, vol. 3772, pp. 127–138. Springer, Heidelberg (2005). https://doi.org/10.1007/11575832_14
13. Park, H.S., Jun, C.H.: A simple and fast algorithm for k-medoids clustering. Expert Syst. Appl. **36**(2, Part 2), 3336–3341 (2009)
14. Shimomura, L.C., Vieira, M.R., Kaster, D.S.: Performance analysis of graph-based methods for exact and approximate similarity search in metric spaces. In: Marchand-Maillet, S., Silva, Y.N., Chávez, E. (eds.) SISAP 2018. LNCS, vol. 11223, pp. 18–32. Springer, Cham (2018). https://doi.org/10.1007/978-3-030-02224-2_2
15. Traina Jr., C., Filho, R.F., Traina, A.J., Vieira, M.R., Faloutsos, C.: The Omni-family of all-purpose access methods: a simple and effective way to make similarity search more efficient. VLDB J. **16**(4), 483–505 (2007)
16. Uhlmann, J.K.: Satisfying general proximity/similarity queries with metric trees. Inf. Process. Lett. **40**(4), 175–179 (1991)

Management and Processing of Knowledge

Statistical Processing of Stopwords on SNS

Yuta Nezu(✉) and Takao Miura

Department of Advanced Sciences, HOSEI University, Kajinocho 3-7-2, Koganei, Tokyo, Japan
yuta.nezu.6f@stu.hosei.ac.jp, miurat@hosei.ac.jp

Abstract. For the purpose of text classification or information retrieval, we apply preprocessing to these texts such as stemming and stopwords removal. Almost all the techniques could be useful only to well-formed text information like textbooks and news articles, but is not true to social network services (SNS) or any other texts in internet world. In this investigation, we propose how to extract *stopwords* in context of social network services. To do that, first we discuss what stopwords mean, how different from conventional ones, and we propose statistical filters TFIG and TFCHI, to identify. We examine categorical estimation to extract characteristic values putting our attention on Kullback Leibler Divergence (KLD) over temporal sequences on SNS data. Moreover we apply several preprocessing to manage *unknown words* and to improve morphological analysis.

Keywords: Natural language processing · Stop words · Social networking services · TFIG · TFCHI

1 Introduction

The internet today is a widespread information infrastructure. Among others, Social Network Services (SNS) play special roles on bi-directional communities in real-time with vast amount of information. Nowadays many contents in SNS become useful to obtain information of interests, thus traditional technologies like information retrieval (IR) or classification should be provided. However, basically the information is semi-structured using *hash tag* and it takes much time to examine and analyse the contents in a reliable manner.

Removal of stopwords provides us with sharp improvement of IR and text classification processing, because they don't help us to do these tasks very much. A stopword means a word that plays no essential nor important role. Typical examples of these words are "stopwords lists" published by GOOGLE and SlothLib.

There have been several investigations of stopwords proposed so far [1,7]. Until now there are few investigation about stopwords in SNS except only introductory ones [4]. In this work, we discuss what stopwords (in SNS) mean and what properties should be provided then introduce statistical filters to extract. We examine categorical estimation to extract characteristic values using Kullback Leibler Divergence (KLD) over temporal sequences on SNS data. And we

S. Hartmann et al. (Eds.): DEXA 2019, LNCS 11706, pp. 125–137, 2019.
https://doi.org/10.1007/978-3-030-27615-7_9

apply several preprocessing to manage *unknown words* and to improve morphological analysis. The rest of the paper is organized as follows. In Sect. 2, we describe stopwords and their properties. In Sect. 3, we introduce statistical filters and describe the relationship with stopwords. In Sect. 4 we propose our ideas to extract stopwords in SNS. Section 5 contains an experimental results to see the effectiveness of this idea, and discussing our results. we conclude this investigation in Sect. 6.

2 Morphological Analysis and Stopwords

Before developing our story, let us see how word structure works in Japanese language. We know the fact that, in English, a word describes grammatical roles such as *case* and *plurality* by means of word order or inflection. For example, we see two sentences.

```
John calls Mary.
Mary calls John.
```

The difference corresponds to the two interpretations of positions, i.e., *who calls whom* over `John` and `Mary`. Such kind of language is called *inflectional.* On the other hand, in Japanese, grammatical relationship can be described by means of postpositional particles, and such kind of languages is called *agglutinative.* For example, let us see the two sentences in Table 1:

Table 1. Morphological analysis in Japanese

`"John/ga/Mary/wo/yobu."`	(*John calls Mary*)
`"John/wo/Mary/ga/yobu."`	(*Mary calls John*)

In the sentences, the positions of `John, Mary` and `yobu` (*call*) are exactly same but the difference of postpositional particles (`ga, wo`). With the postpositional particles, we can put any words to any places[1]. Independent word(s) and a postpositional particle constitute a *clause.* Clearly, in Japanese language, many approach for inflectional languages can't be applied in a straightforward manner[2]. The main reasons come from inherent aspects of Japanese; it is agglutinative while English is inflectional.

[1] One exception is *predicate.* In fact, the predicate should appear as a *last verb* in each sentence.

[2] Morphological analysis means both word segmentation and part of speech processing in Japanese. For example, `"sumomo/mo/momo/mo/momo/no/uchi"` means *Both Plum and Peach are same kind of Peach,* which is a typical tongue twister where you should say "`mo`" many times. There are two nouns "`sumomo`" (*plum*) and "`momo`" (*peach*). There is no delimiter between words (no space, no comma, and no thrash) and everything goes into *one* string as "`sumomomomomomomomonouchi`".

Generally a *stopword* is defined *negatively*, that is, any words which can't be effective for inquiry, classification and feature extraction, or the ones not useful to identify categorical aspects [1]. For examples, a *postpositional particle* (of Japanese) is said very often "hard to understand" since it gives just a *case* to any word usually as we said. Another example is a *symbol* such as operators, Arabic numerals and parenthesis/brackets. These words play common role on documents: (1) appeared many times. (2) appeared independently of document categories. This means the query/classification precision becomes worse and we remove these words in advance for our tasks.

When we do preprocessing with removing stopwords, generally we apply common list to any corpus. These lists work effectively for news articles or novel corpus. However, it is not always effective for SNS documents, because topic of interest keep changing and causing burst words in short period. In other words, we should extract different words depending on time period as stopwords.

They have been several investigations proposed for stopwords removal, including *stopwords list* [1] such as Google [2] and Slothlib [3]. More theoretical is proposed by Feng [7] where statistical values over word frequencies have been examined exhaustively. In twitters, some investigation for stopwords has been discussed by [4] where correlation among words has been utilized in a framework of sentiment analysis. Here in this investigation we propose a sophisticated discussion mainly for text classification.

3 Statistical Filters

Feature selection means techniques to extract suitable words enough to classify texts. They can be seen as the combination of search technique for the purpose of text classification along with an evaluation measure which scores the different candidates [6]. *Filter methods* use a proxy measure which provides efficient computation while capturing the usefulness of our words of interests to examine deep structure of semantics through words and their situation. Since stopwords stand on the opposite side of *feature* words, we can say they are *useless* words.

Sonoda et al. [5] has investigated several statistical filters for mining collocation words in Japanese, and has shown the useful combination of conditional probability and statistical filter. In the following, we examine 3 feature selection. The methods to be examined are Term Frequency (TF), Information Gain (IG) and χ^2 statistics (CHI).

As our running example, let us show words of interests in a Table 2. There are 3 categories, "Figure Skating", "Gymnastics" and "Volleyball" with 5 keywords. The table contains our filter values of TF, IG, CHI.

Table 2. Running example

(keyword)	"Figure Skating"	"Gymnastics"	"Volleyball"	(TF)
YuzuruHanyu	100	10	10	120
KenzoShirai	10	100	10	120
SarinaKoga	10	10	100	120
WorldChampionship	50	40	30	120
FinalGame	5	4	3	12
(total)	175	164	153	492

(keyword)	TF	IG	CHI
YuzuruHanyu	120	0.23226562	119.4654528
KenzoShirai	120	0.259088052	135.0225957
SarinaKoga	120	0.289176081	152.8167134
WorldChampionship	120	0.005267658	2.68907563
FinalGame	12	0.000409702	0.268907563

3.1 Term Frequency (TF)

Term Frequency of a word w means a count how many w appears in a document or a collection of documents, defined formally:

$$tf(i,j) = \frac{f_{i,j}}{N_j} \quad (1)$$

$$f_{w_i} = \sum_j f_{i,j} \quad (2)$$

$$tf(w_i) = \frac{f_{w_i}}{N} \quad (3)$$

$$N_j = \sum_i f_{i,j} \quad (4)$$

$$N = \sum_j N_j \quad (5)$$

Note $f_{i,j}$ means a frequency of a word w_i in a document d_j, N_j a total frequency of words in d_j, f_{w_i} a total frequency of w_i appeared in documents D, N a total sum of frequencies of all the words in D. A formula 1 shows TFs in a document and a formula 3 total TFs in D. Any words appeared many times in a document show more characteristic features, while any words appeared in many documents have bigger TFs but don't always identify specific documents. In Table 2, every TF column contains total frequencies of its keyword over all the categories.

Let us note some comments. Frequent words (big TF words) look like stop-words, but they might carry characteristic features. Certainly removal of frequent words causes dramatic reduction and thus we could improve query overhead within a collection of documents, this means they might help us to improve query efficiency.

3.2 Information Gain (IG)

P(x) means the appearance probability of x, and $P(x|c)$ the conditional probability of x under the category c. Given an event w_i, Information Gain $G(w_i)$ means an amount of information to estimate a suitable category, defined as follows:

$$G(w_i) = H(w_i) - H(w_i|c_K) \tag{6}$$

$$H(w_i) = -P(w_i)\log_2 P(w_i) - P(\overline{w_i})\log_2 P(\overline{w_i}) \tag{7}$$

$$H(w_i|c_K) = \sum_{k \in K} -P(w_i, c_k)\log_2 P(w_i|c_k) - P(\overline{w_i}, c_k)\log_2 P(\overline{w_i}|c_k) \tag{8}$$

Note that $\overline{w_i}$ means a complementary event of w_i, and c_k a category. In our case, given a word w_i, $G(w_i)$ means a difference of entropy $H(w_i)$ in D and entropy of the word in c_k.

For example in Table 2, assume there happens a word *"YuzuruHanyu"* 100 times in "Figure Skating" of 175 words in this category, and only 10 time both in "Gymnastics" and "Volleyball" of 164 and 153 words each. Then we have 0.23226562 of Information Gain. On the other hand, assuming a word *"FinalGame"* appears 5, 4 and 3 times respectively in each category, we have few information gain 0.000409702. As this shows, the less information gain we have, the more similar the probability of occurrence we see, and the word becomes less characteristic to identify category. We expect that stopwords may contain less amount of IG.

3.3 χ^2 Statistics (CHI)

A notion of χ^2 Statistics (CHI) means how independent a word w_i works in each category, defined as follows:

$$\chi^2(w_i) = \sum_k \frac{(P(w_i|c_k) - E(w_i, c_k))^2}{E(w_i, c_k)} \tag{9}$$

$$E(w_i, c_k) = \frac{f_{w_i} \times N_{c_k}}{N_{c_k}} \tag{10}$$

Here N_{c_k} means a total frequency of all the words in a category c_k, and $E(w_i, c_k)$ a frequency of a word w_i in c_k with the assumption that w_i appears independently of c_k.

Yang has shown the strong correlation of IG and CHI [6]. In our example in Table 2, *"Yuzuru Hanyu"* has the value 119.4654528 while *"FinalGame"* has 0.268907563. This means CHI could be one of useful indicators to stopwords.

4 Extracting Stopwords in SNS

In this investigation let us introduce new kinds of statistical filters, *TFIG* and *TFCHI*, to extract stopwords specialized for SNS. Each of our filters consists of two kinds of filters. We also examine categorization of SNS documents. To do that, we introduce a new method based on *Kullback Leibler Divergence* (KLD) over temporal documents.

Whenever we talk about IG and CHI, we assume a category given in advance to each document. So we still need some more notions for measurements.

Among others, let us introduce KLD to evaluate the difference between two probability distributions Given two probability density functions $P(w), Q(w)$ over two sets D_1, D_2 of documents, KLD is defined as follows:

$$KLD(P_W||Q_W) = \sum_{w_i \in W} P(w_i) \log_2 \frac{P(w_i)}{Q(w_i)} \tag{11}$$

Let us note that $KLD(P_W||Q_W)$ is non negative and $KLD(P_W||Q_W) = 0$ if and only if $P_W = Q_W$ for every word. The more KLD we have, the more differently the words are distributed.

Whenever new topics arise, the word distribution fluctuates suddenly before and after the events, the situation is called *burst*. In our case, once such an *edge* is detected, we see very high value of KLD and some category should be changed.

Clearly any words of small TF don't help us to identify documents nor to estimate categorization, because fewer coverage causes decreasing reliability of these tasks. Note we select words as "stopwords" when they happen frequently but useless to categorize.

Zero frequency problem causes a potential issue of zero-frequency symbols as never-arising symbol. That is, zero-frequency means no possibility aspect. To solve this issue, we apply MAP estimation to a conditional probability $P(w_i|c_k)$ given a category c_k and a word w_i:

$$P(w_i|c_k) = \frac{f_{w_i,c_k} + 1}{N + W \times K} \tag{12}$$

Note f_{w_i,c_k} means a frequency of w_i in c_k, and N, W, K a number of total words, a number of entry words and K a number of categories respectively.

Here in this investigation, we define two kinds of statistical filters for stopwords, $TFIG$ and $TFCHI$ defined as below:

$$TFIG(w_i) = \log_2 \frac{tf(w_i)}{G(w_i)} \tag{13}$$

$$TFCHI(w_i) = \log_2 \frac{tf(w_i)}{\chi^2(w_i)} \tag{14}$$

Let us describe our motivation of these filters. All the filters contain TF on its numerator, the bigger TF shows the higher value. The difference comes at the denominators, $G(w_i)$ and $\chi^2(w_i)$. The smaller value shows the higher precision.

In SNS, by *stopwords* we like to remove words that show both the higher TF and the smaller precision (IR or categorization), in other words, we like to remove the words with the higher TFIG or TFCHI.

5 Experimental Results

We show our experimental results to see how well the proposed approach works. We discuss our results and examine whether the candidates are selected correctly or not. As a baseline, we compare GOOGLE/SlothLib stopwords as static list with our approach and examine whether our approach goes better.

5.1 Preliminaries

In this experiment, we examine two kinds of twitter corpus in Japanese[3]. To do that, using R-package `twitteR`, we collect all the twitter documents containing some words in designated period except retweet (RT) documents as shown in a Table 3. We have selected several keywords in trend appeared in `@jptrend` tweet contents as shown in a Table 3 where "Amount" means the number of twitter documents.

Table 3. Corpus detail and keywords

Corpus	Designated Period	Amount
corpus1	2018/7/30/13:00 - 2018/7/31/12:59	65545
corpus2	2018/12/14/4:00 - 2018/12/15/3:59	134182

Corpus1		Corpus2	
Keyword	Amount	Keyword	Amount
ShiningMonday	31034	Home Alone	67930
Crepe with your style	699	Day of the south pole	1946
Staying up late on Monday	2802	LINE Pay	5525
#ijuin	2332	price revision	1044
#musica_od	924	great sword	20132
Yama No Susume	1718	Sasasushi	892
Arc the Lad	6656	Meteor	32878
Day of Umeboshi	1080	Friday	3835
#tama954	897		
Osaka Toin	17776		

We extract text body and the temporal information from them. Considering one tweet as one document, we apply morphological analysis *MeCab* to them.

[3] See https://twitter.com/?lang=ja.

With the help of *mecab-ipadic-NEologd* dictionary and some extension of recent trend words in the period, we examine the following preliminary processes in advance:

(a) Remove the special terms `"URL", "UserName"`
(b) Unify Character-codes, Capital Letters and so
(c) Preprocess verbs: work/worked, go/going, go/went, book/books, ...
(d) Remove specialized expression in Japanese
(e) Treat Unicode as a word: U+1F600, U+FE0F, ...
(f) Concatenate long emoticon/emoji into one
(g) Remove any symbols except (5) and (6)
(h) Unifying long identical characters such as `"aaaaaaa"` into `"a"`
(i) Unify net-slangs such as `"w", "ww", "www", ..` into `"a"`
(j) Unify multiple consonants into one like `"whattt"` into `"what"`
(k) Preprocess adjectives specialized in Japanese
(l) Concatenate names using the part-of-speech information identified by mecab: `"John Smith"` into `"JohnSmith"`
(m) Remove unknown words consisting only one character.

Let us note that all the transformation rules depend on heuristic, local trendy and burst situation as well as Japanese language. In this investigation, we have examined the these local rules but is not the point of our interests.

After preprocessing, we obtain filter values of TF, IG, CHI and KLD where we assume 30 min for unit of temporal information. Then we get TFIG and TFCHI. We make ranking to the high score words to obtain candidates as stopwords. We also examine the words by TF, 1/IG and 1/CHI for the comparison[4]. We remove non-frequent words in advance: a word appeared in less than 4 documents (corpus1) or 5 documents (corpus2).

For evaluation, we examine two measures: reduction ratio and precision ratio based on Naive-Bayes classification. As for the former case, we define the ratio as follows:

$$R = \frac{N - N_d}{N} \tag{15}$$

Note N means the number of total words in corpus and N_d a number of stopword candidate. We examine Naive Bayes classification to see the latter. Assuming multi-nominal probability distribution on corpus, we classify corpus documents. To do that, we divide the whole corpus randomly into training (10%) and test (90%) where we consider keywords extracted as correct answer classes[5]. Then we take an average of the precision ratios.

[4] We say 1/IG instead of IG because we like to smaller value better. So is true for 1/CHI.

[5] For instance, when collecting twitter documents by giving a keyword `"Home Alone"`, then we give a class *"Home Alone"* to the collection.

5.2 Results

In Tables 4, 5, 6, 7 and 8, let us illustrate our results of TF, 1/IG, 1/CHI, TFIG, TFCHI filters. The Table 4 shows a baseline results with GOOGLE/SlothLib lists (322 words). Figure 1 contain some of the stopwords list. Nezu et al. [8] has given details of the list. The other tables contain the reduction ratios and the precision ratios with the top 2% to 30% as stopword candidates with respect to corpus1 and corpus2 respectively.

Table 4. Reduction ratio and precision (baseline)

	Corpus1		Corpus2	
	Reduction (%)	Precision (%)	Reduction	Precision
No reduced	-	78.7799	-	85.0075
Reduced	26.65522	85.56741	24.23754	88.82252

desu (*is*), masu (*is*), ha (*is*), ga (*is*), no (*of*), ni (*into*), wo (*to*), de (*so*), kara (*from*), made (*until*), mo (*also*), to (*and*), yori (*than*), e (*huh*), sorede (*so then*), shikashi (*however*), asoko (*over there*), atari (*per*), achira (*there*), atchi (*over there*), ato (*after*), anata (*you*), are (*that*), ikutsu (*how many*), itsu (*when*), ima (*now*), iya (*no*), iroiro (*various*), uchi (*home*), omaka (*rough*), omae (*your*), ore (*I*), gai (*or*), kaku (*each*), katachi (*form*), kara (*from*), gara (*from a*), kuse (*habit*), koko (*here*), kotchi (*here*), koto (*about*), goto (*per*), kochira (*here*), gotcha (*confuse*), kore (*this*), korera (*these*)

Fig. 1. Some of stopwords proposed by Google and SlothLib

Table 5. Reduction ratio (corpus1)

FreqWords (%)	TF	1/IG	1/CHI	TFIG	TFCHI
2	64.53556123	0.417780664	0.485561887	21.79176509	22.41685859
4	72.87916521	0.803604414	0.973057473	26.18573887	26.98700249
6	77.5376486	1.202455187	1.344225823	29.74058925	30.68341996
8	80.56459927	1.956088759	2.050432894	32.31556332	32.87949089
10	82.82020776	6.350571408	6.61742181	33.78588665	34.30401609
12	84.59218806	6.609076374	7.247400448	35.19463694	35.45456673
14	86.04928082	7.241395805	7.697646892	36.3277843	36.83705943
16	87.28969756	7.530839948	8.618901598	37.55303677	38.1456645
18	88.36055947	7.766445852	8.92381533	38.81380824	39.27107706
20	89.30206712	8.076142943	9.225065211	39.5829114	40.29756568
22	90.13019898	8.345537685	9.462808361	40.59220027	41.37402514
24	90.86510621	8.653809946	9.64976648	41.59182066	42.36591073
26	91.52632934	9.699228454	9.985721163	42.50025189	43.18651133
28	92.11966948	9.967401913	10.4143916	43.2221321	44.20078712
30	92.66151217	10.26386844	10.71218118	43.95408789	45.36324528

As seen easily, in TF, we obtain generally the high reduction ratio but the low precision compared to other filters. On the other hand, in 1/IG and 1/CHI, when we give the higher threshold X, we get the better precision ratio but the poor reduction ratio, in fact, we eventually go under the baseline. Compared to these situation in our approach, both reduction ratios and the precision of TFIG and TFCHI generally go beyond the baseline with appropriate X values.

5.3 Discussion

Let us discuss what our results mean. We generally apply static list as stopwords. We consider static list as baseline in our experiment. Stopwords that we extracted are dynamic. In short, Our result show that our dynamic stopwords work better than the static list in SNS document.

Table 6. Precision (corpus1)

FreqWords (%)	TF	1/IG	1/CHI	TFIG	TFCHI
2	66.05098347	78.96864229	78.96561528	85.36113134	85.61713279
4	62.36276533	79.12759322	79.02520412	86.38548064	86.73345862
6	56.85976232	79.36150652	79.19609344	87.14833688	87.35235747
8	54.73478381	79.69997671	79.49508102	87.49340265	87.5792157
10	51.99666527	80.78893596	80.69829289	87.59887425	87.65927459
12	45.57589672	80.94112965	80.82467652	87.7466111	87.75763509
14	42.90746856	81.18733532	81.02169046	87.79488947	87.86530147
16	40.83059651	81.38506153	81.3061352	87.31501943	87.30132174
18	39.23772874	81.54072827	81.61713396	86.93869208	87.42901402
20	37.20876208	81.80930618	81.78389209	86.86696658	87.37688085
22	35.48136749	81.92539486	82.05079057	86.82524474	87.3736467
24	33.23884925	82.16439626	82.25141092	86.87384853	87.38796379
26	31.76064133	82.53965185	82.46648531	86.89673616	87.41774264
28	30.73940005	82.68269504	82.63570396	86.92760442	86.85731953
30	29.21152876	82.87221207	82.69835408	86.99025875	86.91427377

Table 9 show words which appeared in top 20 words of corpus1 and corpus2 by TFIG and TFCHI. We find that stopwords depend on time period we collect the corpus.

Finally let us examine how TFIG and TFCHI relate with each other. Although some words are ranked very differently, no sharp distinction between TFIG and TFCHI happens in reduction and precision. the Kendall coefficient shows 0.8986 (corpus1) and 0.9462 (corpus2). It show a strong relationship between the words generated by TFIG and TFCHI. We can't see which filter works better for stopwords extraction but the taste of IG and CHI.

Table 7. Reduction (corpus2)

FreqWords (%)	TF	1/IG	1/CHI	TFIG	TFCHI
2	67.79592084	0.225854117	0.226838874	10.31331535	10.31331535
4	76.51028555	0.397989716	0.397989716	21.61385002	20.85317414
6	81.09118017	0.673278656	0.677513113	27.39019093	27.50309337
8	84.01615593	0.802921972	0.79927837	34.01608207	31.18342732
10	86.14958367	1.049308282	1.075355116	34.61338669	34.6843877
12	87.80963851	1.322775421	1.256993624	37.18793652	37.57814666
14	89.10828737	1.50623573	1.535876929	41.20505752	41.29644301
16	90.18792618	1.8516394	1.694619827	42.56141317	42.41192699
18	91.1031105	2.023725761	2.033868762	44.57731011	44.69459472
20	91.88456476	2.376662824	2.381241947	46.02327868	46.25543525
22	92.55572619	2.707935225	2.686615226	47.31242464	48.49433051
24	93.14697455	2.89055849	2.847573828	49.80533807	50.52992259
26	93.6686498	3.183622302	3.044968456	51.87618448	51.24889399
28	94.13375074	3.359007601	3.370283073	53.10077958	52.5678781
30	94.55365131	3.6056401	3.621297743	54.02704242	53.88277546

Table 8. Precision (corpus2)

FreqWords (%)	TF	1/IG	1/CHI	TFIG	TFCHI
2	83.12747268	85.11533939	85.11533939	86.56857785	86.56857785
4	71.48860137	85.20219445	85.20219445	88.53165113	88.3174657
6	60.79242669	85.24647795	85.24898146	89.64811468	89.67587939
8	57.02326666	85.36133946	85.36154388	91.27306434	90.48073555
10	47.33893979	85.56789436	85.57251743	91.38793129	91.38461008
12	44.72489004	85.7560464	85.72580309	91.84780447	91.94545639
14	42.6585451	85.84124225	85.83816735	93.1595112	93.20731609
16	39.4788214	86.01977981	85.98696326	93.40920076	93.43989394
18	37.46658502	86.10889803	86.12324168	94.08269546	94.13186657
20	36.13985343	86.1370522	86.1119204	94.37573391	94.34935803
22	35.26110598	86.29126673	86.29717988	94.50421609	94.81273288
24	33.12731653	86.44030622	86.45235135	95.12131463	95.33087312
26	31.99355093	86.51314284	86.51083376	95.38522048	95.37046366
28	29.95114646	86.68719279	86.68732103	95.54086018	95.44579039
30	28.30743993	86.78154504	86.79714192	95.59829431	95.64159557

Table 9. Stopword of top 20 words

Corpus1		Corpus2	
TF/IG	TF/CHI	TF/IG	TF/CHI
is	is	high	high
and	and	_:(	_:(
so	so	right now	right now
for	for	*koronikuru*	*koronikuru*
or	or	orange	orange
tara	*tara*	*tottoto*	*tottoto*
wa	but	girl	girl
but	*wa*	premise	premise
mashi	*ja*	(-_-;)	(-_-;)
ja	*mashi*	accounting	accounting
nara	*nara*	piano	piano
than	of	door	door
u	than	grandfather	grandfather
of	*u*	U+0001F1FA	U+0001F1FA
terrible	terrible	phraseology	phraseology
ne	*ne*	U+0001F1F8	U+0001F1F8
ne	*ne*	common	common
nante	*nante*	hope	hope
kurai	however	Th	Th
however	inside	center	center

6 Conclusion

In this investigation, we have proposed two kinds of statistical filters, TFIG and TFCHI, to extract stopwords in SNS environment. Although we have utilized traditional filters of TF, IG and CHI as well as KLD, the completely different context has been examined applicable only for SNS environment. Let us note that we have shown the stopword list and the way how to generate the list.

Our experimental results show that the extracted stopwords take advantages to the traditional stopword list: we got 8.90% reduction ratio more and 2.31% precision ratio better compared to the baseline approach.

We expect to apply our approach to other kinds of languages in internet worlds including automatic generation of stopword list over several media languages.

References

1. Manning, C., Raghavan, P.: Introduction to Information Retrieval, 1st edn. Cambridge University Press, Cambridge (2008)
2. Bouge, K.: https://sites.google.com/site/kevinbouge/stopwordslists/stopwordsja-txt. Accessed 28 Dec 2017
3. slothlib - Revision 77. http://svn.sourceforge.jp/svnroot/slothlib/CSharp/Version1/-SlothLib/NLP/Filter/StopWord/word/Japanese.txt. Accessed 19 Jan 2018
4. Saif, H., Fernandez, M., Alani, H.: Automatic stopword generation using contextual semantics for sentiment analysis of Twitter. In: The 13th International Semantic Web Conference (ISCW) (2014)
5. Sonoda, T., Miura, T.: Mining Japanese collocation by statistical indicators. In: 15th International Conference on Enterprise Information Systems (ICEIS), Angers, France (2013)
6. Yang, Y., Pedersen, J.O. : A comparative study on feature selection in text categorization. In: Proceedings of International Conference on Machine Learning (ICML), pp. 412–420 (1997)
7. Zou, F., Wang, F.L., Deng, X., Han, S., Wang, L.S.: Automatic construction of Chinese stop word list. In: Proceedings of the 5th WSEAS International Conference on Applied Computer Science (2006)
8. Nezu, Y., Miura, T.: Extracting stopwords on social network service. In: The 29th International Conference on information Modelling and Knowledge Bases (EJC) (2019)

Multiple Choice Question Answering in the Legal Domain Using Reinforced Co-occurrence

Jorge Martinez-Gil[1(✉)], Bernhard Freudenthaler[1], and A Min Tjoa[1,2]

[1] Software Competence Center Hagenberg GmbH, Softwarepark 21, 4232 Hagenberg, Austria
{jorge.martinez-gil,bernhard.freudenthaler,amin.tjoa}@scch.at
[2] Vienna University of Technology, Favoritenstrasse 9-11/188, 1040 Vienna, Austria

Abstract. Nowadays, the volume of legal information available is continuously growing. As a result, browsing and querying this huge legal corpus in search of specific information is currently a tedious task exacerbated by the fact that data presentation does not usually meet the needs of professionals in the sector. To satisfy these ever-increasing needs, we have designed an appropriate solution to provide an adaptive and intelligent solution for the automatic answer of questions of legal content based on the computation of reinforced co-occurrence, i.e. a very demanding type of co-occurrence that requires large volumes of information but guarantees good results. This solution is based on the pattern-based methods that have been already successfully applied in information extraction research. An empirical evaluation over a dataset of legal questions seems to indicate that this solution is promising.

Keywords: Expert systems · Legal information processing · Knowledge engineering · Information retrieval · Question answering

1 Introduction

An increasing number of professionals from the legal sector agree that the information explosion concerning national and international legislation makes their work more expensive, tedious and even error-prone. The two major reasons for that are: (a) national and international legislation is usually formatted in an unstructured way, and (b) the huge volume and speed at which legislation is published usually lead to information overload in their daily activities.

In this context, working with information concerning legislation and case law has always been attractive to computer scientists and practitioners looking for applying for the latest advances on language and semantic technologies. In fact, these technologies have proven to be very useful for solving a number of problems that have traditionally affected the field of legal information processing. In practice, the daily work of these professionals requires reading a large amount of legal material necessary to identify the relevant documents and to identify the correct

S. Hartmann et al. (Eds.): DEXA 2019, LNCS 11706, pp. 138–148, 2019.
https://doi.org/10.1007/978-3-030-27615-7_10

fragment that they need. One step in the evolution towards the improvement of these processes come from a subfield from the information retrieval (IR) field, and it is called Question Answering (QA) systems. In fact, the design of systems of this kind is presented as an alternative to overcome the traditional processes by trying to provide accurate and understandable answers to specific questions, rather than presenting the user with a list of search-related documents [11].

In the particular case of the legal domain, the research community agrees that a system allowing to generate automatic responses to legal questions could have a strong impact with a lot of practical implications in their daily activities. The degree of usefulness is such that even the reduced version of the problem that we are addressing here (multiple choice, i.e. responding in a scenario where the answers are already given beforehand [1]) can also significantly help to reduce the workload. This is mainly because a QA system would be able to automatically process a huge amount of legal resources to answer a question or doubt in a matter of seconds, and that means that it could save resources in the form of effort, money and time to many professionals in the legal sector.

To tackle this problem, we have focused on computational techniques for co-occurrence analysis. Techniques of this kind have been widely used in various forms of research on content analysis, text mining, thesauri building, and ontology learning. Here, we propose a specific kind of co-occurrence, i.e. reinforced co-occurrence that it is intended to order to discover latent patterns on huge text corpora. And although our field of application in this work is the legal field, some of the conclusions that are drawn can be extrapolated to a wide range of specific domains. Therefore, with this idea in mind, we present here our research from which the following contributions can be highlighted:

- We propose a new method for the automatic answer of multiple choice questions of legal content based on the idea of computing reinforced co-occurrence.
- We have compiled a dataset of legal questions so that the researchers can try and compare their own solutions, and we have empirically evaluated our approach using the aforementioned dataset of legal questions.

The remainder of this work is organized in the following way: Sect. 2 reports the state-of-the-art on question answering methods and tools that have proven to be successful in the legal domain. Section 3 presents the fundamentals of our contribution concerning the computation of the reinforced co-occurrence over huge corpora. Section 4 reports the empirical evaluation of our novel approach over a legal dataset and the analysis of the results that we have achieved. Finally, we outline the conclusions and future lines of research.

2 State-of-the-Art

A QA system is a kind of computer system intended to automatically reply questions by analyzing different sources of either structured or unstructured information. These sources are usually called Knowledge Bases (KBs). In this context, there are basically two different approaches to tackle the problem depending on the KBs to be exploited: working with structured KBs, or working with unstructured KBs. Each of them has different advantages and disadvantages. For example, working with structured KB allows exploiting the knowledge represented by using the so-called inference engines, in order to infer new knowledge and to answer questions. However, at present, there is not an automatic way to introduce a new entity into the KB nor to determine with which existing entities should be related and how [15]. Therefore, finding practical solutions is considered as an important research challenge and its matter of intense research [10].

The fact is that not easy to implement these systems, so they have been progressively replaced by another type of more efficient systems based on lighter knowledge models such as knowledge graphs [6] and other enhanced lexical semantic models [22], but in general, it is widely assumed that building a KB is expensive in terms of resource consumption, it is subject to many errors, it is usually difficult and expensive to maintain, and last but not least, a structured KB is usually hardly reusable.

In contrast, IR systems have more practical benefits as most of them have been specifically designed to efficiently process huge amounts of textual data (usually represented in natural language). These huge amounts of data come from existing documents, databases, websites, and so on. For this reason, the most frequent type of QA system that is mentioned in the literature is the one that uses unstructured KBs including different collections of unstructured natural language KBs. In fact, the current generation of QA systems has evolved to extract answers from a wide range of different plain machine-readable resources. These QA systems exploit the massive set of unstructured information available on some sources to retrieve information about any particular question. It is important to note that these QA systems are possible mainly due to recent advances in the big data and natural language technologies. Moreover, since these novel QA systems are capable of processing questions about different domains and topics, they are now used in a wide range of different scenarios [14].

In this context, IR-based solutions represent words in the form of discrete and atomic units. For example, the first approach (and the simplest) could be to query the number of Google results for a specific question and a given answer together. However, this solution has brought a number of problems like the lack of context. To overcome these problems, word processing models such as LSA [5] and term frequency-inverse document frequency (tf-idf) partially solve these ambiguities by using terms that appear in a similar context based on their vector representation, and then they group the semantic space into the same semantic cluster. In this context, one of the best-known QA systems is IBM Watson [8], that it is very popular for its victory in the televised show *Jeopardy* [9]. Although in recent times, IBM Watson has become a generic umbrella that includes other business analytics capabilities.

If we focus strictly on the legal field, we find that QA technology has been very little used in real information systems, and especially in knowledge management systems [2]. The logic behind these systems is that given a legal issue, the extraction of relevant legal resources and the decision whether or not to use that content to answer the question are two key steps in building a system. In recent times, a number of works have been presented in this context. There are two major branches, (a) with structured KB. For example, Lame et al. [12] and Fawei et al. [7] using ontologies, or Xu et al. [21] by exploiting other KBs such as Freebase. And (b) exploiting unstructured KBs. For example, Brueninghaus and Ashley with a classical IR approach [4], Bennet et al. with strong focus on scalability [2], Maxwell and Schafer paying attention to context [16], Mimouni et al. with the possibility to make use of complex queries [17], or most modern deep learning techniques from Morimoto et al. [18] and Nicula et al. [19], the latter with good results, although with issues concerning the interpretability of the results.

3 Multiple Choice Question Answering Using Reinforced Co-occurrence

To overcome the current limitations of exiting QA approaches in the legal domain, we propose to automatically analyze co-occurrence patterns belonging to different corpora of unstructured text. Therefore, our approach is intended to automatically process huge amounts of legal information in order to look for evidence allowing to infer the most promising answers with regards to the huge range of questions that the legal professionals could potentially make. In this way, our contribution is a novel approach for automatically answering multiple choice questions concerning a wide range of topics belonging to the legal domain. This approach needs to fulfill two stages: first, we need to calculate alignment matrices between the question and the possible choices using textual corpora, and then we need to normalize the results in order to produce a final outcome and associated ranking of possible answers.

It is not difficult to see that the design of such as text mining approach in this context is far from being trivial. However, our experience in rapid prototyping and testing text mining solutions has shown us that it is possible to reach a reasonable level of success [14]. According to our experience, the solution that works best is a method with four levels of co-occurrence depending on the context whereby the question and the choice being evaluated can be found together.

On the other hand, the problem that we are addressing here is based on short answer models. The reason is that these models provide the potentially correct answer in the form of a number, a name, a date, or even a short phrase or text fragment. This makes the work of our text mining engine easier. It is also important to note, that this assumes that there are different ways of asking questions, and most of them are characterized by the formulation of questions expressed by interrogative particles (i.e. what, who, why, when, where, where) or some kind of is-a association. At the same time, the aforementioned possible choices are expressed in natural language.

Although the concept seems to be not to difficult to understand, there are huge technical limitations for its development from a pure engineering perspective. In fact, this approach is limited by an important number of technical issues which should be overcome. These limitations, originally identified by [3], are inherent to the process of massively text mining. In order to facilitate overcoming these limitations, our system is designed in the form of a pipeline, i.e. a workflow whereby the data flow into processes so that the output of one process is the input of the next one. Figure 1 shows us an overall view of our IR pipeline. These components are related to each other and process the textual information available on different levels until the QA process has been completed. The natural language questions formulated to the system are processed initially by the question analysis component.

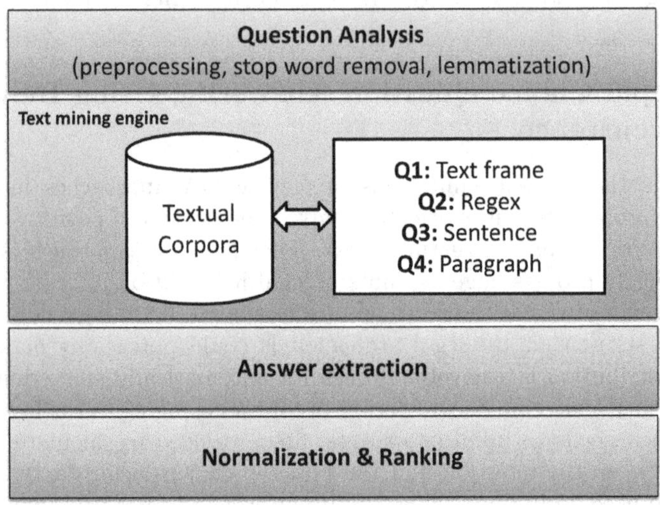

Fig. 1. Pipeline designed to answer the multiple choice tests. First of all, questions and answers need to be pre-processed. After this, a text mining engine is in charge of mining reinforced co-occurrence patterns. Then, these patterns are analyzed. Finally, the results are normalized and a ranking of potential choices is provided

Then, the system continues working by conveniently dividing the information into different parts which will be transferred to the following process which is a text mining engine that looks for the reinforced co-occurrence of the question and each possible answer. Then, the answer extraction module compiles the raw data resulting from the mining phase. Finally, it is necessary to normalize the raw data and create the final ranking to be delivered. The main modules of our QA system could be summarized in the following steps:

- *Question Analysis.* It is in charge of pre-processing both the question and the possible answers. To do that, it is necessary to remove the stop words, to proceed with a lemmatization process, i.e. determining what is the root of the words so irregular forms (e.g. plurals, third person, etc...) does not affect to the co-occurrence, and remove very common adjectives and verbs.

- *Reinforced Co-Occurrence Calculation.* It consists of calculating how many times the pre-processed question and the evaluated answer co-occur together in the same text frame, in the same text expression, in the same sentence, and in the same paragraph.
- *Answer Extraction.* It consists of compiling the results and assign them to each of the possible choices. After this process, we have just raw values that need to be refined.
- *Answer Normalization and Ranking.* Normalization is the process of mitigating the impact of the outliers on the final decision. In this work, we usually work with exponential reductions, but other methods need to be considered in future work. Ranking consists of creating an ordered list of response according to the score obtained after normalization.

3.1 Running Example

In order to illustrate how our approach works, we have designed a running example to better understand how our pipeline processes the information. Let us think in a question whereby we would like to know the kind of document represented by the European Convention on Human Rights. Let us think how the question could be, and how the different choices would look like.

```
What document is the European Convention on Human Rights?
(a) A statute
(b) Delegated legislation
(c) An EU directive
(d) A treaty (correct choice)
```

Then, our system would start by evaluating the suitability of the first answer, i.e. statute. To do that, we can see in Fig. 2 the graphical summary of how this process is performed: The question and the associated choices have to be preprocessed in order to remove non-relevant words, perform lemmatization, etc. Then, this information has to be submitted to text mining engine, where a dispatcher tries to look for the reinforced co-occurrence of the pair question-possible answer by scanning all possible co-occurrences within the corpora. As a result, we get the reinforced co-occurrence values that have to be normalized so the outliers might not have an extreme weight in the final value.

After repeating this process for each of the possible choices, we have that in this case, our solution must discern whether it is about a statute, a delegated legislation, an EU directive or a treaty. In Table 1, a normalization has been applied. In this case, normalization consists of gradually reducing the value associated with the co-occurrence of very general terms since this adds an excessive noise at the time of obtaining a meaningful response. In this case, statute and document have a high degree of co-occurrence that can make other parts of the question lose prominence, so we must proceed to reduce the impact that such co-occurrence has on the final result.

Therefore, the choice that our system would select as the correct one is (d) A treaty, what is also the correct one according to the ground truth. The second

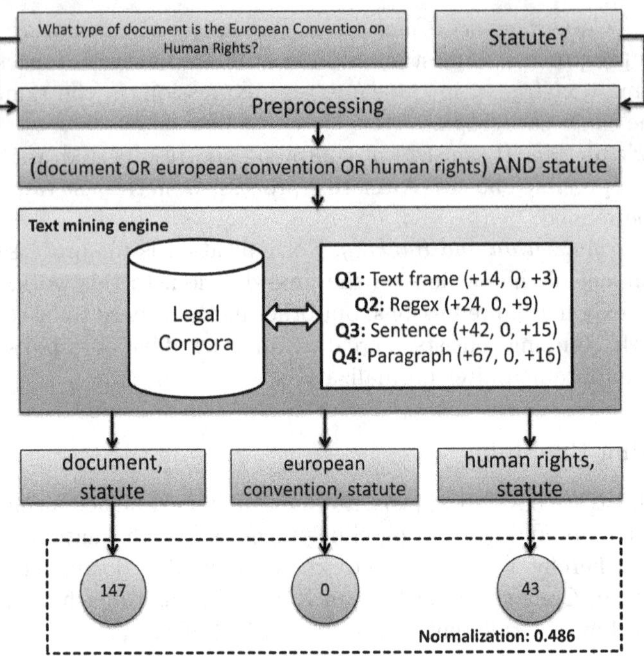

Fig. 2. View of one iteration whereby a question and a potential answer are evaluated

would be (a) A statute. And the other two possible choices has no reinforced co-occurrence, so they would not even be considered as possible answers. Additionally, a heatmap allows to visually inspect the rationale behind the result. This is mainly due to the fact that in some scenarios requiring accountability and/or interpretability, it is not just enough to provide the answer, but also some reasons for helping to interpret that answer.

Table 1. Normalized results obtained for the reinforced co-occurrence, final score and ranking proposal

	Statute	Delegated legislation	EU directive	Treaty
Document	0.376	0.000	0.000	1.000
European convention	0.000	0.000	0.000	0.000
Human rights	0.110	0.000	0.000	0.441
Final score	0.486	0.000	0.000	1.441
Ranking	2	-	-	1

4 Results

We explain here the results. It is important to remark that these results are highly dependent on the base corpus that will be processed. Choosing a relevant, specific base corpus to evaluate each of the possible choices is really important in this context. On the other hand, the task of evaluating the system is of vital importance, as it will assess the performance, as well as the accuracy of the techniques. In this work, we have chosen the strictest methodology to evaluate systems, which consists of binary classification: the answer was right or wrong.

4.1 Solving the Benchmark Dataset

We show here the results that we have obtained when testing our solution. The dataset has been generated by picking randomly legal questions suggested from a number of books from the Oxford University Press[1]. Moreover, it is important to remark that the techniques are applied automatically over the dataset of questions without prior knowledge of the answers (e.g. without machine learning). After comparing the answers with the results, the performance is determined through the accuracy metric.

Therefore, we have obtained 13 correct answers from a total of 20 questions. This means that we got a 65% of accuracy. Table 2 shows a comparison with other approaches. By responding randomly there is a 25% chance of guessing the correct answer, so that is the score we have established as baseline. Please note that, for the sake of fair comparison[2], we just include approaches without machine learning capabilities (as ours). At the same time, we hope that the compilation of this dataset will stimulate the evaluation of more QA systems.

Table 2. Comparison with other approaches

Approach	Correct answers	Accuracy
Baseline	5	25%
Calcipher [20]	7	35%
Li et al. [13]	9	45%
LSA-Classic [5]	9	45%
Our approach	13	65%

QA technology is becoming a very important solution in a wide range of areas overloaded by the constant generation of large amounts of information. In this context, being able to automatically answering specific questions in a correct manner can contribute to alleviating the problem of dealing with those huge amounts of data. Our approach is able to offer good results, at an affordable

[1] http://global.oup.com.

[2] Although we foresee learning the parameters of our system as future work.

cost (in terms of money, time, and effort needed), without the need for training, and with great facilities for interpretability. This technology, however, faces some obstacles in its development related to the amount of engineering work to properly tune the parameters involved along the IR pipeline.

5 Conclusions and Future Work

In the context of the legal domain, methods and techniques for answering specific questions are in high demand, and as a result, a number of solutions for QA have been developed to respond to this need. The major reason for that is that the capability to automatically answer questions by means of computers could help alleviate a problem involving tedious tasks such as an extensive information search what is, in general, time-consuming. By automatically providing hints concerning a wide number of legal topics, lots of resources in the form of effort, money and time can be preserved. In this work, we have presented our research on automatically addressing multiple choice questions and the development of techniques for automatically finding the correct answer by means of IR pipeline that implements reinforced co-occurrence.

We have seen that although approaches based on structured KB often yield good results, it is difficult to use them in practice mainly due to the cost when building such structured KB (i.e. it is expensive in terms of effort, money and time needed) and it is often very difficult to find experts with enough knowledge for curating the KB. In contrast, our approach has a number of practical benefits when selecting the actual right answer from a list of the possible answers due to the advances in big data processing and natural language technology. Moreover, in the present work, we have not yet fully explored the characteristics of legal texts in order to utilize these features for building our legal QA system. In fact, properties such as references between documents or structured relationships in legal statements should be investigated more in depth as part of future work.

As additional future lines of research, we also need to work towards overcoming a number of technical limitations. This includes the capability to work with different multilingual corpora at the same time, the proper processing of verbs when formulating questions and evaluating potential answers, and the proper tuning of the different system parameters by means of a training phase. We think that by successfully addressing these challenges, it is possible to build solutions that can help the legal practitioners to overcome one of the most problematic issues that they have to face in their daily work.

Acknowledgements. This research work has been supported by the Austrian Ministry for Transport, Innovation and Technology, the Federal Ministry of Science, Research and Economy, and the State of Upper Austria in the frame of the COMET center SCCH.

References

1. Aydin, B.I., Yilmaz, Y.S., Li, Y., Li, Q., Gao, J., Demirbas, M.: Crowdsourcing for multiple-choice question answering. In: AAAI, pp. 2946–2953 (2014)

2. Bennett, Z., Russell-Rose, T., Farmer, K.: A scalable approach to legal question answering. In: ICAIL, pp. 269–270 (2017)
3. Blohm, S., Cimiano, P.: Using the web to reduce data sparseness in pattern-based information extraction. In: Kok, J.N., Koronacki, J., Lopez de Mantaras, R., Matwin, S., Mladenič, D., Skowron, A. (eds.) PKDD 2007. LNCS (LNAI), vol. 4702, pp. 18–29. Springer, Heidelberg (2007). https://doi.org/10.1007/978-3-540-74976-9_6
4. Brueninghaus, S., Ashley, K.D.: Improving the representation of legal case texts with information extraction methods. In: ICAIL, pp. 42–51 (2001)
5. Deerwester, S.C., Dumais, S.T., Landauer, T.K., Furnas, G.W., Harshman, R.A.: Indexing by latent semantic analysis. JASIS **41**(6), 391–407 (1990)
6. Ding, J., Wang, Y., Hu, W., Shi, L., Qu, Y.: Answering multiple-choice questions in geographical gaokao with a concept graph. In: Gangemi, A., Navigli, R., Vidal, M.-E., Hitzler, P., Troncy, R., Hollink, L., Tordai, A., Alam, M. (eds.) ESWC 2018. LNCS, vol. 10843, pp. 161–176. Springer, Cham (2018). https://doi.org/10.1007/978-3-319-93417-4_11
7. Fawei, B., Pan, J.Z., Kollingbaum, M., Wyner, A.Z.: A methodology for a criminal law and procedure ontology for legal question answering. In: Ichise, R., Lecue, F., Kawamura, T., Zhao, D., Muggleton, S., Kozaki, K. (eds.) JIST 2018. LNCS, vol. 11341, pp. 198–214. Springer, Cham (2018). https://doi.org/10.1007/978-3-030-04284-4_14
8. Ferrucci, D.A.: Introduction to this is Watson. IBM J. Res. Dev. **56**(3), 1 (2012)
9. Ferrucci, D.A., Levas, A., Bagchi, S., Gondek, D., Mueller, E.T.: Watson: beyond Jeopardy!. Artif. Intell. **199–200**, 93–105 (2013)
10. Hoffner, K., Walter, S., Marx, E., Usbeck, R., Lehmann, J., Ngonga Ngomo, A.C.: Survey on challenges of question answering in the semantic web. Semant. Web **8**(6), 895–920 (2017)
11. Kolomiyets, O., Moens, M.-F.: A survey on question answering technology from an information retrieval perspective. Inf. Sci. **181**(24), 5412–5434 (2011)
12. Lame, G.: Using NLP techniques to identify legal ontology components: concepts and relations. Artif. Intell. Law **12**(4), 379–396 (2004)
13. Li, Y., McLean, D., Bandar, Z., O'Shea, J., Crockett, K.A.: Sentence similarity based on semantic nets and corpus statistics. IEEE Trans. Knowl. Data Eng. **18**(8), 1138–1150 (2006)
14. Martinez-Gil, J., Freudenthaler, B., Natschlaeger, T.: Automatic recommendation of prognosis measures for mechanical components based on massive text mining. IJWIS **14**(4), 480–494 (2018)
15. Martinez-Gil, J.: Automated knowledge base management: a survey. Comput. Sci. Rev. **18**, 1–9 (2015)
16. Maxwell, K.T., Schafer, B.: Concept and context in legal information retrieval. In: JURIX, pp. 63–72 (2008)
17. Mimouni, N., Nazarenko, A., Salotti, S.: Answering complex queries on legal networks: a direct and a structured IR approaches. In: Pagallo, U., Palmirani, M., Casanovas, P., Sartor, G., Villata, S. (eds.) AICOL 2015–2017. LNCS (LNAI), vol. 10791, pp. 451–464. Springer, Cham (2018). https://doi.org/10.1007/978-3-030-00178-0_31
18. Morimoto, A., Kubo, D., Sato, M., Shindo, H., Matsumoto, Y.: Legal question answering system using neural attention. In: COLIEE@ICAIL, pp. 79–89 (2017)
19. Nicula, B., Ruseti, S., Rebedea, T.: Improving deep learning for multiple choice question answering with candidate contexts. In: ECIR, pp. 678–683 (2018)

20. Stam, M.: Calcipher System. https://github.com/matt-stam/calcipher. Accessed 01 Apr 2019
21. Xu, K., Reddy, S., Feng, Y., Huang, S., Zhao, D.: Question answering on freebase via relation extraction and textual evidence. In: ACL, vol. 1 (2016)
22. Yih, W.-T., Chang, M.-W., Meek, C., Pastusiak, A.: Question answering using enhanced lexical semantic models. In: ACL, vol. 1, pp. 1744–1753 (2013)

A Probabilistic Algorithm to Predict Missing Facts from Knowledge Graphs

André Gonzaga(✉), Mirella Moro, and Mário S. Alvim

Universidade Federal de Minas Gerais, Belo Horizonte, Brazil
{andregonzaga,mirella,msalvim}@dcc.ufmg.br

Abstract. Knowledge Graph, as the name says, is a way to represent knowledge using a directed graph structure (nodes and edges). However, such graphs are often incomplete or contain a considerable amount of wrong facts. This work presents ProA: a probabilistic algorithm to predict missing facts from Knowledge Graphs based on the probability distribution over paths between entities. Compared to current state-of-the-art approaches, ProA has the following advantages: simplicity as it considers only the topological structure of a knowledge graph, good performance as it does not require any complex calculations, and readiness as it has no other requirement but the graph itself.

Keywords: Knowledge Graph · Link prediction · Probabilistic solution

1 Introduction

Over time, the Semantic Web [2] has promoted a graph-based representation of knowledge in which entities are graph nodes (e.g., *Shakespeare* and *Hamlet*) and there is an edge between two nodes if they are related (e.g., Shakespeare *has written* Hamlet). Entities may also have types denoted by *is a* relations (e.g., Shakespeare *is a* writer, Hamlet *is a* play). These relations are known as *facts.* In many cases, the sets of possible types and relations are organized in a *schema* or an *ontology*, which defines their interrelations and their usage restrictions.

Such graph representation of knowledge is known as *Knowledge Graphs* (KG), as coined by Google in 2012[1]. Well known knowledge graphs include: Freebase [4], DBpedia [1], Wikidata [21] and YAGO [19]. In a broader perspective, any graph-based representation of knowledge could be considered a Knowledge Graph, which includes any kind of *Resource Description Framework* (RDF) for example – note, RDF organizes knowledge in triples of the form *subject–predicate–object*, which is easily equivalent to node-edge-node in a KG. Also, search engines and e-commerce websites, such as Amazon and Walmart, are trying to improve

[1] https://googleblog.blogspot.com/2012/05/introducing-knowledge-graph-things-not.html.

S. Hartmann et al. (Eds.): DEXA 2019, LNCS 11706, pp. 149–158, 2019.
https://doi.org/10.1007/978-3-030-27615-7_11

user experience by using KGs. Those graphs are often constructed from semi-structured knowledge, such as Wikipedia, or harvested from the Web.

Indeed, Knowledge Graphs on the Web form a backbone of many information systems that require access to structured knowledge. The idea of feeding intelligent systems and agents with formalized knowledge of the world dates back to classic Artificial Intelligence research in the 1980s. Recently, the advent of Linked Open Data [3] has spurred interest over representations of general world knowledge as graphs from completely fresh perspectives, for example [16]. Nonetheless, constructing a knowledge graph presents several issues.

Specifically, no matter the building procedure used, the result is expected not to be perfect. The reason is simple: as a model of the real world, formalized knowledge cannot reach full coverage (i.e. cover all information about every entity in the universe). Also, achieving 100% of both completion and correctness is unlikely, which defines a trade-off between coverage and correctness that is addressed differently in each knowledge graph, for example [5] and [20]. Therefore, the problem here is summarized as identifying missing facts in Knowledge Graphs.

Taking a closer look at such a problem, its solution may borrow from other known issues. Specifically, *link prediction* is the task of predicting the existence (or probability of correctness) of edges in graphs [11]. In our context, link prediction is critical for KG not achieving 100% of completion and correctness. Even if KGs could reach a perfect representation of the world, it may change over time, and then link prediction still remains important. Furthermore, link prediction can be easily mapped to not only adding missing data to Knowledge Graphs [13,17], but also discovering links in the context of linked data [9], predicting new relationships over professional networks [6], and even applied to predicting drug-target and protein-protein interactions in biomedical graphs [8].

Following such a link prediction trend, this work proposes a probabilistic algorithm, called ProA, based on path frequency of edges to infer a possible missing edges from an entity source to an entity target. Since this probabilistic approach uses only the distribution over paths between nodes in the graph, there are no entities properties or any other external dependencies to predict facts. Our evaluation also shows that using the probability distribution provides good performance when compared to other algorithms. Our advantages over the state-of-the-art are simplicity in terms of operation and computational complexity, and ProA is ready to be used in any Knowledge Graph.

The paper is organized as follows. Section 2 presents a comprehensive overview of related work, most of which is used as baseline. Section 3 presents the new probabilistic algorithm. Then, Sect. 4 presents our evaluation and results, whereas Sect. 5 concludes this work.

2 Related Work

Link prediction is usually an important task since graphs are often incomplete or need improvement. For example, a social network developer may want to predict relationships between users or study the possible evolution of a professional network [6]. Now, there are many works improving link prediction over Knowledge Graphs. This section follows with a comprehensive overview of relevant approaches that are used as our baselines later on.

In a different perspective, Chekol el al. [7] describe how to combine uncertainty and time in Knowledge Graphs. Since facts are usually accompanied by a confidence score that witnesses how likely is for them to hold, the authors show a solution for the management of uncertain and temporal data in KGs. The work relies on one main step: use Markov Logic Network (MLN) that provides the necessary underpinning to formalize the syntax and semantics of uncertain temporal KGs. Then, they explore the usage of Probabilistic Soft Logics (PSL) and compare the results against the MLN approach.

From yet another perspective, RESCAL [15] is a relational latent-feature model that explains triples via pairwise interactions of latent features. In particular, it models the score of a triple x_{ijk} and a weight matrix $\mathbf{W}$ whose entries w_{abk} specify how much the latent features a and b interact in the k-th relation. This is a bilinear model, since it captures the interactions between two entity vectors using multiplicative terms. In general, it models block structure patterns via the magnitude of entries in matrix $\mathbf{W}$, while it uses homophily patterns via the magnitude of its diagonal entries.

Lastly, there are other tensor factorization methods for learning from Knowledge Graphs. ER-MLP [12] uses a global weight vector for all relations based on multi-layer perceptrons (MLPs). TransE [5] translates the latent feature representations via a relation-specific offset instead of transforming them via matrix multiplications. HolE [14] learns compositional vector space representations of an entire Knowledge Graph. This method is related to holographic models of associative memory in that it employs circular correlation to create compositional representations. DistMult [22] is a special case of RESCAL, in which the relation matrix is assumed to be diagonal. Complex [20] performs sparse tensor factorization of KG in the complex domain. Specifically, nodes and relations are modeled by d dimensional vectors with a real and imaginary part ($Re(x)$, $Im(x)$). This allows to model anti-symmetric relations since the three way dot product (inner product) in the complex domain is not symmetric.

Here, we propose a new way to predict edges in Knowledge Graphs based on path information in the topological structure of the graph. Given the probability distribution over all paths between two entities and the probability distribution of those paths in the whole graph, the probabilistic algorithm learns about it and generates a new probability distribution to infer if there is an edge to be predicted or not. The main difference between this approach to others is the input. ProA only takes the path information over entities to predict facts, without any complex mathematical calculation. Moreover, the results show competitive numbers against current state-of-the-art, specially for dense graphs.

3 Missing Facts Prediction

We now introduce concepts necessary to understand our proposed solution, followed by our algorithm called *ProA*.

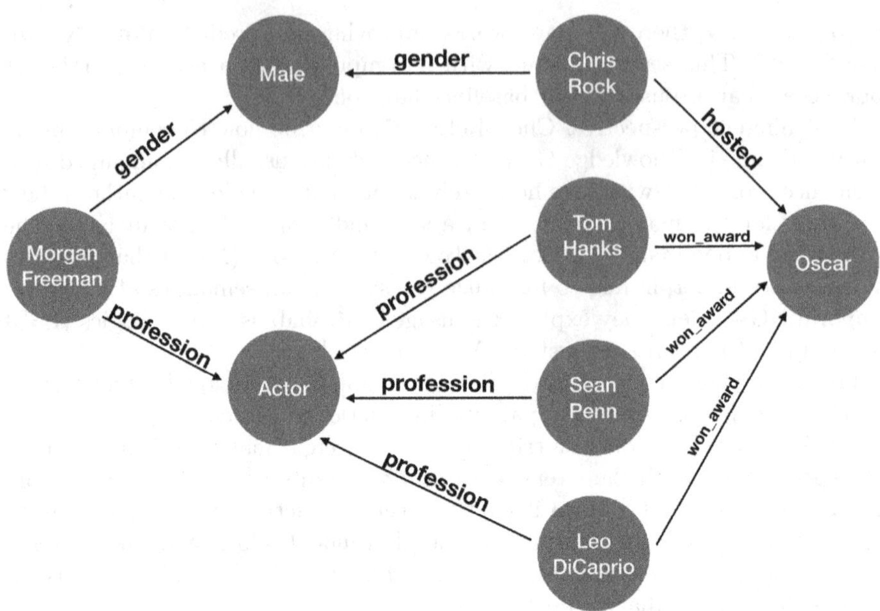

Fig. 1. Probability dependencies between entities. There are two ways to reach `Oscar` from `Morgan Freeman`: through [`gender`, `gender`$^{-1}$, `hosted`], and through [`profession`, `profession`$^{-1}$, `won_award`].

3.1 Basic Concepts

Knowledge Graphs typically have statistical patterns or regularities, which are not universally true but nevertheless allow for useful predictive power. They also have some deterministic rules, such as type constraints and transitivity (e.g., if Barack Obama was born in Honolulu, and Honolulu is located in the USA, then we can infer that Barack Obama was born in the USA). One example of such statistical pattern is homophily, that is, the tendency of entities to be related to other entities with similar characteristics. This has been widely observed in various social networks. For example, Brazil born actors are more likely to star in Brazil made movies.

Formally, we model each possible triple $x_{ijk} = (e_i, r_k, e_j)$, indicating that entity e_i is related to entity e_j by relation type r_k, over these sets of entities and relations as a binary random variable $Y(ijk) \in \{0, 1\}$, which indicates its existence. While $Y(ijk) = 1$ indicates the existence of a triple, the interpretation of $Y(ijk) = 0$ depends on whether the open world or closed world assumption is considered. In this work, we assume it is open world, which means if a random variable is $Y(ijk) = 0$ then this fact is unknown.

A path D is a sequence of relations, in which the last entity of a relation is the first entity of the following one, with length greater than 0, connecting two or more entities. Moreover, random variables $Y(ijk)$ are conditionally dependent on each other because we consider paths and their dependencies. In other words, if the probability to reach a node x, from z node, is $P(x)$ and the node y depends on x, then the probability of reaching y is $P(y|x)$.

Figure 1 illustrates an example of a Knowledge Graph. The graph represents real facts regarding `profession`, `gender`, `hosted_oscar` and `won_oscar` relations. Then, the probability of reaching the target node `Oscar` from source node `Morgan Freeman` depends on the probability of each relation in the path. Since this graph has four paths connecting `Morgan Freeman` to `Oscar`, the question here is: "Can we infer the existence of any direct relation between `Morgan Freeman` and `Oscar`"?

Let D be the set of paths of length at most ℓ between any two nodes. We only consider paths in which the random variable Y is 1, i.e., there is a path between two nodes. There is also a probability distribution over D, because some paths can be more likely than others. There is a probability distribution for every valid path in D in the whole topological structure in the graph which means how often the path is in the graph structure. Note it is a valid probability distribution since the sum of values is 1.

Regarding the distribution math, a distribution over all edges in the graph can be defined by counting the number of each type of relation divided by the total of edges in the KG. For instance, consider the Freebase train dataset with 483,142 facts. One of its popular edge is `award_nominee`, as it appears almost 16,000 representing 0.03311 of total. Then, the probability of an edge being `award_nominee` in KG is about 3.3%.

Formally, the distribution to predict a fact is defined by Eq. 1.

$$P(Y(ijk) = 1) = \sum_{m_k}^{|D|} \frac{d_{m_k}}{|D|}, \tag{1}$$

where $|D|$ is the number of paths in D, and each d_{m_k} is a path in D ending with a relation k. Then, the final distribution after this calculation can be sorted and ranked. Intuitively, the most probable relation after this calculation should be the correct fact between those entities.

In summary, the problem tackled in this work is defined as: "Given a source and a target entities, what steps are necessary to predict a fact between them?"

3.2 ProA Algorithm

Our solution, the ProA algorithm, gets all possible paths from a source node to a target node to infer which relation is most likely to connect between those two nodes. It satisfies the Markov property of the probability conditional distribution, since it depends upon the present state of the graph. In other words, the algorithm has the memoryless property of a stochastic process.

Again using Fig. 1 with a piece of real Knowledge Graph that focuses on Morgan Freeman and the Oscars. Entity `Morgan Freeman` is connected to `actor` and `male` entities through relations `profession` and `gender`, respectively. Applying ProA algorithm for entities `Morgan Freeman` and `Oscar` finds all drawn paths to generate D. Then, for each path in D, a final probability distribution of facts

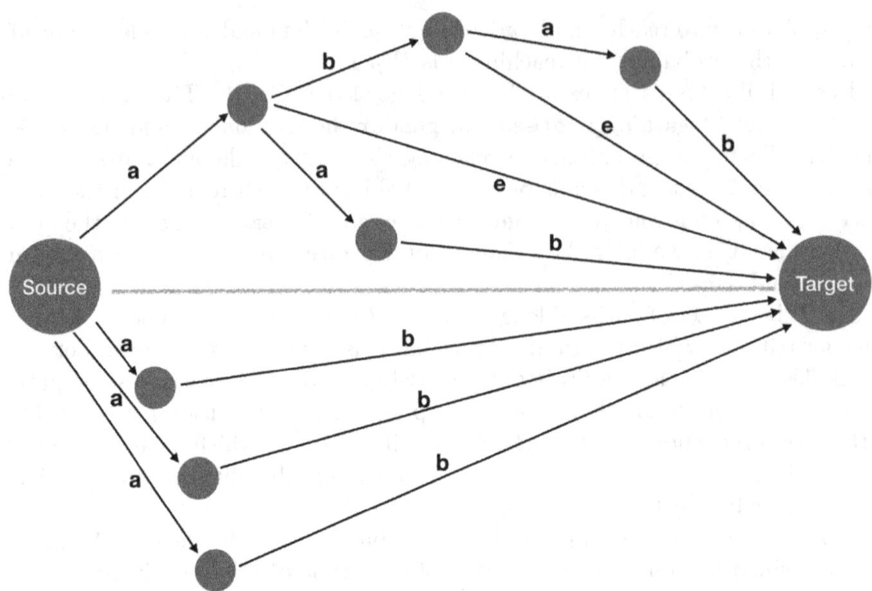

Fig. 2. Link prediction based on path probabilistic distribution. To predict the red fact between source and target, ProA finds all possible paths between two nodes at most length ℓ (in this scenario $\ell = 3$). The set of possible paths is $\{(a, b), (a, e), (a, a, b), (a, b, e)\}$, and the most common path is (a, b). Usually, the fact $\{e\}$ exists when (a, b) is a path between two nodes; hence, ProA infers with a certain probability that $\{e\}$ could be the missing red fact. (Color figure online)

that connect entities is calculated. In summary, ProA learns that `won_award` is more probable to occur for an actor that `hosted` from gender relation. The final distribution is about all relations found between source and target, and ProA can properly infer each missing relation (Fig. 2).

ProA does not only infer missing relations between entities, but also predicts whether there is no relation between them. Considering the Knowledge Graphs Freebase and Wordnet, an initial empirical evaluation (performed while testing ProA) shows that `None` is the most probable relation to occur between entities connected by a path at most length three. Therefore, ProA learns about this special relation and infers whether there is a relation or not.

The calculation of time complexity for ProA is similar to well known algorithms in graphs such as Breadth-first search and Depth-first search. Since in the worst case every node and relation will be explored, ProA is at least $O(|V|+|E|)$. However ProA takes ℓ as argument, so the final time complexity is $O((|V|+|E|)^{\ell})$, which means if ProA considers the number of paths at most 3 it will have a time complexity of $O((|V| + |E|)^3)$ and so on.

Table 1. Dataset details

Dataset	Entities	Relations	Train	Test	Valid
FB15K	14,951	1,345	483,142	59,071	50,000
WN18	40,943	18	141,442	5,000	5,000

Table 2. MMR and Hits@3 on FB15K and WN18 datasets

Algorithm	MMR		Hits@1		Hits@3		Hits@10	
	FB15K	WN18	FB15K	WN18	FB15K	WN18	FB15K	WN18
[12] ER-MLP	0.288	0.712	0.317	0.775	-	-	-	-
[5] TRANSE	0.380	0.454	0.231	0.089	0.472	0.823	0.641	0.934
[15] RESCAL	0.354	0.890	0.409	0.904	-	-	-	-
[14] HOLE	0.524	0.938	0.402	0.93	0.613	**0.945**	0.739	**0.949**
[22] DISTMULT	0.654	0.938	0.546	0.728	0.733	0.914	0.824	0.936
[20] COMPLEX	0.692	**0.941**	**0.599**	**0.936**	0.759	**0.945**	0.840	0.947
PROA	**0.695**	0.483	0.54	0.304	**0.8**	0.701	**0.96**	0.703

4 Experiments and Results

In order to evaluate ProA performance, we first detail the dataset over which we build a knowledge graph. Then, we present two of the most common metrics for evaluating prediction on KG. Last, we present our evaluation and results.

4.1 Datasets

Following [5][2], this work evaluates the probabilistic approach sampling on the Freebase dataset (FB15k) and WordNet dataset (WN18). They are very different in coverage: FB15k contains mostly named entities connected through strongly typed relations, whereas WN18 contains mostly common nouns connected through lexical and semantic relations. Table 1 presents the dataset statistics.

The data contain relations with high variation in the number of instances – 39% of the relations have at most 10 instances, while the most frequent relation has almost 16,000. This disparity is also reflected in the distribution of node degrees: 12% of the entities have degree equal or less than 10 (appear in at most 10 times). The average degree of a node in FB15k is approximately 13.2 overall, and 32.4 on the training data [10] (Fig. 3).

[2] A very well cited paper with over 680 citations in June, 2018.

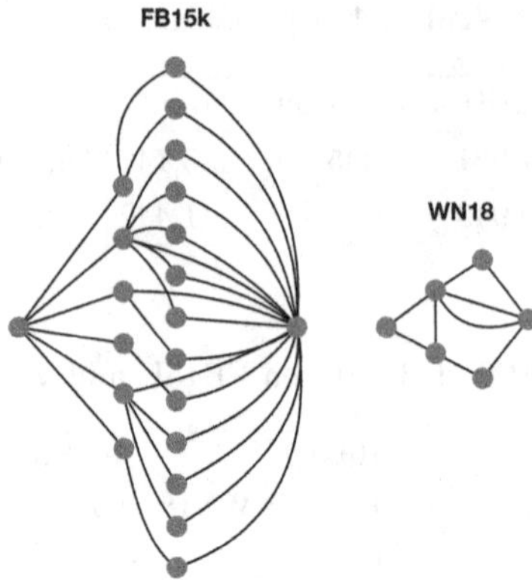

Fig. 3. Differences in the number of paths between FB15k and WN18: there are four paths from entity source to entity target in WN18 and 36 paths in FK18k.

4.2 Results

Table 2 presents the results for ProA and each algorithm of the state-of-the-art tested, as discussed in Sect. 2. Specifically, the columns are grouped by the metrics (previous section) calculated over each of the two datasets. *ER-MLP* and *RESCAL* do not provide information about Hits@3 and Hits@10. It shows that ProA probabilistic approach has great results over baselines for FB15K. The main reason is that dense graphs have multiple paths between entities, and ProA learns from those and properly infers the missing facts. Specifically, the Freebase dataset has an average degree of 32.4 that indicates a high level of relation between entities, which could classify it as a dense graph.

The *MMR* measure results show a small improvement over the state-of-the-art, but without significant statistical difference over *Complex* and *DistMult*. However there is a significant improvement for Hits@3 and Hits@10 indicating that ProA often has a good precision under ranking of possible facts. On the other hand, Wordnet has average degree of three. Then, Table 2 shows that ProA does not fully learn the paths. In our complementary evaluations of this dataset, often the number of paths between two entities is only one, and then ProA cannot predict properly based solely in one relation. Moreover, ProA has a considerable hits rate in sparse graphs indicating that in 70% of cases, the number of facts predicted is in the top three and top 10 results.

Overall, ProA works well with dense Knowledge Graphs (such as Freebase) because it requires a certain amount of paths to generate the paths probability distribution. For KGs with less paths between entities, the probabilistic algorithm could not learn properly and then the prediction is just reasonable. Wordnet is an example of sparse Knowledge Graph, which means there is a few number of paths between nodes, making it difficult for ProA.

5 Conclusion

This work presents ProA, a probabilistic algorithm to infer missing facts in Knowledge Graphs. The approach generates probability distributions over paths between entities and predicts missing facts. In ProA, the paths are exploited to create probabilistic model that can capture rich interactions in relational data.

Our experimental evaluation shows ProA provides a good performance against the state-of-the-art, specially in dense graphs, then solving a complex problem as link prediction. Moreover, ProA is publicly on GitHub[3] and it is easy to use. Intuitively, user can setup a knowledge graph dataset as input and define parameters such as the most allowed length of paths, and then ProA predicts every possible fact.

As future work, we plan to further exploit the probability distribution over sparse graphs and apply ProA to complex networks, as the topological structure is important. A related idea is to explore the presence of relations `None` and correlate it with the concept of *dull nodes* as proposed in [18].

Acknowledgements. Work partially funded by CNPq and FAPEMIG, Brazil.

References

1. Auer, S., Bizer, C., Kobilarov, G., Lehmann, J., Cyganiak, R., Ives, Z.: DBpedia: a nucleus for a web of open data. In: Aberer, K., et al. (eds.) ASWC/ISWC -2007. LNCS, vol. 4825, pp. 722–735. Springer, Heidelberg (2007). https://doi.org/10.1007/978-3-540-76298-0_52
2. Berners-Lee, T., Hendler, J., Lassila, O.: The semantic web. Sci. Am. **284**(5), 29–37 (2001)
3. Bizer, C., Heath, T., Berners-Lee, T.: Linked data - the story so far. Int. J. Seman. Web Inf. Syst. **5**(3), 1–22 (2009)
4. Bollacker, K., Evans, C., Paritosh, P., Sturge, T., Taylor, J.: Freebase: a collaboratively created graph database for structuring human knowledge. In: Proceedings of the 2008 ACM SIGMOD International Conference on Management of Data, pp. 1247–1250. ACM, Vancouver (2008).https://doi.org/10.1145/1376616.1376746
5. Bordes, A., Usunier, N., Garcia-Durán, A., Weston, J., Yakhnenko, O.: Translating embeddings for modeling multi-relational data. In: Proceedings of the 26th International Conference on Neural Information Processing Systems - Volume 2, USA, pp. 2787–2795 (2013)
6. Brandão, M.A., Moro, M.M.: Social professional networks: a survey and taxonomy. Comput. Commun. **100**, 20–31 (2017)
7. Chekol, M.W., Pirrò, G., Schoenfisch, J., Stuckenschmidt, H.: Marrying uncertainty and time in knowledge graphs. In: Proceedings of the Thirty-First AAAI Conference on Artificial Intelligence, San Francisco, CA, USA, pp. 88–94 (2017)
8. Crichton, G., Guo, Y., Pyysalo, S., Korhonen, A.: Neural networks for link prediction in realistic biomedical graphs: a multi-dimensional evaluation of graph embedding-based approaches. BMC Bioinform. **19**(176) (2018)

[3] https://github.com/andrehigher/ProA.

9. Georgala, K., Hoffmann, M., Ngomo, A.N.: An evaluation of models for runtime approximation in link discovery. In: IEEE/WIC/ACM International Conference on Web Intelligence, pp. 57–64 (2017)
10. Kotnis, B., Nastase, V.: Analysis of the impact of negative sampling on link prediction in knowledge graphs. CoRR (2017)
11. Liben-Nowell, D., Kleinberg, J.: The link-prediction problem for social networks. J. Am. Soc. Inf. Sci. Technol. **58**(7), 1019–1031 (2007)
12. Mikolov, T., Chen, K., Corrado, G., Dean, J.: Efficient estimation of word representations in vector space. CoRR abs/1301.3781 (2013). http://arxiv.org/abs/1301.3781
13. Nickel, M., Murphy, K., Tresp, V., Gabrilovich, E.: A review of relational machine learning for knowledge graphs. Proc. IEEE **104**(1), 11–33 (2016). https://doi.org/10.1109/JPROC.2015.2483592
14. Nickel, M., Rosasco, L., Poggio, T.A.: Holographic embeddings of knowledge graphs. CoRR abs/1510.04935 (2015). http://arxiv.org/abs/1510.04935
15. Nickel, M., Tresp, V., Kriegel, H.P.: A three-way model for collective learning on multi-relational data. In: Proceedings of the 28th International Conference on International Conference on Machine Learning, pp. 809–816 (2011)
16. Nishioka, C., Scherp, A.: Keeping linked open data caches up-to-date by predicting the life-time of RDF triples. In: IEEE/WIC/ACM International Conference on Web Intelligence, pp. 73–80 (2017)
17. Paulheim, H.: Knowledge graph refinement: a survey of approaches and evaluation methods. Semant. Web **8**, 489–508 (2017)
18. Sett, N., Chattopadhyay, S., Singh, S.R., Nandi, S.: A time aware method for predicting dull nodes and links in evolving networks for data cleaning. In: IEEE/WIC/ACM International Conference on Web Intelligence, pp. 304–310 (2016)
19. Suchanek, F.M., Kasneci, G., Weikum, G.: Yago: a core of semantic knowledge. In: Proceedings of International Conference on World Wide Web, Banff, Canada, pp. 697–706 (2007). https://doi.org/10.1145/1242572.1242667
20. Trouillon, T., Welbl, J., Riedel, S., Gaussier, E., Bouchard, G.: Complex embeddings for simple link prediction. Int. Conf. Mach. Learn. (ICML) **48**, 2071–2080 (2016)
21. Vrandečić, D., Krötzsch, M.: Wikidata: a free collaborative knowledgebase. Commun. ACM **57**(10), 78–85 (2014). https://doi.org/10.1145/2629489
22. Yang, B., Yih, W., He, X., Gao, J., Deng, L.: Embedding entities and relations for learning and inference in knowledge bases. CoRR abs/1412.6575 (2014). http://arxiv.org/abs/1412.6575

Semantic Oppositeness Embedding Using an Autoencoder-Based Learning Model

Nisansa de Silva(✉) and Dejing Dou

Department of Computer and Information Science, University of Oregon, Eugene, USA
{nisansa,dou}@cs.uoregon.edu

Abstract. Semantic oppositeness is the natural counterpart of the much popular natural language processing concept, semantic similarity. Much like how semantic similarity is a measure of the degree to which two concepts are similar, semantic oppositeness yields the degree to which two concepts would oppose each other. This complementary nature has resulted in most applications and studies incorrectly assuming semantic oppositeness to be the inverse of semantic similarity. In other trivializations, "semantic oppositeness" is used interchangeably with "antonymy", which is as inaccurate as replacing semantic similarity with simple synonymy. These erroneous assumptions and over-simplifications exist due, mainly, to either lack of information, or the computational complexity of calculation of semantic oppositeness. The objective of this research is to prove that it is possible to extend the idea of word vector embedding to incorporate semantic oppositeness, so that an effective mapping of semantic oppositeness can be obtained in a given vector space. In the experiments we present in this paper, we show that our proposed method achieves a training accuracy of 97.91% and a test accuracy of 97.82%, proving the applicability of this method even in potentially highly sensitive applications and dispelling doubts of over-fitting. Further, this work also introduces a novel, unanchored vector embedding method and a novel, inductive transfer learning process.

Keywords: Semantic oppositeness · Autoencoder · Transfer learning · Unanchored Learning

1 Introduction

Semantic similarity measures are widely used in Natural Language Processing (NLP) applications [1–3]. The reason for this popularity is that NLP methods built around the simple exact matching approach would yield results with a weaker recall, in comparison with the gold standard. Free text has a tendency to use synonyms and similar text in substitution, which may go unnoticed if an exact match method is used in NLP. Semantic oppositeness is the natural counterpart of the semantic similarity function [4]. While semantic similarity

S. Hartmann et al. (Eds.): DEXA 2019, LNCS 11706, pp. 159–174, 2019.
https://doi.org/10.1007/978-3-030-27615-7_12

yields the degree to which two concepts are similar in a given domain, to be used in the purpose of confidence calculation in applications, semantic oppositeness yields the degree to which two concepts oppose each other in a given domain for the same purpose.

The use of an oppositeness measure in NLP is relevant in the case of contradiction finding, or in the case of extracting negative (negation) rules from a corpus. This, in turn, helps in building reasoning chains and other utilities for various front-end applications, such as question-answering systems and chat bots. It can also be used in the NLP applications which concern fake news or propaganda, given that the candidate text that is being analyzed would contain concepts opposing the generally accepted corpus of knowledge. In fact, one previous work on this domain [4] was an implementation of an oppositeness measure to discover contradictions from medical abstracts in the PubMed archive [5]. Some later considerations of this algorithm were in the legal domain [6] and in text classification [7].

Despite being both innovative and useful, the oppositeness calculation algorithm of the PubMed Study [4] is very computationally intensive. Therefore, in large tasks it would significantly slow down the process. It is in an attempt to avoid such computational complexity that most Natural Language Processing (NLP) tasks involve the incorrect generalization of reducing semantic oppositeness to antonymy [8] or inverse of semantic similarity. In the case of semantic similarity, this problem was overcome with the rise of the word embedding systems. Tasks which used to be a complex set of word similarity calculations [9,10] were reduced to simple K-NN look-ups in vector spaces [11,12]. The objective of this paper is to obtain such an embedding for semantic oppositeness, so that NLP applications that involve semantic oppositeness can become more efficient and cost effective. The proposed method first autoencodes word vectors, then it transfer learns the decode half of the deep neural network by using values obtained by the previous oppositeness algorithm [4] as the target.

In addition to the main research contribution of introducing an embedding for the semantic oppositeness function, in this paper we also introduce a novel, unanchored vector embedding approach and a novel, *inductive transfer learning* [13] process based on autoencoders [14], which utilizes both the learnt embeddings and the learnt latent representation.

2 Related Work

Even though not as extensively as its counterpart, semantic similarity [9,10], there have been a few studies on the derivation and uses of semantic oppositeness [4,8,15–17]. However, almost all of these studies reduce oppositeness from a scale to bipolar scales [16] or anonymity [8]. The study by [4] proves that the reduction of oppositeness to a binary function or defining it as the inverse of semantic similarity function is incorrect. The oppositeness function that we use in this study is heavily influenced by the alternative oppositeness function that is proposed in their study. The said study is an NLP application of the PubMed archive to find contradictions in medical research paper abstracts.

Word embedding has recently risen as an emerging field in the domain of NLP. The leading algorithms for this task are: Word2vec [11,18], GloVe [12], and Latent Dirichlet Allocation (LDA) [19]. In considering the flexibility, ease of customization, and wide usage, in this study we use word2vec as the starting point for our embedding system. Even though this study is focused on embedding oppositeness rather than embedding words, given that oppositeness is an emergent property between pairs of words, the points of embedding, in this study, remain as words. This is the reason it is possible to use word2vec as a reasonable starting point.

Autoencoders are one of the simplest forms of the representation learning algorithms. They consist of two components, an encoder and a decoder. While autoencoders are trained to preserve as much information as possible, special steps are taken to prevent them from learning the identity function [14]. Autoencoders are fairly common in contemporary research [20–22]. A study by [20] proved that stacked autoencoders can out-perform older models.

The proposed unanchored approach to word vector embedding, while being a novel idea introduced in this paper, has some similarity to studies by [23–25]. Transfer learning is a machine learning technique mainly employed when there is a task in a domain of interest with a scarcity of data, while another related domain exists with sufficient training data [13]. Given that the task employed in this work uses transfer learning in such a way that source and target domains are the same, while the source and target tasks are different but related, by the definition given by [13] it is possible to declare that this methodology is based on the principals of *inductive transfer learning*. In the NLP domain, transfer learning is commonly used for the task of document classification [26–28] and sentiment analysis [29]. The novelty in this paper pertaining to transfer learning on autoencoders is how the system is first trained as an autoencoder, and then how the transfer learning transforms it into a neural model, where the encoder toward latent space is kept intact, while the decoder is retrained into a mapper to a different vector space.

3 Methodology

The methodology of this work consists of two components. The first component is calculating the oppositeness value from the algorithm adapted from the work of [4]. For ease of reading, this method shall be referred to as the *original oppositeness measure (OOM)* in the remainder of this paper. The second component is embedding the oppositeness values obtained from the above step.

3.1 Linguistic Measures

The first component, which is needed to calculate the oppositeness between two given words, w_1 and w_2, in the algorithm proposed by the *OOM*, is the weighted semantic similarity. Among the various semantic similarity measures available as the *sim* function, the *OOM* picks the method proposed by [10] which gives the similarity between two words in the 0 to 1 range. For their reasoning of selecting of this method for semantic similarity over the other methods, they refer to their earlier work [30] in which they comprehensively compared various semantic similarity measures. We, in this paper, decided to follow the same progression and use the Wu and Palmer method as the semantic similarity measure for this work. However, it should be noted here that the *OOM* was intended for finding oppositeness between relationships in triples, whereas the work discussed in this paper is interested in embedding oppositeness between individual words. Thus the length constant values in equation for $simil_{w_1,w_2}$ and all subsequent equations would be trivially collapsed to carry the value 1. Therefore, in this study, $simil_{w_1,w_2}$ simplifies to [10]'s $sim(w_1, w_2)$. However, we discuss the possibility of extending this algorithm to incorporate embedding oppositeness between phrases in Sect. 5 by reintegrating the above constants.

For the *difference* component of the oppositeness calculation, it is needed to calculate the lemma of the given words and then obtain the antonym set of all the possible senses of the given word. The *OOM* does not provide a formal mathematical expression on this step; but we define L_w as $lemma(w)$ and A_w as $antonyms(w)$. The final relative difference equation we use is almost identical to the equation proposed by the *OOM*. However, a few alterations are made to accommodate the cumulative changes we have performed above. The final relative difference equation is shown in Eq. 1 where $P = \{(w_1, w_2), (w_2, w_1)\}$.

$$reldif_{w_1,w_2} = \underset{(i,j)\in P}{avg}\left[\underset{a_k \in A_i}{\arg\max}\left(sim(L_j, a_k)\right)\right] \tag{1}$$

Original Oppositeness Model. The *OOM* is built on the principle that the oppositeness value of two words that are highly similar should be more correlated with their difference value than oppositeness value of two words that are less similar. This property is obtained by the Eq. 2. To explain this property, they use an example where they calculate the oppositeness values of *expand*, *decrease*, *change*, and *cat* against the word *increase*. The expectation is that the high difference values of the words *change* and *cat* would be augmented by the similarity value in such a way that the relevant word *change* would be put in the correct place on the oppositeness scale (between *expand* and *decrease*), while the irrelevant word *cat* would be easily distinguishable for removal. We continue to use this same example set of words (or extensions thereof) in our subsequent comparisons, for ease of comparison and preservation of flow. The calculated values for similarity, difference, and finally the oppositeness by Eq. 2 are shown in Table 1. The power scaling constant K is determined by the instructions from *OOM*. Further, we use the same visualization structure as used by the *OOM* in Fig. 1(a). However, we add an overlay of the placement of the four example words (*expand*, *decrease*, *change*, and *cat*) in relation to the word *increase*, for the ease of explanation of the subsequent alterations and additions we perform upon the basic algorithm.

$$oppo\_ori_{w_1,w_2} = reldif_{w_1,w_2}^{\left(K * simil_{w_1,w_2} + 1\right)} \tag{2}$$

Table 1. Oppositness with $w_1 = \textit{increase}$

	expand	*decrease*	*change*	*cat*
$simil_{w1,w2}$	0.80	0.75	0.46	0.25
$reldif_{w1,w2}$	0.63	1.00	0.72	0.25
$oppo\_ori_{w1,w2}$	0.25	1.00	0.49	0.11

The main weakness of this original model is the heavy dependence on the antonym property of the candidate words to calculate the difference. This weakness did not affect the performance of the application discussed by the *OOM*, because they were comparing the oppositeness between the relationship component extracted from triples from medical abstracts. In that application, the relationship component always returns one or more action verbs. Coupled with the fact that they are comparing relationship strings which contain more than one word, most of the instances translate to a high probability of encountering words with antonyms. But in the application of this study, not only should the algorithm handle single word instances, it also has to handle the possibility of that word not having an antonym. In such cases where one or both considered words do not have antonyms, the difference value calculated by Eq. 1 collapses to zero. This in turn further collapses the final oppositeness value calculated by Eq. 2 to zero; thus effectively rendering the particular data point obtained by the word pair in question, unusable.

Proposed Balancing Model Derived from the Naïve Oppositeness Measure. In the attempt to solve the problem of incalculable difference values, we turn to the naïve oppositeness measure that was replaced by the oppositeness measure proposed by the *OOM*. This naïve oppositeness measure is the simple operation of declaring the complement of similarity as oppositeness. The Eq. 3 shows the definition of this measure. The Table 2 contains a comparative analysis of the above mentioned oppositeness measure and the naïve method. It is obvious from this data that the naïve method on its own does not achieve the desired properties of an oppositeness scale as stipulated in the *OOM*. While it is obvious that the naïve oppositeness measure is independent of the difference measure, we plot it on the same axis as the above oppositeness measure for the sake of comparison in Fig. 1(b). However, at this point it should be noted that this model does not suffer from the weakness to words without antonyms that impacts the improved model proposed by the *OOM*.

$$oppo\_nai_{w_1,w_2} = (1 - simil_{w_1,w_2}) \tag{3}$$

Combined Oppositeness Model. We have proposed that the solution to the weakness of the *OOM* is to augment it with the naïve model discussed in Sect. 3.1. However, given that the original oppositeness model is far superior to

Table 2. Oppositness with $w_1 = increase$

	expand	*decrease*	*change*	*cat*
$oppo\_ori_{w_1,w_2}$	0.25	1.00	0.49	0.11
$oppo\_nai_{w_1,w_2}$	0.20	0.25	0.54	0.75

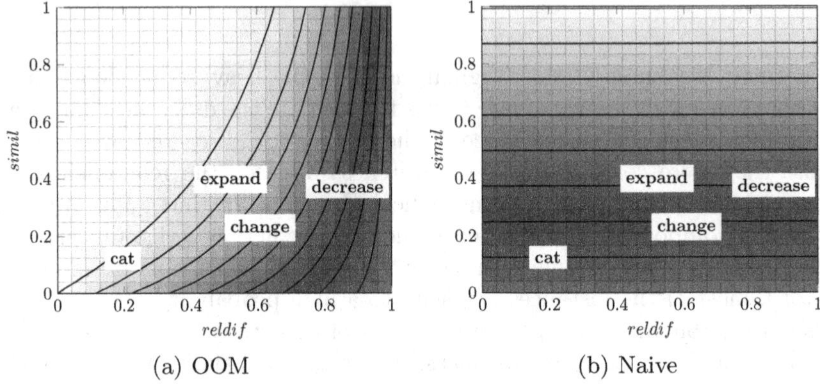

(a) OOM (b) Naive

Fig. 1. Contour plots

the naïve model, and that the original model is weak only at the specific instance where the difference measure is valued zero, it is vital that the two models are combined in a way that the naïve model would only take over at points where the original oppositeness model is weak. We achieved this by multiplying the naive oppositeness function with the term $(1 - reldif)$. Note here that there is no need to further multiply OOM with $reldif$, given that it is already positively correlated with $reldif$. To further fine tune the balance between the OOM and the naïve oppositeness measure, we introduced a hyper parameter α. The final combined oppositeness measure is shown in Eq. 4. Finally, we show the subtle alteration brought about by this improvement in the familiar visualization in Fig. 2. In the example visualization, we have set α to 0.9. While it was needed to set α to this value for the purpose of showing a difference in the graphs discernible to the human eye, in practice, it was observed that α value should be kept at 0.99 or higher, to prevent the naïve oppositeness measure from negatively affecting the overall calculation at points where the difference value is greater than zero. At this point it should be noted that the reason for employing this continuous method to aggregate the two methods, rather than using a case-based approach, where the naïve oppositeness measure is only used at points where the difference measure is zero, is to make-sure that the active surface of the oppositeness curve would be continuous and smooth at all points. Further, note the slight curvature present in Fig. 2(b) in comparison with Fig. 1(a), due to this addition.

$$oppo(w_1, w_2) = \alpha * oppo\_ori + (1 - \alpha)(1 - reldif) * oppo\_nai \tag{4}$$

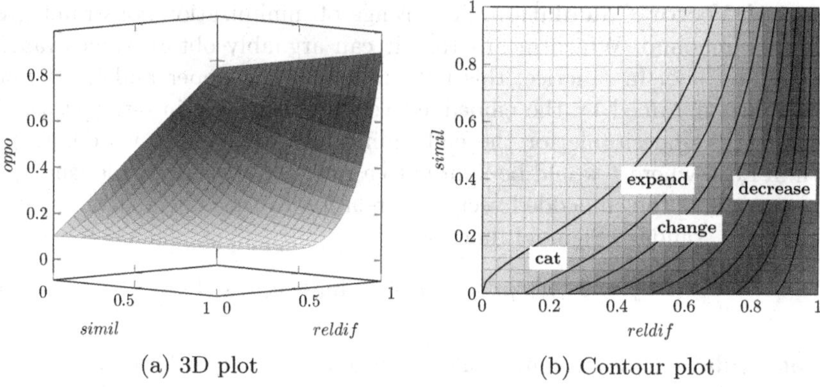

(a) 3D plot (b) Contour plot

Fig. 2. Oppositeness function with $\alpha = 0.9$

3.2 Embedding Semantic Oppositeness

Once the algorithm described in Sect. 3.1 is used to calculate the oppositeness measures for word pairs, the next step of the process is to embed them in a vector space. Embedding the oppositeness gives applications the ability to do simple K-NN queries on the vector space, instead of running the costly algorithm *OOM* each time. This section discusses the process of embedding the said oppositeness values in a vector space.

Minimization Constraint. In an embedding process it is important to first define what the minimization constraint is. In almost all cases it is defined as a function to calculate the distance between the vector currently obtained by the embedded object and the vector expected to be obtained by the embedded object. However, in this study our objective is novel in the sense that for this algorithm, it does not matter where the individual word vectors map to. All that matters is the difference between given two embedded word vectors approaching the oppositeness value calculated above. Therefore, in the learning process, instead of anchoring a vector (or a context) and trying to move the target vector close to it, we can employ an algorithm to push both vectors together with no contextual attachments. Thus the minimization constraint becomes a matching of two distance scalars, rather than minimizing the distance between two vectors. The proposed minimization constraint is given in Eq. 5, where $||a - b||$ denotes the Euclidean distance between vectors a and b. Note that we need to preserve the sign of the difference to use the unanchored training. Thus, the absolute value function (abs) is deconstructed into three cases in later steps.

$$min\Big[abs\Big(oppo(w_t, w_s) - ||y_t - y_s||\Big)\Big] \quad (5)$$

Expected Vector Calculation. The range of minimization constraint given in Eq. 5 is unbound. Which means that it can arguably obtain values ranging from $-\infty$ to $+\infty$. In practice, this is bounded by the upper and lower limits of the values obtained by the embedded vectors. However, in either case, this large range is undesirable for the embedding task. Therefore, we define f as shown in Eq. 6 where f would be limited to a range of $+1$ and -1, depending on the placement of the embedded vectors in relation to the expected value, and σ indicates the standard Sigmoid function.

$$f = 2 * \sigma\big[oppo(w_t, w_s) - ||y_t - y_s||\big] - 1 \tag{6}$$

Target Update Rule. As mentioned in Sect. 3.2, the embedding in this study does not confirm to the idea of anchoring one vector (or context) and pushing the candidate to match the expected vector. Instead, both the vectors in question are moved to make sure the oppositeness is defined by the distance between the said vectors are embedded. It should be noted that to the best of our knowledge, this study is the first to utilize such an unanchored approach to word vector embedding. In this section we derive the update rule for each of the two vectors. For the simplicity of subsequent calculations, we define ΔY as $y_t - y_s$, while the expected shifts are $y_{t'} - y_t$ and $y_{s'} - y_s$. There are three possible cases of vector placement, as shown by Fig. 3. Each of these cases are uniquely identifiable by the f value calculated by Eq. 6. The Table 3 summarizes all update cases.

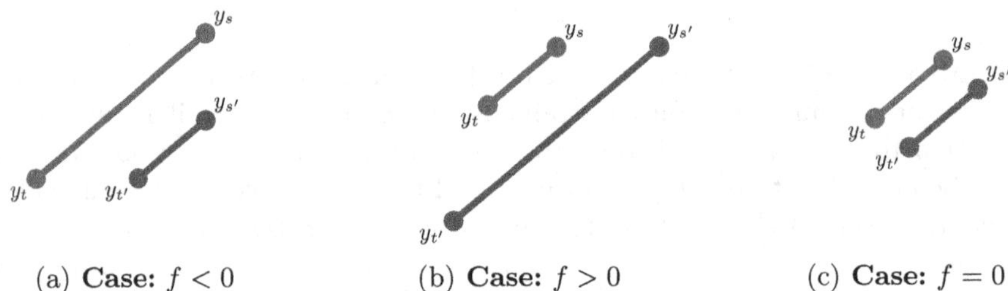

(a) **Case:** $f < 0$ (b) **Case:** $f > 0$ (c) **Case:** $f = 0$

Fig. 3. Possible cases of vector placement

Table 3. Case-based update rules

	$f < 0$	$f > 0$	$f = 0$
$y_{t'} - y_t$	$-\eta_1 \Delta Y$	$+\eta_2 \Delta Y$	0
$y_{s'} - y_s$	$+\eta_1 \Delta Y$	$-\eta_2 \Delta Y$	0

Finally, it is possible to combine all the above embedding target update rules together, based on the fact that they are uniquely mapped to the value of f, as shown in Eqs. 7 and 8.

$$y_{t'} = y_t + f\eta \Delta Y \tag{7}$$

$$y_{s'} = y_s - f\eta \Delta Y \tag{8}$$

Autoencoder-Based Learning. While the above algorithm is sound as a solution to embed words in a vector space guided by the oppositeness values, starting with fully empty or fully randomized word vectors would be counterproductive. In such an approach, our system will implicitly have to learn the word embeddings that are achieved by word embedding systems such as word2vec [11] or GloVe [12]. Further, there is the initial hurdle of declaring the input (x_s, x_t) in an unambiguous manner. The solution to both of these problems is to involve an already trained word embedding model as the starting point. In this study we decided to use word2vec. This solves the second problem outright. The declaration of input (x_s, x_t) in an unambiguous manner is now a simple matter of querying the trained Word2vec model with the expectant word string.

The second step is not so straightforward. The objective of this step is to utilize the already existing embedding of words in word2vec to make the oppositeness embedding faster. The rationale here is the fact that word2vec already clusters words by similarity and thus following the naïve method we discussed in above sections, it is reasonable to predict that the oppositeness embedding would be comparatively closer to achieve when starting from a similarity embedding than by a random or a zero embedding. Here, note the fact that the naïve assumption was to assume that the similarity embedding trivially translates to the oppositeness embedding. We do not confirm to that naïve assumption. We only claim that the similarity embedding would be reasonably closer to the expected oppositeness embedding rather than a zero or random starting point. Therefore we propose the novel idea of applying transfer learning [13] on the decoder portion of the autoencoder. The learning model proposed here is shown in Fig. 4.

First of all, to employ the proposed model, we should obtain a mapping from words to vectors. Among the various algorithms and models available to map words to vectors such as Word2Vec [11] and GloVe [12], we propose to use Word2Vec based on the wider support (especially the availability of large Google-trained data set[1]). This component of the ensemble would map a given word to a vector. Incidentally, this would become the input of our neural network.

The proposed model has two learning phases. The first phase of the proposed model is called the *Autoencoding Phase*. In this, we keep the word2vec model locked and the weights of the encoder and the decoder unlocked. The formal representation of an autoencoder is given in Eq. 9 where, the section $\sigma_1(W_1 x_i + b_1)$ correlates to the *encode*(X) function where the W_1 and b_1 are weights and biases of the *encode* function. The $\sigma_1'(W_1' l) + b_1'$ portion, where l represents $\sigma_1(W_1 x_i + b_1)$ discussed above, correlates to the *decode*(l) function where the W_1' and b_1' are weights and biases of the *decode* function and l is a

[1] https://goo.gl/yV57W3.

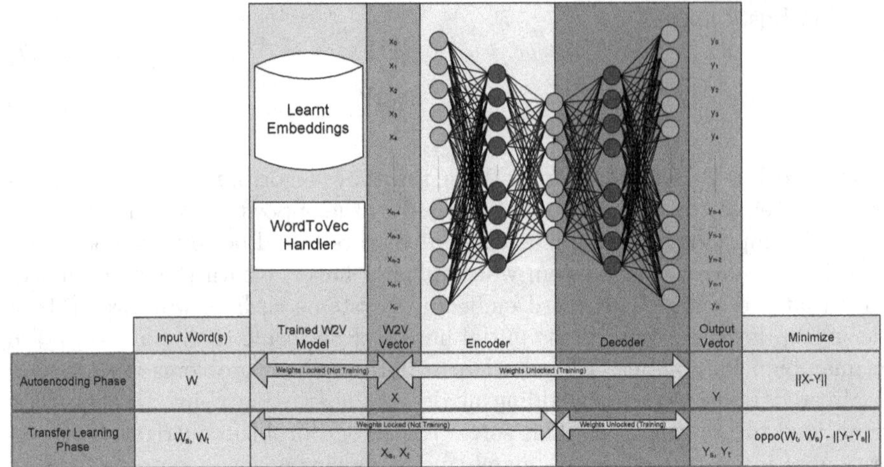

Fig. 4. Overall autoencoder-based learning model

latent representation output by the $encode(X)$ function. The learning objective of the autoencoder is to $minimize(||X - Y||)$ where X is the input vector of the encoder and Y is the output vector of the decoder. Note here that in literature, Y is commonly refereed as X' to showcase the fact that it is supposed to be a reconstruction of the original X. However, in this study we opted to use Y for the sake of clarity of the subsequent steps where we use transfer learning instead of reconstruction (autoencoding).

$$Y_i = (\sigma_1'(W_1'\sigma_1(W_1X_i + b_1)) + b_1') \tag{9}$$

In summary, during the *Autoencoding Phase* of the proposed model, the neural network learns to reconstruct a given word vector. As mentioned above, this is an attempt to utilize the learnt artifacts of a word embedding system, where related words are clustered together while unrelated words are embedded far apart.

The second phase of the proposed model is the *Transfer Learning Phase*. The transfer learning proposed in this work differs from prior work in literature for three facts. Firstly, the transfer process done on the autoencoder is not done to train yet another model, but to map the same inputs to a different vector space. Thereby, at the end of the training process, the trained neural network is *not* an autoencoder. However, for the sake of readability of the paper, we would continue to refer to the two components of the neural network as *encoder* and *decoder*. It is imperative that after the training, the output of the *decoder* no longer tries to reconstruct the input to the *encoder*. However, given the linguistic properties, that vector will still be *reasonably close* to the input vector. The second difference is the fact that we lock the weights of the *encoder* along with the word2vec model in the training process of this phase. While it is a given property of transfer learning applications to lock a certain number of initial layers and train only a certain number of layers close to the output layer, usually, that choice is open-ended and unrestricted. In this study however, we specifically lock all the layers that were previously in the *encoder* and keep the layers that were previously in the *decoder* unlocked. The rationale for this decision is as follows: the autoencoder has already learnt a latent representation of the word vectors, by using the autoencoding process, where the output is the input itself. Therefore, we can be sure of the accuracy of the learnt latent representation. The latent representation of a given vector need not be altered when the application is changed. For the best of our knowledge, this study is the first to propose this autoencoder-based *inductive transfer learning* [13] process to utilize both learnt embeddings and the learnt latent representation. The third aspect that distinguishes this model from the traditional learning processes is the fact that this phase uses two forward passes to calculate the error for a single back-propagation pass. This is due to the fact that, as discussed in Sect. 3.2, the learning objective of the neural network is to achieve the minimization proposed in Eq. 5.

4 Experiments and Results

4.1 Calculating the Oppositeness Data

For the purpose of obtaining an adequate collection of words for experimentation, we used the list available in the Linux dictionary[2]. The dictionary contained 72186 total strings. However, it was observed that a certain portion of the strings were non-words. Further, for the sake of preserving the variety of the sample set, it was decided to replace words by their lemmas in cases where there are multiple morphological forms. This process was achieved by passing the potential word strings through the WordNet [31] lemmatizer. This yielded a reduced word list of 65167.

Next the methodology discussed in Sect. 3.1 was used on the 65167 words taken as pairs. Given that each word was considered against all other words, this resulted in 4246737889 pairs of words. For each pair of words, the *similarity*, *difference*, and *oppositeness* were calculated. A few sample lines from the *increase* file are shown in Example 1.1. Note the minimal value of the *similarity* slot for *advents*. This implies that the similarity measure could not give a similarity value to the pair (i.e., the pair is disjoint).

Example 1.1. Sample Oppositeness Lines

```
increase ,adrian:0.125 ,0.1 ,0.0313516
increase ,adriatic:0.125 ,0.1 ,0.0313516
increase ,advent:0.1875 ,0.2777778 ,0.088623986
increase ,advents:1.4E−45 ,1.4E−45 ,0.01
```

The above files were then processed by applying the irrelevancy threshold proposed by *OOM*. Thus, for each word, all pairs with oppositeness value *less than* the oppositeness value of the most similar pair were eliminated. After this

[2] /usr/share/dict/words.

reduction step, only 76084553 pairs out of the original 4246737889 were left. This means, on average, each word contained 1168 pairs after this step, showing a reduction of 98.21%. As an example, the corresponding file to the word *increase* has only 1107 lines. This is a significant reduction in the case of potential computational load. Given that the file format stays the same as shown in Example 1.1, we do not provide a separate example here.

4.2 Autoencoding on Word2Vec Data

We used a TensorFlow [32, 33] based implementation of the two-layer autoencoder proposed by [34] for the purpose of training on the MNIST data set [35]. The input layer was altered to have the size of 300 to match the trained model of Google's word2vec embedding. The middle layer was of size 256, and the latent layer was of 128. By definition, the decoder layer had mirrored counts. Following the precedence set by [34] for this particular configuration of auto-encoder, we found 30000 epochs to balance accuracy against the threat of over-fitting. By employing the multiple random restart method, we obtained a trained autoencoder with a validation accuracy of 94.83%. This is the model that we used for the next step of transfer learning.

4.3 Learning of Oppositeness Data

Here it was decided to use a 3 : 2 split for the training-validation set vs. test set for the transfer learning of oppositeness data. It is reasonable to assume that the word2vec vector and the expected oppositeness vector will be close in the vector space. Thus, transferring the decoder weights to the system gives a more efficient starting point, compared to initiating with zero weights or random weights.

The training-validation set was divided to 30 equal parts, and each were used to train a clone of the implementation. Here inverse-n-fold cross-validation was used where for each clone, a single portion of data is used as the training data, and then the remaining $n-1$ portions of data are used for validation. We report in Fig. 5 the accuracies of the separate clones resulting from the cross-validation process. The Y axis (rows) of the Matrix corresponds to each transfer learning clone and the x axis (columns) corresponds to the portion of data set. Hence, the 30 entries on the diagonal of the matrix correspond to the training accuracies, and the 870 entries on the remainder of the matrix correspond to the validation accuracies. It is observable that while the diagonal is sightly distinguishable, some clones seem to be performing better than the others across the board. We claim that this is because of the linguistic property that some words are more central in a lexicon than others. These words might distinguish themselves by having more synonyms or by having polysemy. When the data set given to a clone has a majority of such words, it is possible to claim that the trained model would generalize better than in the case where the data set given to a clone has a minority of such words.

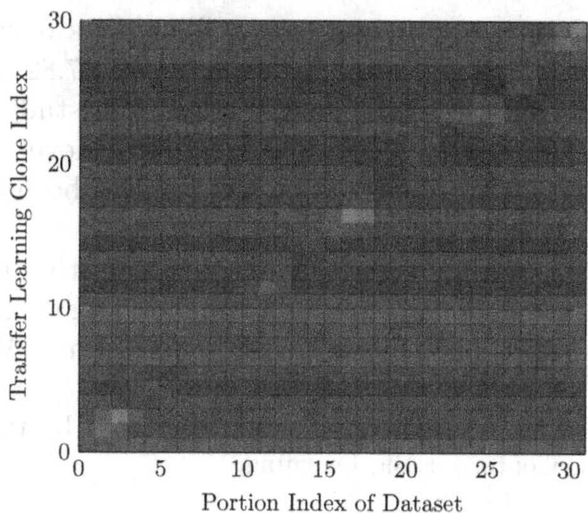

Fig. 5. Training/Validation Matrix of the Clones

Finally, the relevant vectors from the 30 separate embeddings were averaged together to produce the final singular embedding. For that combined singular embedding, on the complete 14820 training-word set (i.e., all 30 portions together), we obtained a mean training accuracy of **97.91%** with a standard deviation of 0.38. The reason for this re-calculation is the fact that we anticipated that since we merged the trained models, the performance of the merged model would not be the average of the separate components. The observable change in accuracy is proof that the said assumption is justified. Also, it is possible to note here that the above rigorous validation has cleared, in the case of the merged model, any possibility of over-fitting which may have threatened individual transfer learning clones. Following that, on the 9910 test words, we obtained a mean test accuracy of **97.82%** with a standard deviation of 0.43. Yet again, we present the closeness of the training accuracy and test accuracy as proof that the system has not over-fitted to the data despite obtaining very good training accuracy which is higher than 97%. Note that all accuracy values at this point are calculated by taking *OOM* output as the gold standard.

The embedding enables practical NLP applications to do a simple k-NN look-up as opposed to the slow and costly oppositeness calculation from scratch. Further, the algorithm by [4] completely fails if even one of the words does not have antonyms. The embedding approach proposed in this study works for all word pairs and gives the closest approximation. Therefore, we claim that this approach provides a more constantly reliable source of oppositeness look-up for practical NLP applications.

5 Conclusion

The main research contribution of this study was the introduction of semantic oppositeness embedding. This study successfully proposed and demonstrated an embedding methodology on more than 49 million pairs of words, to obtain a training accuracy of 97.91% and a testing accuracy of 97.82%.

In addition to this main research contribution, this study also introduced a novel, unanchored vector-embedding approach and a novel, *inductive transfer learning* process based on auto encoders which utilizes both learnt embeddings and the learnt latent representation.

As for future work, we propose extending this algorithm to embed semantic oppositeness of phrases, similar to the phrase embedding extensions done to the implementations of other word embedding systems, such as Word2Vec.

Acknowledgement. This research is partially supported by the NSF grant CNS-1747798 to the IUCRC Center for Big Learning.

References

1. Gomaa, W.H., Fahmy, A.A.: A survey of text similarity approaches. Int. J. Comput. Appl. **68**(13), 13–18 (2013)
2. Stavrianou, A., Andritsos, P., Nicoloyannis, N.: Overview and semantic issues of text mining. ACM Sigmod Rec. **36**(3), 23–34 (2007)
3. Turney, P.D.: Mining the web for synonyms: PMI-IR versus LSA on TOEFL. In: De Raedt, L., Flach, P. (eds.) ECML 2001. LNCS (LNAI), vol. 2167, pp. 491–502. Springer, Heidelberg (2001). https://doi.org/10.1007/3-540-44795-4_42
4. de Silva, N., Dou, D., Huang, J.: Discovering inconsistencies in PubMed abstracts through ontology-based information extraction. In: Proceedings of the 8th ACM International Conference on Bioinformatics, Computational Biology, and Health Informatics, pp. 362–371. ACM (2017)
5. National Center for Biotechnology Information: PubMed Help, March 2017
6. Ratnayaka, G., Rupasinghe, T., de Silva, N., Gamage, V.S., Warushavithana, M., Perera, A.S.: Shift-of-perspective identification within legal cases. In: Proceedings of the 3rd Workshop on Automated Detection, Extraction and Analysis of Semantic Information in Legal Texts (2019)
7. de Silva, N.: Sinhala Text Classification: Observations from the Perspective of a Resource Poor Language (2019)
8. Paradis, M., Goldblum, M.C., Abidi, R.: Alternate antagonism with paradoxical translation behavior in two bilingual aphasic patients. Brain Lang. **15**(1), 55–69 (1982)
9. Jiang, J.J., Conrath, D.W.: Semantic similarity based on corpus statistics and lexical taxonomy. In: 10th International Conference on Research in Computational Linguistics, ROCLING 1997 (1997)
10. Wu, Z., Palmer, M.: Verbs semantics and lexical selection. In: Proceedings of the 32nd Annual Meeting on Association for Computational Linguistics. ACL 1994, pp. 133–138. Association for Computational Linguistics, Stroudsburg (1994)
11. Mikolov, T., Sutskever, I., et al.: Efficient estimation of word representations in vector space. arXiv preprint arXiv:1301.3781 (2013)
12. Pennington, J., Socher, R., Manning, C.D.: Glove: global vectors for word representation. In: EMNLP, vol. 14, pp. 1532–1543 (2014)
13. Pan, S.J., Yang, Q.: A survey on transfer learning. IEEE Trans. Knowl. Data Eng. **22**(10), 1345–1359 (2010)

14. Goodfellow, I., Bengio, Y., Courville, A.: Deep Learning. MIT Press (2016). http://www.deeplearningbook.org
15. Mettinger, A.: Aspects of Semantic Opposition in English. Oxford University Press, New York (1994)
16. Schimmack, U.: Pleasure, displeasure, and mixed feelings: are semantic opposites mutually exclusive? Cogn. Emotion **15**(1), 81–97 (2001)
17. Rothman, L., Parker, M.: Just-about-right (jar) Scales. ASTM International, West Conshohocken (2009)
18. Mikolov, T., Sutskever, I., et al.: Distributed representations of words and phrases and their compositionality. In: Advances in Neural Information Processing Systems, pp. 3111–3119 (2013)
19. Das, R., Zaheer, M., Dyer, C.: Gaussian LDA for topic models with word embeddings. In: ACL, vol. 1, pp. 795–804 (2015)
20. Lv, Y., Duan, Y., et al.: Traffic flow prediction with big data: a deep learning approach. IEEE Trans. Intell. Transp. Syst. **16**(2), 865–873 (2015)
21. Alsheikh, M.A., Niyato, D., et al.: Mobile big data analytics using deep learning and apache spark. IEEE Network **30**(3), 22–29 (2016)
22. Hinton, G.E., Salakhutdinov, R.R.: Reducing the dimensionality of data with neural networks. Science **313**(5786), 504–507 (2006)
23. Hinton, G.E., Roweis, S.T.: Stochastic neighbor embedding. In: Advances in Neural Information Processing Systems, pp. 857–864 (2003)
24. Ono, M., Miwa, M., Sasaki, Y.: Word embedding-based antonym detection using thesauri and distributional information. In: Proceedings of the 2015 Conference of the North American Chapter of the Association for Computational Linguistics: Human Language Technologies, pp. 984–989 (2015)
25. Chen, Z., Lin, W., et al.: Revisiting word embedding for contrasting meaning. In: Proceedings of the 53rd Annual Meeting of the Association for Computational Linguistics, vol. 1, pp. 106–115 (2015)
26. Fung, G.P.C., Yu, J.X., et al.: Text classification without negative examples revisit. IEEE Trans. Knowl. Data Eng. **18**(1), 6–20 (2006)
27. Al-Mubaid, H., Umair, S.A.: A new text categorization technique using distributional clustering and learning logic. IEEE Trans. Knowl. Data Eng. **18**(9), 1156–1165 (2006)
28. Sarinnapakorn, K., Kubat, M.: Combining subclassifiers in text categorization: a DST-based solution and a case study. IEEE Trans. Knowl. Data Eng. **19**(12), 1638–1651 (2007)
29. Blitzer, J., Dredze, M., Pereira, F.: Biographies, bollywood, boom-boxes and blenders: domain adaptation for sentiment classification. In: Proceedings of the 45th Annual Meeting of the Association of Computational Linguistics, pp. 440–447 (2007)
30. de Silva, N.H.N.D.: SAFS3 algorithm: frequency statistic and semantic similarity based semantic classification use case. In: 2015 Fifteenth International Conference on Proceedings of Advances in ICT for Emerging Regions (ICTer), pp. 77–83. IEEE (2015)
31. Miller, G.A., Beckwith, R., et al.: Introduction to wordnet: an on-line lexical database. Int. J. Lexicography **3**(4), 235–244 (1990)
32. Abadi, M., Barham, P., et al.: Tensorflow: a system for large-scale machine learning. In: OSDI, vol. 16, pp. 265–283 (2016)
33. Abadi, M., Agarwal, A., et al.: TensorFlow: Large-scale machine learning on heterogeneous systems (2015). tensorflow.org

34. Damien, A.: Auto-Encoder Example. https://goo.gl/wiBspX (2017). Accessed 06 June 2018
35. LeCun, Y., Bottou, L., et al.: Gradient-based learning applied to document recognition. Proc. IEEE **86**(11), 2278–2324 (1998)

COMET: A Contextualized Molecule-Based Matching Technique

Mayesha Tasnim[1,3], Diego Collarana[2,3(✉)], Damien Graux[3], Mikhail Galkin[3], and Maria-Esther Vidal[4,5,6]

[1] RWTH Aachen, Aachen, Germany
mayesha.tasnim@rwth-aachen.de
[2] University of Bonn, Bonn, Germany
collaran@cs.uni-bonn.de
[3] Fraunhofer Institute for Intelligent Analysis and Information Systems, Sankt Augustin, Germany
{damien.graux,mikhail.galkin}@iais.fraunhofer.de
[4] TIB Leibniz Information Centre for Science and Technology, Hannover, Germany
maria.vidal@tib.eu
[5] Universidad Simón Bolívar, Caracas, Venezuela
[6] L3S Research Centre, Leibniz University of Hannover, Hannover, Germany

Abstract. Context-specific description of entities –expressed in RDF– poses challenges during data-driven tasks, e.g., data integration, and context-aware entity matching represents a building-block for these tasks. However, existing approaches only consider inter-schema mapping of data sources, and are not able to manage several contexts during entity matching. We devise COMET, an entity matching technique that relies on both the knowledge stated in RDF vocabularies and context-based similarity metrics to match *contextually equivalent* entities. COMET executes a novel 1-1 perfect matching algorithm for matching contextually equivalent entities based on the combined scores of semantic similarity and context similarity. COMET employs the Formal Concept Analysis algorithm in order to compute the context similarity of RDF entities. We empirically evaluate the performance of COMET on a testbed from DBpedia. The experimental results suggest that COMET is able to accurately match equivalent RDF graphs in a context-dependent manner.

Keywords: Data integration · RDF knowledge graphs · RDF entities

1 Introduction

The semi-structuredness nature of the RDF data model allows for naturally representing entities of the same type with different properties, as well as for the encoding of multiple contexts of an entity within the same graph. This feature of

This research was supported by the European project QualiChain (number 822404).

S. Hartmann et al. (Eds.): DEXA 2019, LNCS 11706, pp. 175–185, 2019.
https://doi.org/10.1007/978-3-030-27615-7_13

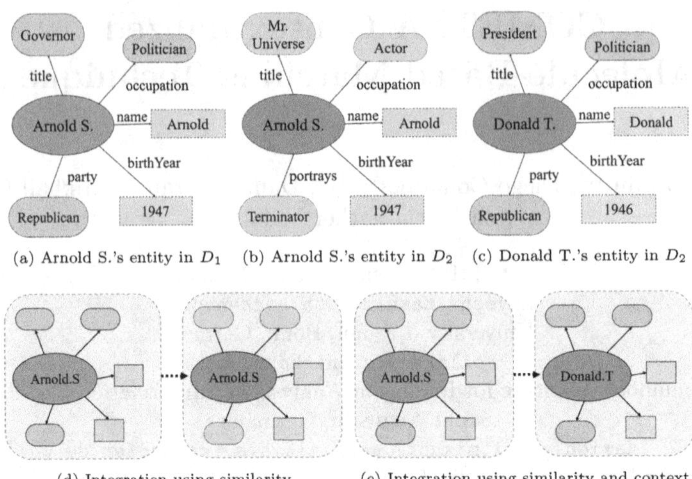

(a) Arnold S.'s entity in D_1 (b) Arnold S.'s entity in D_2 (c) Donald T.'s entity in D_2

(d) Integration using similarity (e) Integration using similarity and context

Fig. 1. Motivation Example. Top row shows three entities across two datasets. The bottom row shows two matching scenarios, the left one not considering context during entity matching, and the right one taking context into consideration.

RDF is of paramount importance for addressing the data complexity challenge of variety—a dominant dimension of data in the Big Data era [6]. Nevertheless, enabling diverse representations of the same entity poses new challenges during the matching of RDF graphs, particularly, because specific contexts need to be considered for an effective identification of similar entities [2]. To perform a contextually RDF entity matching, the specific context under which a matching is being performed must be taken into account.

We motivate our work using an example of a context-based entity matching scenario using RDF entities representing persons. *Arnold Schwarzenegger* is a person with an extensive career in both politics and acting. Consequently, there is data available regarding both his career in politics and his achievements in the movie industry. Consider an integration system aiming to match data about American politics. Dataset D_1 contains information about Arnold Schwarzenegger and his political career. In another dataset D_2, there exists information about Arnold's acting career, e.g., the movies he has acted in and the roles he has portrayed. The same dataset D_2 also contains information about other celebrities, like Donald Trump, President of the United States. These entities are presented in Fig. 1a, b and c, respectively. In a typical integration scenario where context is not considered, entities are matched to the ones that are most similar to them. In such a case, Arnold Schwarzenegger's entity from D_1 will be matched with the entity in D_2 containing information about his acting career, as shown in Fig. 1d. However, in the context of politics, Arnold's political career is more similar to Donald Trump's than his own career in acting. They are politicians of almost the same age who are both supporting the Republican party. In a political context, their careers are far more similar than when Arnold's post as the Governor of

California is compared with his portrayal of the Terminator in Terminator 2. Therefore, when context of American politics is considered, the entity of Arnold S. from D_1 should be matched with Donald T. entity from D_2.

To support scenarios where the inclusion of *context* in entity matching makes results more relevant for real-world scenarios, we present the **CO**ntextualized **M**olecul**E**-based matching **T**echnique (COMET). COMET is a context-aware entity matching method that employs a two-fold approach for both: (1) identifying contextually similar entities, and (2) matching them into a 1-1 set of entities. We present the results of an empirical evaluation that illustrate the benefits of the techniques implemented in COMET.

The remainder of the paper is structured as follows: Sect. 2 summarizes the related work and compares COMET with the state of the art. Then, we define the COMET approach in terms of its main characteristics and architecture in Sect. 3. A comprehensive evaluation of the COMET approach is reported in Sect. 4. Finally, Sect. 5 summarizes our main conclusions and future work.

2 Related Work

The problem of entity matching between RDF graphs has been extensively treated by the Semantic Web community. As a result, a vast amount of approaches and frameworks have been developed. In the task of identifying whether the given entities refer to the same real-world entity, growing attention in the relational databases field is given to crowdsourcing mechanisms [11]. Reporting impressive results, such approaches, however, might struggle in sophisticated domains with multiple contexts due to a lack of human experts who could reliably provide necessary example or training data.

Entity matching is particularly important in data integration scenarios and integration frameworks tackle this problem. Knoblock et al. [7] propose KARMA, a framework for integrating a variety of data sources including databases, spreadsheets, XML, JSON, and Web APIs. KARMA implements a hybrid approach that relies on supervised machine algorithms for identifying mapping rules from structured sources to ontologies; these mapping rules should solve entity matching problems. Schultz et al. [10] describe the Linked Data Integration Framework (LDIF). LDIF is oriented to integrate RDF datasets from the Web and provides a set of independent tools to support interlinking tasks. LDIF provides an expressive mapping language for translating data from various vocabularies to a unified ontology. LDIF tackles the problem of identity resolution by defining linking rules using the SILK tool [5]. Based on the defined rules, SILK identifies `owl:sameAs` links among entities of two datasets. ODCleanStore [8] relies as well on SILK to perform instance matching and provides a custom data fusion modules to merge the data of the discovered matches. Other efforts to produce a unified view from RDF graphs are often combined with federated SPARQL query engines, e.g., ANAPSID [1], MULDER [4]. Albeit effective in query planning, such engines process raw tuples coming from endpoints and therefore employ only basic entity matching of those raw tuples according to join operators.

The above-mentioned entity matching and integration approaches aim at mapping different data sources with possibly varying schema, i.e., they perform inter-schema mapping. Context-based entity matching could only be supported on a superficial level via filtering query results without applying much of inherent semantics. Contrary, COMET considers diverse criteria during entity matching, e.g., entity similarity and contextual knowledge. In consequence, COMET provides the building blocks for context-based semantic integration mechanisms.

3 The COMET Approach

Grounded on the entity matching component from the data integration technique proposed by Collarana et al. [3], we propose COMET, an entity matching approach to identify and synthesize contextually equivalent RDF entities. Thus, a solution to the *problem of contextually matching entities* is provided.

Problem Definition

RDF Molecule [3] – If $\Phi(G)$ is a given RDF Graph, we define RDF Molecule M as a subgraph of $\Phi(G)$ such that,

$$M = \{t_1, \ldots, t_n\} \quad \forall i, j \in \{1, \ldots, n\}\big(subject(t_i) = subject(t_j)\big)$$

Where $t_1, t_2, \ldots, t_n$ denote the triples in M. In other words, an RDF Molecule M consists of triples which have the same subject. That is, it can be represented by a tuple $M = (R, T)$, where R denotes the URI of the molecule's subject, and T denotes a set of property and value pairs $p = (prop, val)$ such that the triple *(R, prop, val)* belongs to *M*. For example, the RDF molecule for *Arnold Schwarzenegger* is (dbr:Arnold-Schwarzenegger, {(`dbo:occupation`, *Politician*), (`dbp:title`, *Governor*)}). An RDF Graph $\Phi(G)$ described in terms of RDF molecules is defined as follows:

$$\Phi(G) = \{M = (R, T) | t = (R, prop, val) \in G \wedge (prop, val) \in T\}$$

Context – We define a context C as any Boolean expression which represents the criteria of a system. Two entities, such as an RDF molecule M_1 and M_2, can be either similar or not similar with respect to a given context. That is, C is a Boolean function that takes as input two molecules M_1 and M_2 and returns `true` if they are similar according to system context, and `false` otherwise. Below is an example of context C, modeled after the example presented in Fig. 1, where two molecules are similar if they have the same occupation. If $P = (p, v)$ is the predicate representing the occupation property of a molecule, then context

$$C(M_1, M_2) = \begin{cases} \texttt{true}, & \text{if } P \in M_1 \wedge P \in M_2. \\ \texttt{false}, & \text{otherwise.} \end{cases}$$

Depending on the requirements of the integration scenario, this context can be any Boolean expression.

Semantic Similarity Function – Let M_1 and M_2 be any two RDF molecules. Then *semantic similarity function* Sim_f is a function that measures the *semantic similarity* between these two molecules and returns a value between [0, 1]. A 0 expresses that the two molecules are completely dissimilar and 1 expresses that the molecules are identical. Such a similarity function is defined in GADES [9].

Contextually Equivalent RDF Molecule – Let $\Phi(G)$ and $\Phi(D)$ be two sets of RDF molecules. Let M_G and M_D be two RDF molecules from $\Phi(G)$ and $\Phi(D)$ respectively. Then, M_G and M_D are defined as contextually equivalent iff (i) They are in the same context. That is, $C(M_1, M_2) = \texttt{true}$ and, (ii) They have the highest similarity value, i.e., $Sim_f(M_G, M_D) = max(\forall_{m \in \Phi(D)} Sim_f(M_G, m))$.

Let F_c be an idealized set of *contextually integrated* RDF molecules from $\Phi(G)$ and $\Phi(D)$. Let θ_C be a homomorphism such that $\theta_C : \Phi(G) \cup \Phi(D) \rightarrow F_c$. Then there is an RDF Molecule M_F from F_c such that $\theta(M_D) = \theta(M_G) = M_F$. From the motivation example, this means that the molecule of *Arnold Schwarzenegger*, the politician is *contextually equivalent* to the molecule of *Donald Trump* as they are similar *and* they satisfy the context condition of having the same occupation.

In this work, we tackle the problem of explicitly modeling the context and then, matching RDF molecules from RDF graphs that are both highly-similar and equivalent in terms of this context. This problem is defined as follows: given RDF graphs $\Phi(G)$ and $\Phi(D)$, let M_G and M_D be two RDF molecules such that $M_G \in \Phi(G)$ and $M_D \in \Phi(D)$. The system is supplied a context parameter C, which is a boolean function evaluating if two molecules are in the same context. It is also supplied a similarity function Sim_f, which evaluates the semantic similarity between M_G and M_D.

The problem of creating a contextualized graph Φ_C consists of building a homomorphism $\theta_C : \Phi(G) \cup \Phi(D) \rightarrow F_c$, such that for every pair of RDF molecules belonging to Φ_C there are none that are *contextually equivalent* according to system context C. If M_G and M_D are contextually equivalent molecules belonging to F_c, then $\theta_C(M_G) = \theta_C(M_D)$, otherwise $\theta_C(M_G) \neq \theta_C(M_D)$.

Architecture. We propose COMET, an approach to match contextually equivalent RDF graphs according to a given context, thus providing a solution to the problem of *contextually matching* RDF graphs. Figure 2 depicts the main components of the COMET architecture. COMET follows a two-fold approach to solve the problem of entity matching in RDF graphs in a context-aware manner: First, COMET computes the similarity measures across RDF entities to create a bipartite graph; Finally, COMET executes a context-aware 1-1 perfect matching algorithm for matching RDF entities based on the combined scores of similarity and context, by employing Formal Concept Analysis to validate context.

Building a Bipartite Graph. The COMET pipeline receives two RDF graphs $\Phi(G), \Phi(D)$ as input, along with context parameter C, and a similarity function Sim_f. COMET first constructs a bipartite graph between the sets $\phi(G)$ and $\phi(D)$. The *Dataset Partitioner* employs a similarity function Sim_f and ontology

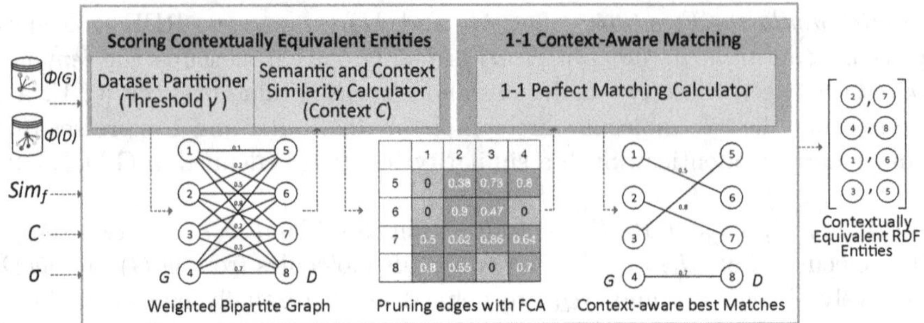

Fig. 2. The COMET Architecture. COMET receives datasets G and D, similarity function Sim_f and context C. Output is a set of matched RDF entities.

O to compute the similarity between RDF molecules in $\phi(G)$ and $\phi(D)$ assigning the similarity score as edge weight in the bipartite graph. COMET supports a variety of similarity functions including simple string similarity. However, as showed in [3], semantic similarity measures are advocated (in the implementation of this work we particularly use GADES [9]).

After similarity computation of the RDF molecules, the result of the similarity function is tested against a threshold γ to determine entity similarity (the similarity threshold's minimum acceptable score). Thus, edges are discarded from the bipartite graph whose weights are lower than γ.

1-1 Context-Aware Perfect Matching Calculator. The main contribution of the COMET pipeline is a novel 1-1 context-aware perfect matching calculator, which validates and prunes pairs of RDF molecules that do not comply with the input context C, making COMET a context-aware approach. For identifying contextually equivalent RDF entities, the *Context Validator* component employs the Formal Concept Analysis (FCA) algorithm. FCA[1] is the study of binary data tables that describe the relationship between objects and their attributes. Applying this context validation step over the RDF molecules ensures only contextually relevant tuples are kept. In COMET, context is modeled as any boolean function. Two molecules are matched if they satisfy this condition, otherwise they are not matched. The algorithm by Vychodil [12] is applied in COMET; it performs formal concept analysis to compute formal concepts within a set of objects and their attributes. This algorithm is extended in our approach for validating complex *boolean conditions*. A typical formal concept analysis takes as input a binary data table. The rows of this table correspond to object, while the columns denote if an object contains a certain attribute.

In our approach, instead of using objects and attributes, we replace the attributes with a *boolean condition* C. This is the same as the context condition C used in our approach. For example, the context C from the motivating

[1] https://en.wikipedia.org/wiki/Formal_concept_analysis.

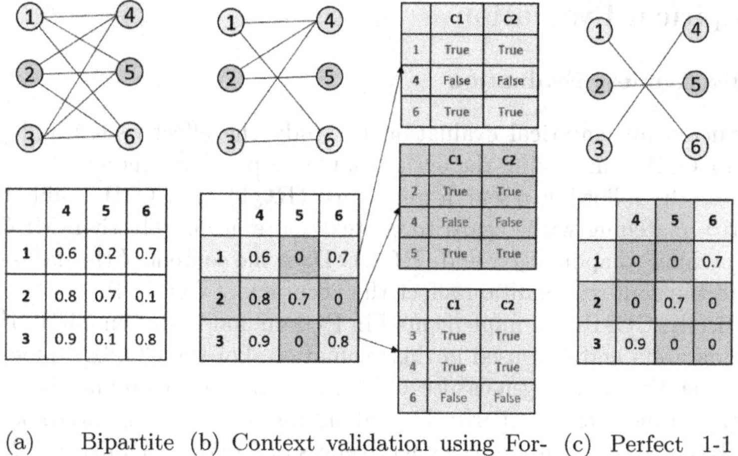

(a) Bipartite graph (b) Context validation using Formal Concept Analysis (c) Perfect 1-1 matches

Fig. 3. Context Validation. A 1-1 matching algorithm over a bipartite graph that validates context of entities using FCA; 1-1 perfect matches are generated.

example can be broken down into $C = C_1 \wedge C_2$ where C_1 = "contains property `dbo:occupation`", and C_2 = "has the same value for property `dbo:occupation`". The execution of the FCA algorithm remains unchanged by this adaptation since the format of the input to FCA is still a binary matrix.

When applied to RDF molecules, formal concept analysis returns a set of formal concepts $< M, C >$ where M is a set of all the molecules that contain all conditions contained in C. Thus, the molecules that do not meet the context condition are pruned. In Fig. 3 an example of context validation is demonstrated. Edges in a bipartite graph are filtered according to a threshold value γ as detailed in the previous section. Next, the remaining edges are validated by constructing an FCA matrix according to context condition C. The FCA algorithm returns the edge satisfying the context conditions. The edges that do not satisfy the context condition are discarded.

COMET solves the problem of *context-aware entity matching* by computing a 1-1 weighted perfect matching between the sets of RDF molecules. The input of the 1-1 weighted perfect matching component is the weighted bipartite graph created on the previous step. Since each weight of an edge between two RDF molecules corresponds to a combined score of semantic similarity and context equivalence value, we call this a 1-1 context-aware matching calculator. Finally, the Hungarian algorithm is utilized to compute the matching.

4 Empirical Evaluation

4.1 Effectiveness Evaluation

We conducted an empirical evaluation to study the effectiveness and performance of COMET in solving the entity matching problem among RDF graphs. We address the following research questions: **(RQ1)** "Is COMET able to perform entity matching with regard to context more accurately than MINTE [3] entity matching component?", and **(RQ2)** "Does the content of the dataset with respect to the context condition affect the accuracy of COMET?"

Practically, COMET is implemented in Python and hosted in GitHub[2] along with the datasets and logs used in this evaluation. For the COMET pipeline we use the semantic similarity measure *GADES* [9] *GADES* examines both string similarity and hierarchy similarity by making use of graph neighbourhoods.

As a baseline, we compare the effectiveness of COMET against the MINTE pipeline proposed by Collarana et al. [3]. Towards **(RQ1)** and **(RQ2)** we design an experiment to measure the *precision*, *recall* and *f-measure* of COMET in comparison to MINTE, while supplying both the pipelines with datasets of different compositions of molecules with respect to context to observe the effect of contextual content on the effectiveness of COMET.

Table 1. Benchmark Description. Datasets used in the evaluation including: number of RDF molecules (M), number of triples (T), evaluated contexts (C).

Configuration	Experiment: Effectiveness					
	A		B		C	
Datasets	*A1*	*A2*	*B1*	*B2*	*C1*	*C2*
Molecules	1000	1000	1000	1000	1000	1000
Triples	70,660	70,660	70,776	70,776	71,124	71,124
Context	$C(M_{D1}, M_{D2}) = \texttt{true}$, if `dbo:occupation` match					

Although each experiment has different datasets and gold standards, we use the same metrics for all the experiments: *Precision*, *Recall*, and *F-meaure*. *Precision* measures what proportion of the performed integrations are actually correct. That is, *precision* is the fraction of RDF molecules that has been identified as contextually equivalent by COMET (C), which intersects with the Gold Standard (GS). On the other hand, *recall* measures the overall proportion of integrated RDF molecules that were identified correctly. That is, *recall* is measured by the fraction of correctly identified similar molecules with respect to the Gold Standard, i.e., $Precision = \frac{|C \cap GS|}{|C|}$ and $Recall = \frac{|C \cap GS|}{|GS|}$. *F-measure* is the harmonic mean of *Precision* and *Recall.*

[2] https://github.com/RDF-Molecules/COMET.

Table 2. Effectiveness evaluation of COMET.

Configuration	COMET			MINTE		
	Precision	Recall	F-Measure	Precision	Recall	F-Measure
A	1.0	1.0	1.0	0.54	0.54	0.54
B	0.708	0.708	0.708	0.449	0.449	0.449
C	0.558	0.558	0.558	0.408	0.408	0.408

The datasets contain 1,000 people entities from DBpedia. In order to test the effect of contextual data content on the accuracy of COMET, three pairs of datasets *(A1, A2)*, *(B1, B2)* and *(C1, C2)* are generated using configurations A, B and C: In configuration A, every molecule *a1* in dataset A1 has **2** *highly similar* molecules *a2* and *a3* in dataset A2, such that *a2* satisfies context condition, but *a3* does not. That is, $C(a1, a2) = \texttt{true}$ and $C(a1, a3) = \texttt{false}$. In configuration B, every molecule *b1* in dataset B1 has **3** *highly similar* molecules *b2*, *b3* and *b4* in dataset B2, such that *b2* and *b3* satisfy the context but *b4* does not. For configuration C, every molecule *c1* in dataset C1 has **4** *highly similar* molecules in dataset C2, two of which satisfy context, and two that do not.

Every pair of datasets are synthesized as follows: First, molecules from the original set of 1,000 DBpedia person entities are duplicated according to the configuration condition to create n number of highly similar molecules in the second dataset. Then predicates inside the similar molecules are randomly edited and deleted to create some variation of similarity. The predicates are then edited to ensure that the correct number of similar molecules in the second dataset satisfy the context according to the original dataset.

Similar to the motivation example shown in Fig. 1, the context C used in this experiment checks if two molecules have the same value for the predicate `dbo:occupation`. The Gold Standard contains matches between molecules that (i) Satisfy the context condition and, (ii) Are highest in similarity among all other molecules. For every pair of datasets belonging to the three configurations (i.e., A, B and C), there is a corresponding Gold Standard G_A, G_B and G_C. The datasets, gold standard and the experiment code are all available on GitHub.

Table 1 describes the dataset used during our evaluations. This experiment was conducted on MINTE and COMET once for each pair of datasets *(A1, A2)*, *(B1, B2)* and *(C1, C2)*, with the context condition requiring that every pair of matched molecules must have the same value for `dbo:occupation` property. The threshold value γ for this experiment is applied at 97th percentile in every case. Then, we compare against the Gold Standard G_A, G_B and G_C for configurations A, B and C, respectively; Precision and Recall are calculated each time.

4.2 Discussion of Observed Results

Based on the values of Precision, Recall, and F-measure reported in Experiment 1 (Table 2), we can positively answer **(RQ1)**, i.e., COMET is able to perform contextual entity matching more effectively than MINTE, and answer **(RQ2)** by illustrating how the contextual content of the dataset affects accuracy. In every case, COMET performs better than MINTE, since MINTE does not take context into consideration during its 1-1 perfect matching whereas COMET does. Moreover, a decrease in Precision and Recall of COMET with the increase of highly similar molecules occurs since COMET has a lesser chance of identifying the perfect match with a higher number of options to choose from. On the other hand, in case of *configuration A*, Precision and Recall is perfect since the dataset supplies only one highly similar molecule that also meets the context. Thus, the highest values of precision and recall demonstrate that in an ideal condition with only one perfect option, COMET will always find the correct match.

5 Conclusions and Future Work

We presented COMET, an approach to match contextually equivalent RDF entities from different sources. COMET executes a 1-1 perfect matching where contextually equivalent RDF molecules are identified according to a combined score of semantic and context similarity. COMET utilizes the Formal Concept Analysis algorithm to decide whenever two RDF molecules are contextually equivalent. The behavior of COMET was empirically studied on two real-world RDF graphs under different context configurations. Observed results suggest that COMET is able to effectively identify and match contextually equivalent entities. COMET makes use of a very simple definition of context conditions, modeling context as a boolean function of entities. In future, context can be modeled in a more generalized way, e.g., a probabilistic function. We plan to evaluate a multi-threading version of the FCA algorithm that may enable the implementation of this work on large datasets. Finally, FCA context parameters will be empowered by materializing implicit facts in RDF knowledge graphs via a reasoner.

References

1. Acosta, M., Vidal, M.-E., Lampo, T., Castillo, J., Ruckhaus, E.: ANAPSID: an adaptive query processing engine for SPARQL endpoints. In: Aroyo, L., et al. (eds.) ISWC 2011. LNCS, vol. 7031, pp. 18–34. Springer, Heidelberg (2011). https://doi.org/10.1007/978-3-642-25073-6_2
2. Beek, W., Schlobach, S., van Harmelen, F.: A contextualised semantics for `owl:sameAs`. In: Sack, H., Blomqvist, E., d'Aquin, M., Ghidini, C., Ponzetto, S.P., Lange, C. (eds.) ESWC 2016. LNCS, vol. 9678, pp. 405–419. Springer, Cham (2016). https://doi.org/10.1007/978-3-319-34129-3_25
3. Collarana, D., Galkin, M., Ribón, I.T., Vidal, M., Lange, C., Auer, S.: MINTE: semantically integrating RDF graphs. In: Proceedings of the 7th International Conference on Web Intelligence, Mining and Semantics, WIMS, pp. 22:1–22:11 (2017)
4. Endris, K.M., Galkin, M., Lytra, I., Mami, M.N., Vidal, M.-E., Auer, S.: MULDER: querying the linked data web by bridging RDF molecule templates. In: Benslimane, D., Damiani, E., Grosky, W.I., Hameurlain, A., Sheth, A., Wagner, R.R. (eds.) DEXA 2017. LNCS, vol. 10438, pp. 3–18. Springer, Cham (2017). https://doi.org/10.1007/978-3-319-64468-4_1

5. Isele, R., Bizer, C.: Active learning of expressive linkage rules using genetic programming. J. Web Semant. **23**, 2–15 (2013)
6. Jagadish, H.V., et al.: Big data and its technical challenges. Commun. ACM **57**(7), 86–94 (2014)
7. Knoblock, C.A., et al.: Semi-automatically mapping structured sources into the Semantic Web. In: Simperl, E., Cimiano, P., Polleres, A., Corcho, O., Presutti, V. (eds.) ESWC 2012. LNCS, vol. 7295, pp. 375–390. Springer, Heidelberg (2012). https://doi.org/10.1007/978-3-642-30284-8_32
8. Michelfeit, J., Knap, T.: Linked data fusion in ODCleanStore. In: ISWC Posters and Demonstrations Track (2012)
9. Ribón, I.T., Vidal, M., Kämpgen, B., Sure-Vetter, Y.: GADES: a graph-based semantic similarity measure. In: SEMANTICS - 12th International Conference on Semantic Systems, Leipzig, Germany, pp. 101–104 (2016)
10. Schultz, A., Matteini, A., Isele, R., Mendes, P.N., Bizer, C., Becker, C.: LDIF - a framework for large-scale linked data integration. In: International World Wide Web Conference (2012)
11. Verroios, V., Garcia-Molina, H., Papakonstantinou, Y.: Waldo: an adaptive human interface for crowd entity resolution. In: International Conference on Management of Data, pp. 1133–1148 (2017)
12. Vychodil, V.: A new algorithm for computing formal concepts. na (2008)

Authenticity, Privacy, Security and Trust

Differentially Private Non-parametric Machine Learning as a Service

Ashish Dandekar(✉), Debabrota Basu, and Stéphane Bressan

School of Computing, National University of Singapore,
13, Computing Drive, Singapore 117417, Singapore
{ashishdandekar,debabrota.basu}@u.nus.edu, steph@nus.edu.sg

Abstract. Machine learning algorithms create models from training data for the purpose of estimation, prediction and classification. While releasing parametric machine learning models requires the release of the parameters of the model, releasing non-parametric machine learning models requires the release of the training dataset along with the parameters. The release of the training dataset creates a risk of breach of privacy. An alternative to the release of the training dataset is the presentation of the non-parametric model as a service. Still, the non-parametric model as a service may leak information about the training dataset.

We study how to provide differential privacy guarantees for non-parametric models as a service. We show how to apply the perturbation to the model functions of histogram, kernel density estimator, kernel SVM and Gaussian process regression in order to provide (ϵ, δ)-differential privacy. We empirically evaluate the trade-off between the privacy guarantee and the error incurred for each of these non-parametric machine learning algorithms on benchmarks and real-world datasets.

Our contribution is twofold. We show that functional perturbation is not only pragmatic for releasing machine learning models as a service but also yields higher effectiveness than output perturbation mechanisms for specified privacy parameters. We show a practical step to perturbate the model functions of histogram, kernel SVM, Gaussian process regression along with kernel density estimator and perform evaluation on a real-world dataset as well as a selection of benchmarks.

Keywords: Differential privacy · Data privacy · Non-parametric models · Functional perturbation

1 Introduction

Organisations are amassing data at an unprecedented scale and granularity. They release either raw data or machine learning models that are trained on the raw data. All machine learning models do not fit the choice of releasing only the models. A parametric machine learning model [22] assumes a parametric model

S. Hartmann et al. (Eds.): DEXA 2019, LNCS 11706, pp. 189–204, 2019.
https://doi.org/10.1007/978-3-030-27615-7_14

function[1] that maps a new data to the corresponding output. A non-parametric machine learning model [22] does not assume a parametric model function but calculates some form of correlation between new data and the training data to compute the corresponding output. For instance, kernel density estimation [24] computes the probability density of new data by assimilating probabilities of the new data with reference to the assumed probability distributions centred at every data-point in the training data. Kernel SVM [6] and Gaussian process regression [26] compute kernel Gram matrix between new data and the training data. Thus, while releasing parametric machine learning models requires the release of the parameters of the model function, releasing non-parametric machine learning models requires the release of the training dataset along with the parameters. An alternative to the release of the training dataset is utilising non-parametric models as a service. While using a non-parametric model as a service, user would send a new data-point to the model to obtain the output of estimation, prediction, or classification.

Publication of raw data without any processing leads to a violation of the privacy of users. Not only raw data but also publication of a 'non-private' machine learning model as a service leads to a violation of the privacy of users. For instance, experiments by Shokri et al. [28] show that models created using popular machine-learning-as-a-service platforms, such as Google and Amazon, can leak identity of a data-point in the training dataset with accuracy up to 94%. In order to reduce the risk of breach of privacy, we need to take preemptive steps and provide quantifiable privacy guarantees for the released machine learning model. Differential privacy [14] is one of such privacy definitions to quantify the privacy guarantees.

We study how to provide differential privacy guarantees for non-parametric models as a service. This cannot be achieved by differentially private output perturbation mechanisms, such as Laplace mechanism and Gaussian mechanism [14]. Due to sequential composition [14] of differential privacy, the privacy guarantee of a mechanism linearly degrades with the number of times the noise is added from a given noise distribution. Output perturbation requires addition of calibrated noise in the output for every new data input. Therefore, it suffers from the degradation of privacy guarantee. When machine learning is provided as a service one can not limit the number of queries. If the noise is added to a model function, further evaluations are performed on the noisy model and we do not need to introduce noise for every evaluation. We adopt functional perturbation proposed in [18] that adds a scaled noise sampled from a Gaussian process to the model function in order to provide a robust privacy guarantee. [18] proves that an appropriate calibration of this mechanism provides (ϵ, δ)-differential privacy. We show how to calibrate the functional perturbation for histogram, kernel density estimator, kernel SVM, and Gaussian process regression. We evaluate the trade-off between the privacy guarantee and the error incurred for each of these non-parametric machine learning algorithms on benchmarks and US census dataset.

[1] Model function refers to the mapping from input to output that is learned by the corresponding machine learning algorithm.

Our contribution is twofold. Firstly, we show that functional perturbation is a viable alternative to output perturbation to provide privacy guarantees for machine learning models as a service. We also hypothesise as well as experimentally validate that output perturbation is less effective than functional perturbation for a given privacy level and a given test set. Additionally, output perturbation is not directly applicable for machine learning models with categorical outputs, such as classification, where functional perturbation naturally operates. Secondly, we show the practical step to perturb the model functions of histogram, kernel SVM, Gaussian process regression, and the kernel density estimator. We evaluate the trade-off between the privacy guarantee and the error incurred for each of these non-parametric machine learning algorithms for US census dataset [1] and a comprehensive range of benchmarks. The results validate that the error decreases for nonparametric machine learning as a service with increase in the size of training dataset and the values of privacy parameters ϵ and δ.

2 Related Work

Most of the big technology companies offer machine learning as a service on their cloud platforms, such as Google's Cloud Machine Learning Engine[2], Microsoft's Azure Learning Studio[3], Amazon's Machine Learning on AWS[4], IBM's Bluemix[5]. These apps provide machine learning models as easy to use APIs for data scientists. Cloud services also provide storage space to host training datasets. For an extensive survey of such platforms, readers can refer to [25].

Privacy of machine learning models is a well-studied topic. Ateniese et al. [4] show the ability to learn statistical information about the training data through parameters of a trained machine learning model. They show a successful attack on support vector machine and hidden Markov model. Homer et al. [19] identify the presence of a certain genome in a publicly released highly complex genomic mixture microarray dataset. They do so by comparing distributions of genomes from the released sample to available statistics of the population. Fredrikson et al. [16] propose the model inversion attack on machine learning models wherein they learn some sensitive attribute in the training dataset. Given black-box access to the model and access to demographic information about patients, they successfully learn genomic markers of patients. In the follow-up work, Fredrikson et al. [17] show instantiation of a successful model inversion attack on decision trees and neural networks that are implemented on machine learning as a service platform. Shokri et al. [28] propose membership inference attack that infers the presence of a data-point in the training dataset based on the outputs machine learning models. They perform attacks on classification models provided by commercial platforms from Google and Amazon. We have enlisted the attacks that

[2] https://cloud.google.com/ml-engine/.
[3] https://azure.microsoft.com/en-us/services/machine-learning-studio/.
[4] https://aws.amazon.com/machine-learning/.
[5] https://www.ibm.com/cloud/.

are pertinent to research in this work. For an extensive survey of attacks on various machine learning models, readers can refer to [15].

Differential privacy [13] has become a popular privacy definition to provide privacy guarantees for machine learning algorithms. Researchers have devised privacy-preserving mechanisms to provide differential privacy guarantees for linear regression [8,32], logistic regression [7,31], support vector machines [27], deep learning [2,28]. Chaudhuri et al. [8] propose differentially private empirical risk minimisation, which lies at the heart of training of machine learning models. They propose output perturbation and objective perturbation. These mechanisms are helpful for releasing parametric machine learning models. Zhang et al. [32] propose the functional mechanism that introduces noise in the loss function of a machine learning model. The functional mechanism is useful for parametric machine learning models that estimate parameters of the model by minimising its loss function. Hall et al. [18] propose the use of functional perturbation that induced noise in the coefficient of expansion of a function in a functional basis. Functions of non-parametric models that use kernels lie in the RKHS spanned by the kernel. Therefore, it is possible to apply the functional perturbation to provide privacy guarantees for non-parametric models. Smith et al. [29] apply functional perturbation by Hall et al. to provide differential privacy Gaussian process regression. Aldá and Rubinstein [3] propose Bernstein mechanism that provides a differentially private way of releasing functions of machine learning models in a non-interactive way. Balog et al. [5] provide a functional perturbation that ensures the closure of functions in a finite dimensional RKHS under the appropriate perturbations. Nozari et al. [23] propose a functional perturbation algorithm that is catered to distributed machine learning task.

Jain and Thakurta [21] propose three ways, which are interactive, non-interactive and semi-interactive, of using machine learning models with differential privacy guarantees. This work is an instance of interactive use of non-parametric machine learning model wherein we provide differential privacy guarantees using the functional perturbation proposed in the work of Hall et al. [18].

3 Methodology

In this section, we discuss release of trained machine learning models. We argue that non-parametric machine learning models need to be released as a service to users. We further instantiate functional perturbation by Hall et al. [18], which provides (ϵ, δ)-differential privacy guarantee, to four non-parametric models.

Our work stands at the crossroads of machine learning and differential privacy. We present a brief background to each of these fields in the technical report [10] for providing privacy operators as service [11].

3.1 Non-parametric Machine Learning Models as a Service

Jain and Thakurta [21] propose three ways in which an organisation can use machine learning models. Firstly, they propose a non-interactive model release wherein an organisation releases a model with quantifiable privacy guarantees. Non-interactive model release is plausible for parametric machine learning models since values of the parameters are sufficient to compute outputs for a new data. Non-parametric machine learning models require training dataset along with the parameters to compute outputs for new data. Secondly, they propose a semi-interactive model release wherein an organisation releases model that provides quantifiable privacy guarantees for a specified set of test data. A priori knowledge of test data is not an assumption that can be realised in every business scenario. Lastly, they propose interactive model release wherein an organisation provides machine learning model as a service. It keeps trained model on the server and users send queries to the server. For non-parametric models, release of training dataset violates the privacy of the users. Therefore, interactive model release, i.e. release of machine learning as a service is a viable alternative.

Differential privacy [14] is a privacy definition for randomised algorithms. In order to provide quantifiable differential privacy guarantees, we need to introduce randomisation while using machine learning models as a service. A privacy-preserving mechanism introduces randomisation to avoid the release of true outputs. Under appropriately calibrated randomisation, privacy-preserving mechanisms provide differential privacy guarantees.

Firstly, randomisation can be introduced by adding an appropriately calibrated random noise to the output of the query. These privacy-preserving mechanisms are called as *output perturbation mechanisms*. For instance, Laplace mechanism [14] adds noise drawn from Laplace distribution whereas Gaussian mechanism [14] adds noise drawn from Gaussian distribution to the output of the model. Multiple evaluations of such mechanisms result in a sequential composition [14]. Privacy guarantee of the sequential composition of privacy-preserving mechanisms linearly degrades with the number of evaluations of privacy preserving mechanisms. Secondly, randomisation can be introduced by adding an appropriately calibrated random noise to the model function. Unlike output perturbation mechanisms, which add calibrated noise to every output of the query, the privacy-preserving mechanisms that perturb functions are *one-shot* privacy-preserving mechanisms. They add calibrated noise in a function leading to change in its functional form. The noisy functional form is used for computing outputs. Therefore, functional perturbation does not suffer from the degradation in the differential privacy guarantee with increasing the number of queries.

When a machine learning model is provided as a service, one cannot strictly control the number of times a user accesses the service. Therefore, we choose functional perturbation based privacy-preserving mechanism. Hall et al. [18] propose the functional mechanism that adds calibrated noise to the expansion of the model function in an appropriate functional basis. Functions of non-parametric machine learning models, especially the ones that use kernels, lie in Reproducing Kernel Hilbert Space (RKHS) [30] associated with the kernel. Thus, RKHS readily provides a functional basis for functions of non-parametric models. Zhang et al. [32] propose the functional mechanism that adds calibrated noise to the loss function of machine learning model. Loss functions are akin to parametric models that train their parameters using some appropriate loss function. Therefore, we choose to use functional perturbation as proposed by Hall et al. [18] to provide differential privacy guarantees for non-parametric models released as a service.

3.2 Functional Perturbation in RKHS

Hall et al. [18] propose a mechanism that provides a calibrated functional perturbation that provides quantifiable (ϵ, δ)-differential privacy guarantee. We briefly explain functional perturbation of a function that lies in a reproducing kernel Hilbert space (RKHS).

Suppose that a function $f : \mathbb{R}^d \rightarrow \mathbb{R}$ lies in RKHS, $\mathcal{H}_k$, associated with a kernel $k : \mathbb{R}^d \times \mathbb{R}^d \rightarrow \mathbb{R}$. For a given dataset $D = \{x_i\}_{i=1}^n$ where each $x_i \in \mathbb{R}^d$, let $\{k(\cdot, x_i)\}_{i=1}^n$ denotes a basis of $\mathcal{H}_k$. In this basis, any function $f \in \mathcal{H}_k$ is expanded as:

$$f(\cdot) = \sum_{i=1}^{n} w_i^f k(\cdot, x_i)$$

where each $w_i^f \in \mathbb{R}$. Inner product between two functions $f, g \in \mathcal{H}_k$ is defined as:

$$\langle f, g \rangle = \sum_{i=1}^{n} \sum_{j=1}^{n} w_i^f w_j^g k(x_i, x_j)$$

The inner product is used to define norm of any function in $\mathcal{H}_k$ as $\|f\|_{\mathcal{H}_k} = \sqrt{\langle f, f \rangle}$.

Functional perturbation adds calibrated noise sampled from a Gaussian process to a model function. Gaussian process uses the kernel that is associated with RKHS where the function lies. We formally define functional perturbation in Definition 1.

Definition 1 (Functional perturbation [18]**).** *Let f_D denotes a model function, whose parameters (or hyperparameters) are estimated on a dataset $D \in \mathcal{D}$. Assume that f_D lies in a reproducing kernel Hilbert space, $\mathcal{H}_k$, with an associated kernel k. Functional perturbation is a privacy-preserving mechanism that perturbs f_D as follows:*

$$f'_D = f_D + \Delta \frac{c(\delta)}{\epsilon} G. \tag{1}$$

where G is a sample path of a Gaussian process with mean zero and covariance function k and $\Delta, \epsilon, \delta > 0$.

Functional perturbation in Definition 1 satisfies (ϵ, δ)-differential privacy [14] when parameters are calibrated as $\Delta \geq \max_{D,D'} \|f_D - f_{D'}\|_{\mathcal{H}_k}$ and $c(\delta) \geq \sqrt{2 \log \frac{2}{\delta}}$. Δ is *sensitivity* of the functions in RKHS $\mathcal{H}_k$. Sensitivity is the maximum deviation of model functions that are trained on any two neighbouring datasets D and D'. In order to apply functional perturbation for machine learning tasks, we need to compute the sensitivity of respective model functions.

3.3 Applications to Four Non-parametric Machine Learning Models

We now illustrate application of functional perturbation for four non-parametric machine learning models. We use non-parametric models that are based on kernel methods. For such non-parametric models, model functions lie in the RKHS associated with the specified kernel.

Histogram. Histogram [22] is used for solving discretised probability density estimation problem. It discretises the domain of a given dataset into a finite number of *bins*. Each bin defines an interval in the domain of the training dataset. Probability of a data-point inside an interval is commensurate to the number of training data-points that lie in the interval.

For a fixed number of bins b, histogram is a vector in $\mathbb{R}^b$. Therefore, we can consider histogram estimation as a function $f : \mathcal{D} \rightarrow \mathbb{R}^b$ wherein $\mathcal{D}$ is a universe of datasets. Let, $\{e_i\}_{i=1}^b$ be the standard basis of Euclidean space $\mathbb{R}^b$. Standard basis spans RKHS associated with the dot product kernel, i.e. $k(x, y) = x^T y$. For a pair of neighbouring datasets, L_1 norm between two histograms is two in the case when the distinct data-points occupy two different bins. Therefore, sensitivity of the histogram function is 2. Let, f_D denotes the histogram for a dataset D with the number of bins b. Thus, the functional perturbation of Eq. 1 for histograms takes the form

$$f'_D = f_D + \frac{2}{n}\frac{c(\delta)}{\epsilon}G.$$

Kernel Density Estimation. Kernel density estimation [22] is a probability density estimation problem that estimates the probability density function of a training dataset. It assumes a probability density function centred at every data-point in the training dataset. Probability of a new data-point is computed as weighted average of the probabilities computed using the probability densities centred at every data-point.

We consider the kernel function namely Gaussian kernel that outputs values in the range $[0, 1]$. It acts as a probability density function. Let k denotes a Gaussian kernel with bandwidth h. Estimate of the probability density function for a dataset $D = \{x_i\}_{i=1}^n$ for the Gaussian kernel k is presented as

$$f_D(\cdot) = \frac{1}{n}\sum_{x_i \in D} k(\cdot, x_i) = \frac{1}{n}\sum_{x_i \in D} \frac{1}{(2\pi h^2)^d} \exp\left(-\frac{\langle \cdot, x_i \rangle}{2h^2}\right).$$

Hall et al. [18] compute the sensitivity Δ of kernel density estimator with a Gaussian kernel as $\frac{\sqrt{2}}{n(2\pi h^2)^{d/2}}$. Thus, from Eq. 1 the functional perturbation for kernel density estimate with Gaussian kernel is

$$f'_D = f_D + \left(\frac{\sqrt{2}}{n(2\pi h^2)^{d/2}}\right)\frac{c(\delta)}{\epsilon}G.$$

Gaussian Process Regression. Gaussian process [26] is a collection of Gaussian random variables such that any subset follows a multivariate Gaussian distribution. Covariance function for the multivariate Gaussian distribution is calculated using a kernel function k. Gaussian process regression outputs a response sampled from posterior distribution of a test data-point given the training dataset. Mean function $\bar{f_D}$ and variance function $Var(f_D)$ of the posterior distribution computed on a training dataset D are given in Eq. 2.

$$\begin{aligned}\bar{f_D}(\cdot) &= \sum_{d_i \in D} \sum_{d_j \in D} (K_D + \sigma_n^2 \mathbb{I})_{ij}^{-1} y_j k(\cdot, x_i) \\ V\bar{a}r(f_D)(\cdot) &= k(\cdot,\cdot) - \sum_{d_i \in D} \sum_{d_j \in D} (K_D + \sigma_n^2 \mathbb{I})_{ij}^{-1} k(\cdot, x_i) k(\cdot, x_j)\end{aligned} \tag{2}$$

K_D is the Gram matrix computed using kernel k on the training dataset and d is the dimension of each training data-point.

Smith et al. [29] use the functional perturbation to provide differential privacy guarantee to Gaussian process regression. Equation 2 shows that the posterior covariance function does not require responses y_j's in the training data. Since only the responses are sensitive towards the disclosure, Smith et al. [29] proposed to perturb only the posterior mean function. Since the sensitivity of the posterior mean function with Gram matrix K_D is $d\|(K_D + \sigma_n^2 \mathbb{I})^{-1}\|_\infty$, they apply the functional perturbation to the posterior mean function as

$$\bar{f_D}' = \bar{f_D} + (d\|(K_D + \sigma_n^2 \mathbb{I})^{-1}\|_\infty) \frac{c(\delta)}{\epsilon} G.$$

Kernel Support Vector Machine. Support vector machine (SVM) [9] is used for solving a classification problem. SVM outputs the class label of a data-point that is specified as the input. Linear SVM is a parametric machine learning model whereas kernel SVM is a non-parametric machine learning model.

Let us consider a data-point $d = (x, y)$ where $x \in \mathbb{R}^d$ are the predictors and $y \in \{-1, 1\}$ is the associated class label. Let $\mathcal{D}$ denotes universe of datasets with n data-points each. We fit a support vector machine classifier with a kernel k on a training dataset $D \in \mathcal{D}$ with n data-points. Kernel support vector machine assumes the form $f(\cdot) = \langle w, \phi(\cdot) \rangle$ where $w \in \mathbb{R}^F$ and $\phi : \mathbb{R}^d \to \mathbb{R}^F$. w is estimated by solving the optimisation problem in Eq. 3. In Eq. 3, C denotes the regularisation constant and l denotes the loss function.

$$\max_{w \in \mathbb{R}^F} \frac{\|w\|^2}{2} + C \sum_{d \in D} l(y_i, f_D(x_i)) \tag{3}$$

Using *hinge loss*, $l_{hinge}(x, y) = max(0, xy)$, as the loss function we obtain a closed form solution. It is presented in Eq. 4. In the solution, α^*'s are called support vectors that are solutions to *dual* of the optimisation problem in Eq. 3.

$$w_D = \sum_{i=1}^{n} \alpha_i^* y_i k(\cdot, x_i) \tag{4}$$

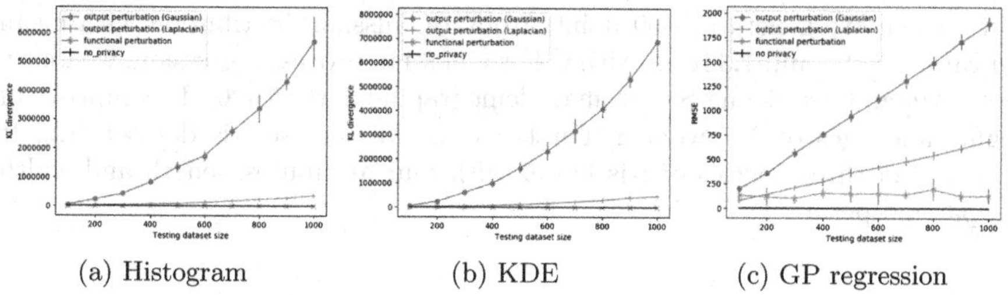

(a) Histogram (b) KDE (c) GP regression

Fig. 1. Comparative evaluation of functional and output perturbation mechanisms for varying size of test datasets. We compare (0.4, 0.001)-differentially private functional perturbation, (0.4, 0.001)-differentially private Gaussian mechanism and (0.4, 0.0)-differentially private Laplace mechanism.

Hall et al. [18] compute the sensitivity of the minimisers of regularised functionals in RKHS. Equation 3 represents an instance of the same problem. Since the sensitivity of w_D is $\frac{2C}{n}$, following Eq. 1 the functional perturbation for kernel SVM takes the form

$$w'_D = w_D + \left(\frac{2C}{n}\right)\frac{c(\delta)}{\epsilon}G.$$

4 Performance Evaluation

In this section, we present effectiveness and efficiency evaluation of functional perturbation for four non-parametric models, *viz.* histogram, kernel density estimation (KDE), Gaussian process regression (GP regression) and kernel support vector machine (kernel SVM), as a service. We comparatively evaluate output perturbation and functional perturbation mechanism. We observe that output perturbation mechanism are less effective than functional perturbation mechanism for a specified setting of differential privacy parameters.

4.1 Dataset

Real World Dataset. We conduct experiments on a subset of the 2000 US census dataset provided by Minnesota Population Center in its Integrated Public Use Microdata Series [1]. The census dataset consists of 1% sample of the original census data. It spans over 1.23 million households with records of 2.8 million people. The value of several attributes is not necessarily available for every household. We have therefore selected 212, 605 records, corresponding to the household heads, and 6 attributes, namely, *Age, Gender, Race, Marital Status, Education, Income.* We treat this dataset as the population from which we draw samples of desired sizes.

Benchmark Datasets. For histogram and kernel density estimation, we follow Hall et al. [18] and synthetically generate a dataset from a known probability distribution. We generate 5000 points from a Gaussian distribution with mean and variance of 2 and 1.3 respectively. For Gaussian process regression, we follow Smith [29] and use !Kung San woman demographic dataset [20]. It comprises of heights and ages of 287 women. For kernel SVM, we use Iris dataset [12]. It comprises of three species of Iris flower with four attributes: length and width of sepal and petal.

4.2 Experimental Setup

All experiments are run on Linux machine with 12-core 3.60 GHz Intel® Core i7™processor with 64 GB memory. Python® 2.7.6 is used as the scripting language. We use RBF kernel for the experiments. Hyperparameters of the kernel are tuned by performing cross-validation on respective dataset.

4.3 Evaluation Metrics

We perform experiments on four non-parametric models solving the problems of estimation, prediction, and classification. Therefore, we use different metrics of effectiveness for the evaluation. Histogram and kernel density estimation are used for estimating probability density of a given data-point and we use Kullback-Leibler divergence (KL divergence) as the metric of effectiveness. Gaussian process regression is used for predicting real-valued attribute and we use root mean squared error (RMSE) as the metric of effectiveness. Kernel SVM is used for classification and we use classification error as the metric of effectiveness. Smaller the value of any of these metrics higher is the effectiveness of the model. In order to evaluate efficiency, we compute query execution time, i.e. the time required to compute output of the model.

4.4 Effectiveness Evaluation

In this section, we present the results on the real-world census dataset.

We start by the comparative evaluation of the functional perturbation and output perturbation mechanisms, namely Gaussian mechanism and Laplace mechanism. Output perturbation mechanisms are not directly applicable for machine learning models with categorical outputs, such as SVMs. Therefore, we perform comparative study for histograms, KDE and GP regression. We also plot the effectiveness of the model without any application of privacy-preserving mechanism. We denote it by "no privacy". In case of histogram and KDE, we do not have the true distributions of the attributes from the census dataset. Therefore, we compute effectiveness by computing KL divergence between functionally perturbed estimators and their non-private counterparts.

In Fig. 1, we comparatively evaluate effectiveness for varying size of testing datasets. Across three models, we observe that effectiveness of the output perturbation mechanisms degrades as the testing dataset size increases. We do not observe such a phenomenon with the functional perturbation. Due to sequential

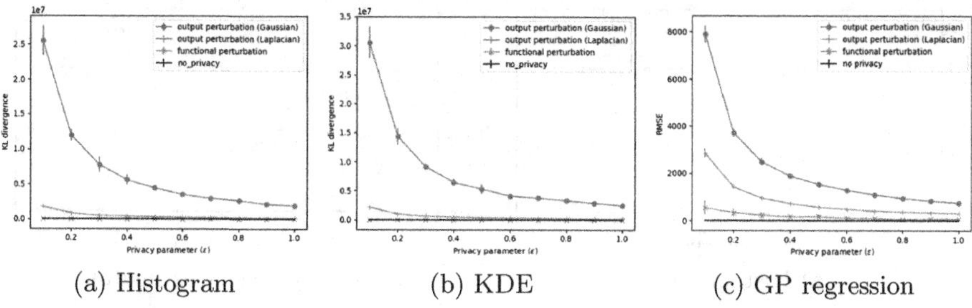

(a) Histogram (b) KDE (c) GP regression

Fig. 2. Comparative evaluation of functional and output perturbation mechanisms for varying privacy parameter ϵ and $\delta = 0.001$. We use dataset of size 5000 to train the models.

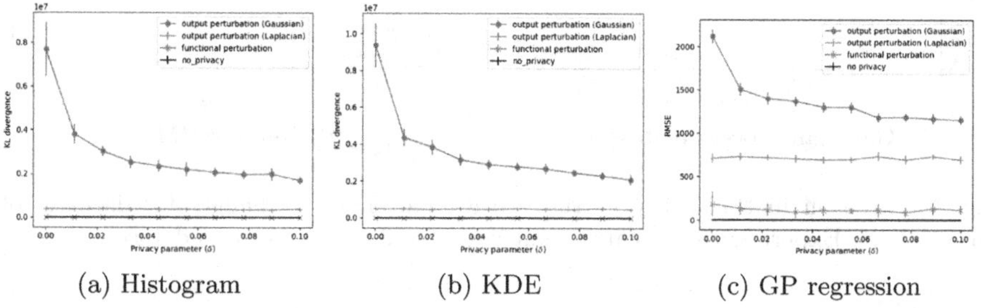

(a) Histogram (b) KDE (c) GP regression

Fig. 3. Comparative evaluation of functional and output perturbation mechanisms for varying privacy parameter δ and $\epsilon = 0.4$. We use dataset of size 5000 to train the models.

composition [14], privacy guarantee of output perturbation mechanisms linearly degrades with the number of evaluations. In order to attain differential privacy with specified privacy parameters, output perturbation mechanisms introduce higher amount of noise for testing datasets of large sizes. Higher amount of noise results in reduction in the effectiveness.

In Figs. 2 and 3, we comparatively evaluate effectiveness for varying privacy parameters ϵ and δ respectively. Across three models, we observe that the effectiveness of the output perturbation mechanisms increases as values of privacy parameters increase. Privacy parameter ϵ quantifies the privacy guarantee of differential privacy. Higher values of ϵ provides weaker privacy guarantees. Weaker privacy guarantees require less amount of noise and hence yield higher effectiveness. Privacy parameter δ is a quantifier of the extent of slack provided in the privacy guarantee of ϵ-differential privacy. In order to provide a robust differential privacy guarantee, we require the value of δ to be as small as possible. Thus, with increasing value of δ the amount of perturbation in the function reduces and hence, the effectiveness increases.

We continue our evaluation of functional perturbation for four non-parametric models on the census dataset. In Fig. 4, we present the effectiveness as privacy parameter ϵ varies between 0 to 1 keeping $\delta = 0.0001$ for different sizes

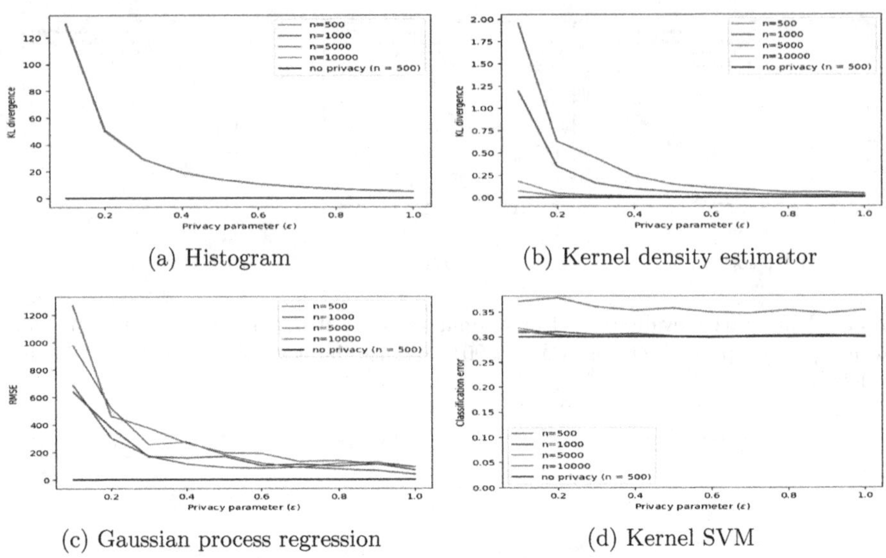

(a) Histogram (b) Kernel density estimator

(c) Gaussian process regression (d) Kernel SVM

Fig. 4. Variation in the utility as the privacy parameter ϵ changes for datasets of varying sizes. Experiments are carried out with $\delta = 0.0001$ on census dataset.

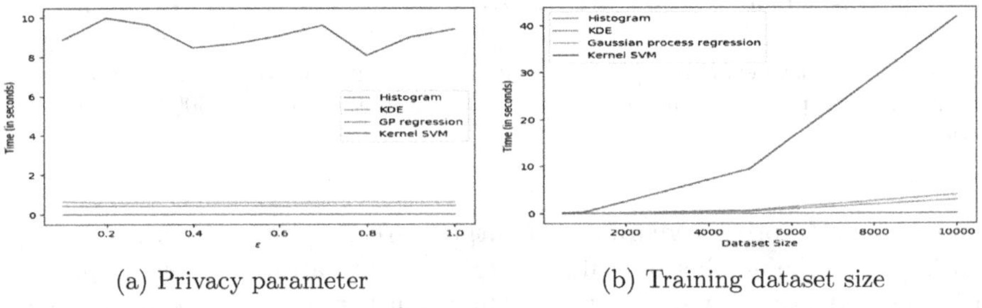

(a) Privacy parameter (b) Training dataset size

Fig. 5. Evaluation of efficiency of functional perturbation for various four non-parametric machine learning models. Figure (a) plots query execution time versus privacy level. Figure (b) plots query execution time versus training dataset size. For both experiments, we set $\delta = 0.001$. We set $\epsilon = 0.2$ for the plot in Figure (b).

of training dataset sizes. We observe that effectiveness of the models increases with increasing the size of dataset. The reason for this is twofold. Firstly, effectiveness of non-parametric models increases with increasing size of the training dataset [22]. Secondly, closer inspection of equations of functional perturbation for each of the four models tells that the amount of noise is inversely proportional to the number of training data-points. Thus, the functional perturbation adds lesser amount of noise for specified privacy parameters as the size of training dataset increases. We make similar observations while evaluating the effectiveness under variation in privacy parameter δ for a fixed value of ϵ. Due to lack of space, we do not provide these results.

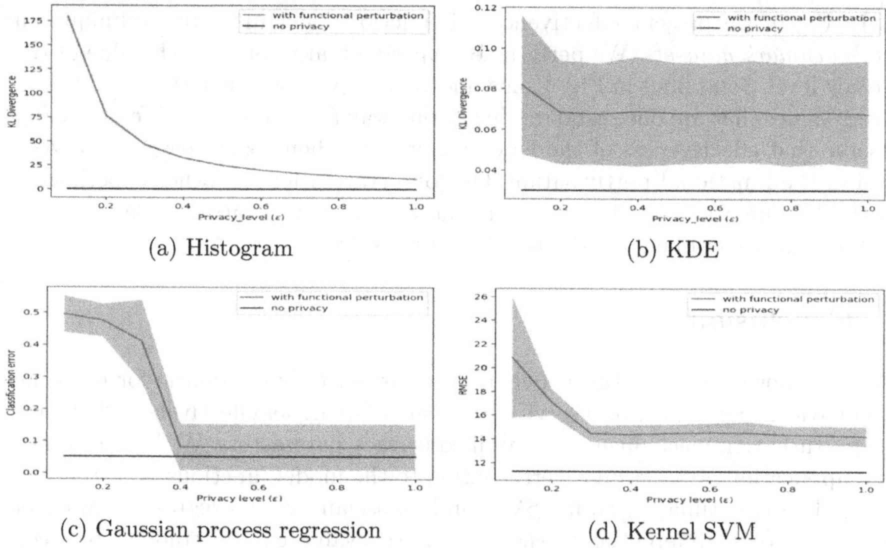

(a) Histogram (b) KDE

(c) Gaussian process regression (d) Kernel SVM

Fig. 6. Variation in the utility as the privacy level changes for datasets of varying sizes. Experiments are carried out with $\delta = 0.0001$ on benchmark datasets.

4.5 Efficiency Evaluation

In Fig. 5(a), we plot the query execution time that is the time required to compute the output for non-parametric models as a service, on a dataset of size 5000 with varying privacy levels. For a given non-parametric model, we observe that query execution time does not depend on the value of the privacy parameter ϵ. Functional perturbation involves sampling a path from the Gaussian process with zero mean function and covariance function computed using the kernel function used in the non-parametric model. The computation of covariance functions requires a significant computational time. This computation time is not affected by any particular value of privacy level. We make similar observation for the privacy parameter δ, which we do not include in the paper due to lack of space.

In Fig. 5(b), we plot query evaluation time for varying size of the training datasets. For this experiment, we set privacy parameters ϵ and δ to 0.2 and 0.001 respectively. We observe that evaluation time increases with increasing size of the training dataset. Large training datasets require large amount of correlations to be computed for every new data-point. Therefore, larger training datasets incur higher amount of time.

4.6 Experiments on the Benchmark Datasets

For reproducibility of the results, we also conduct experiments on the datasets that are either synthetic or publicly available. We observe results that are consistent with the results on the real-world dataset.

In Fig. 6, we present effectiveness of functional perturbation technique on the *benchmark datasets.* We perform 10 experimental runs for each value of the privacy level. Solid lines in Fig. 4 show mean effectiveness whereas shaded region covers values that are one standard deviation away from the mean. We invariably observe that effectiveness of the models increases when we increase the privacy level in the functional perturbation. Our observation for the other experiments on the benchmark datasets are consistent with the observations that we make for the same experiment on the census datasets.

5 Conclusion

We have shown that functional perturbation is not only pragmatic for releasing machine learning models as a service but also yields higher effectiveness than output perturbation mechanisms for specified privacy parameters. We have instantiated application of functional perturbation to the model functions of histogram, kernel density estimator, kernel SVM and Gaussian process regression in order to provide (ϵ, δ)-differential privacy. We have evaluated the tradeoff between the privacy guarantee and the error incurred for each of these non-parametric machine learning algorithms for a real-world dataset as well as a selection of benchmarks.

We are now studying functional perturbation for non-parametric machine learning methods such as k-nearest neighbour density estimation and kernel Bayesian optimisation. We are also interested in studying a step by step functional perturbation method that perturbs a model function in adaptive way balancing the specified privacy and utility requirements.

Acknowledgement. This project is supported by the National Research Foundation, Singapore Prime Minister's Office under its Corporate Laboratory@University Scheme between National University of Singapore and Singapore Telecommunications Ltd.

References

1. Minnesota population center. Integrated public use microdata series – international: Version 5.0 (2009). https://international.ipums.org
2. Abadi, M., et al.: Deep learning with differential privacy. In: Proceedings of the 2016 ACM SIGSAC Conference on Computer and Communications Security, pp. 308–318. ACM (2016)
3. Aldà, F., Rubinstein, B.: The Bernstein mechanism: function release under differential privacy (2017)
4. Ateniese, G., Mancini, L.V., Spognardi, A., Villani, A., Vitali, D., Felici, G.: Hacking smart machines with smarter ones: how to extract meaningful data from machine learning classifiers. Int. J. Secure. Network. **10**(3), 137–150 (2015)
5. Balog, M., Tolstikhin, I., Schölkopf, B.: Differentially private database release via kernel mean embeddings. arXiv preprint arXiv:1710.01641 (2017)
6. Boser, B.E., Guyon, I.M., Vapnik, V.N.: A training algorithm for optimal margin classifiers. In: Proceedings of the Fifth Annual Workshop on Computational Learning Theory, pp. 144–152. ACM (1992)

7. Chaudhuri, K., Monteleoni, C.: Privacy-preserving logistic regression. In: Advances in Neural Information Processing Systems, pp. 289–296 (2009)
8. Chaudhuri, K., Monteleoni, C., Sarwate, A.D.: Differentially private empirical risk minimization. J. Mach. Learn. Res. **12**, 1069–1109 (2011)
9. Cortes, C., Vapnik, V.: Support-vector networks. Mach. Learn. **20**(3), 273–297 (1995)
10. Dandekar, A., Basu, D., Bressan, S.: Evaluation of differentially private non-parametric machine learning as a service. Technical report TRA3/19, National University of Singapore, March 2019
11. Dandekar, A., Basu, D., Kister, T., Poh, G.S., Xu, J., Bressan, S.: Privacy as a service: publishing data and models. In: Li, G., Yang, J., Gama, J., Natwichai, J., Tong, Y. (eds.) DASFAA 2019. LNCS, vol. 11448, pp. 557–561. Springer, Cham (2019). https://doi.org/10.1007/978-3-030-18590-9_86
12. Dheeru, D., Karra Taniskidou, E.: UCI machine learning repository (2017). http://archive.ics.uci.edu/ml
13. Dwork, C.: Differential privacy. In: Bugliesi, M., Preneel, B., Sassone, V., Wegener, I. (eds.) ICALP 2006. LNCS, vol. 4052, pp. 1–12. Springer, Heidelberg (2006). https://doi.org/10.1007/11787006_1
14. Dwork, C., Roth, A., et al.: The algorithmic foundations of differential privacy. Found. Trends® Theor. Comput. Sci. **9**(3–4), 211–407 (2014)
15. Dwork, C., Smith, A., Steinke, T., Ullman, J.: Exposed! a survey of attacks on private data. Ann. Rev. Stat. Appl. **4**, 61–84 (2017)
16. Fredrikson, M., Lantz, E., Jha, S., Lin, S., Page, D., Ristenpart, T.: Privacy in pharmacogenetics: an end-to-end case study of personalized warfarin dosing. In: Proceedings of the USENIX Security Symposium. UNIX Security Symposium, vol. 2014, pp. 17–32. NIH Public Access (2014)
17. Fredrikson, M., Jha, S., Ristenpart, T.: Model inversion attacks that exploit confidence information and basic countermeasures. In: Proceedings of the 22nd ACM SIGSAC Conference on Computer and Communications Security, pp. 1322–1333. ACM (2015)
18. Hall, R., Rinaldo, A., Wasserman, L.: Differential privacy for functions and functional data. J. Mach. Learn. Res. **14**, 703–727 (2013)
19. Homer, N., et al.: Resolving individuals contributing trace amounts of DNA to highly complex mixtures using high-density SNP genotyping microarrays. PLoS Genet. **4**(8), e1000167 (2008)
20. Howell, N.: Data from a partial census of the !kung san, dobe. (1967). https://public.tableau.com/profile/john.marriott#!/vizhome/kung-san/Attributes
21. Jain, P., Thakurta, A.: Differentially private learning with kernels. In: Dasgupta, S., McAllester, D. (eds.) Proceedings of the 30th International Conference on Machine Learning. Proceedings of Machine Learning Research, vol. 28, pp. 118–126. PMLR (2013)
22. Murphy, K.P.: Machine Learning: A Probabilistic Perspective. The MIT Press, Cambridge (2012)
23. Nozari, E., Tallapragada, P., Cortés, J.: Differentially private distributed convex optimization via functional perturbation. IEEE Trans. Control Netw. Syst. **5**(1), 395–408 (2018)
24. Parzen, E.: On estimation of a probability density function and mode. Ann. Math. Stat. **33**(3), 1065–1076 (1962)
25. Pop, D.: Machine learning and cloud computing: survey of distributed and SaaS solutions. arXiv preprint arXiv:1603.08767 (2016)

26. Rasmussen, C.E.: Gaussian processes in machine learning. In: Bousquet, O., von Luxburg, U., Rätsch, G. (eds.) ML 2003. LNCS (LNAI), vol. 3176, pp. 63–71. Springer, Heidelberg (2004). https://doi.org/10.1007/978-3-540-28650-9_4
27. Rubinstein, B.I., Bartlett, P.L., Huang, L., Taft, N.: Learning in a large function space: privacy-preserving mechanisms for SVM learning. J. Priv. Confid. **4**(1) (2012)
28. Shokri, R., Stronati, M., Song, C., Shmatikov, V.: Membership inference attacks against machine learning models. In: 2017 IEEE Symposium on Security and Privacy (SP), pp. 3–18. IEEE (2017)
29. Smith, M.T., Zwiessele, M., Lawrence, N.D.: Differentially private Gaussian processes. arXiv preprint arXiv:1606.00720 (2016)
30. Smola, A.J., Schölkopf, B.: Learning with kernels, vol. 4. Citeseer (1998)
31. Yu, F., Rybar, M., Uhler, C., Fienberg, S.E.: Differentially-private logistic regression for detecting multiple-SNP association in GWAS databases. In: Domingo-Ferrer, J. (ed.) PSD 2014. LNCS, vol. 8744, pp. 170–184. Springer, Cham (2014). https://doi.org/10.1007/978-3-319-11257-2_14
32. Zhang, J., Zhang, Z., Xiao, X., Yang, Y., Winslett, M.: Functional mechanism: regression analysis under differential privacy. Proc. VLDB Endow. **5**(11), 1364–1375 (2012)

PURE: A Privacy Aware Rule-Based Framework over Knowledge Graphs

Marlene Goncalves[1(✉)], Maria-Esther Vidal[1,2,3], and Kemele M. Endris[2,3]

[1] Universidad Simón Bolívar, Caracas, Venezuela
mgoncalves@usb.ve
[2] TIB Leibniz Information Centre for Science and Technology, Hanover, Germany
Maria.Vidal@tib.eu
[3] L3S Research Center, Hanover, Germany
endris@L3S.de

Abstract. Open data initiatives and FAIR data principles have encouraged the publication of large volumes of data, encoding knowledge relevant for the advance of science and technology. However, to mine knowledge, it is usually required the processing of data collected from sources regulated by diverse access and privacy policies. We address the problem of enforcing data privacy and access regulations (EDPR) and propose PURE, a framework able to solve this problem during query processing. PURE relies on the local as view approach for defining the rules that represent the access control policies imposed over a federation of RDF knowledge graphs. Moreover, PURE maps the problem of checking if a query meets the privacy regulations to the problem of query rewriting (QRP) using views; it resorts to state-of-the-art QRP solutions for determining if a query violates or not the defined policies. We have evaluated the efficiency of PURE over the Berlin SPARQL Benchmark (BSBM). Observed results suggest that PURE is able to scale up to complex scenarios where a large number of rules represents diverse types of policies.

1 Introduction

A larger volume of the data published on the Web and accessible over the Internet is on the form of linked data. While a large number of linked data sources are open, some of them may contain sensitive information that are preserved with access restrictions. Thus, whenever a user requires to combine these two sources to mine meaningful knowledge by creating RDF knowledge graphs or by processing a given query, he/she has to individually deploy access methods to the sources. In the literature, several approaches have been proposed for addressing the problem of access control; exemplary approaches include: access control ontologies for RDF data [6,27], access enforcement on centralized or distributed RDF stores (e.g., [2]) or federated RDF sources (e.g., [10,16]). These approaches have shown to be effective in the representation and management of privacy-aware regulations. Nevertheless, they are not expressive enough to

S. Hartmann et al. (Eds.): DEXA 2019, LNCS 11706, pp. 205–214, 2019.
https://doi.org/10.1007/978-3-030-27615-7_15

consider complex privacy-aware regulations during the execution of federated queries. As a consequence, some queries may fail during query answer or produce results that violate data access regulations.

Problem and Research Objective: In this paper, we tackle the problem of enforcing data privacy and access regulations (EDPR) during query processing. This problem is important because the size of available data in domains like biomedical and astronomic, or from social networks, keeps increasing at exponential size [25], and scalable solutions are demanded. The main research goal of this work is to devise an expressive formalism able to describe privacy and access control policies while query processing efficiency is preserved.

Approach: The main idea of this paper is to present PURE, an approach capable of collecting only the data that is relevant for query answering and respects the data access regulations of the sources relevant for a query. PURE resorts to ontologies to describe the main classes and properties of the sources in a federation. It also exploits the *Local-As-View* (LAV) approach [20] to express the privacy and access regulations using conjunctive rules. Furthermore, basic graph patterns of SPARQL queries are represented as conjunctive queries involving the classes and properties included in the federation ontology. The problem of deciding if a set of regulations can be met during the execution of a SPARQL query is afterwards cast to the LAV *Query Rewriting Problem* (QRP) [13]. Then, state-of-the art SAT solvers [12] are used for generating solution models in the combinatorial space of all solutions which eventually correspond to the re-writings of the query. Once a query passes the test of rule satisfiability, a federated query engine like BOUNCER is utilized to execute the query [10]. Using the well-known Berlin SPARQL Benchmark (BSBM), and rules of several complexity, we demonstrate the impact on query execution of checking and respecting the data access regulations represented using a set of LAV rules. The results of the experiments suggest that the number of rules and shape of the rules impact on the total query execution time. However, even in complex scenarios with a large number of rules (up to 150) and queries with up to 25 triple patterns, PURE checks privacy satisfiability in less than 50% of the total query execution time.

Contributions: Our main contribution is a model for enforcing data privacy and access regulations. Specific contributions are summarized as follows: *(i)* The casting of the problem of enforcing data privacy and access regulations (EDPR) to the QRP; *(ii)* PURE, a framework able to exploit state-of-the-art solvers for QRP in order to decide if a query meets or not the privacy and access control regulations; *(iii)* An empirical evaluation of the behavior of PURE over a benchmark of queries and views of different complexity.

The remainder of this article is structured as follows: Sect. 2 motivates the problem of privacy aware federated query processing over a set of biomedical collections, Sect. 3 describes the rule-based approach, PURE, and Sect. 5 reports on the results of the empirical study. Related work is presented in Sect. 6, and finally, Sect. 7 concludes and give insights for future work.

2 Motivation Example

Figure 1 describes an example where data sources from Electronic Health Records (EHR) are integrated into a knowledge graph and linked to DBpedia [19]. Figure 1a shows a patient information published by the first data source $DS1$; name and birth date are protected in consonance with the EU General Data Protection Regulation (GDPR). Figure 1b depicts information from DBpedia, as well as mortality rates and side effects. Data is represented as RDF molecules, where a set of RDF triples that share the same subject correspond to an RDF molecule.

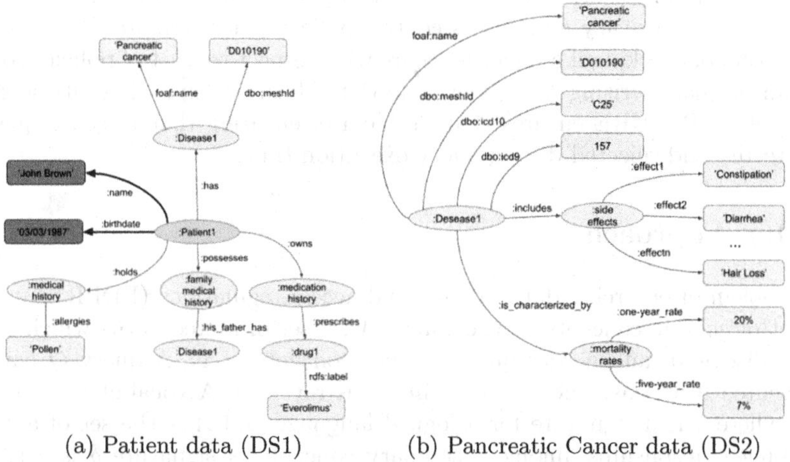

(a) Patient data (DS1) (b) Pancreatic Cancer data (DS2)

Fig. 1. Motivating Example. A federation of two RDF knowledge graphs: DS1 and DS2. (a) An instance of DS1 or RDF molecule representing a pancreatic cancer patient; thicker arrows correspond to sensitive data. (b) An RDF molecule representing information about pancreatic cancer data.

Depression is one of the adverse events in cancer patients; particularly, it has been reported for pancreatic cancer patients [15] who are at a risk of suicide [23,26]. Clinicians, family members, and caregivers should be trained to provide support and monitor the suicide risk of these patients, and thus mitigating risk and saving lives, especially during the first months after diagnosis. In this sense, accessing a complete EHR with all a patient's information, including conditions, indications, side effects, mortality rates can be counterproductive for the patient. Specially, pancreatic cancer is one of the cancers with the highest mortality where the five-year survival rate after diagnosis is 7% for all stages of pancreatic cancer combined [1]. A doctor may decide that a patient with previous serious emotional problems in his/her medical history should not have access to information in his/her EHR that is not easy to handle, such as the mortality rates. A doctor may want to give this information personally accompanied by mental health professionals. In this case, a doctor may decide that the mortality rates cannot be consulted by his/her pancreatic cancer patients by means of EHRs. It is important to note that an EHR can retrieve data over an RDF graph from a federated query engine which can join patients' data of $DS1$ with the second data source, $DS2$. In this context, access control policies over the RDF graph must be defined and enforced to avoid the access of restrictive information, e.g., mortality rates from a join between medical histories and diseases. Several approaches have been defined to address either the problem of describing access control policies [14,17] or enforcing these policies during query processing [10,16]. However, none of these approaches proposes management techniques that enable efficient checking of these policies during query execution. PURE addresses this problem and implements a privacy-aware component that resorts to LAV rules to describe data privacy and access control policies. Furthermore, PURE casts the problem of checking if a query is secure with respect to a set of policies to the problem of query writing using views (QRP). Thus, by exploiting off-the-shelf solvers of QRP, PURE ensures that problem of ensuring that insecure queries are detected and rejected during query execution time.

3 Our Approach

The problem of enforcing data privacy and access regulations (EDPR) consists of a set of generic concepts or vocabulary V, a set of secrecies SS which must not be disclosed, and a user query Q over concepts of V. A query Q cannot be executed if at least one secrecy s in SS is revealed. A vocabulary is a pair $\langle \delta, A \rangle$ where δ is a signature for a logical language and A is the set of axioms describing relationships among vocabulary concepts; a signature δ is a set of predicate and constants from which logical formulas can be constructed. Data privacy and access policies are expressed using rules or assertions that must be enforced. They are modelled in terms of another signature $Sg = \langle S_1^{s1} \dots S_n^{sn} \rangle$ where each symbol S_i^{si} represents the secrecy i of arity si. These rules assure the secrecies on sensitive data will not be revealed and are defined following the Local-As-View (LAV) approach based on mappings that describe them in terms of vocabulary concepts. Thus, for each rule on a secrecy S_i^{si}, there is a mapping that describes S_i^{Si} as a conjunctive query on the vocabulary concepts that also restricts which attributes can be projected in a user query. For example, consider the following two secrecies represented as rules:

$S_1(X,W)$:-*has*(X,Y),*disease*$(Y,$*'Pancreatic cancer'*$)$, *is_characterized_by*(Y,Z), *five_year_rate*(Z,W).
$S_2(Zip, Birthdate, Gender)$:-*patient*$(\_, Zip, Birthdate, Gender)$.

The first secrecy $S_1(X,W)$ assures that a pancreatic cancer patient X and the five-year mortality rate are not projected out or processed in a user query. Further, the second secrecy $S_2(Zip, Birthdate, Gender)$ holds that the quasi-identifiers *Zip*, *Birthdate*, and *Gender* should not be projected together in a query; a quasi-identifier is a set of attributes that can be combined to reveal identity of an entity and reidentify a record [11]. The interpretation of a mapping among secrecies and conjunctive queries are as follows: (i) a head of a secrecy S_i provides information in the form of tuples, e.g. $(Zip, Birthdate, Gender)$, on which attributes should not be projected or processed together in a query; (ii) a secrecy body is a conjunctive query that represents the condition to be checked corresponding to a data policy. In addition to the set SP of secrecies, for each predicate in the vocabulary, there is a LAV rule; we call them, *basic LAV rules*. A user query Q is a conjunctive query over the vocabulary concepts:

$$Q(Name, Birthdate, Zip, Gender, Rate) \text{:-} has(Name, Y), \\ patient(Name, Birthdate, Zip, Gender),\ is\_characterized\_by(Y, Z), \\ disease(Y, \text{'Pancreatic cancer'}),\ five\_year\_rate(Z, Rate).$$

The evaluation of this query returns the name, birthdate, zip, gender, and five-year mortality rate from those patients who have pancreatic cancer. Any rewriting of the query Q where all the predicates in the body of the secrecies S_1 or S_2 are selected in the rewriting lead to a query that violates the regulations– insecure rewritings, e.g., the following rewriting makes the query Q to be *insecure*:

$$Q'(Name, Zip, Birthdate, Gender, Rate) \text{:-}\ \boldsymbol{S_1(Name, Rate)}, \\ patient(Name, Zip, Birthdate, Gender),\ \boldsymbol{S_2(Zip, Birthdate, Gender)}.$$

A rewriting Q' of a query Q is *insecure* with respect to a set of secrecies SS if the following conditions are met: (i) All the subgoals between secrecies in SS and the query match, i.e., each subgoal s in the body of a secrecy S in SS appears in the body of the query Q or there exists any subsumption relationship entailed between s and any subgoal in the body of the query. For example, for the secrecy $S1$, *has* and *is_characterized_by* appears in the query Q; (ii) For each matched subgoal, the variables of the secrecy body can be instantiated with the variables or constants of the query subgoals in a manner that the resulting terms are equal; (iii) The attributes in the head of the secrecy S_1 appear or are instantiated with the attributes in the body of the query Q. For example, the attributes (X, W) in the head of the secrecy S_1 appears in the query Q; (iv) All attributes in the head of the secrecy S_2 appears or are instantiated with the attributes in the head of the query Q.

Formally, EDPR receives a set of LAV secrecies SS on a set of predicates P that define sources in the vocabulary V, and a conjunctive query Q over predicates in P, and a set SP of *basic* LAV rules. Q is *insecure* if there is at least one *insecure* rewriting Q' of Q with respect to secrecies S. Contrary, if all the rewritings of Q include views that are in SP, i.e., basic predicates, the query can be safely evaluated following any of these rewritings. A rewriting Q' of Q on SP is secure and satisfies the following conditions: (i) Each subgoal of a rule in SP corresponds to a subgoal in Q or can be implied based on subsumption relationships, i.e., subclass; (ii) All variables in the head of a rule in SP appears or are instantiated with variables in the body of the query Q. In this work, we assume that all the LAV rules are *safe*, i.e., all variables in the head of the query (resp. in the head of each LAV rule) appear in the query body (resp. in the body of each LAV rule). In addition, we deal with queries and LAV rule without arithmetic predicates, e.g., predicates likes $x \leq y$ are forbidden.

4 The PURE Architecture

Figure 2 depicts the PURE architecture. Given a SPARQL query, the access policy engine component retrieves all the access control policies that must be respected by the query. As PURE follows a LAV approach, access control policy rules are stored in a data access policy catalog. For example, the secrecies S_1 and S_2, and the predicate $patient(Name, Zip, Birthdate, Gender)$ are defined as views in the access policy database and they are retrieved by the access policy engine component when the user's query Q is received. To check if a user's query is secure or not, the query rewriter component obtains the input SPARQL query and the policies retrieved by the access policy engine component. If the user's query is *insecure*, the query rewriter component identifies subqueries that are secure rewritings in terms of the basic LAV rules, i.e., the rewritten query will identify a subquery of the original query that do not violate any secrecy. For instance, if there is a rule restricting information about the set of cancer patients who live in Hannover and this set of data is subset of a user's query response, then the records corresponding to these cancer patients will be extracted from the final query answer. Lastly, the rewritten query and secure queries are evaluated on the graph knowledge by a federated query engine (e.g., BOUNCER); the graph knowledge unifies heterogeneous information from different data sources.

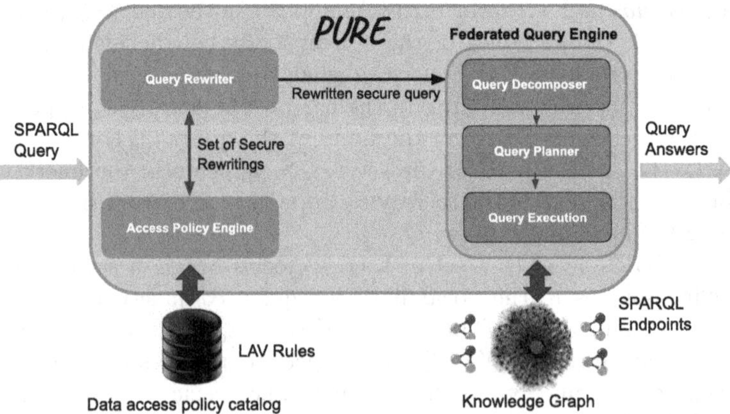

Fig. 2. The PURE Architecture. PURE receives a SPARQL query and outputs the results of executing the SPARQL query over a federation of SPARQL endpoints in terms of access policy rules. PURE resorts to solvers of the query rewriting problem (QRP) to identify secure rewritings of a query.

5 Experimental Study

We study the impact of query rewriting on the query execution of BSBM queries.

Benchmarks: The Berlin SPARQL Benchmark (BSBM) generates a dataset of 200M triples and 12 queries; answer size is limited to 10,000 per query. The third query (Q3) of BSBM was omitted for our experimental study because it is a union query and it is not a conjunctive query. Also, the set of rules was randomly generated and each one corresponds to a star-join between 1 and 3 predicates. The number of rules varies from 10 to 150.

Metrics: We measure performance as *query execution time*; it is computed as the elapsed time in seconds between the submission of a query and the delivery of the answers. The *time* command of the operating system is utilized.

Implementation: PURE techniques are implemented in Python 3.5. The BSBM dataset is partitioned into eight parts (one part per RDF type) and deployed on one machine as SPARQL endpoints using Virtuoso 6.01.3127, where each dataset resides in a dedicated Virtuoso docker container. Experiments are executed on a Dell Optiplex 5050, i5-6600 (4x 3,3 GHz), 24 GB RAM, 500 GB NVMe. Also, MCDSAT [4] is used for the query rewriting problem.

Figure 3 shows the average query rewriting time with respect to the average total query execution time for several configurations of number of rules: 40, 80, 120, and 150 (time in seconds). As observed, for each query, the average query rewriting time is approximately 50% of average total execution time. Although the time complexity of the tasks of query rewriting and execution are different, i.e., QRP is an NP-problem [4] while the evaluation of a conjunctive query is polynomial on the size of the RDF graph and the number of query predicates [22], both problems are impacted by query complexity. Being expected that even in complex queries like Q4, Q11, and Q12, query rewriting and execution time maintain a similar relation. We can also observe that for each query, the execution time grows as the number of rules increases. Since the number of rewritings performed by the rewriter for conjunctive queries blows up exponentially with the number of views (rules) [3], the execution time shows a growth pattern. Thus, execution time is affected by the search space of query rewritings.

6 Related Work

In [24], authors have developed a system for mapping secured ontologies using graph similarity concept and three phase based ontology merge method has been proposed by [5]. Due to the domain this method has been proposed, no privacy exploitation is concerned when merging ontologies. Also, with the current architectures of EHR data management systems, it is evident they maintain a local database at every institute and data access is limited to local users. Dewri et al. [8] explore the problem of privately detecting and tracking the health records of an individual in a distributed infrastructure. They supported federated querying

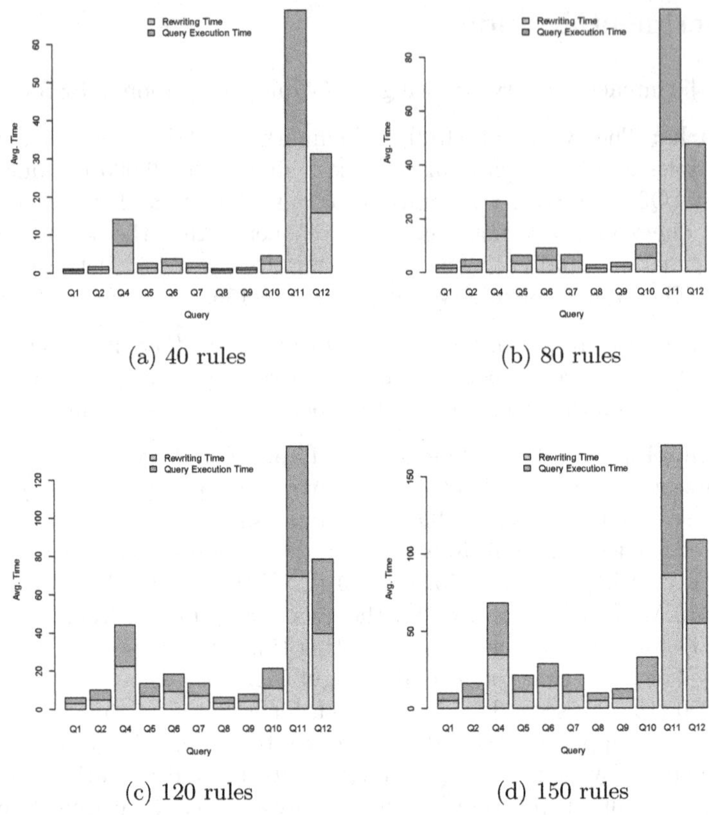

Fig. 3. Impact of Privacy Validation on Total Execution Time. Rewriting time and query execution time for each BSBM query and several configurations of number of rules (40, 80, 120, and 150). In all the cases, the time of privacy validation is on average 50% of the total execution time. Reported results suggest that time for the validation and query execution remain proportional.

through enhancing privacy in underlying data sources. Server/client and decentralized architectures to perform privacy-preserving federated data analysis are proposed in [7]. Authors also introduce secure multiparty computation protocols to protect the intermediary results in communication. Even though authors do not focus on linked data in their research, this system can be used as a comparison to our proposed architecture for privacy preserved federated querying on distributed knowledge graphs. The Informatics for Integrating Biology and the Bedside (i2b2) [18,21], an open source platform and software implementation, allows clinical researchers to search across multiple i2b2 sites, find sets of interesting patients, and reuse medical data while preserving patient privacy and ensuring data integrity. While i2b2 created an analytic platform for a single clinical data repository, the Shared Pathology Information Network (SPIN) [9] has tackled the problem of cross-institution data sharing across a peer-to-peer network, in which each participating institution maintains autonomy and control of its own data. The Linked Medical Data Access Control (LiMDAC) framework [14] consists of three Linked Data models. The proposed architecture exploits and orchestrates these models to enable controlling access to medical data. Although these approaches enable the modeling of privacy and access regulations, none of them is able to validate these policies during execution time. Finally, approaches like BOUNCER [10] or SAFE [16] enforce privacy regulations during the execution of federated queries. However, they rely on formalisms to represent data access policies where LAV rules cannot be expressed and validated.

7 Conclusions and Future Work

In this paper, the problem of enforcing data privacy and access regulations is presented. The PURE framework is devised for enabling the detecting of conjunctive queries that violate the privacy regulations. PURE resorts to LAV rules for expressing data privacy and access regulations and to state-of-the-art solvers of the problem of query rewriting to provide an efficient solution. Thus, our work broadens the repertoire of techniques available to privacy-aware query processing, and we hope that our techniques will help in ensuring privacy over knowledge graphs. In the future, we will extend the formalism to represent more complex rules. Further, the modeling of real-world regulations is part of our future plans.

Acknowledgement. This work has been partially supported by the EU H2020 RIA funded project iASiS with grant agreement No. 727658.

References

1. Hirshberg foundation for pancreatic cancer research. Prognosis. http://pancreatic.org/pancreatic-cancer/about-the-pancreas/prognosis/. Accessed 29 Mar 2019
2. Amini, M., Jalili, R.: Multi-level authorisation model and framework for distributed semantic-aware environments. IET Inf. Secur. **4**(4), 301–321 (2010)
3. Arvelo, Y., Bonet, B., Vidal, M.: Compilation of query-rewriting problems into tractable fragments of propositional logic. In: Proceedings, The Twenty-First National Conference on Artificial Intelligence and the Eighteenth Innovative Applications of Artificial Intelligence Conference, Boston, Massachusetts, USA, 16–20 July 2006, pp. 225–230 (2006)
4. Blai Bonet: MCDSAT. https://github.com/bonetblai/mcdsat. Accessed 15 Sept 2019
5. Châabane, S., Jaziri, W., Gargouri, F.: A proposal for a geographic ontology merging methodology. In: 2009 International Conference on the Current Trends in Information Technology (CTIT), pp. 1–6. IEEE (2009)
6. Costabello, L., Villata, S., Gandon, F.: Context-aware access control for RDF graph stores. In: ECAI-20th European Conference on Artificial Intelligence (2012)
7. Dai, W., Wang, S., Xiong, H., Jiang, X.: Privacy preserving federated big data analysis. In: Srinivasan, S. (ed.) Guide to Big Data Applications. SBD, vol. 26, pp. 49–82. Springer, Cham (2018). https://doi.org/10.1007/978-3-319-53817-4_3

8. Dewri, R., Ong, T., Thurimella, R.: Linking health records for federated query processing. Proc. Priv. Enhanc. Technol. **2016**(3), 4–23 (2016)
9. Drake, T.A., et al.: A system for sharing routine surgical pathology specimens across institutions: the shared pathology informatics network. Hum. Pathol. **38**(8), 1212–1225 (2007)
10. Endris, K.M., Almhithawi, Z., Lytra, I., Vidal, M.-E., Auer, S.: BOUNCER: privacy-aware query processing over federations of RDF datasets. In: Hartmann, S., Ma, H., Hameurlain, A., Pernul, G., Wagner, R.R. (eds.) DEXA 2018, Part I. LNCS, vol. 11029, pp. 69–84. Springer, Cham (2018). https://doi.org/10.1007/978-3-319-98809-2_5
11. Gkoulalas-Divanis, A., Loukides, G., Sun, J.: Publishing data from electronic health records while preserving privacy: a survey of algorithms. J. Biomed. Inform. **50**, 4–19 (2014)
12. Gomes, C.P., Kautz, H.A., Sabharwal, A., Selman, B.: Satisfiability solvers (2008)
13. Halevy, A.Y.: Answering queries using views: a survey. VLDB J. **10**(4), 270–294 (2001)
14. Kamateri, E., Kalampokis, E., Tambouris, E., Tarabanis, K.: The linked medical data access control framework. J. Biomed. Inform. **50**, 213–225 (2014)
15. Kenner, B.J.: Early detection of pancreatic cancer: the role of depression and anxiety as a precursor for disease. Pancreas **47**(4), 363 (2018)
16. Khan, Y., et al.: SAFE: SPARQL federation over RDF data cubes with access control. J. Biomed. Semant. **8**(1), 5 (2017)
17. Kirrane, S., Villata, S., d'Aquin, M.: Privacy, security and policies: a review of problems and solutions with semantic web technologies. Semant. Web **9**(2), 153–161 (2018)
18. Kohane, I.S., Churchill, S.E., Murphy, S.N.: A translational engine at the national scale: informatics for integrating biology and the bedside. JAMIA **19**(2), 181–185 (2012)
19. Lehmann, J., et al.: DBpedia - a large-scale, multilingual knowledge base extracted from Wikipedia. Semant. Web **6**(2), 167–195 (2015)
20. Levy, A.Y., Rajaraman, A., Ordille, J.J.: Querying heterogeneous information sources using source descriptions. In: Proceedings of 22nd International Conference on Very Large Data Bases (1996)
21. Malin, B., Karp, D., Scheuermann, R.H.: Technical and policy approaches to balancing patient privacy and data sharing in clinical and translational research. J. Investig. Med. **58**(1), 11–18 (2010)
22. Pérez, J., Arenas, M., Gutiérrez, C.: Semantics and complexity of SPARQL. ACM Trans. Database Syst. **34**(3), 16:1–16:45 (2009)
23. Saad, A.M., et al.: Suicidal death within a year of a cancer diagnosis: a population-based study. Cancer **125**(6), 972–979 (2019)
24. Shenoy, K.M., Shet, K.C., Acharya, U.D.: Secured ontology matching using graph matching. In: Meghanathan, N., Nagamalai, D., Chaki, N. (eds.) Advances in Computing and Information Technology. Advances in Intelligent Systems and Computing, vol. 177, pp. 11–18. Springer, Heidelberg (2013). https://doi.org/10.1007/978-3-642-31552-7_2
25. Stephens, Z.D., et al.: Big data: astronomical or genomical? PLoS ONE **13**(7), e1002195 (2015)
26. Turaga, K.K., Malafa, M.P., Jacobsen, P.B., Schell, M.J., Sarr, M.G.: Suicide in patients with pancreatic cancer. Cancer **117**(3), 642–647 (2011)
27. Unbehauen, J., Frommhold, M., Martin, M.: Enforcing scalable authorization on SPARQL queries. In: SEMANTiCS (Posters, Demos, SuCCESS) (2016)

FFT-2PCA: A New Feature Extraction Method for Data-Based Fault Detection

Matheus Maia de Souza, João Cesar Netto, and Renata Galante(✉)

Instituto de Informática, Universidade Federal do Rio Grande do Sul (UFRGS), Porto Alegre, Brazil
{mmsouza,netto,galante}@inf.ufrgs.br

Abstract. The industrial environment requires constant attention for faults on processes. This concern has central importance both for workers safety and process efficiency. Modern Process Automation Systems are capable of produce a large amount of data; upon this data, machine learning algorithms can be trained to detect faults. However, this data high complexity and dimensionality causes a decrease in these algorithms quality metrics. In this work, we introduce a new feature extraction method to improve the quality metrics of data-based fault detection. Our method uses a Fast Fourier Transform (FFT) to extract a temporal signature from the input data, to reduce the feature dimensionality generated by signature extraction, we apply a sequence of Principal Component Analysis (PCA). Then, the feature extraction output feeds a classification algorithm. We achieve an overall improvement of 17.4% on F1 metric for the ANN classifier. Also, due to intrinsic FFT characteristics, we verified a meaningful reduction in development time for data-based fault detection solution.

Keywords: Feature extraction · Fault detection · Tennessee Eastman process

1 Introduction

The industrial automation level achieved in the last decades allow improvements in production and safety. Increase in the sophistication of process control systems has not eliminated abnormal situations. Consequently, faults in a complex environment, like the factory floor, are inevitable. These faults may cause from a reduction in the process performance to injuries in workers [2]. A single fault can cause many process variables to exceed their limits. According to [7], faults occur in the industrial processes that produce undesired or unacceptable system behavior. For instance, in hazardous processes as, e.g., chemical plants, the consequences of a fault can be disastrous.

In this work, we discuss only data-based techniques. Data-based techniques are more generic, as there is no need for in-depth knowledge of the target system. Two groups divide data-based techniques, those that use statistical methods and those that use machine learning methods. This paper presents a new feature

This work is supported by the BNDES under the FUNTEC-SDCD project.

S. Hartmann et al. (Eds.): DEXA 2019, LNCS 11706, pp. 215–224, 2019.
https://doi.org/10.1007/978-3-030-27615-7_16

extraction method for machine learning fault detection. The new method is based on the frequency spectrum extraction of input signals, using a discrete Fourier Transform. Input signals will later be classified according to their states, fault or normal. Given the natural cyclical behavior of industrial environments, the frequency spectrum functions as a signature of behavior that can be classified as normal behavior or fault behavior. To verify if the use of these behavioral signatures produces a performance gain, compared to the use of data in the time domain, we performed experiments using datasets generated by the *Tennessee Eastman* industrial control simulation [5]. The simulation *Tennessee Eastman* was chosen due to the control of the environment that the simulation offers. The measure used to compare the techniques will be the precision with which the algorithm, or set of algorithms, can detect a failure.

The rest of the paper is structured as follows. Section 3 provides a review of the literature. Section 2 describes the background. In Sect. 4, we discuss the method FFT-2PCA, showing the functional steps and components. Section 5 presents the experiments to implement the proposed method along with the acquired results, while discusses these results and their application to actual setups. Section 7 concludes the paper, providing directions for future work.

2 Background

We begin with our background section with data source; the Tennessee Eastman process (TEP), an industrial process simulation which has an academic use in the industrial control field. The TEP is a suitable data source due to its capability to provide a structured data output of a complex system.

The Fast Fourier Transform (FFT) is used to determine the frequencies and phase content of local sections of a signal as it changes over time [3]. Its uses extend from data compression to image processing. In our work, we apply the FFT to obtain a frequency signature that intends to extract the time series features created by the TEP.

Besides our method, FFT-2PCA, we use two other machine learning algorithms to extract features. The first one, Principal Component Analysis (PCA), is a well-known method for dimensionality reduction, it does that by extracting a new set of orthogonal variables from interdependent ones. To do so, it applies singular value decomposition to find these orthogonal variables, also called principal components [1]. The PCs are the new dimensional space in which the original variables will be represented. The second one, Independent Component Analysis (ICA), aim to maximize independence. It finds a linear transformation, which applied to the original set of variables, results in a new space where the independence between variables is maximized [4].

3 Related Work

In [12], authors present a comparison between a set of statistical tools to perform process monitoring and fault detection. The tools chosen were: PCA, Partial Least Squares (PLS), independent component analysis (ICA), and variations of each technique, and a method based on the model of the process to be monitored. The statistical tests T^2 and squared prediction error (SPE) are used to classify a sample as a normal state or fault state. The results show that the techniques that use PLS obtained higher accuracy and lowered false positive rate. However, these statistical measures, with thresholds manually selected, makes it challenging to use on a large scale since each class of failure requires individual analysis.

In [8], the authors compare two techniques for fault diagnosis, a process to identify a fault class. One based on PCA and another one based on Support Vector Machines (SVM). Like in [12], PCA is a failure detection method having as a deciding factor a threshold for a T^2 test that defines if a sample represents a failure situation and its class. The results show the PCA as a better option since it has less computational effort to obtain better results. To evaluate performance, the authors chose to measure just accuracy. Since accuracy only computes the ratio of correct predictions over total predictions, the results of false negatives are neglected.

In [11], authors use a combination of ANN and PCA as an approach to fault detection and fault diagnosis. The paper hypothesis is that a dimensionality reduction improves classification performance. They apply fault detection for each fault class getting the best overall detection rate of 84.32%. Authors also tested the approach in fault diagnosis obtained the best overall diagnosis rate of 48.29%. In this work again, authors use a fragile methodology to evaluate performance by using only the missed detection rate as a metric. Due to a substantial similarity with our work and being state of the art in our domain, we define this work as our baseline.

The raw input data that comes from multiple sensors in an industrial environment are complex and noisy. All cited works face this problem and make an effort to extract useful information from the input data to perform fault detection with better results.

4 FFT-2PCA

With the rise of industrial automation, processes that have been pure analog now are part of the digital world. This transformation brought the challenge to keep the environment safe while dealing with a sheer amount of data. Our contribution to face this challenge is a feature extraction method designed for data-based fault detection. The data collected in the industrial environment, used as input for the classification, carry some inherent problems, such as noisy and redundant information. Our method aims to mitigate these problems by extracting relevant features in the time domain and taking them to a representation space where the classifiers can more easily distinguish the states of operation.

The Algorithm 1 aims to extract proper signatures from all-time series, providing the time-related knowledge to the classifiers. Our input data comprise of time series, organized in a matrix form. As the name time series implies, each data point is directly dependent on the previous ones. Simple classification algorithms do not take into account this connection, assessing only one data point at a time. The classifier receives the extracted signatures as a matrix that now expresses a representation space created by our algorithm.

Algorithm 1: FFT-2PCA algorithm

Input : $A[m, n]$, k, pcs1, pcs2
Output: $E[(\frac{m}{k}), n \cdot pcs2]$
$B[n(\frac{m}{k}), \frac{k}{2}] \leftarrow$ `Dff(`$A[m, n], k$`);`
$Bnorm[n(\frac{m}{k}), \frac{k}{2}] \leftarrow$ `Normalize(`$B[(\frac{m}{k})n, \frac{k}{2}]$`);`
$C[(\frac{m}{k}), n(\frac{k}{2})] \leftarrow$ `Pivoting(`$B[n(\frac{m}{k}), \frac{k}{2}]$`);`
pca = `setPCA(`$pcs1$`);`
`pca.fit(`$C[(\frac{m}{k}), n(\frac{k}{2})]$`);`
$D[(\frac{m}{k}), n \cdot pcs1] \leftarrow$ `pca.transform(`$C[(\frac{m}{k}), n(\frac{k}{2})]$`);`
pca = `setPCA(`$pcs2$`);`
`pca.fit(`$D[(\frac{m}{k}), n \cdot pcs1]$`);`
$E[(\frac{m}{k}), pcs2] \leftarrow$ `pca.transform(`$D[(\frac{m}{k}), n \cdot pcs1]$`);`

The first step is data entry; the FFT-2PCA receives a matrix $A_{m,n}$ with m being the number of observations and n the number of system features. To apply the FFT to the matrix $A_{m,n}$, we have to use the transform in its discrete form. This transform requires a parameter k that defines how many observations are going to be used in one transformation. Then we can define arrays with length k for each of n features in the input matrix, these arrays are, separately, transformed by the discrete FFT to the frequency domain. The matrix $B_{(\frac{m}{k})\cdot n, \frac{k}{2}}$ shows the result of the transformation on the previous step, the second matrix has $\frac{k}{2}$ columns due characteristics of the transformation, and each observation became n rows in the second matrix. This goes on for all block of k rows in the matrix $A_{m,n}$ resulting in a matrix $B_{(\frac{m}{k})\cdot n, \frac{k}{2}}$.

To get again a matrix with rows as observations, we rearrange the matrix $B_{(\frac{m}{k})\cdot n, \frac{k}{2}}$, each block of n rows is concatenated in one row, resulting in matrix $C_{(\frac{m}{k}), n(\frac{k}{2})}$. To reduce the dimensionality we apply the first PCA, selecting the more representative FFT weights. The result is the matrix $D_{(\frac{m}{k}), n\cdot pcs1}$ that has its columns as the principal components. So the number of columns will be given by pc1 multiplied by n system features. The last step selects relevant Feature-PC values applying a second PCA. The parameter pcs2 controls the number of PCs; consequently, the number of columns in the last matrix. Finally, we obtain a matrix $E_{(\frac{m}{K}), pcs2}$ that is the input for the classification algorithms.

5 Experiments

Our experiments aim to evaluate how each feature extractors impact classification performance. Our main assumption is that the FFT-2PCA method generates a better input for the classifiers, by extracting the frequency signatures and latter creating a more efficient representation space. Beyond the quality performance, we assess how the feature extractors impact the classifiers regarding training time.

Datasets. We create seven datasets were each one represents a TEP configuration. The structure that keeps this data is a matrix where the columns are the 53 simulation parameters, for instance, internal reactor temperature or the apertures that control the elements mixture. The rows are an observation, also called data point. All seven datasets have the data points count balanced between the two states: faulty or normal. The datapoints count in each dataset respectively is: 261 k; 261 K; 522 K; 783 K; 1.02 G; 1.3 G; 1.56 G; 1.82 G.

Baseline Model. We choose the work "Integration of principal component analysis and neural classifier for fault detection and diagnosis of Tennessee Eastman process" [11], present in Sect. 3, as our baseline. This choice is based on the similarity and relevance to our work. The primary goal of our baseline is to test the impact of the PCA algorithm as the feature extractor for fault classification and diagnosis using ANN. For our work, we choose to focus only on the fault detection problem, to test with two more classifiers (Key-Nearest Neighbor (KNN) and SVM) and bring another algorithm for feature extraction (ICA).

Metrics. To evaluate performance, we use the F1 score that is a harmonic mean of precision and recall. The positive concept, in our case, is related to a fault. For instance, we called a false positive when the classification algorithm gives us a fault indication for a given instance of data, and in the real world, that same instance represents a normal behavior.

Performance Experiments. To evaluate the performance we combine all three feature extractors (PCA, ICA, and FFT-2PCA) with all three classifiers. Also, to test if the use of a given feature extractor results in a significant improvement, we test the classifier without any feature extractor. We set up the FFT-2PCA with parameters as follows: $K = 200$, $pc1 = 100$. The $pc2$ parameter assumes the values: 1000, 2000, 3000, 4000, 5200. These values were chosen to maintain the number of pcs consistently spaced, as in our baseline that configures the PCA with pc values: 10, 20, 30, 40, 52. The ICA and PCA components follow our baseline setup. Each feature extractor configuration paired with a classifier execute all seven datasets 30 times. Then, the F1 results are summarized by dataset calculating the mean for all configurations. To test the statistical significance, we use the Kruskal-Wallis test with a 99% confidence interval.

The ANN was built using a similar architecture of our baseline, with an input layer that matches the dimensionality of PCA or FFT-2PCA output, two hidden layers with 40 and 15 neurons, respectively, and one neuron as the output layer. Except for the output layer that uses a sigmoid, all layers use Relu [10] as the activation function. For the loss function, we use binary cross-entropy. For the KNN classifier, we chose a k-value as five, uniform weights and Minkowski [9] as the distance metric. The SVM classifier is used in Linear [6] form, to handle the amount of data, with the l2 penalty and squared hinge as the loss function. We use two libraries as resources for our method, Keras to build and train the Neural Networks, and Scikit-learn for FFT, PCA, ICA, SVM, and KNN.

Timing Experiments. The experiments regarding the timing characteristics for the pairs feature extractors/classifiers follow the same configuration used in the performance setup. We begin measuring the required time for the feature extraction phase. Also, we perform a time testing for the loading phase, to isolate the weight of the feature extraction in this first phase. Following the timing experiments, we capture the training time spent by each pair feature extractor/classifier. Then, we calculate the mean for each dataset, to verify the impact of the size of the different datasets on the timing results. The statistical test used was the Wilcoxon test, with a 99% confidence interval. The machine used in all timing tests has a processor Intel I5-6500 and 16 GB of DDR3 memory.

6 Results

Performance. Figure 1 shows the results of the experiments for each pair of feature extractor and classifier. Each box-plot inform the distribution, median, and mean of experiments using one of our seven datasets. The KNN classifier had the best performance between our three classifiers. Being a parametric algorithm, the KNN benefits from the dataset size and balance. It is worth notice that in a real fault detection application, these conditions are hardly achieved, but in this work, we use these artificial datasets to establish a parameter to comparison to our baseline and future works. Regarding feature extraction, the PCA had a small impact on the KNN performance; improving only 0.93% the overall performance. Our method, FFT-2PCA, did not improve the KNN classifier.

For the SVM and ANN, the FFT-2PCA improve the overall performance. As we can see on Fig. 1, only when applied to the dataset one the FFT-2PCA, in conjunction with the respective classifier, do not improve the F1 performance. More precisely, we achieve the best improvements when applying the FFT-2PCA with the ANN classifier. The increase in the F1 metric was 17.4%; in comparison with our baseline, the improvement was 48.7%.

Table 1. Overall F1 values for each pair Feature Extractor and Classifier.

ANN	F1	KNN	F1	SVM	F1
FFT-2PCA	**0.75**	PCA	**0.86**	FFT-2PCA	**0.73**
Plain	0.64	Plain	0.85	Plain	0.70
ICA	0.51	ICA	0.84	PCA	0.65
PCA	0.50	FFT-2PCA	0.82	ICA	0.62

Table 1 shows the overall results for each pair feature extractor and classifier. The overall F1 score is the mean of all experiments results across the seven datasets. Each column represents a classifier with rows populated by the feature extractor method; they are ordered by the F1 overall score presented in the right side of each classifier column.

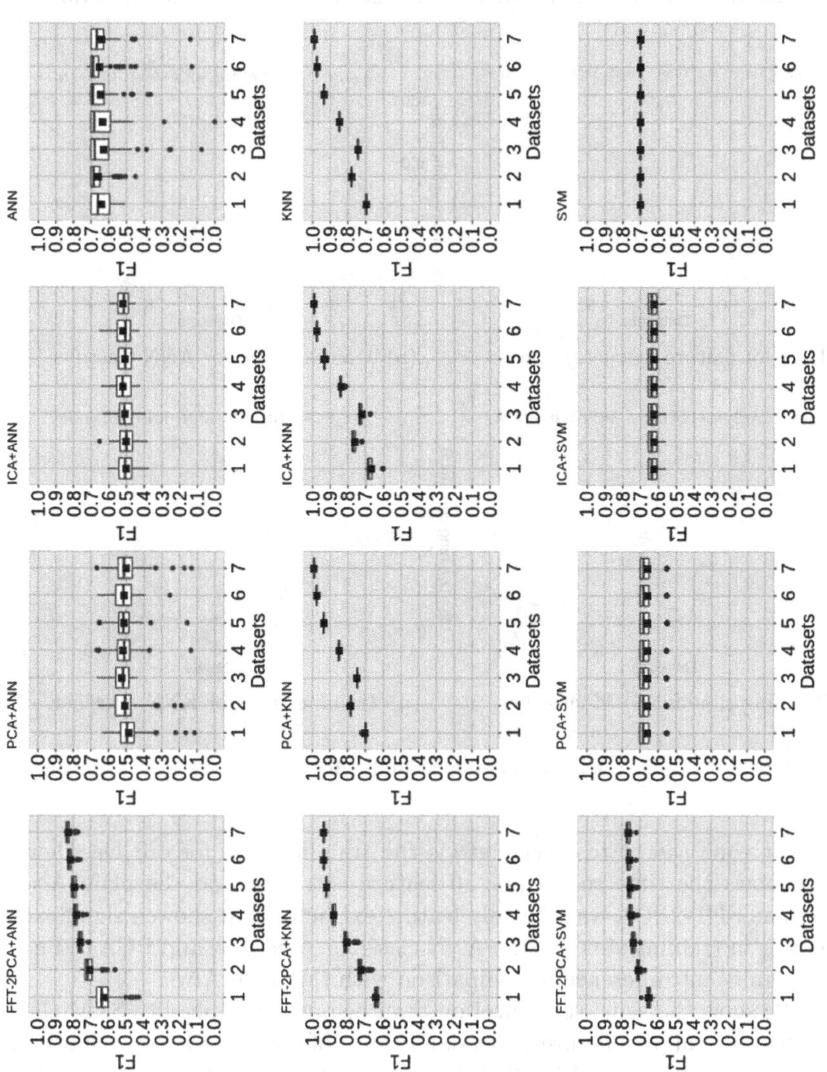

Fig. 1. Box plot graph of F1 measurement for all combinations of feature extraction setup and classifier.

Timing. Figure 2a shows the results for the data load time and feature extraction time applied on all seven datasets. The result for features extraction includes the data loading time too. The graph shows close time results for FFT-2PCA and the other two feature extractors, but, as the data size grows, they drifted. Figure 2a also show that the timing for PCA and ICA feature extractors are close. The PCA and ICA are the only pair that did not achieve statistical relevance.

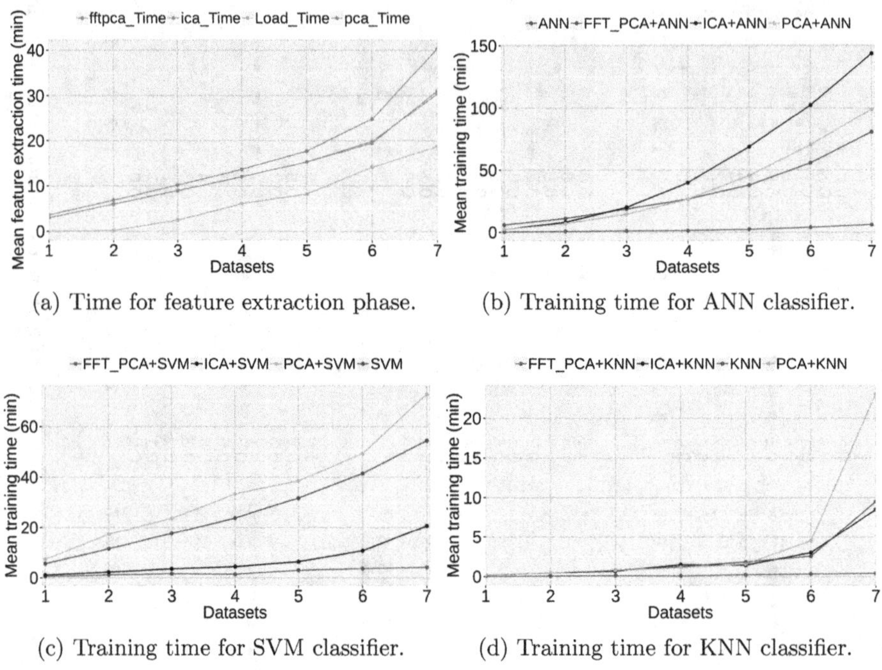

(a) Time for feature extraction phase. (b) Training time for ANN classifier.

(c) Training time for SVM classifier. (d) Training time for KNN classifier.

Fig. 2. Time measurements.

The next time aspect to investigate is the training time for our classifiers. Figure 2b shows the time results for all feature extractor and the plain ANN classifier. The FFT-2PCA was the only feature extractor that showed significant improvement in training time for the ANN classifier. The ICA and PCA did not present a significant difference in comparison with the plain ANN.

Figure 2c show the same configuration than before but applied to SVM classifier. In this case, all results present a significant difference between each other. It means that the ICA and FFT-2PCA can improve the training time for the SVM classifier.

Last training time graph, showed in Fig. 2d, shows the values for the KNN classifier. Again the only feature extractor that achieves a statistical improvement was the FFT-2PCA.

Table 2 shows the average time (in minutes) for the entire process, feature extraction, and training, for all seven datasets. In this table, we can see how

Table 2. Average feature extraction time + average training time (in minutes).

Datasets	FFT-2PCA			ICA			PCA			Plain		
	ANN	KNN	SVM	ANN	KNN	SVM	ANN	KNN	SVM	ANN	KNN	SVM
1	3.90	3.66	4.15	5.20	3.19	4.08	5.27	3.21	10.35	5.88	0.26	5.66
2	7.45	6.90	7.91	13.88	6.43	8.13	12.94	6.45	23.18	11.13	0.65	11.58
3	11.24	10.31	11.73	28.98	9.64	12.44	23.53	9.76	32.58	21.23	3.25	19.97
4	15.00	13.75	15.02	51.79	13.50	16.37	38.68	13.18	45.33	32.76	7.39	29.82
5	19.99	17.91	20.61	83.93	16.72	21.62	60.71	16.85	53.79	46.09	10.05	39.89
6	28.72	25.07	28.36	122.22	22.80	30.45	89.73	23.91	68.67	69.85	16.67	55.49
7	**46.51**	40.71	44.46	174.23	38.82	50.92	130.10	54.05	103.83	**99.38**	28.20	73.11

FFT-2PCA improve the development time of fault detection solutions. Again, our method, in conjunction with KNN, does not result in an improvement. However, for the SVM and ANN, the FTT-2PCA improve the total time. For the ANN case, the improvement is particularly interesting. As the dataset size increases, the impact of the training times becomes more relevant in the total time, causing the total time reductions brought by the FFT-2PCA to be more prominent. In the best case, a 53% reduction in the total development time of the solution.

Discussion. We begin our discussion reasoning about our classifiers nature. Two of our classifier, KNN, and SVM, are non-parametric classifiers. This trait has a few implications; One benefit is that the classifier can be more flexible, fitting a large number of functions. However, to achieve good performance, a large amount of data must be provided. We can see this behavior in Fig. 1, which is showing a major increase in F1 for the KNN classifier as the dataset size increases. With the fifth dataset, the gain reaches a plateau, and there is not a major increase. The SVM reach its limits right on the first dataset, except when using the FFT-2PCA as a feature extractor that provides a small gain.

Our third classifier, ANN, is a parametric algorithm. This property imposes less dependency on large quantities of training data but also brings more difficulties to fit the hidden function. We can see this caveat in Fig. 1, by looking to results for the ANN classifier. In this case, the F1 metric has more dispersion and more outliers when compared with the SVM and KNN.

With all these concepts in mind, we can think about the applicability of a given fault detection solution beyond the artificial datasets. In real applications, well-annotated fault behavior can be scarce, imposing a limit in size for the training datasets. This factor can limit the use of classifiers, such as KNN and SVM.

As our results showed, the FFT-2PCA brought the most critical improvement were achieved when combined with the ANN classifier. Not address only the problem of the better fitting, but also bring a large improvement in training times. These results show us a promising path for the development of a solution for fault detection based in ANN with our FFT-2PCA as a feature extractor.

7 Conclusion

This work introduces a feature extraction method for fault detection. This method called FFT-2PCA applies Fast Fourier Transform to a range of data points, after, a sequence of PCA reduces the dimensionality, then the resulting data entered in the classification algorithm. Our method was tested against other two feature extractors, ICA and PCA, and the classifiers without any preprocessing. The FFT-2PCA showed the best results when applied with the ANN classifier, a 17.4% improvement in performance and a 54% reduction in training time when compared with the plain ANN classifier. With these results, we can consider the FFT-2PCA a good option in the development of a fault detection solution based in ANN.

In our future work, we plan to expand our tests on the fault diagnosis problem. A fault diagnosis solution consists in not only detect a fault but identify the class that fault belongs. We intend to explore other aspects that we do not address in this work, such as the fault classes balance in our datasets.

References

1. Abdi, H., Williams, L.J.: Principal component analysis. Wiley Interdiscip. Rev. Comput. Stat. **2**(4), 433–459 (2010)
2. Blanke, M., Kinnaert, M., Lunze, J., Staroswiecki, M., Schröder, J.: Diagnosis and Fault-Tolerant Control, vol. 691. Springer, Heidelberg (2006). https://doi.org/10.1007/978-3-540-35653-0
3. Brigham, E.O., Brigham, E.: The Fast Fourier Transform and Its Applications, vol. 1. Prentice Hall, Englewood Cliffs (1988)
4. Comon, P.: Independent component analysis, a new concept? Signal Process. **36**(3), 287–314 (1994)
5. Downs, J.J., Vogel, E.F.: A plant-wide industrial process control problem. Comput. Chem. Eng. **17**(3), 245–255 (1993)
6. Fan, R.E., Chang, K.W., Hsieh, C.J., Wang, X.R., Lin, C.J.: LIBLINEAR: a library for large linear classification. J. Mach. Learn. Res. **9**, 1871–1874 (2008)
7. Frank, P.M., Blanke, M.: Fault diagnosis and fault-tolerant control. In: Control Systems, Robotics and Automation XVI (2007)
8. Jing, C., Hou, J.: SVM and PCA based fault classification approaches for complicated industrial process. Neurocomputing **167**, 636–642 (2015)
9. Kruskal, J.B.: Nonmetric multidimensional scaling: a numerical method. Psychometrika **29**(2), 115–129 (1964)
10. Nair, V., Hinton, G.E.: Rectified linear units improve restricted Boltzmann machines. In: Proceedings of the 27th International Conference on Machine Learning (ICML 2010), pp. 807–814 (2010)
11. Nashalji, M.N., Arvand, S., Norouzifard, M.: Integration of principal component analysis and neural classifier for fault detection and diagnosis of Tennessee Eastman process. In: 2014 4th International Conference on Engineering Technology and Technopreneuship (ICE2T), pp. 166–170. IEEE (2014)
12. Yin, S., Ding, S.X., Haghani, A., Hao, H., Zhang, P.: A comparison study of basic data-driven fault diagnosis and process monitoring methods on the benchmark Tennessee Eastman process. J. Process Control **22**(9), 1567–1581 (2012)

Consistency, Integrity, Quality of Data

A DaQL to Monitor Data Quality in Machine Learning Applications

Lisa Ehrlinger[1,2(✉)], Verena Haunschmid[1,3], Davide Palazzini[4], and Christian Lettner[1]

[1] Software Competence Center Hagenberg GmbH, Hagenberg, Austria
{lisa.ehrlinger,christian.lettner}@scch.at
[2] Johannes Kepler University Linz, Linz, Austria
lisa.ehrlinger@jku.at
[3] Institute of Computational Perception, Johannes Kepler University Linz, Linz, Austria
verena.haunschmid@jku.at
[4] University of Bergamo, Bergamo, Italy
d.palazzini@studenti.unibg.it

Abstract. Machine learning models can only be as good as the data used to train them. Despite this obvious correlation, there is little research about data quality measurement to ensure the reliability and trustworthiness of machine learning models. Especially in industrial settings, where sensors produce large amounts of highly volatile data, a one-time measurement of the data quality is not sufficient since errors in new data should be detected as early as possible. Thus, in this paper, we present DaQL (Data Quality Library), a generally-applicable tool to continuously monitor the quality of data to increase the prediction accuracy of machine learning models. We demonstrate and evaluate DaQL within an industrial real-world machine learning application at Siemens.

Keywords: Data quality · Machine learning · Trust

1 Introduction

Poor data in companies can lead to effects like incorrect decision making, cost increase, customer dissatisfaction, and organizational mistrust [9]. Although the importance of data quality (DQ) has been recognized in the business community already in the 90s [13], it has been largely ignored in the machine learning (ML) and artificial intelligence (AI) community in favor of focusing on the algorithms and methods themselves [12]. DQ is often described from a consumer point of view, who can judge whether the data is *fit for use* [13]. In the context of ML, *fitness for use* can be associated with the performance of a trained ML model. If poor-quality data (e.g., values are missing or incorrect) is used for ML, the employed models will also perform poorly and generate unreliable predictions.

S. Hartmann et al. (Eds.): DEXA 2019, LNCS 11706, pp. 227–237, 2019.
https://doi.org/10.1007/978-3-030-27615-7_17

Thus, the key to high-quality ML are the three principles of DQ: *prevention*, *detection* and *correction* [2]. In the frame of a real-world ML application at Siemens (previously introduced in [3]), we have deployed a framework for DQ-driven ML that allows to follow those three principles. *Prevention* of known DQ issues is achieved with an ETL (extract, transform, load) process during the integration of the data into a central repository. This central repository holds all data used for the ML application. The *detection* of new problems within the repository is performed with DaQL (Data Quality Library), a novel library to continuously monitor DQ. Eventually, and based on the DQ measurements, DQ issues are corrected on-the-fly and depending on the respective ML model.

The focus and main contribution of this paper is the introduction of DaQL, a generally-applicable library for *continuous data quality measurement* (CDQM), which constitutes the basis for comprehensive DQ-driven ML. CDQM describes the calculation and storage of DQ metrics over time in order to ensure that the qualitative condition of the data remains stable [5,10]. The applicability and usefulness of DaQL is demonstrated by its application to an existing ML application at Siemens. We show that the ML model performance increases when DQ is monitored with DaQL and corrected according to the measurement results.

In Sect. 2, we describe the design decisions and architecture of DaQL, and demonstrate it in Sect. 3. Section 4 presents a case study that shows the impact of poor DQ on ML and how model performance could be increased with DaQL. In Sect. 5 we distinguish DaQL from related DQ tools.

2 DaQL: A Library for Data Quality Monitoring

DaQL has been implemented along the lines of [5] and can be divided into three components shown in Fig. 1: (1) the core Python library with the newly developed DaQL language, (2) the Jupyter-based user interface, and (3) two different engines for adjusting the scalability: `pandas` and `spark`.

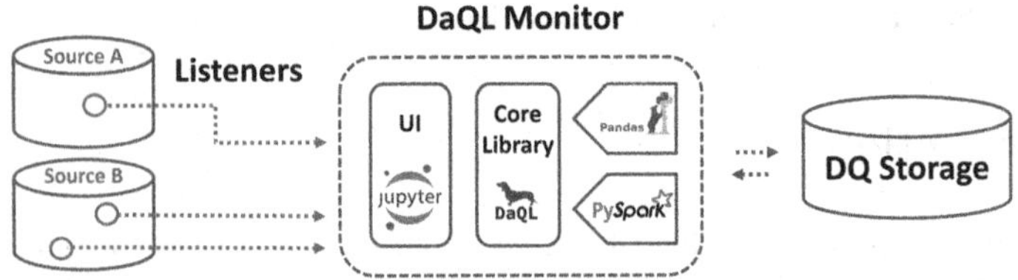

Fig. 1. Architecture of DaQL

There are four core requirements a CDQM framework should fulfill, which have been identified in [5]: (1) DQ measurement capabilities, (2) storage of DQ measurement results, (3) analysis of DQ measurement results, and (4) the automation of DQ measurement. In the following paragraphs, we describe each of those requirements shortly and describe the adaptions to our use case along with implementation details for DaQL. We also added an additional requirement, which is specifically important for industrial applications: scalability.

(R1) DQ measurement. The question "how DQ should actually be measured" is one of the biggest challenges for practitioners [10]. One reason is that DQ is typically described with "dimensions", where not all are well defined and no standard is commonly agreed-on [10]. We refer to [10] for an objective discussion of this circumstance. We, in contrast, aim at a practical approach by focusing on relevant DQ errors that have been detected at our industry partner and are of special interest for ML data (cf. [1,7]): missing values (e.g., `null`, 0.0, `Inf`) or missing records, outliers, and values outside a given range.

(R2) Storage. Following the suggestion from [5], DaQL stores CDQM results (`Mes`) with the respective metric name (`MetricName`) and a timestamp (`DQTime`) for later analysis. Since in industrial environments, databases (DBs) from the company's IT infrastructure can usually be reused, DaQL allows an exchange of the DQ storage. At Siemens, DQ measurements are stored in an MS SQL server, which was already used for the ML application we applied DaQL for.

(R3) Analysis. Visual analytics (e.g., tables, histograms) are required to present the DQ measurements and help a user to detect patterns and sudden changes in DQ [5]. The requirement in DaQL was an intuitive display of the DQ measurements to allow a user to identify the records violating a given constraint. We used Jupyther notebooks for the implementation of a flexible user interface.

(R4) Automation. The CDQM tool should be executed automatically and continuously store measurement results [5]. Since the exact schedule of the DQ measurements depends on the customer environment (typically on the ETL job), DaQL allows application-specific configuration with `crontab` to schedule the execution of the Jupyter notebooks. At Siemens, we schedule DQ checks daily.

(R5) Scalability An additional requirement to [5] was the efficient processing of small and large amounts of data, which justifies the design decision for the two different engines `pandas` and `spark`.

2.1 DaQL Core and DaQL Language

The core library of DaQL is implemented in Python and provides a novel high-level language to formulate DQ checks, which can also be used by non-data-engineering experts. DaQL language follows the lazy evaluation strategy, which avoids repeated execution since it delays DQ check evaluation to the time when the results are required. It also acts as abstraction layer for the two data engines.

The basic concept is a *dataset*, which can, e.g., be a table in a relational DB or an SQL statement. DaQL accesses datasets through *listener* objects, which are registered to a specific dataset. To perform referential integrity checks, *lookups* to other listeners can be applied. DQ checks can be implemented by applying one or several *constraints* to these listeners, to which the observed data entities must conform. The resulting DQ measurements, for example, the number of records that violate a given constraint, can be plotted or stored in the DQ storage.

It is either possible to define a new constraint in form of a lambda expression, or to reuse a predefined constraint function for very common data errors, e.g.,

Constraints.ISNULL, Constraints.ISZERO, or Constraints.ISNEGATIVE. We implemented predefined constraints for the three error types listed in Sect. 2:

1. A certain value must be *present*. Here null checks are not sufficient, since, depending on the measuring device, also 0.0 or Inf might be transmitted. Further, if a record is missing, referential checks are needed.
2. A certain value must not lie too far away from the rest of the values. For detecting outliers, we apply a proximity-based method as described in [1] by computing a range for "normal" values. The range is defined by the mean and 3 times the standard deviation ($t_{min} = \bar{x} - 3 * s_x$ and $t_{max} = \bar{x} + 3 * s_x$), where the multiplier 3 is a standard value used in statistic process control [10].
3. A certain value must be within a user-specified range (between t_{min} and t_{max}) or it is bounded below (t_{min}) or bounded above (t_{max}).

The following operations are currently supported by the DaQL language:

- applyConstraint(column1, [column2,] func) applies a constraint function (func) upon one or two columns of the dataset.
- lookup(keyColumn, lookupDataset) applies a referential integrity check against a lookup dataset. A lookup dataset can be defined on a dataset by specifying a certain lookup column: Lookup(dataset, lookupColumn).
- success() sets the mode of the previous operation to success, i.e., the resulting dataset contains all rows for which the previous operation evaluated to true (e.g., for applyConstraint) or was successful (e.g., for lookup). If not specified, success is considered to be the default mode for every operation.
- failure() sets the mode of the previous operation to failure, i.e., the resulting dataset contains only rows for which the previous operation evaluated to false or no lookup could be found.
- union(otherDataset) merges two datasets that must be from one listener.
- count() executes a DQ operation and returns the resulting number of rows.
- collect() executes the defined DQ operations and returns the result set as a pandas DataFrame or spark RDD (Resilient Distributed Dataset), depending on the chosen engine.
- store(metricName) stores the value of the DQ metric result in the DQ storage, which can then be analyzed using a TimeSeries object.

The DaQL language also allows to define temporal window functions over data streams. Time constraints (which are based on the time when the monitored data entity is created or modified) can be applied to those window functions. The field that determines the temporal ordering of the data instances must be specified when the listener is defined. The full use of temporal windows can be achieved with the implementation of stream-based DaQL engines, which is planned for future work. In the meanwhile it represents a convenient way to specify temporal constraints, although they are executed in batch-form.

2.2 DaQL User Interface

The implementation of the user interface (UI) is realized with Jupyter Notebooks since they can easily be copied and modified. DQ measurements can be displayed as scalars or formatted as tables as shown in Sect. 3. We developed two types of notebooks: (1) notebooks for computing DQ measurements, which provide a view on current results and store these measurements in the DQ storage, and (2) notebooks to visualize the history of DQ measurement results over time. An observation of the change of DQ metrics over time can help detecting issues in the data processing pipeline, e.g., a sudden increase in missing values.

2.3 DaQL Data Processing Engines

In many industrial applications, the amount of data is continuously increasing and thus, the scalability of DaQL is essential for successful data management. DaQL allows to choose how data should be handled by switching between two engines: `pandas` and `spark`. Currently, batch-based implementations exist for `pandas`, which is better suited for smaller datasets that fit into main memory, and for `Apache Spark`, which is designed to distribute Big Data across several nodes. In DaQL, `pandas` is set as default engine.

3 Application and Demonstration

We demonstrate the application of DaQL to monitor DQ for an existing ML application at Siemens. The aim of the ML application (previously described in [3]) is to optimize the slitting of metal coils used for building power transformers. The implemented optimization algorithm can be seen as a generalization of the cutting-stock problem described in [6] with multiple objectives and constraints. The objectives of the optimization are the reduction of production times, manual machine adjustments, and the optimization of physical power transformer properties. These physical transformer properties are predicted on-the-fly from the current slitting plan during optimization, using a previously learned ML model [3]. The models are trained on transformer features (e.g., size), the used coils (in terms of quality and amount) and the measured coil properties. Those coil properties are aggregated to create a new transformer feature [3]. For example, an input feature of the ML model $\tilde{P}$ can be computed as the weighted (w_j) sum over the measured property (p_j) for each of the used coils j: $\tilde{P} = \sum_{j=1}^{c} w_j p_j$.

Unfortunately, we are not allowed to publish details about the measured properties and the amount of data at Siemens due to privacy concerns. Thus, we created an anonymized dataset, which closely resembles the structure of the original data, but has only one generic type of `MeasuredValue`. For the code listings in this section, we use a table `CoilStock` that contains 1,000 coils with the attributes `BookingDate` (when the coil was "booked" into the system) and `Weight`. The table `Measurement` stores about 30,000 measurement values (`MeasuredValue`) along with the date when the measurement was taken (`MeasureDate`) and the respective `CoilID`. In our dataset, only 992 coils have assigned measurement values. We also inserted measurements, where the measured value is 0.0 or `null` instead of a proper value.

Missing Values. A common fault in our ML application are `MeasuredValue` entries of 0.0 or `null` instead of proper values. Listing 1 shows the definition of a listener for the table `Measurement`, along with a single constraint to observe 0.0. Here, we use one of the predefined constraints `Constraints.ISZERO`, which implements the lambda expression $x : x == 0$. There are two possibilities to display the results of the listener: (line 3) to count the number of rows that violate the constraint, or (line 4) to display the violating rows in table-form. In addition, the DQ results can be stored in the DQ storage (line 5). Here, the name of the DQ metric is used as identifier to retrieve the stored records later on. If 0.0 and `null` entries do not have to be considered separately, the `union` function can be used to combine two or more constraints as shown in line 7. The count can be used to calculate higher-level DQ metrics, e.g., 95% of values are complete and do not contain 0.0 or `null` respectively.

```
1 | lst0 = Listener('Measurement')
2 | trn0 = lst0.applyConstraint('MeasuredValue', Constraints.ISZERO)
3 | trn0.count() // output: '24'
4 | trn0.collect().head()
5 | trn0.store('Measurement_MeasuredValue_0')
6 | trn1 = lst0.applyConstraint('MeasuredValue', Constraints.ISNULL)
7 | trn2 = trn1.union(trn0)
```

Listing 1. Monitor '0' and `null` values

Missing Records. Another DQ problem are missing records. This occurs when no measurements are made for a certain coil, which requires a check against a reference table (in our application the table `CoilStock`, which contains all coils). For such scenarios, a `lookup()` can be applied to a listener on the reference table. A second listener needs to be defined for the table with potentially missing measurements and the columns that define the relationship. The result set contains all CoilIDs (e.g., "91354") with their booking date (e.g., "2018-07-22").

```
lst1 = Listener('CoilStock')
lkp = lst1.lookup('CoilID', Lookup(lst0, 'CoilID'))
trn = lkp.failure()
```

Listing 2. Check missing records with a lookup

Consistency Checks. DaQL also allows to formulate and monitor different kinds of consistency checks using lambda expressions. In our application, it is for example important to ensure that the measurement for a coil is performed after the coil was booked into the system. To verify such a consistency rule, a lookup is performed on the table `Measurement` to get all rows for which a reference in the table `CoilStock` is available (i.e., invoking the `success` method). Listing 3 shows the definition of the constraint, which returns all records where the measure date is smaller than the booking date.

```
lkp = lst0.lookup('CoilID', Lookup(lst1, 'CoilID', 'BookingDate'))
trn0 = lkp.success()
trn1 = trn0.applyConstraint('BookingDate', 'MeasureDate', lambda x,y: x >
    ↪ y)
```

Listing 3. Constraint definition on two columns

Outlier Detection. To identify outliers, we calculate predefined lower and upper limits and extract the values that fall outside this range. The constraint in line 1 from Listing 4 is necessary to compute the mean without taking into account '0' measurements. The resulting interval after computing the lower and upper limits is $[0.869, 1.131]$. A subset of the resulting outliers (from which all exceed the upper limit) is: [1.154563, 1.139064, 1.170858, 1.158262, 1.168459].

```
1 trn0 = lst0.applyConstraint('MeasuredValue', lambda x: x != 0)
2 df = trn0.collect()
3 up = df['MeasuredValue'].mean() + 3*df['MeasuredValue'].std()
4 lo = df['MeasuredValue'].mean() - 3*df['MeasuredValue'].std()
5 trn1 = trn0.applyConstraint('MeasuredValue', lambda x: x < lo or x > up)
6 trn1.collect().head()
```

Listing 4. Outlier monitoring

Time Series Analysis. As shown in Listing 1, DQ measurement results can be stored in the DQ storage. DaQL provides `TimeSeries` objects, which can be used to query and visualize historic DQ measurements by providing the name of the DQ metric. Listing 5 shows how to query the respective data from the DQ storage and Fig. 2 visualizes the respective DQ measurement over time.

```
ts2 = TimeSeries('Measurement_MeasuredValue_OUTLIER')
ts2.plot('2018-07-21')
```

Listing 5. Querying and plotting of TimeSeries objects

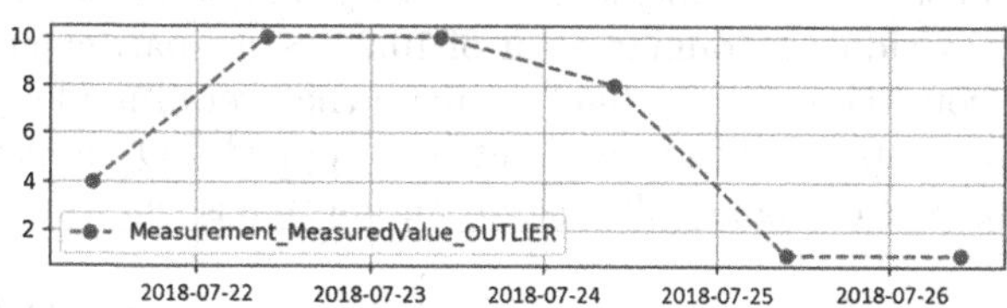

Fig. 2. Plotting the DQ metric *Mesurement_MeasuredValue_OUTLIER* over time

4 DaQL Evaluation and the Impact of DQ on ML

To assess the efficacy of DaQL at Siemens, we conducted a case study within the frame of the ML application, which was described in [3]. The results of this study demonstrate the impact of poor DQ on ML model performance.

For the evaluation, we observed a transmission error in the measurement device, which resulted in storing 0.0 instead of proper measurement values in the DB. In our example, this measurement value refers to p_j of the ML model $\tilde{P}$. To demonstrate the impact of this DQ problem on the performance of the resulting formula, we built three models with different strategies to deal with faulty data: (1) ignoring the problem, continuously measuring faulty values with DaQL and (2) removing faulty samples, or (3) correcting them. The three strategies are discussed below. Figure 3 shows the predictions of each model plotted against the measured true values with the training error and the number of samples used for training. The training error is calculated as a weighted root mean squared error (RMSE), where underestimated predictions imply a higher cost than overestimated predictions. Samples with faulty values are marked by red crosses.

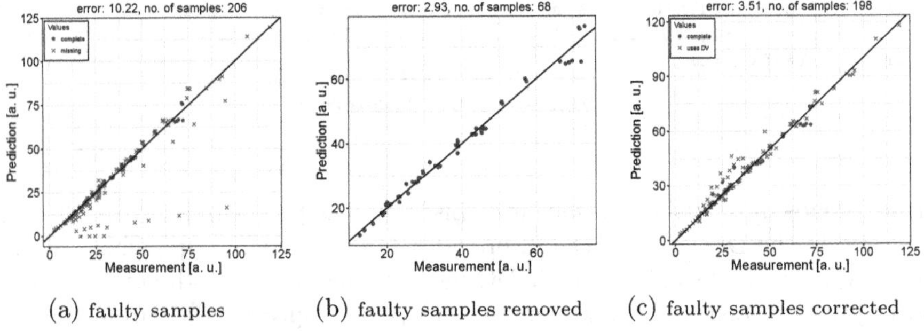

(a) faulty samples (b) faulty samples removed (c) faulty samples corrected

Fig. 3. Error for models trained on datasets with different quality. (Color figure online)

Strategy 1: Ignoring the DQ Problem. Figure 3a shows that it is generally possible to train a model using all data including the faulty samples, but the performance is not satisfying, yielding a training error of 10.22. Some particularly poor samples (far away from the reference line) decrease the overall model performance. This model represents a solution without any applied DQ checks. This naive trust that the data is of high quality is not an unusual scenario in practice [10].

Strategy 2: Removing Faulty Samples. In this scenario, we used DaQL to detect all faulty samples, where at least one coil is 0.0 and simply removed those samples from the training data. In the following, Fig. 3b shows that a model with satisfying performance (error: 2.93) can be trained. However, no predictions are possible for transformers with a lot of faulty and consequently removed data later on. Data deletion is the easiest method to deal with faulty data in the context of ML, and is called complete-case analysis [7] in terms of missing values.

Strategy 3: Correcting Faulty Values. Figure 3c shows the performance of a model trained on data where the faulty 0.0 values have been detected with DaQL and corrected with the DQ framework surrounding DaQL. The implemented data correction component uses domain knowledge about the typical transformer properties from experts, which has been encoded in rules. The corrected values are estimated slightly pessimistically to prevent underestimation of the transformer property. This fact can be observed in Fig. 3c since several samples with corrected values (indicated by the red crosses) lie above the reference line and are predicted a little bit too high. A small amount of samples (8 of 206) could not be included in this model since no suitable rule could be found by the data correction component. However, in general, the model shows with a training error of 3.51 a performance similar to the model based on strategy 2, while using the majority of samples for training. The solution with strategy 3 leads to high quality ML models and can also be used to predict the majority of samples in contrast to the slightly better performing model in Fig. 3b.

This analysis clearly demonstrates the impact of poor data on ML model performance, and consequently the need to *detect* and *correct* DQ problems in ML applications. In order to understand the degree and effectiveness of data correction and to define goals for further DQ improvement activities, it is previously necessary to measure and know the quality of the data [10]. Thus, without a defined DQ measurement strategy (in our case using DaQL), it is not possible to identify and correct DQ issues properly. However, with this evaluation, we also want to highlight the importance of defining the DQ improvement process. Simply excluding data will in some cases result in models with desirable performance, but preclude the further application of ML models. In strategy 2 of our case study at Siemens, many excluded transformer samples had only few faulty measurement values for the assigned coils.

5 Related Work

None of the DQ tools observed in [8] offers monitoring capabilities, and the focus of most tools is on data profiling and data cleansing. Gartner [11] indicates that monitoring (i.e., the deployment of controls to ensure that data conforms to rules) is one of the core functionalities of a DQ tool. The majority of existing tools that support DQ monitoring refers to a specific implementation.

DaQL in contrast, is domain-independent, and can be applied to monitor the DQ within any DB. To the best of our knowledge, we found three tools that can be compared to DaQL: Apache Griffin[1], MobyDQ[2], and QualIe [4]. While QualIe implements DQ dimensions and metrics as proposed in literature [4], we chose a more practical approach with DaQL by implementing DQ checks with constraints (both: predefined constraints and leaving the user to define new ones). Likewise, Apache Griffin implements a DQ dimension for "accuracy", which needs to be further specified by the user. MobyDQ aligns with DaQL in the type of DQ checks, but focuses on DQ checks along the data pipeline

[1] http://griffin.apache.org (June 2019).
[2] https://github.com/mobydq/mobydq (June 2019).

(between DBs) in contrast to the focus on one centralized ML application by DaQL. Thus, to the best of our knowledge, DaQL is the first library to offer continuous DQ measurement that is dedicated to improve the performance of ML models and tackles DQ issues from the practice perspective.

6 Conclusion and Outlook

In this paper, we introduced DaQL, a library to monitor the quality of data employed for ML applications. We demonstrated the efficacy of DaQL by deploying it to a ML application at Siemens and achieved increasing model performance.

In future work, we are going to add duplicate detection and extend outlier detection with more sophisticated methods. The DaQL language will be expanded from batch-based to stream-based execution. In addition, we are going to extend the time-series plots with ML methods to automatically detect patterns and sudden changes to support early fault detection, as proposed by [5,10].

Acknowledgments. The research reported in this paper has been supported by the Austrian Ministry for Transport, Innovation and Technology, the Federal Ministry for Digital and Economic Affairs, and the Province of Upper Austria in the frame of the COMET center SCCH.

References

1. Aggarwal, C.C.: Outlier Analysis, 2nd edn. Springer, Heidelberg (2017). https://doi.org/10.1007/978-3-319-47578-3
2. Chapman, A.D.: Principles of data quality. Technical report, Global Biodiversity Information Facility Material (2005)
3. Chasparis, G., Zellinger, W., Haunschmid, V., Riedenbauer, M., Stumptner, R.: On the optimization of material usage in power transformer manufacturing. In: Proceedings of the 8th International Conference on Intelligent Systems. IEEE (2016)
4. Ehrlinger, L., Werth, B., Wöß, W.: Automated continuous data quality measurement with QuaIIe. Int. J. Adv. Softw. **11**(3 & 4), 400–417 (2018)
5. Ehrlinger, L., Wöß, W.: Automated data quality monitoring. In: Proceedings of the 22nd MIT International Conference on Information Quality (ICIQ 2017), pp. 15.1–15.9 (2017)
6. Gerstl, A., Karisch, S.E.: Cost optimization for the slitting of core laminations for power transformers. Ann. Oper. Res. **69**, 157–169 (1997)
7. Pigott, T.D.: A review of methods for missing data. Educ. Res. Eval. **7**(4), 353–383 (2001)
8. Pushkarev, V., Neumann, H., Varol, C., Talburt, J.R.: An overview of open source data quality tools. In: Proceedings of Information and Knowledge Engineering Conference, pp. 370–376 (2010)
9. Redman, T.C.: The impact of poor data quality on the typical enterprise. Commun. ACM **41**(2), 79–82 (1998)
10. Sebastian-Coleman, L.: Measuring Data Quality for Ongoing Improvement: A Data Quality Assessment Framework. Newnes, New York (2012)

11. Selvege, M.Y., Judah, S., Jain, A.: Magic quadrant for data quality tools. Technical report, Gartner, October 2017
12. Sessions, V., Valtorta, M.: The effects of data quality on machine learning algorithms. In: Proceedings of the 11th International Conference on Information Quality (ICIQ 2006), vol. 6, pp. 485–498 (2006)
13. Wang, R.Y., Strong, D.M.: Beyond accuracy: what data quality means to data consumers. J. Manage. Inf. Syst. **12**(4), 5–33 (1996)

Automated Detection and Monitoring of Advanced Data Quality Rules

Felix Heine[1], Carsten Kleiner[1(✉)], and Thomas Oelsner[2]

[1] University of Applied Sciences and Arts Hannover, Hannover, Germany
{felix.heine,carsten.kleiner}@hs-hannover.de
[2] mVISE AG, Hamburg, Germany
thomas.oelsner@mvise.de

Abstract. Nowadays business decisions heavily rely on data in data warehouse systems (DWH), thus data quality (DQ) in DWH is a highly relevant topic. Consequently, sophisticated yet still easy to use solutions for monitoring and ensuring high data quality are needed. This paper is based on the IQM4HD project in which a prototype of an automated data quality monitoring system has been designed and implemented. Specifically, we focus on the aspect of expressing advanced data quality rules such as checking whether data conforms to a certain time series or whether data deviates significantly in any of the dimensions within a data cube. We show how such types of data quality rules can be expressed in our domain specific language (DSL) RADAR which has been introduced in [10]. Since manual specification of such rules tends to be complex, it is particularly important to support the DQ manager in detecting and creating potential rules by profiling of historic data. Thus we also explain the data profiling component of our prototype and illustrate how advanced rules can be semi-automatically detected and suggested to the DQ manager.

Keywords: Data quality · Domain specific language · Advanced data quality monitoring · Rule based data quality · Time series rules · Automated rule generation · Data heterogeneity

1 Introduction

Data is the core of many modern businesses. Data is typically stored in different distributed databases and business decisions rely on an integrated view on this data. In order to take correct decisions, it is important that data at the foundation is correct. In line with requirements of our partners, we focus on data quality in integrated data warehouses. Yet, the DSL presented in this paper is also applicable to heterogeneous, distributed data sources. An important part of data quality monitoring is checking internal consistency of databases and conformance with external data sets. These checks can

---

The project IQM4HD has been funded by the German Federal Ministry of Education and Research under grant no. 01IS15053A. We would also like to thank our partners SHS Viveon/mVise for implementing the prototype and CTS Eventim for providing important requirements and reviewing practical applicability of the prototype and concepts.

S. Hartmann et al. (Eds.): DEXA 2019, LNCS 11706, pp. 238–247, 2019.
https://doi.org/10.1007/978-3-030-27615-7_18

be rather simple checks like NOT NULL, structural checks like referential integrity, or complex statistical checks like conformance of the data to a certain distribution or time series model. We integrate all these kinds of checks in our data quality DSL RADAR.

Whereas the general design of the language and the specification and execution of rather simple quality rules has been discussed in [10], this paper will present advanced data quality rules such as time series and multidimensional cube checks. These advanced rules are of particular interest for the DQ manager as they facilitate to detect deeply rooted data quality issues such as erroneous data provisioning processes or malfunctioning external data sources. Specification of advanced rules requires more effort than for simple rules as many more parameters need to be defined (e. g. a histogram of the desired target distribution of a certain attribute). Consequently, it is even more important to support the DQ manager as much as possible in order to make definition of rules as easy as possible. In addition to the separation of concerns in RADAR (cf. Sect. 3.1) we also provide a data profiling component which can suggest data quality rules to the DQ manager based on analyzing historical data.

The prototype has been developed in the research project IQM4HD[1]. The goal of this project was creating a solution for continuous data quality monitoring of heterogeneous data in form of a rule based system with the addition of automated rule detection.

An important novel contribution of this paper is the specification of advanced data quality rules such as time series checks and multidimensional cube checks that can be monitored automatically. Moreover these rules cannot only be checked in relational data sources but also in other types of data source (e. g. document databases) as our quality language operates on an intermediary internal data model abstracting from the concrete source. In addition, data profiling in combination with the (semi-)automated generation of data quality rules is another aspect that has not been subject of any other publication to the best of our knowledge. Finally, the prototypical implementation showing the feasibility of the approach is another important contribution.

In this paper Sect. 2 covers work related to our approach. The main concepts of the DSL are presented in Sect. 3.1, whereas Sect. 3.2 presents the specification of advanced quality rules such as time series rules together with an illustrative example. The following Sect. 4 presents the profiling component of our DQ system with a specific focus on generation of advanced DQ rules based on profiling. Finally, the paper closes with Sect. 5 where a conclusion and an outlook is given.

2 Related Work

Data Quality and Monitoring Systems. Data quality is a topic with many facets, a lot of different dimensions of data quality exist. The predecessor of this project, the Data Checking Engine (DCE) [7], had also been developed in our group. It is an SQL-based system for complex data quality rules. The DCE provides an interface for data quality managers to compose data quality rules in form of SQL statements. Even though templates were introduced to reduce the complexity of rule composition, especially

[1] http://iqm4hd.wp.hs-hannover.de/english.html.

regarding the creation of multiple rules of the same type for a lot of different data, the complexity of SQL and the underlying database schema still has to be known by the domain expert. Also, once written, rules were hard to read and to fully comprehend. Therefore, this project, as a successor, and especially the DSL copes with the mentioned problems the DCE has, in form of the separation of concerns via the split into Sources, Checks and Actions.

Recently, [12] presented an approach with a very similar goal to our work, namely automating data quality checks in large data sets. Whereas the goal and some of the features are similar to our approach (e. g. presence of a wide range of basic types of checks, methods to automatically suggest potential quality rules, integrating heterogeneous data sources), significant differences are also present. They use a more sophisticated optimization procedure to efficiently process the checks based on Spark and address differential quality checking in detail (which is not applicable to all checks in our system). On the other hand we provide more advanced data quality checks such as ARIMA time series model based checks or multidimensional checks, require less programming knowledge by the DQ manager by separating concerns in the rule definition, have a comprehensive profiling component that can also detect complex quality rule candidates and facilitate what Jack Olsen calls "value rules" [11], which are checks that output a score instead of a clear correct/error decision. To this end both approaches complement each other. All in all, the extension of compatibility for heterogeneous data sets based on a well-defined internal data model, the split between technical and subject-specific knowledge and the advanced types of quality checks such as ARIMA set our approach apart from existing solutions.

Advanced Rules and Profiling. An overview of data profiling can be found in [1]. There is a wide range of research activities about various constraint types and efficient methods to detect them. As an example, there has been research about extensions to FD in order to improve them. One result are conditional functional dependencies (CFD) as described in [3] or even in the context of data quality rules in [5]. Many techniques for discovering CFDs have been presented, an overview can be found in [4]. Such methods are useful for our rule detection engine. All kinds of outlier detection algorithms [2] are interesting for our work as well. Both, unsupervised algorithms can be implemented and provided as rule types, as well as supervised algorithms that use some kind of model. Here, we are specifically interested in time series outlier detection [6] and outlier detection in multidimensional data (see e.g. [9]), as this fits perfect with typical data warehouses.

3 Advanced Data Quality Rules in RADAR

3.1 Brief Overview of RADAR

This chapter briefly introduces the proposed language RADAR to specify data quality rules. We only present the most important components that are required to understand the example of specifying advanced data quality rules thereafter. More details of the language itself can be found in [10] for the interested reader.

In order to compose data quality rules one needs to have information about (i) *what data should be checked*, (ii) *what quality aspect should the data be checked for* and

Listing 1. DSL code for a Customer Source from a Document DB with Projection.

```
SOURCE Customer TYPE LIST QUERY ROLES (_id: IDENTIFIER):
NATIVE
  db.customer.aggregate( [ {
    $project : { "name" : 1, "dob" : 1,
                 "email" : 1, "ssn" : 1 } } ] )
END
```

Listing 2. DSL code for a unique check.

```
CHECK UniqueCheck ON LIST src(val):
  RETURN
    SELECT val ROLE value, COUNT(*) ROLE value
    FROM src
    GROUP BY val
    HAVING COUNT(*) > 1;
END
```

lastly (iii) *what should be done with the result.* These aspects are typically handled by different roles, namely database expert and DQ manager (depending on the complexity of the desired specification a software developer might also be required). These three aspects are reflected in the quality rules in RADAR by three different components, namely Sources, Checks and Actions.

Sources define the data to be checked by mapping the relevant external data entities to a RADAR internal representation used within the other components of the language based on the external data source's native language. These are specified by a database expert. Checks define what and how data should be checked in order to assess data quality. A simple example would be whether two different data sources consist of same number of distinct elements. Checks are generic in that they only define what comprises the check and not which data it shall be applied to. In addition Checks can have an arbitrary number of parameters to configure them. Implementing completely new Checks requires software development knowledge, but many generic Checks are already predefined in our prototype and should be sufficient for many cases. Finally, Actions are used to define what data should be subject to which Check and how the results should be interpreted. These require domain knowledge and are thus specified by the domain expert. Note that results of a quality check are not always of binary nature but may be a numeric quality measure, e. g. how different a certain distribution is from its reference distribution. The DQ manager has to decide which values of this measure are critical or worth a warning.

As a first practical, purposely very simple (in order to focus on the DQ language aspect), example consider checking whether data in a certain value field of customer documents from a MongoDB contains a unique SSN and in another field a unique e-mail address. We have to define the data to be checked as Source (cf. Listing 1) in the

Listing 3. DSL code for checking MongoDB customer attributes for uniqueness.

```
ACTION UniqueCheckCustomerSsnMail:
  EXECUTE UniqueCheck ON Customer(ssn: val)
  RESULT IN Error
END
```

MongoDB query language. The Check analyzing whether the attribute value in the SSN field is unique over all rows in the data source is displayed in Listing 2. Note that while the code inside the check looks similar to an SQL statement, it is not implemented in SQL, but rather in our own specification language. An important difference is that our language operates on sets of records in our internal data model and not on relations in a relational database. Thus, `src` in this case is a set of records from a document database as shown in Listing 1. The implementation counts the number of occurrences of each attribute value by a grouping function, detecting non-unique values by having more than a single occurrence. This type of Check is shown for completeness purposes only; it would be provided with the product and not be implemented by the DQ manager or a software developer in his enterprise. Finally, Listing 3 shows how the source data is assigned to the Check in an Action. The same Check could be used for a different attribute (e.g. `Customer(email)`) in another Action.

All in all, RADAR satisfies the requirements of reusability of quality checks, allowing a separation of concerns for different user groups, extensibility by parametrization and support for different heterogeneous data sources. The optimizability of the execution of data quality checks has been presented in [10] and the possibility to define complex data quality rules will be shown in the next section.

3.2 Advanced Data Quality Rules

This section presents an example of a value rule, showing how more complex data quality rules are specified in RADAR. The basic idea is to use time series modeling to find data quality problems in time-related data. I.e. the sales volume of a company normally behaves according to some time series model. Unusual values not matching the model could indicate quality problems, like missing data or duplicate data fed into the data warehouse. A basic type of time series models are SARIMA models.

A time series in SARIMA models (see e.g. [8]) is described by multiple components. They are driven by a series of random components e_t with zero mean and equal variance σ^2. The AR component is a linear combination of the previous values of the time series and the current random component. The MA component is a moving average of the previous random components. The "I" stands for "Integrated" meaning that the differenced series is looked at. The seasonal part references back to the values in the previous season, for example $S = 7$ for the same day in the previous week. The seasonal part can also be described using AR, MA or I components. Overall, the order of an SARIMA model is described by $(p, d, q) \times (P, D, Q)_S$, where the first triple defines the order of the AR, I, and MA components in the non-seasonal part, and the second triple defines these orders for the seasonal component.

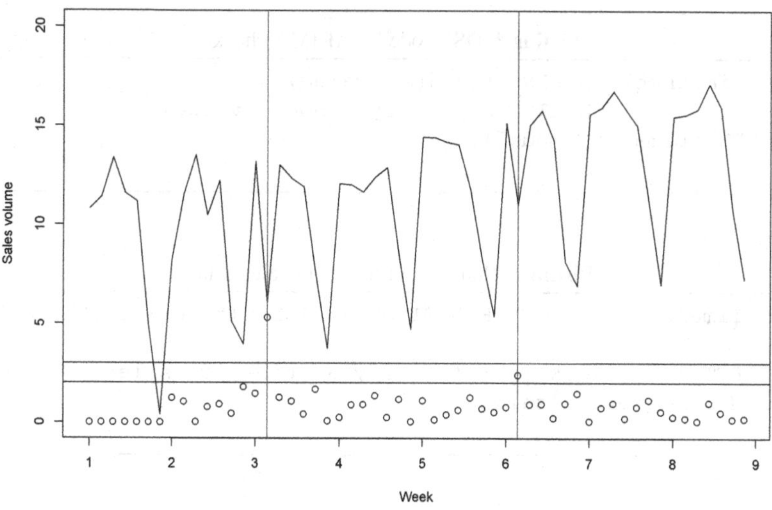

Fig. 1. Sales time series with daily ARIMA scores.

For example, the model we use for the time series in Fig. 1 is an $(1, 0, 0) \times (2, 1, 0)_7$ model, which means that we have an AR model that only uses the last value and an $S = 7$ weekly seasonal component that uses 2 AR components. Furthermore, the seasonal model is differentiated once as $D = 1$. We have no MA parts here. The equation for this model series is:

$$(1 - \text{ar}_1 B)(1 - \text{sar}_1 B^7 - \text{sar}_2 B^{14})(1 - B^7) y_t = e_t \quad (1)$$

Here, B^l is the backshift operator, meaning $B^l y_t = y_{t-l}$. The parameter values are $\text{ar}_1 = 0.6301$, $\text{sar}_1 = -0.5609$, and $\text{sar}_2 = -0.1503$. The variance of e_t is estimated to $\sigma^2 = 2.12$. Using this model we can calculate a forecast $\hat{y}_t$ by resolving (1) for y_t and setting $e_t := 0$. The forecast interval is calculated using the variance of e_t.

Our check now computes $\hat{y}_t$ for every time step t using the previous values. The absolute difference between the forecast and the actual value, normalized with σ, defines the score for each point:

$$\text{score}_t = \frac{|y_t - \hat{y}_t|}{\sigma}$$

For the forecast calculation of later points, we replace outliers where the score is larger than 2 with the forecasted value to avoid that a single error influences the forthcoming forecasts and produces consecutive errors.

As the check code uses a built-in method `arima` that internally calls a suitable standard function of the R statistics package, the code is rather simple, cf. Listing 4. Data to be checked is specified using a relational data source, see Listing 5.

The model itself is described in RADAR using a constant Source, that is not computed by executing an external query, but rather uses a fixed structure throughout. This constant Source lists the order of the model, the coefficients and the variance of e_t.

Listing 4. DSL code for ARIMA check.

```
CHECK ArimaCheck ON LIST ts(time, value)
               WITH LIST model(key, index, value):
   RETURN arima(ts, model);
END
```

Listing 5. Source for time series sales data.

```
SOURCE TimeSeriesSals TYPE LIST QUERY ROLES(time_id: IDENTIFIER):
NATIVE
    SELECT '' || week || '-' || day as time_id, sales_volume
    FROM time_series_sales
END
```

An example matching the above model is shown in Listing 6. Since it is very difficult to find this model manually, especially for the DQ manager, we will use a profiling component described in Sect. 4 to automatically suggest a model from analyzing historical data.

Finally, an action puts it all together. It calls the ARIMA check using the model and the time series data source, and assigns the appropriate roles, as shown in Listing 7.

Continuing the example Fig. 1, we see that the sales volume data has a weekly seasonal pattern. The circles at the bottom denote the computed scores; the upper horizontal line indicates the error threshold of 3, the lower horizontal line indicates the warning threshold of 2. Note that we don't have scores for the first week as we need historic data for forecasting. At time 3-2 (second day of week 3), there is an error, the value of the sales volume is too small for a non-weekend day. Due to the seasonal model, this error can be detected. On the other hand, at time 6-2 there is a warning as the value is rather low but not as exceptional compared to 3-2. Yet, the data steward may know that these values are due to some shops having been closed for a public holiday. On the other hand, it might also be due to an error in data provisioning and in that case the error and the warning should be observed. Thus, days 3-2 and 6-2 will be reported as suspicious.

Listing 6. Source for sales data ARIMA model.

```
SOURCE ArimaModelSales TYPE LIST CONST ROLES (key,index,value):
   ["coef",1,0.6301], ["coef",2,-0.5609], ["coef",3,-0.1503],
   ["coefnames",1,"ar1"], ["coefnames",2,"sar1"],
   ["coefnames",3,"sar2"], ["var",1,2.12],
   ["arma",1,1], ["arma",2,0], ["arma",3,2], ["arma",4,0],
   ["arma",5,7], ["arma",6,0], ["arma",7,1]
END
```

Listing 7. Action for sales time series data quality rule.

```
ACTION ArimaCheckSales:
  EXECUTE ArimaCheck
  ON TimeSeriesSals(time_id: time, sales_volume: value)
  WITH ArimaModelSales RESULT IN ERROR ABOVE 3 WARNING ABOVE 2
END
```

It would be impossible to detect both types without using a seasonal time series, as both values are within the normal range of values.

Similar to these time series checks, we have implemented multidimensional quality checks. Syntactically, those are pretty similar to the time series checks, with the difference, that the Action provides the Check with information as to which attributes contain the cube metric/fact value, cube dimensions and cube time dimension, respectively. The Check itself is rather simple, basically invoking a built-in function computing a score for the passed data cube by comparing it against a desired cube model. This model is specified as a constant data source similar to the one presented in Listing 6 for time series, just with different parameters. Based on the computed score the Action can issue warnings or errors, as before. Since the constant data sources needed to provide the models for the advanced quality checks tend to be lengthy and difficult to discover, it is important to assist the DQ manager in specifying such sources. This is done by our profiling component as will be explained below.

4 Semi-automated Detection of Quality Rules

In order to ensure data quality, a huge set of rules is necessary to regularly test as much quality aspects and parts of the data set as possible. When writing rules, two challenges arise. First, the effort to write huge numbers is high and thus a good coverage of the data with rules might not be achieved. Second, finding good parameters and thresholds for complex rule types by hand is extremely difficult. This holds true especially for rules that define a statistical model for the data and check the conformance of the current data with the model (see Sect. 3.2). To simplify these tasks, we developed a profiling component that examines existing data and suggests rules and parameters for these rules. The data quality steward can then inspect these suggestions and decide, using additional business knowledge, which candidates should be switched to active.

We follow a modular approach and provide various profiling modules. Each module is focused on one type of profiling and outputs specific types of rules. As a simple example, the NOT NULL profiling module counts the current number of NULL and NOT NULL entries in a column. Based on these statistics, the module suggests either a hard or a soft NOT NULL rule, stating that not more than a certain fraction of the values may be NULL, or both. For the soft rule, we need to find reasonable parameter values. Another example is a profiling module for a time series rule, that has to estimate the parameters of the time series model.

The UI to steer the profiling is two-fold. First, we have a basic profiling UI that targets the generation of huge amounts of basic rules. This UI shows the data sources and their attributes in a grid-like view and automatically suggests basic profiling modules that match the data type of the attributes, i.e. numeric range profiling for numeric columns, NULL profiling for every column, etc. The goal is to quickly generate simple rules for individual column values with a good coverage. The profiling functionality here is similar to commercial data profiling tools, but with the extension, that quality rule candidates are automatically generated based on the profiling results. The other UI, called advanced profiling, targets more complicated rules where the profiling needs more information to run. An example is the time series profiling, which needs to know which attribute of a data source contains the time and which contains the value of the time series. Here, running automatically over every combination would produce extreme load and generate lots of meaningless rules. Thus we assume the data steward can steer the process better using business knowledge about data. For the basic profiling UI, we use one generic kind of dialog. For the advanced profiling UI, each module defines its own UI components and parameters that are needed to steer this module.

Due to space limitations of the final version, details on profiling to detect different types of quality rules such as range, time series or cube rules are omitted here. A detailed discussion can be found in [10].

5 Conclusion and Future Work

5.1 Conclusion

In this paper we have presented a DSL to specify data quality rules. In particular we have focused on specifying advanced quality rules such as time series and multidimensional rules. We have also illustrated how simple (unique check) as well as more complex (time series, distribution check) quality rules are expressed in this language. Also, we have described how (semi-)automated generation of data quality rules is performed in our prototype by means of profiling. For simple rules the profiling works similar to commercial data profiling tools, but in addition offers the data steward the option to directly adopt data quality rules in our rule language based on the results. Configuration of advanced rules by specifying the necessary model parameters is tedious; this task is also significantly simplified by our profiling component, providing the data steward with potential settings for these parameters based on a statistical analysis of historic data. Our prototype is able to execute these quality checks on heterogeneous types of databases providing the source data. The implementation of the DSL allows for easy extensibility in case additional functionality is required. In summary, our DSL together with the prototypical implementation and the integrated profiling functionality provide an important step forward in efficiently managing data quality in data warehouses.

5.2 Future Work

Whereas this paper's focus lies on the specification of advanced rules and their profiling, one part of future work is to improve execution efficiency with an optimization step.

The general approach is to push logic toward the database and only execute logic inside of our engine when it is really necessary. We already have implemented some optimization steps, however there is still room for improvements.

While the applicability to different types of data sources has been shown in principal by using relational as well as document databases in the prototype, an extension to more types of data sources needs to be implemented. Nevertheless the DSL itself will remain unchanged even for different types of sources.

The profiling for advanced quality rule candidates is still in a prototypical state. More advanced UIs for the advanced quality rules will have to be implemented for a production system. For instance, the time series rule candidates should be presented in a graphical way to the data steward to be able to decide whether to adopt the rule more easily. Similarly, advanced UIs for other complex rule candidates will also be needed.

Finally, future extensions of the DSL capabilities, particularly in the body of Checks might be necessary in order to provide functionality that has not been necessary so far. Based on our language user-defined Checks are already possible, yet additional conceptual requirements might arise that need more advanced language constructs.

References

1. Abedjan, Z., Golab, L., Naumann, F.: Profiling relational data: a survey. VLDB J. **24**(4), 557–581 (2015). https://doi.org/10.1007/s00778-015-0389-y
2. Aggarwal, C.C.: Outlier Analysis, 1st edn. Springer, New York (2013). https://doi.org/10.1007/978-1-4614-6396-2
3. Bohannon, P., Fan, W., Geerts, F., Jia, X., Kementsietsidis, A.: Conditional functional dependencies for data cleaning. In: 2007 IEEE 23rd International Conference on Data Engineering, pp. 746–755. IEEE (2007). http://ieeexplore.ieee.org/xpls/abs_all.jsp?arnumber=4221723
4. Caruccio, L., Deufemia, V., Polese, G.: Relaxed functional dependencies - a survey of approaches. IEEE Trans. Knowl. Data Eng. **28**(1), 147–165 (2016)
5. Chiang, F., Miller, R.J.: Discovering data quality rules. Proc. VLDB Endow. **1**(1), 1166–1177 (2008). http://dl.acm.org/citation.cfm?id=1453980
6. Gupta, M., Gao, J., Aggarwal, C., Han, J.: Outlier detection for temporal data: a survey. IEEE Trans. Knowl. Data Eng. **26**(9), 2250–2267 (2014). https://doi.org/10.1109/TKDE.2013.184
7. Heine, F., Kleiner, C., Koschel, A., Westermayer, J.: The data checking engine: complex rules for data quality monitoring (2014). http://citeseerx.ist.psu.edu/viewdoc/download?doi=10.1.1.682.7950&rep=rep1&type=pdf
8. Hyndman, R.J., Athanasopoulos, G.: Forecasting: Principles and Practice. OTexts, Melbourne (2018)
9. Li, X., Han, J.: Mining approximate top-k subspace anomalies in multi-dimensional time-series data. In: Proceedings of the 33rd International Conference on Very Large Data Bases, pp. 447–458. VLDB Endowment (2007)
10. Oelsner, T., Heine, F., Kleiner, C.: IQM4HD concepts. Technical report, University of Applied Sciences and Arts Hannover, Germany (2018). http://iqm4hd.wp.hs-hannover.de/ConceptsIQM4HD.pdf
11. Olson, J.E.: Data Quality: The Accuracy Dimension. Morgan Kaufmann, San Francisco (2003)
12. Schelter, S., Lange, D., Schmidt, P., Celikel, M., Biessmann, F., Grafberger, A.: Automating large-scale data quality verification. Proc. VLDB Endow. **11**(12), 1781–1794 (2018). https://doi.org/10.14778/3229863.3229867. http://www.vldb.org/pvldb/vol11/p1781-schelter.pdf

Effect of Imprecise Data Income-Flow Variability on Harvest Stability: A Quantile-Based Approach

Zied ben Othmane[1,2], Cyril de Runz[2(✉)], Amine Ait Younes[2], and Vincent Mercelot[1]

[1] Kantar Media Company, Advertising Intelligence, Paris, France
{zied.benothmane,vincent.mercelot}@kantarmedia.com
[2] Université de Reims Champagne-Ardenne, CReSTIC, Reims cedex 2, France
{cyril.de-runz,amine.ait-younes}@univ-reims.fr
https://kantarmedia.com, https://crestic.univ-reims.fr/en/equipe/modeco

Abstract. Retrieved data from sensors may have a high level of quality to ensure crucial decisions and determine effective strategies. Nowadays, in view of the mass of generated information from these data, there is a real need to handle their quality. This paper propose new indices for quantifying the variability/stability of a data flow according to a data modeling that handles data imperfection. To deal with the data imprecision, we adopt a quantile-based approach. Our index definitions use parameters. Hence, to obtain an efficient judgments by this approach, we examine the choice of the appropriate parameters, and how it can affect the judgment on the harvest stability.

Keywords: Variability · Harvest stability · Temporal series · Imprecision · Quantile-based · Approach

1 Introduction

Retrieved data from sensors may have a high level of quality to make efficient analysis and take the right decisions. This paper deals with stream data harvested from web scraping robots, considered as sensors. The problem is that the credibility of the records is not guaranteed, due to the absence of a way to evaluate them.

Indeed, our data are imprecise and incomplete. For instance, a sensor may provide no value for a period of time, producing a lack of the information. Moreover, collected information are more tend information than real and exhaustive information. Finally, since the recorded values vary from one timestamp to another, the recording process present exceptional behavior.

We propose an approach handling these constraints with the aim to provide quality information for the future use of the crawled data. In our context, a crucial aspect of data quality is the variability/stability of the harvest for which we propose new indices.

S. Hartmann et al. (Eds.): DEXA 2019, LNCS 11706, pp. 248–257, 2019.
https://doi.org/10.1007/978-3-030-27615-7_19

This article is structured as follows. Section 2 introduces the main related works. Section 3 gives an overview of our data. Section 4 is devoted to explaining our approach. In Sect. 5, an examination to choose the appropriate parameters is done by varying these parameters, in Sect. 6, we study the effect to make the better judgment on the harvest. Finally, Sect. 7 presents the conclusion of this work.

2 Related Work

With the continuous rise of mass data harvested from sensors, traditional techniques evaluating the quality of the harvest are no longer convenient to generate timely valuable information and find suitable knowledge [2]. To improve the value of web information, Capiello et al. [3] discussed the role of data quality and cited techniques to highlight the importance of tracking data. Therefore, and based on different expert feedback, an approach regarding the quality of recordings is proposed in [6], given a method to identify and validate a new data quality dimension. Morishima et al. [8] adopt a test of a selection of harvested data, that accord an evaluation due to dependencies and relational schemes between attributes.

Working on evaluating data quality is an important complementary work that must be realized [3]. [12] identifies different data quality dimensions. In the frame of our work and regarding harvested temporal data, data quality is considered as a perception or an assessment informing the consistency, efficiency, and reliability of the data. [1] estimates that veracity over web data is demonstrated by analyzing the space of conflicting information. The embedded mechanism extracts entities and relations and provides a label of the veracity. Those approach needs an expert evaluation which is not possible in our context. The veracity definition is related to the presence of uncertain or imprecise data due to errors or missing or invalid data, and these issues can compromise the usefulness of the collected values [4].

To extract realistic information from data, the presence of uncertain, incomplete and imprecise data must be verified by handling their credibility and ensuring perfection in the data [4]. Information is considered imperfect when it is incomplete, imprecise, and/or uncertain. In our context, the data imperfection is primarily due to the imprecision resulting from the recording process.

Many techniques and approaches deal with imprecision in data. A usual one affects intervals of the measurement [7]. according to the specification of our values, these approaches will be discarded due to the impossibility to handle each element apart. We then use a model based on quantile classification that allows us to encompass the imprecision.

Quantile-based systems define quantiles as system parameters to continuously differentiate and compute convenient simulations to make decisions. For instance, Painter et al. [9] used quantile regression to improve the comprehension of the data and report predictions. In view of quantile flexibility and its range of applicability, [10] adopted an analysis to verify the division validation.

Other approaches elicit expert knowledge on a continuous uncertain quantity to define the number of specific proportions [11].

In this paper, our data are characterized by their imperfection, especially imprecision, timeliness, volume, and lacunarity. However, we want to verify the variability of the harvest and determine its effectiveness on the stability of the generated information, as they can highlight several issues such as extreme events [5].

In view of web-harvested data, measuring the variability and flow stability is a part of data quality assessments. In order to measure them, this paper proposes indices to quantify them. Those indices allows highlighting sensor behaviors in a web data harvesting system.

3 Data Description

Our harvesting system falls in the working process of scraping web data. Several robots, considered as sensors in this paper, are deployed to scrape website data. Each of them is affected to a unique website with a specific and unknown harvest strategy.

In fact, different web scrapers capture various log data about advertisements (content, banner format, advertiser, etc.). Those data are aggregated according to several timestamp to compute diverse variables, such as: total number of banners scraped by website, total number of URLs scraped by website, total number of advertiser detect by website, etc. These kind of information are temporal, i.e its measures differ from one timestamp to another.

Over m successive timestamps, interlinks between harvested data dynamics may hide diverse nuances in data. Them, we aim to understand their variability. Nevertheless, visually or using simple techniques, it is quite hard to understand the crowded behavior of one website or category, meta-category, etc.

To make the approach generic, the followed notations are considered:

- S is the set of n sensors: $S = \{s_1, \ldots, s_n\}$.
- T is the set of m timestamps (a timestamp represent a period for which data are aggregated): $T = \{t_1, \ldots, t_m\}$.
- V is the set of p studied variables: $V = \{v_1, \ldots, v_p\}$.
- $v_i^{t_k}(s_j)$ is the value of v_i at t_k for s_j.

4 Data Quality Prototype

To treat the cited imperfections, in this section, we begin by explaining the base mechanism of the work. Then, we show how it leads to constructing appropriate data quality indices. These indices are the main issues to measure the quality of the harvest.

Our goal is to evaluate temporal data when these ends are imprecise, uncertain, lacunar and come with irregular flow. In view of the gap in the literature, we present an approach in data quality context that makes in evidence a comparison and a classification of temporal data in view of previous cited features.

The quantile-based approach aims to create quality indicators that consider the uncertainty in the volume of the on-time upcoming data, the significant fluctuation of recordings, the discontinuity, and the lack of data on different variables of study.

Principle Steps. The prototype began by harvesting and classifying data into variables of study. Next, since our temporal data present imperfections, discontinuities and imprecision in values, we have applied a quantile-based approach to handle this. First, we range data into a chosen number of quantiles, which decreases imprecision since it surrounds the saved values and locates data behavior.

The approach is based on two computation types. Internal computation uses indicators measuring the internal variability of the data (i.e., judging whether they always exhibit the same behavior over time). However, the external computation judges the temporal positioning toward the rest. The approach considers the lacunarity in data and adopts a jump quantification method to measure the variability in its two sides. A stability index is defined as an aggregation of the two variability indices.

4.1 Quantile-Based Approach

Since our data represent volumes, it is important to keep a notion of the order in the data modeling. The projection of a data into its quantile could therefore give it. Indeed, a quantile-based projection allows to partition data into subsets, equilibrated in terms of cardinality, according to their values. Moreover, it can provide a meta-understanding of their placement across a period, by understanding data according to other values. Thus, placing data into their quantiles involves a certain endorsement of comparison between the data distributions.

As the absence of harvested data does not mean the absence of advertisements on the website, the absence of crawled data should not be considered a classic 0 and should not be considered in the quantile computation. For this, we place null data into a specific level called quantile 0.

In the view of other data imperfection, representing data by the appropriate quantile provides a method to encompass imprecision in the measures. This allows us to obtain new information about the harvest, mostly the movement of each sensor.

We first project sensors' values, at each timestamp, into their quantiles according to the whole set of data collected from sensors at the same timestamp. This is called external quantile superposition and illustrated in Fig. 1 (right).

Moreover, to inform about the functionality of a sensor, we propose to project into quantile its harvested data at each timestamp according to data collected by the same sensors at other timestamp. This is called internal quantile superposition and Fig. 1 (left) presents the chronological position of the data according to previous one of the sensor.

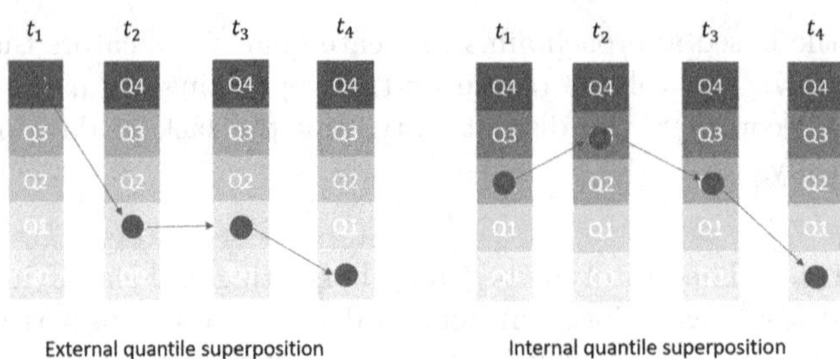

Fig. 1. Example of a relative sensor's harvest into the quantile division (where q = 4).

Computing Principle. In our context, the quantile of $v_i^{t_k}(s_j)$ can be computed over the set of values from different sensors at t_k and produce an external point of view, or from the same sensor s_j at different times.

Let Q be the quantile function that associates a value in a finite set Ω to its quantile position according to this set. Let r be the chosen number of quantiles. Using a quantile-based projection we will obtain r possible values (1 to r) that define subsets equilibrated in terms of cardinality. By construction, a quantile equal to 0 represents a lack of data. Thus:

Therefore, $Q_{ext}(v_i^{t_k}(s_j), \{v_i^{t_k}(s_z), \forall z \in [1, n]\})$, the external quantile of $v_i^{t_k}(s_j)$, is computed over the set of values $\{v_i^{t_k}(s_z), z \in [1, n]\}$.

With the same logic, $Q_{int}, (v_i^{t_k}(s_j), \{v_i^{t_z}(s_j), \forall z \in [1, o]\})$, the internal quantile of $v_i^{t_k}(s_j)$, is computed over the set of values $\{v_i^{t_z}(s_j), z \in [1, m]\}$.

In the following, $Q_{int}, (v_i^{t_k}(s_j), \{v_i^{t_z}(s_j), \forall z \in [1, o]\})$ is noted as $Q_{ext}(i, k, j)$ and $Q_{int}, (v_i^{t_k}(s_j), \{v_i^{t_z}(s_j), \forall z \in [1, o]\}))$ is noted as $(Q_{int}(i, k, j)$.

4.2 Quantile-Based Variability

To obtain more information about the different quantile-based distributions, this section examines the meaning of the variability and shows how to compute it in the quantile-based context. In our approach, the word "variability" of temporal data refers to the method of providing measurement to temporal data x_t along T. Higher varying values of x_t during T lead to a higher variable x_T. According to our context, a data flow is projected into two data flows: internal quantile-based representation and external quantile-based representation. Therefore, there are two variability indices. According to a given sensor s_j, the internal variability is the variability relative to the temporal data (i.e., the variability of the internal trajectory of the data harvested by this sensor). The objective of the study of this variability is to answer the question: "Do I have the same harvest over time?" or "Do I have the same behavior during T?". For the external variability that is related to the external quantile computation, it is important to ask: "Am I volatile against the rest? How much? Do I seem to have the same behavior as they have?". Hence, the computation is related to the positioning of the rest of the selected elements. If a specific temporal series of data is kept along T in a quantile q and the rest of the selection in the quantile $(q + i)$, this end is not considered variable at all, which is why the quantile-based approach quantifies jumps over time to compute the variability index and to determine appropriate indicators relevant to the parameter choice. Let Sc be a score that measures variability for a temporal series and q be the number of levels where the data can be placed. In the following equation (Eq. 1), the score Sc_{ext} is given for Q_{ext}. The same computation can be done for Sc_{int} based on Q_{int}.

$$Sc_{ext}(i,k,j) = \begin{cases} 1, \text{ if} \\ \quad Q_{ext}(i,k,j) - Q_{ext}(i,k-1,j) \geq 2 \\ or, \text{ if } Q_{ext}(i,k-1,j) > 0 \\ \quad \text{and } Q_{ext}(i,k,j) = 0 \\ 0, \text{ else} \end{cases} \tag{1}$$

Consequently, to calculate the index in a period between t_k and t_{k+z}:

$$Sc_{ext}(i,k,k+z,j) = \frac{\sum_{x=k}^{x=t+z} Sc(i,x,j)}{z - IPMD(i,j)} \tag{2}$$

where $IPMD(i,j)$ is the number of inter-periods without received data.

The functionality of the variability index of a sensor s_j is based on counting the significant jumps between timestamps and dividing them by the effective period. If the jump between t_k and t_{k+1} moves over $jp = 2$, the count takes 1 in addition, or else no value will be added. We note that we are on quantile scales since the data are placed in their quantiles. Score variability also considers break periods (b). If we have more than one consecutive period (b = 1) with no saved data, the index will not progress by 1, even in Q_0, because the absence of data is not stabile since we consider that we have no idea about what was gone behind.

Figure 2 gives an example where there are four jumps into an active period of five active inter-periods (over eight inter-periods and nine timestamps). Consequently, we deduce that if $Sc(i,k,j) \rightarrow 1$, the propagation is well varied for the Δt (i.e., many exceptions have been detected on that sensor).

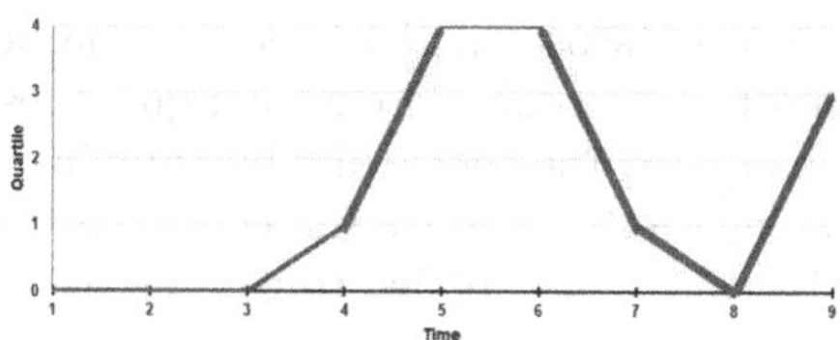

Fig. 2. Sample of sensor data-flow evolution with $r = 4$, $jp = 2$, $b = 1$, and $var = \{v_i\}$.

An index that measures the variability detection of a sensor s_j allows the user to understand its reliability. This key factor in turn (e.g., the mean of that value of group of sensors) helps the user to determine certain categories. The calculation could be applied toward external or internal s_j superpositions, which allows the validation of such distributions. The variability index can be static or dynamic (i.e., a single measure in one timestamp or a chronological signal over a period, respectively).

4.3 Instability and Stability Indices

To conclude some visions to judge the crawling algorithm's functionality, we have unified some calculations by associating a measure that groups the internal and external variability of a sensor or group. This measure is called the instability index.

Previously, for a variable v_i and a source s_j at a time t_k, we defined an internal Sc_{int} and external Sc_{ext} variability scores. According to these scores, we define the instability index of a variable v_i of a source S_j at t_k as follows:

$$ISt(i,k,j) = \frac{\sqrt{Sc_{ext}(i,k,j)^2 + Sc_{int}(i,k,j)^2}}{\sqrt{2}} \tag{3}$$

Therefore, for a given period between t_k and t_{k+z}, it is defined as follows:

$$ISt(i,k,k+z,j) = \frac{\sqrt{Sc_{ext}(i,k,k+z,j)^2 + Sc_{int}(i,k,k+z,j)^2}}{\sqrt{2}} \tag{4}$$

The stability index St is defined as follows:

$$St(i,k,j) = 1 - ISt(i,k,j) \tag{5}$$

$$St(i,k,k+z,j) = 1 - ISt(i,k,k+z,j) \tag{6}$$

The sense of the instability index is to give significant indicators that provide visual inspection to expand user expertise.

5 Experimental Sensibility Evaluation

The objective of this section is to examine the choice of the parameters' values and treat the effect of the variability features on the harvest stability. The different results are computed on the basis of 702 sensors, three variables of study, the summarized data flow by sensor and by month, the period of study of 36 consecutive months, and the internal and external quantiles.

First, we propose examining the choice of r by applying the Silhouette Index and the Dunn Index as methods to verify the consistency of the quantile sets. Table 1 shows the statistical results applied to gain an overview of the variation of r.

In view of our dataset, the general results show when the jp increases, the data are poorly matched to their associated quantile sets. We can note minor variations, such as the passage from $r = 4$ to $r = 6$ in the Dunn Index. According to our dataset, this does not have a great effect because the general tendency gives $r = 4$ as the best number of the quantile set.

To examine the number of the choice of jp, we choose a greater number for the quantile set: $r = 9$.

Table 2 illustrates the variability indices according to $r = 9$ and $b = 2$. By this table we understand that the high variability (on both sides) depends on the choice of jp. However, if jp increases then the variability decreases.

Table 1. Influence of the quantile number on the data quantile distribution quality according to the Silhouette Index and Dunn Index.

r	Silhouette Index				Dunn Index			
	Min	1Q	Mean	Max	Min	1Q	Mean	Max
$r = 4$	−0.2864	0.4566	0.4593	0.7540	0.0004234	0.4603	0.4462	1.6703
$r = 5$	−0.2974	0.4356	0.4506	0.6278	0.28640	0.4603	0.4645	1.6700
$r = 6$	−0.3816	0.4165	0.4452	0.7061	0.00079	0.4603	0.4276	1.6990
$r = 9$	−0.2790	0.3794	0.4322	0.5532	0.00031	0.3298	0.3801	2.7190
$r = 10$	−0.2740	0.380	0.4307	0.586	0.0007039	0.285	0.3710	1.108

Table 2. External and internal variability indices calculated regarding several jumps within $r^* = 9$, $b = 2$ and $T = 36$.

$r_{b=2} = 9$	Median(V_{ext})	Mean(V_{ext})	Median(V_{int})	Mean(V_{int})
$jp = 2$	0.7996737	0.7732519	0.7444	0.5358
$jp = 3$	0.7996240	0.7732425	0.7444	0.5255
$jp = 4$	0.7995924	0.7732330	0.7412	0.5179
$jp = 5$	0.7995292	0.7732235	0.7355	0.5113

On other hand, Table 3 shows that the high variability depends on b. If b is bigger, then the variability is bigger. Accordingly, a decision maker must find an adjustment between these two parameters.

Table 3. Example of $r = 9$ external and internal variability by the mean. Indices are calculated by varying the number of breaks b within $T = 36$.

mean	Internal variability				External variability			
	jp-2	jp-3	jp-4	jp-5	jp-2	jp-3	jp-4	jp-5
b=2	0.5358	0.5255	0.5179	0.5113	0.7732	0.7732	0.7732	0.7732
b=3	0.7685	0.7685	0.7685	0.7685	0.77110	0.7710	0.7710	0.7710
b=4	0.7625	0.7623	0.7622	0.7621	0.7684	0.7684	0.7684	0.7684

The choice of b can be extracted from measuring the dissimilarity between correspondent rows in Table 3 or be defined due to system knowledge or other techniques. We continue by admitting $b = 2$ as the best time break since the first row of external variability is the lower one and its internal variability is accepted. Nevertheless, the variation of jp can affect the judgment on the stability of the harvest. To avoid wrong judgments, the choice of jp must be reasonable.

6 Significance and Effects

Regarding the significance of the modeled indicators, such as judging the data quality, measuring the variability of sensor functionality, measuring the stability of the harvest, etc., in this section, we show some new experiments. The results are presented by varying the choice of the variables of study, period, and timestamp.

By combining internal and external variability using the stability index on three variables of study, we can determine a global overview about the stability of the harvest over 36 months, as seen in the Fig. 3. Without any analytical method, we can easily recognize two different large groups and an outlier. We can then adopt deeper data mining techniques. This kind of experiment allows the generation of a confidence sample (i.e., the most stable sensor functionality). Alternatively, we can also dismiss others and verify the back-end harvest mechanism.

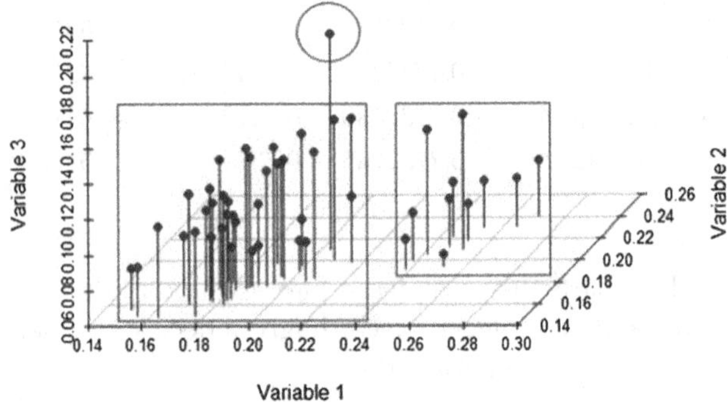

Fig. 3. 3D vision of the stability indicators of 702 sensors. This information is detected due to averaging the indices of the stability indicator, and is produced on the basis of r = 4, jp = 2, b = 1, and $var = \{v_i\}$ in a monthly timestamp (i.e., variability is calculated on a daily scale every month).

7 Conclusion

In this paper, we focused on mining aspects that facilitate managers' awareness of the quality of a flow data from sensors. As a crucial aspect is the data volatility, we proposed a new variability index that considers the complex nature of data and mainly its imperfection. To model data in accordance with their imperfection, we, first, differentiated the case of the lack or presence of data and projected the data into quantiles to deal with imprecision. Based on this modeling, we defined a variability and instability indices. Experimentation shows the interest of our approach for quantifying variability over imprecise data flows. In our future work, we will study the generalization of our approach by using fuzzy sets.

References

1. Ba, M.L., Berti-Equille, L., Shah, K., Hammady, H.M.: Vera: a platform for veracity estimation over web data. In: Proceedings of the 25th International Conference Companion on World Wide Web, WWW 2016 Companion, pp. 159–162 (2016)
2. Ben Othmane, Z., Bodenes, D., de Runz, C., Ait younes, A.: A multi-sensor visualization tool for harvested web information: insights on data quality. In: International Conference on Information Visualisation vol. 22, pp. 10–13 (2018)
3. Cappiello, C.: On the role of data quality in improving web information value. In: Proceedings of the 24th International Conference on World Wide Web Companion, WWW (2015)
4. Cappiello, C., Samá, W., Vitali, M.: Quality awareness for a successful big data exploitation. In: Proceedings of the 22nd International Database Engineering & Applications Symposium, pp. 37–44. ACM (2018)
5. Coelho, C., Ferro, C., Stephenson, D., Steinskog, D.: Methods for exploring spatial and temporal variability of extreme events in climate data. J. Clim. **21**(10), 2072–2092 (2008)
6. Held, J., Lenz, R.: Towards measuring test data quality. In: Proceedings of the 2012 Joint EDBT/ICDT Workshops, pp. 233–238. ACM (2012)
7. Meany-Daboul, M.G., Roscoe, E.M., Bourret, J.C., Ahearn, W.H.: A comparison of momentary time sampling and partial-interval recording for evaluating functional relations. J. Appl. Behav. Anal. **40**(3), 501–514 (2007)
8. Morishima, A., Yumiya, E., Takahashi, M., Sugimoto, S., Kitagawa, H.: Efficient filtering and ranking schemes for finding inclusion dependencies on the web. In: Proceedings of the 22nd ACM international conference on Information & Knowledge Management, pp. 763–768. ACM (2013)
9. Painter, I., Eaton, J., Olson, D., Revere, D., Lober, B.: Generation of prediction intervals to assess data quality in the distribute system using quantile regression. In: JSM Proceedings, Statistics in Defense and National Security Section (2011)
10. Sankaran, P., Sunoj, S.: Quantile-based cumulative entropies. Commun. Stat.-Theory Methods **46**(2), 805–814 (2017)
11. Selikhovkin, I.A.: An imprecise model of combining expert judgments about quantiles. Eur. J. Technol. Des. **3**(1), 49–60 (2014)
12. Sidi, F., Panahy, P.H.S., Affendey, L.S., Jabar, M.A., Ibrahim, H., Mustapha, A.: Data quality: a survey of data quality dimensions. In: 2012 International Conference on Information Retrieval & Knowledge Management, pp. 300–304. IEEE (2012)

Decision Support Systems

Fairness-Enhancing Interventions in Stream Classification

Vasileios Iosifidis[1,2(✉)], Thi Ngoc Han Tran[1], and Eirini Ntoutsi[1,2]

[1] Leibniz University, Hannover, Germany
tranthingochan.03@gmail.com
[2] L3S Research Center, Hannover, Germany
{iosifidis,ntoutsi}@L3S.de

Abstract. The wide spread usage of automated data-driven decision support systems has raised a lot of concerns regarding accountability and fairness of the employed models in the absence of human supervision. Existing fairness-aware approaches tackle fairness as a batch learning problem and aim at learning a fair model which can then be applied to future instances of the problem. In many applications, however, the data comes sequentially and its characteristics might evolve with time. In such a setting, it is counter-intuitive to "fix" a (fair) model over the data stream as changes in the data might incur changes in the underlying model therefore, affecting its fairness. In this work, we propose fairness-enhancing interventions that modify the input data so that the outcome of any stream classifier applied to that data will be fair. Experiments on real and synthetic data show that our approach achieves good predictive performance and low discrimination scores over the course of the stream.

Keywords: Data mining · Fairness-aware learning · Stream classification

1 Introduction

Despite the wide spread belief that data-driven decision making is objective in contrast to human-based decision making that is subject to biases and prejudices, several cases have been documented, e.g., [9,13], in which data-driven decision making incurs discrimination. As a recent example, a Bloomberg report has suggested signs of racial discrimination in Amazon's same-day delivery service [13]. The sensitive attribute race was not employed as a predictive attribute in Amazon's model(s), however the location of the users might have acted as a proxy for race. As a result, predominantly black ZIP codes were excluded from services and amenities. Therefore, the wide spread usage of automated data-driven decision support systems has raised a lot of concerns regarding accountability and fairness of the employed models in the absence of human supervision [1,24,28]. Such issues result in societal and legal implications, therefore, recently the domain of discrimination-aware data mining [23] has attracted a lot of attention and several

S. Hartmann et al. (Eds.): DEXA 2019, LNCS 11706, pp. 261–276, 2019.
https://doi.org/10.1007/978-3-030-27615-7_20

methods have been proposed ranging from discrimination discovery to discrimination elimination and explanation of model decisions.

Most of these methods, however, tackle fairness as a batch learning problem aiming at learning a "fair" model which can be then used for predicting future instances of the population. In many modern applications, however, data is generated sequentially and its characteristics might change with time, i.e., the data is non-stationary. Such dynamic environments (or, data streams) call for model adaptation [12]. As an example, in the EU, the non-native population has significantly changed in the last years due to European refugee crisis and internal EU migration with a potential effect on the racial discrimination in the labor market. In such *non-stationary* environments, the main challenge for supervised learning is the so-called concept drifts, i.e., changes in the underlying data distribution which affect the learning model as the relationships between input and class variables might evolve with time [26]. Existing solutions from the data stream mining domain tackle this issue by adapting the learning models online. However, as the decision boundary of the classifier changes as a result of model adaptation, the fairness of the model might get hurt.

An example of an evolving stream with discrimination is shown in Fig. 1; one can see the deprived and favored communities (w.r.t. some sensitive attribute) over time as well as their class assignments. The favored community dominates the stream. The decision boundary of the classifier (solid line) changes in response to changes in the underlying data. As a result, the associated fairness of the model also changes, calling for "fairness-enhancing interventions" (dashed line). It is important, therefore, model adaptation to also consider fairness‘ to ensure that a valid fairness-aware classifier is maintained over the stream. In this work, we propose fairness-enhancing interventions that modify the input data before updating the classifier. Our method belongs to the category of pre-processing approaches to fairness, investigated that far only in the context of static learning [5,6,14–17]. Our contributions are: (a) we introduce the fairness-aware classification problem for streams (b) we propose pre-processing fairness-enhancing interventions for streams (c) we propose a synthetic generator for simulating different drift and fairness behaviors in a stream and (d) we present an extensive experimental evaluation with different stream learners and on different datasets.

The rest of the paper is as follows: In Sect. 2, we overview the related work. Our approach is presented in Sect. 3. Experimental results are discussed in Sect. 4. Finally, conclusions and outlook are presented in Sect. 5.

2 Related Work

Although more than twenty different notions of fairness have been proposed in the last few years [25,29], still there is no agreement on which measure to apply in each situation. The most popular is that of statistical parity [29] that checks whether the favored and deprived communities have equal probability of being assigned to the positive class. This is the measure we also adopt in this work.

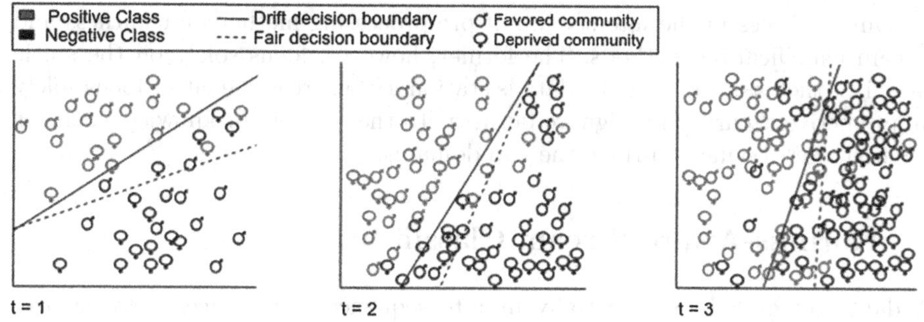

Fig. 1. An evolving stream with discrimination: At each timepoint, the current decision boundary (solid line) and the "fairness-corrected" boundary (dashed line) are displayed.

Pre-processing Fairness-Enhancing Interventions: Methods in this category work under the assumption that in order to learn a fair classifier, the training data should be discrimination-free. To this end, they try to balance the representation of the different groups in the population. Massaging [15] modifies the data distribution by re-labeling some of the instances which reside close to the decision boundary in order to neutralize discriminatory effects. Re-weighting [5] assigns different weights to the different group, e.g., the deprived group will receive a higher score comparing to the favored group. These methods are typically model-agnostic and therefore, any classifier is applicable after the pre-processing phase.

In-processing Fairness-Enhancing Interventions: Methods in this category directly modify the learning algorithm to ensure that it will produce fair results. As such, they are algorithm-specific; e.g., [18] proposes a decision tree that encodes fairness by employing a modified entropy-based attribute splitting criterion and [8] includes sensitive attributes in the learning process by utilizing a joint loss function that makes explicit trade-off between fairness and accuracy.

Post-processing Fairness-Enhancing Interventions: Post-processing methods modify the results of a trained classifier to ensure the chosen fairness criterion is met; e.g., [18] modifies the leaf labels of a decision tree, [22] changes the confidence values of classification rules and [10] shifts the decision boundary of an AdaBoost learner until the fairness criterion is fulfilled.

Stream Classification: Data stream algorithms must be able to adapt to concept drifts in order to maintain a good performance over the stream [12]. Model adaptation is typically enabled by: (i) incorporating new instances from the stream into the model and (ii) forgetting or downgrading outdated information from the model. The former calls for online/incremental algorithms, whereas the latter calls for methods that are able to forget e.g., [11,19]. We discuss several stream classifiers in the experiments (Sect. 4).

Sequential Fairness: When a sequence of decisions has to be taken, the notion of sequential fairness is relevant. For example, [27], studies fair online item ranking for groups and [20] how fairness criteria interact with temporal indicators of well-being and affect discriminated populations on the long-term.

Our work lies in the intersection of pre-processing methods for fairness and stream classification methods. The former, however, focus solely on the static case, i.e., they assume that the data is stationary, whereas the latter focus solely on predictive accuracy and ignore fairness. To the best of our knowledge, this is the first work trying to bridge the two domains.

3 Fairness-Aware Stream Classification

A data stream S is a potentially infinite sequence of instances arriving over time, each instance described in a feature space $A = (A_1, A_2 \cdots A_d)$. One of the attributes is the *sensitive attribute*, denoted by SA, with values $SA = \{s, \overline{s}\}$; we refer to s and $\overline{s}$ as "deprived" and "favored", respectively. We also assume a binary class attribute $C = \{rejected, granted\}$. We refer to "granted" class value as *target class*. We process the stream in chunks of fixed size, $S_1, \cdots, S_t$ with S_t being the most recent chunk. We assume the fully supervised learning setting, where the labels of the instances are available shortly after their arrival. Therefore, the goal is to make a prediction for the instances based on the current classifier and use the labels later on for update (the so-called prequential evaluation [12]). The underlying stream population is subject to changes, which might incur concept drifts, i.e., the decision boundary might change overtime (c.f., solid line in Fig. 1) and therefore, fairness implications may take place (c.f., dashed line in Fig. 1). A stream classifier typically takes care of concept drifts, but does not consider fairness.

The *discrimination aware stream classification problem* therefore is to maintain a classifier that performs well (i.e., the predictive accuracy is high) and does not discriminate (i.e., the discrimination score is low, c.f. Eq. 1) over the course of the stream. In this work, we follow the pre-processing approaches to fairness-aware learning that intervene at the input data to ensure a fair representation of the different communities. In particular, we monitor the discrimination in each incoming chunk from the stream (Sect. 3.1) and if the discrimination score exceeds a user defined threshold ϵ, we "correct" the chunk for fairness (Sect. 3.2) before feeding it into the learner (Sect. 3.3). We assume an initialization phase at the beginning of the stream for which an initial fairness-aware classifier F_0 is trained upon an initial dataset S_0 from the stream. An overview of our approach is depicted in Fig. 2, where M1–M4 are the adaptation strategies introduced in Sect. 3.3.

3.1 Detecting Classifier Discrimination in Data Streams

Let F be the current (non-discriminating) stream classifier and S_t be the current chunk received from the stream. We evaluate the discriminative behavior of F over S_t, i.e., $disc_S(F, S_t)$ by evaluating the predictions of F over instances of S_t. First, we define four communities in each chunk S_t by combining the sensitive attribute SA with the (predicted) class attribute C (both binary attributes):

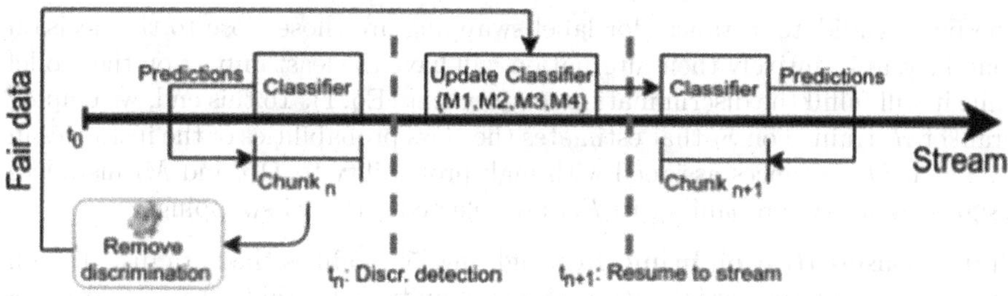

Fig. 2. Fairness-aware stream classification overview

Table 1. (Chunk S_t) Communities

Sensitive Attribute SA	(predicted) class	
	Rejected	Granted
s (Female)	DR_t *(deprived rejected)*	DG_t *(deprived granted)*
$\overline{s}$ (Male)	FR_t *(favored rejected)*	FG_t *(favored granted)*

As discrimination measure, we employ statistical parity that evaluates whether the favored and deprived groups have equal probabilities of being granted [25]:

$$disc_S(F, S_t) = \frac{FG_t}{FG_t + FR_t} - \frac{DG_t}{DG_t + DR_t} \tag{1}$$

If the discrimination value exceeds the threshold ϵ, i.e., $disc_S(F, S_t) > \epsilon$, the discrimination performance of the model degrades, due to, e.g., changes in the distribution that reside in the newly arrived chunk S_t. A typical stream classifier would update F based on S_t to adapt to the incoming stream. However, to also account for fairness, we first "correct" S_t for fairness (c.f., Sect. 3.2), before employing its instances for updating the model (c.f., Sect. 3.3).

3.2 Fairness-Enhancing Data Interventions in Data Streams

If discrimination is detected, S_t is "corrected" for fairness before being used for model update (Sect. 3.3). To this end, we employ two different data intervention techniques: massaging and re-weighting.

Chunk-Based Massaging. Massaging [15] modifies the data distribution by swapping the class labels of certain instances (from "granted" into "rejected" or vise versa) from each of the deprived rejected (DR) and favoured granted (FG) communities. The amount of affected instances, M_t, from each community is derived by Eq. 1 and is as follows:

$$M_t = \frac{FG_{S_t} * (DG_{S_t} + DR_{S_t}) - DG_{S_t} * (FG_{S_t} + FR_{S_t})}{|S_t|} \tag{2}$$

The best candidate instances for label swapping are those close to the decision boundary, as intuitively their alternation will have the least impact on the model while it will fulfill the discrimination requirement (Eq. 1). To this end, we employ a ranker R_t trained on S_t that estimates the class probabilities of the instances in S_t. Then, M_t instances assigned with high probability to DR and M_t instances assigned with low probability to FG are selected for label swapping.

Chunk-Based Re-weighting. Re-weighting [5] modifies the data distribution by assigning different weights to each community (c.f., Table 1) to "enforce" a fair allocation of deprived and favored instances w.r.t the target class. Similarly to massaging, the exact weights depend on the S_t distribution in the different communities. Below we provide the weight for the favoured granted community, same rationale holds for the other communities:

$$W_t^{FG} = \frac{|\overline{s}_{S_t}| * |\{x \in S_t | (x.C = \text{"}granted\text{"})\}|}{|S_t| * |FG_{S_t}|} \tag{3}$$

Each instance $x \in S_t$ is weighted by "inheriting" the weight of its community.

Massaging vs Re-weighting. Both methods modify the data distribution to equalize the number of deprived and favored communities in the target class. However, there are fundamental differences between the two approaches: massaging interferes at the instance level by altering single instances, whereas re-weighting affects a whole community by lowering/increasing its weight. Moreover, massaging is more intrusive than re-weighting as it alters the class labels. Both interventions result in a "corrected" chunk S_t' ($|S_t| = |S_t'|$) used for updating the classifier (c.f., Sect. 3.3).

3.3 Fairness-Aware Classifier Adaptation in Data Streams

The update of a classifier should take into account both concept drifts and fairness. For the former, we work with stream classifiers, like, Hoeffding Trees, Accuracy Updated Ensembles and Naive Bayes that already adapt to concept drifts. In that sense, the concept drift problem is directly tackled by the learner. For the latter, we "correct" the input stream per chunk, using either massaging or re-weighting, to ensure that learners are trained on "fair" data (c.f., Sect. 3.2). In particular, we propose update strategies for fairness-aware stream classification:

- ACCUM&FULLTRAIN (shortly $M1$): F is continuously updated over the stream using the original current chunk S_t, if no discrimination is detected, or its "corrected" counterpart S_t', if discrimination is detected.
- RESET&FULLTRAIN (shortly $M2$): Similar to M1, but if discrimination is detected, F is reset and a new model is created from the "corrected" S_t'.

The underlying assumption for ACCUM&FULLTRAIN is that if trained with "fair" chunks, the classifier F should be fair. In practice, though and due to the complex interaction between input data and learning algorithms, this might not be true (c.f., [17]); therefore, we also propose the RESET&FULLTRAIN model that resets the learner once its predictions incur discrimination.

In addition, we propose two variations that focus more on fairness. The rationale is similar to the previous approaches but the model is updated only if discrimination is detected and only via "corrected" data. Therefore, these two models adapt slower to concept drifts comparing to the first two models, as their adaptation occurs only if discrimination is detected. In particular:

- ACCUM&CORRECTEDTRAIN (shortly $M3$): F is updated over the stream only if discrimination is detected. Update is based on "corrected" chunks S'_t.
- RESET&CORRECTEDTRAIN (shortly $M4$): Similar to M3, but once discrimination is detected F is reset and a new model is created from the corrected chunk S'_t. Thus, $M4$ adapts only via reset, when discrimination is detected.

4 Experiments

We evaluate the performance of our methods for discrimination elimination in data streams using both real and synthetic datasets (Sect. 4.1). As evaluation measures, we use the performance of the model, in terms of accuracy and discrimination, over the new coming chunk from the stream. We report on the performance of the different methods over the stream but also on the overall accuracy-vs-fairness behavior of the different methods. We experiment with a variety of stream classifiers such as Naive Bayes (NB), Hoeffding Tree (HT), Accuracy Updated Ensemble (AUE) and k-Nearest Neighbors (KNN). The aforementioned models are updated based on the new incoming chunk from the stream, however they differ w.r.t how they handle historical information. NB and HT classifiers do not forget, whereas AUE forgets by replacing old learners with new ones. kNNs on the other hand, rely solely on the last chunk for the predictions, due to its internal buffer. An overview of each classifier is given below:

- Naive Bayes (NB): A probabilistic classifier that makes a simplistic assumption on the class-conditional independence of the attributes. The stream version of NBs [2] is an online algorithm, i.e., the model is updated based on new instances from the stream, but does not forget historical information.
- Hoeffding Tree (HT): A decision tree classifier for streams that uses the Hoeffding bound to make a reliable decision on the best splitting attribute from a small data sample [7]. HT is an online algorithm (so, it is updated based on new instances from the stream) but does not forget.
- Accuracy Updated Ensemble (AUE): An ensemble model that adapts to concept drifts by updating its base-learners based on the current data distribution, tuning their weights according to their predictive power on the most recent chunk [4]. The model replaces old learners with newer ones trained upon more recent chunks. We used HTs as base learners for the ensemble and we set the maximum number of base learners to 10.
- KNN: A lazy learner which predicts based on the class labels of the neighboring instances [3]. In particular, the previous chunk instances and their labels are used to make predictions for the instances of the current chunk. The neighborhood is set to $k = 10$.

We evaluate our strategies M1–M4 (c.f., Sect. 3.3) for the different classifiers, as well as against the following baselines that do not explicitly handle discrimination:

B1 **B.NoSA** (Baseline NoSensitiveAttribute): The classifier F does not employ SA neither in training nor in testing. The model is continuously updated over the stream from the original chunks S_t. Intuitively, the model tackles discrimination by omitting SA.

B2 **B.RESET** (Baseline Reset): If discrimination is detected, the old model F is deleted and a new model is learned on S_t. The model is updated over the stream, but without any correction. Discrimination is being monitored and if it is detected again, the whole procedure starts over. Intuitively, this approach tackles discrimination by resetting the model when discrimination is detected.

For the massaging techniques, we use NB as a ranker which according to [5] is the best ranker. We implemented our methods[1] in MOA [3]. For all of our reported experiments, we consider a discrimination threshold of $\epsilon = 0.0$, that is, we do not tolerate any discrimination, and a chunk size of $|S| = 1,000$ instances. The effect of these parameters is discussed in Sect. 4.2.

4.1 Datasets

As real dataset we employ the census-income (or adult-census) dataset, which comprises one of the most popular datasets in this area; we simulate the stream using the file order. Due to lack of stream data for fairness, we extend an existing stream generator to simulate different discrimination scenarios in data streams.

Census-Income [21]: The learning task is to predict whether a person earns more than 50K/year using demographic features. We consider gender as the sensitive attribute with females being the deprived community and males being the favored community. In addition, we consider an annual income of more than 50K as the target class. The dataset consists of 48,842 records and has an overall discrimination of 19.45%. In Fig. 3a, the discrimination score and the different community volumes (DR_t, DG_t, FR_t, FG_t) are shown over time using a chunk size of $|S| = 1,000$ instances. The discrimination score ranges between 15%–25% overtime.

Synthetic Generator: Our generator comprises an extension of the static fairness generator of [30] that represents each community using a Gaussian distribution. It forces the DG community to be closer to the negative class whereas the FG community is placed further away from the negative class. An example can be already seen in Fig. 1. We extend this idea in a stream setting by varying the amount of discrimination over the stream while introducing concept drifts.

In particular, we initialize four Gaussians, as follows, similarly to the static generator: $p(DG) = N([2; 2], [3, 1; 1, 3])$, $p(FG) = N([2.5; 2.5], [3, 1; 1, 3])$,

[1] Code will be made available online.

$p(DR) = N([0.5; 0.5], [3, 3; 1, 3])$ and $p(FR) = N([-2; -2], [3, 1; 1, 3])$. In the initialization phase, all Gaussians contribute equally to each community with n instances each, giving a total of $N = 4n$ instances for the initial chunk. With respect to discrimination, we introduce a parameter SPP that controls the statistical parity by controlling the number of generated instances x in the DG community over the stream. The exact amount of instances x can be derived from Eq. 1 as follows:

$$SPP = \frac{n}{2*n} - \frac{x}{x+n} \Rightarrow x = n * \frac{1 - 2*SPP}{1 + 2*SPP} \tag{4}$$

where n is the amount of instances generated by each Gaussian in a chunk and x is the amount of instances for the DG community based on the desired SPP value; the rest $n - x$ instances generated originally by its corresponding Gaussian are evenly distributed to the FG and FR communities. This way, the ratio of positive instances in the favored community remains the same. To simulate concept drifts in the population, we change the means of the Gaussians at random points over the stream. To maintain the initial problem (unfair treatment of one community), we shift the means all together at a random direction *up, down, left or right* by a random value $k \in [0, 2]$.

For evaluation purposes, we generate a synthetic dataset of 200,000 instances (200 chunks, $N = 1,000$ instances per chunk), 4 numerical attributes, 1 binary sensitive attribute and 1 binary class. We inserted 20 concept drift at random points and vary SPP randomly over time from 0% to 30%. The dataset characteristics are shown in Fig. 3b.

4.2 Evaluation Results

For each dataset, we report on the discrimination-vs-accuracy behavior of the different classifiers under the different adaptation strategies. The discrimination-vs-accuracy plot (an example is shown in Fig. 4a) allows for a quick evaluation of the different behaviors. Values close to 0 in the x-axis mean fair models, whereas as the values increase the corresponding classifier become more discriminating. w.r.t accuracy (y-axis), good predictive power models reside close to 100%, whereas low y values indicate poor performing models. The ideal models are located on the up left region which indicates high accuracy and low discrimination performance models. The worst models are located in the bottom right region where low accuracy and high discriminating behavior take place. Up right and bottom left regions indicate unfair but accurate models and fair but inaccurate models, respectively.

Census-Income. For the massaging, c.f., Fig. 4a, our strategies achieve lower discrimination comparing to the baselines (our values are closer to 0 in the x-axis). As expected, the improvement w.r.t discrimination incurs a drop in accuracy, i.e., baselines have better accuracy comparing to our strategies (baseline values are closer to 100% in the y-axis). We also observe that some strategies depict very similar performance, e.g., $M2$ and $M4$ when combined with HT.

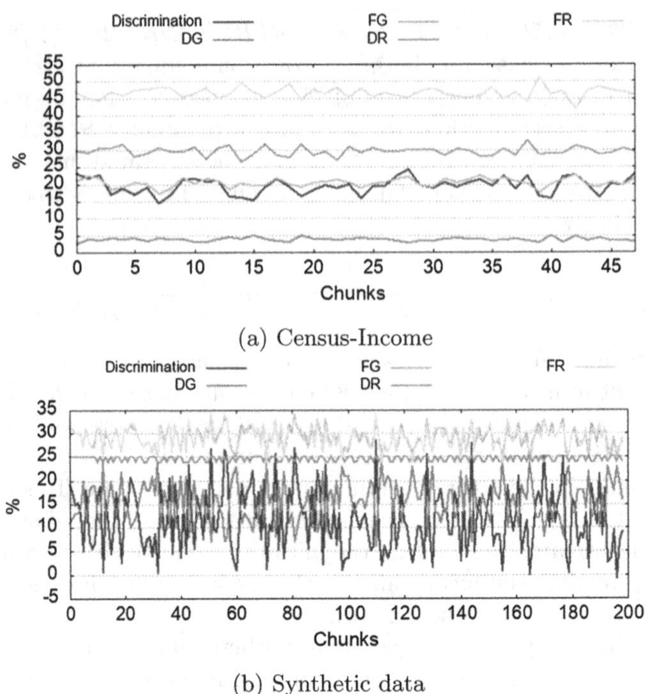

(a) Census-Income

(b) Synthetic data

Fig. 3. Discrimination and community size over the stream ($|S| = 1,000$ instances/chunk)

The reason is that since $\epsilon = 0$, our discrimination detector is activated on almost every chunk from the stream and therefore strategies like $M2$ and $M4$ will both reset the model on each chunk. Accumulative strategies, $M1$ and $M3$, perform better than reset strategies, $M2$ and $M4$; the reason is probably that the latter ones forget too fast. Regarding the different classifiers employed by our strategies, we can see that the best performing ones in terms of both accuracy and discrimination are KNN and AUE. AUE and HT models yield better accuracy and less discrimination when they do not discard previous knowledge. Although KNN is not the best performing model in terms of accuracy, it yields the lowest discrimination score with the smallest drop in accuracy when compared to its baselines, namely $B.RESET$ and $B.NoSA$. In particular, discrimination drops from 19% to 4% while accuracy drops by almost 1%, when KNN is employed by $M3$ and $M4$ strategies.

In Fig. 4b, we compare the discrimination-vs-accuracy behavior of the different classifiers under re-weighting. Same as in massaging, our strategies reduce discrimination in predictions. Classifiers such as HT behave similarly under different strategies since the detector detects discrimination in almost every chunk. KNN on the other hand, doesn't take into consideration weights, hence all the strategies perform identically.

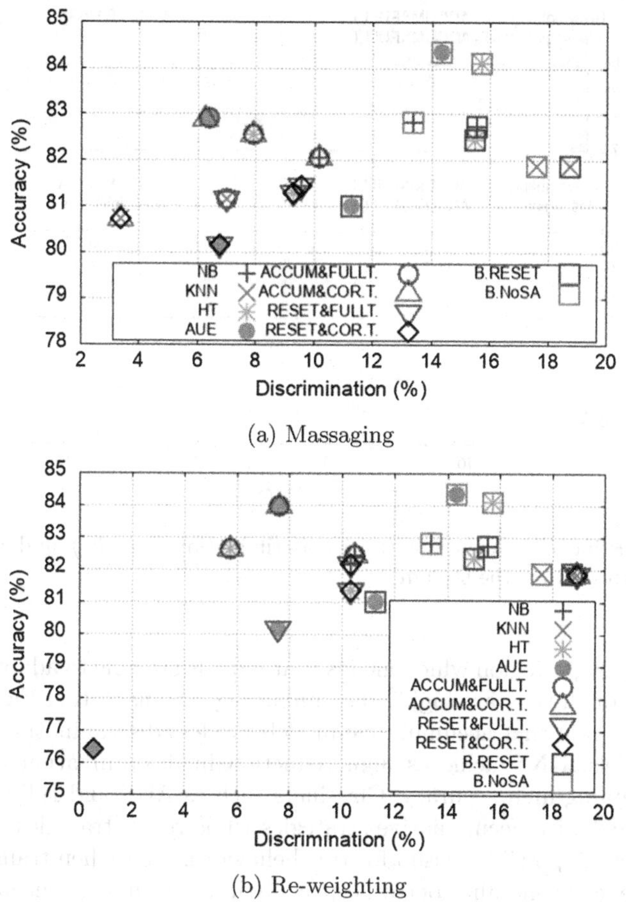

(a) Massaging

(b) Re-weighting

Fig. 4. Census-income (small): Discrimination-vs-accuracy of the different strategies

We also compare models overtime in Fig. 5. We have selected one model for each method (massaging/re-weighting) based on the discrimination-accuracy trade off (points which are closer to (0, 1) based on Euclidean distance) and the best baseline of those two models. Although HT's baseline has the best accuracy overtime, its discrimination score is close to stream's discrimination. On the other hand, re-weighting and massaging methods result in a significant drop in discrimination. The number of massaged instances M varies over the stream, based on how discriminative the KNN's predictions are.

Synthetic Data. The dataset contains 20 concept drifts while its discrimination score varies overtime, as seen in Fig. 3b. The majority of baselines, in Figs. 6a and b, are able to adapt to concept drifts (i.e., they achieve high accuracy), however they cannot handle discrimination, which in some cases is even amplified comparing to the original stream overall discrimination. The vast majority of baselines

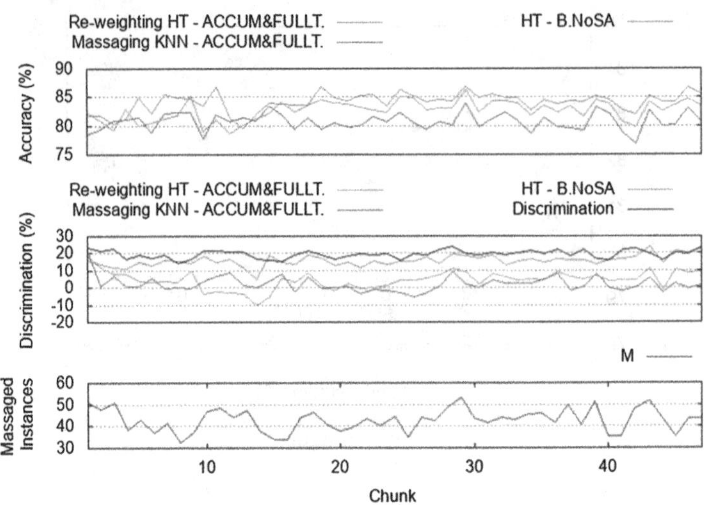

Fig. 5. Census-income: Accuracy (top), discrimination (middle) and # massaged instances (bottom) over the stream

occupy the up right region which means that models are able to adapt to concept drifts even though they are highly discriminating. By inspecting Fig. 6a, we can observe once again that accumulative models are less discriminating in contrast to reset models. KNN achieves high reduction in discrimination (up to 6%), while maintaining high accuracy. Classifiers such as AUE and HT perform well when combined with accumulative strategies while reset strategies incur higher discrimination. A possible reason for this behavior is that when trained on more data, a model can generalize better, especially in re-occurring concepts, comparing to reset strategies that rely solely on recent chunks. KNN is an exception as it performs well despite relying on an internal sliding window for its predictions. A possible reason is that a kNN learner is an instance-based learned and does not perform explicit generalization like HT and AUE. Similarly to census-income dataset, NB is not able to tackle discrimination.

In Fig. 6b, we observe that almost all baselines, same as in massaging, cover the up right region area. AUE's performance is increasing while it becomes more discriminating in contrast to AUE in massaging. Same as before, HT and KNN have the least discriminating behavior while NB performs poorly. Again, reset strategies produce good accuracy models but fail to reduce discrimination.

In Fig. 7, we compare the best "correction" methods and the best baseline. $M1$ combined with KNN has the lowest discrimination score overtime. Its accuracy is slightly worse than its baseline. However, discrimination is lower than stream's and baseline's discrimination overtime. HT's overall performance w.r.t accuracy is relatively good except the interval between 77th and 90th chunk where four concept drifts occurred incurring accuracy loss. Despite the accuracy degradation, HT achieved lower discrimination compared to other classifiers.

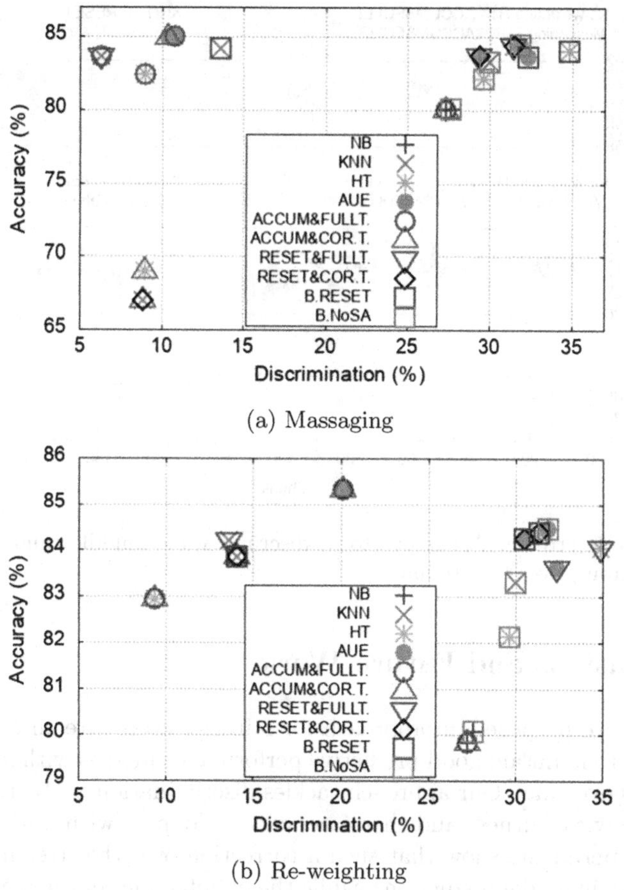

Fig. 6. Synthetic stream: Discrimination-vs-accuracy of the different strategies

Parameter effect: Due to lack of space we omit the time execution charts. A derived conclusion is that our strategies are executed slightly slower compared to the baselines and moreover, that the reset strategies are faster than the accumulative strategies. We have also experimented with different chunk sizes $|S|$ and discrimination thresholds ϵ. Based on our experiments, increasing ϵ results in better accuracy models but their discrimination also increases. With respect to the chunk size effect, there was no clear effect on the performance except for the execution time that decreases with chunk size as less operations take place.

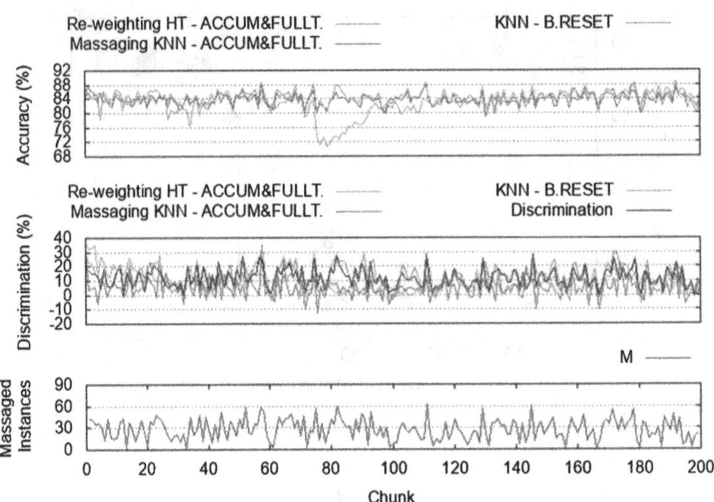

Fig. 7. Synthetic stream: Accuracy (top), discrimination (middle) and # massaged instances (bottom) over the stream

5 Conclusions and Future Work

In this work, we proposed an approach for fairness-aware stream classification, which is able to maintain good predictive performance models with low discrimination scores overtime. Our approach tackles discrimination by "correcting" the input stream w.r.t fairness and therefore, can be coupled with any stream classifier. Our experiments show that such a correction over the stream can reduce discrimination in model predictions, while the maintenance of the model over the stream allows for adaptation to underlying concept drifts. Comparing the different fairness-intervention methods, our experiments show that massaging performs better than re-weighting. A possible explanation is that massaging works at the individual instance level by swapping its class label, whereas re-weighting works at a group level by applying different weights to different communities. Moreover, massaging affects selected instances, which are closer to the boundary.

Our approach is model-agnostic, however our experiments show that the effect of "data correction for discrimination" depends on the classifier and therefore, how to "best correct" for specific classifiers is an interesting research direction. Moreover, we want to investigate in-processing fairness-aware stream classifiers that incorporate fairness notion directly in the classification algorithm.

Acknowledgements. The work is inspired by the German Research Foundation (DFG) project OSCAR (Opinion Stream Classification with Ensembles and Active leaRners) for which the last author is Co-Principal Investigator.

References

1. Bhandari, E.: Big data can be used to violate civil rights laws, and the FTC agrees (2016)
2. Bifet, A., Frank, E.: Sentiment knowledge discovery in twitter streaming data. In: Pfahringer, B., Holmes, G., Hoffmann, A. (eds.) DS 2010. LNCS (LNAI), vol. 6332, pp. 1–15. Springer, Heidelberg (2010). https://doi.org/10.1007/978-3-642-16184-1_1
3. Bifet, A., Holmes, G., Kirkby, R., Pfahringer, B.: MOA: massive online analysis. JMLR **11**, 1601–1604 (2010)
4. Brzeziński, D., Stefanowski, J.: Accuracy updated ensemble for data streams with concept drift. In: Corchado, E., Kurzyński, M., Woźniak, M. (eds.) HAIS 2011. LNCS (LNAI), vol. 6679, pp. 155–163. Springer, Heidelberg (2011). https://doi.org/10.1007/978-3-642-21222-2_19
5. Calders, T., Kamiran, F., Pechenizkiy, M.: Building classifiers with independency constraints. In: ICDMW 2009, pp. 13–18. IEEE (2009)
6. Calmon, F., Wei, D., Vinzamuri, B., Ramamurthy, K.N., Varshney, K.R.: Optimized pre-processing for discrimination prevention. In: NIPS, pp. 3992–4001 (2017)
7. Domingos, P., Hulten, G.: Mining high-speed data streams. In: SIGKDD, pp. 71–80. ACM (2000)
8. Dwork, C., Immorlica, N., Kalai, A.T., Leiserson, M.D.: Decoupled classifiers for group-fair and efficient machine learning. In: FAT, pp. 119–133 (2018)
9. Edelman, B.G., Luca, M.: Digital discrimination: the case of airbnb.com (2014)
10. Fish, B., Kun, J., Lelkes, Á.D.: A confidence-based approach for balancing fairness and accuracy. In: SIAM, pp. 144–152 (2016)
11. Forman, G.: Tackling concept drift by temporal inductive transfer. In: SIGIR, pp. 252–259. ACM (2006)
12. Gama, J.: Knowledge Discovery from Data Streams. CRC Press, Boca Raton (2010)
13. Ingold, D., Soper, S.: Amazon doesn't consider the race of its customers. Should it. Bloomberg, April 2016
14. Iosifidis, V., Ntoutsi, E.: Dealing with bias via data augmentation in supervised learning scenarios. Jo Bates Paul D. Clough Robert Jäschke, p. 24 (2018)
15. Kamiran, F., Calders, T.: Classifying without discriminating. In: IC4. IEEE (2009)
16. Kamiran, F., Calders, T.: Classification with no discrimination by preferential sampling. In: BeneLearn, pp. 1–6. Citeseer (2010)
17. Kamiran, F., Calders, T.: Data preprocessing techniques for classification without discrimination. KAIS **33**(1), 1–33 (2012)
18. Kamiran, F., Calders, T., Pechenizkiy, M.: Discrimination aware decision tree learning. In: ICDM, pp. 869–874. IEEE (2010)
19. Klinkenberg, R.: Learning drifting concepts: example selection vs. example weighting. IDA **8**, 281–300 (2004)
20. Liu, L.T., Dean, S., Rolf, E., Simchowitz, M., Hardt, M.: Delayed impact of fair machine learning. arXiv preprint arXiv:1803.04383 (2018)
21. Merz, C.J., Murphy, P.M.: {UCI} repository of machine learning databases (1998)
22. Pedreschi, D., Ruggieri, S., Turini, F.: Measuring discrimination in socially-sensitive decision records. In: SDM, pp. 581–592. SIAM (2009)
23. Pedreshi, D., Ruggieri, S., Turini, F.: Discrimination-aware data mining. In: SIGKDD, pp. 560–568. ACM (2008)

24. USA: Executive Office of the President, Podesta, J.: Big data: seizing opportunities, preserving values. White House, Executive Office of the President (2014)
25. Romei, A., Ruggieri, S.: A multidisciplinary survey on discrimination analysis. TKER **29**(5), 582–638 (2014)
26. Schlimmer, J.C., Granger, R.H.: Beyond incremental processing: tracking concept drift. In: AAAI, pp. 502–507 (1986)
27. Stoyanovich, J., Yang, K., Jagadish, H.: Online set selection with fairness and diversity constraints. In: EDBT (2018)
28. Sweeney, L.: Discrimination in online ad delivery. Queue **11**(3), 10 (2013)
29. Verma, S., Rubin, J.: Fairness definitions explained (2018)
30. Zafar, M.B., Valera, I., Gomez Rodriguez, M., Gummadi, K.P.: Fairness constraints: mechanisms for fair classification. arXiv preprint arXiv:1507.05259 (2017)

Early Turnover Prediction of New Restaurant Employees from Their Attendance Records and Attributes

Koya Sato(✉), Mizuki Oka, and Kazuhiko Kato

Graduate School of Systems and Information Engineering,
University of Tsukuba, Tsukuba, Japan
{koya,mizuki,kato}@osss.cs.tsukuba.ac.jp

Abstract. It is widely known that the turnover rate of new employees is high. Several studies have been conducted on filtering candidates during the recruitment process to avoid hiring employees that are likely to leave early. However, studies on the prediction of early turnover of new employees, which might enable appropriate interventions, are scarce. In the restaurant industry, which suffers from labor shortages, filtering candidates is unrealistic, and it is important to maintain newly hired employees. In this study, we propose a new model, based on recurrent neural networks, that predicts the early turnover of new restaurant employees by using their attendance records and attributes. We have evaluated the effectiveness of the proposed model by using anonymized data from a restaurant chain in Japan, and we confirmed that the proposed model performs better than baseline models. Furthermore, our analysis revealed that gender and hiring channel had little influence on early turnover and decreased prediction performance. We believe that these results will help in designing efficient interventions to prevent new restaurant employees from leaving early.

Keywords: Turnover prediction · Restaurant industry · Deep learning

1 Introduction

Employee turnover leads to an increase in hiring costs and a decrease in the quality of services [12]. Therefore, it is viewed as one of the most critical issues in human resource management, and has been the focus of many studies. The turnover rate of new employees is known to be particularly high [7]. Many studies have been conducted on filtering candidates during the hiring process to avoid hiring employees [12] that will leave early. However, few studies have attempted to predict the early turnover of new employees.

In industries that suffer from labor shortages, filtering candidates is unrealistic, and it is more important to maintain newly hired employees. The turnover rate of the restaurant industry is high, and labor shortages are a serious problem [14]. For example, in the restaurant industry in Japan, the ratio of job offers

S. Hartmann et al. (Eds.): DEXA 2019, LNCS 11706, pp. 277–286, 2019.
https://doi.org/10.1007/978-3-030-27615-7_21

to job applicants was 3.28–3.36 in January 2019 [13]. Because of an aging population, labor shortages are becoming a severe problem globally. For this reason, we propose a new early turnover prediction model in the restaurant industry to enable efficient interventions for new employees that are likely to leave early.

Because of technological advances, the restaurant industry has adopted attendance management systems, which allow access to employees' daily attendance records. These capture how employees contribute to the workplace, which affects the likelihood of turnover [1,9]. Therefore, by using employees' attendance records, which have recently become available, and attributes that are often used for employee turnover prediction, our hypothesis is that it is possible to predict the early turnover of employees.

In this research, we propose a new model to predict the early turnover of new employees. We constructed this model using recurrent neural networks (RNNs), which are known to exhibit high classification performance on temporal data [10]. We evaluated the effectiveness of the proposed model by using anonymized data from a restaurant chain in Japan, which is at the forefront of the problem of population aging. Furthermore, we conducted a feature analysis to attempt to discover how each feature affects the predictions. The contributions of this research are as follows:

1. We proposed a model to predict the early turnover of new employees from their attendance records and attributes.
2. We confirmed the effectiveness of the proposed model on employee data from a restaurant chain in Japan.
3. We investigated how each feature affects the prediction of early turnover.

2 Related Research

Many studies have been conducted on employee turnover prediction. Machine learning methods, such as logistic regression, random forest, and support vector machine, have been used to predict employee turnover accurately; a multi-layered neural network has also been applied [12].

Despite the volume of previous research on employee turnover prediction, few studies have attempted early turnover prediction specifically for new employees. According to a review by Strohmeier et al. [12], in the field of human resource management, most studies have been conducted on recruitment. With respect to new employees, research has not been conducted with the strategy of maintaining newly hired employees, but of avoiding hiring candidates who have a high turnover potential. In the restaurant industry, which has labor shortages [1], filtering at the recruitment stage is impractical, and it is important to maintain newly hired employees.

Many studies have been conducted to discover the reasons for turnover in the restaurant industry. Cantrell et al. suggested that low wages are the primary reasons for turnover, and inconvenient working hours and inflexible shifts are the secondary reason [1]. McFillen et al. also demonstrated that job satisfaction and commitment have a large effect on turnover [9]. Because attendance records contain information on how an employee contributes to the work, we believe that employees likely to leave early can be predicted by using this information.

Several studies have focused on the early turnover of new employees. Genu et al. investigated the factors leading to the early turnover of foreign nurses through a questionnaire survey [5]. Thoms et al. revealed a relationship between immediate turnover and theft in the restaurant industry [14]. These studies aimed to identify the cause of early turnover, not to predict it. To the best of our knowledge, there is no model that predicts the early turnover of new employees based on employees' attendance records and attributes.

It is well known that RNNs show high classification performance on data with temporal correlation [10]. For example, in the field of video classification and speech recognition, RNNs showed the highest classification performance [3,6]. Therefore, we consider encoding the employees' attendance records by using an RNN to predict the early turnover of new employees.

3 Data

3.1 Employee Data of Restaurant Chain in Japan

We analyzed anonymized data from a restaurant chain in Japan. The staff consist of permanent, part-time, and contractual employees. The working style of permanent employees differs from the others, and the impact of turnover on employers is higher for permanent employees than for the others. Therefore, we only targeted permanent employees.

The data cover the period from January 1, 2016 to March 21, 2018. They cover 220 restaurants and 1260 working employees. Of the working employees, 617 were newly hired during this period. The target employers utilize shift work, which means that the working style varies among employees. The attendance management system records each employee's daily attendance (starting and finishing time) by reading the employee card at the terminal installed at each restaurant. Each employee has an anonymized unique id, and we can access each employee's attendance records and attributes.

To measure the degree of early turnover in the data, we considered an employee who did not attend for 30 consecutive days as a *turnover employee*. In Japan, it is not so common for employees to take a vacation lasting one month. Thus, we considered 30 days as a reasonable threshold. Figure 1a shows the relationship between the elapsed days since employment and the turnover rate. The turnover rate linearly increases, and approximately 19% of new employees left within 150 days. This confirms that early turnover is a problem for this restaurant chain. Because this value is the mean for each restaurant on same chain, it is considered that the early turnover is also a problem for competitor restaurants to almost the same extent.

3.2 Early Turnover Prediction

Figure 1b shows an overview of our early turnover prediction method. We defined an employee who does not attend within 30 days after C days as an *early turnover*

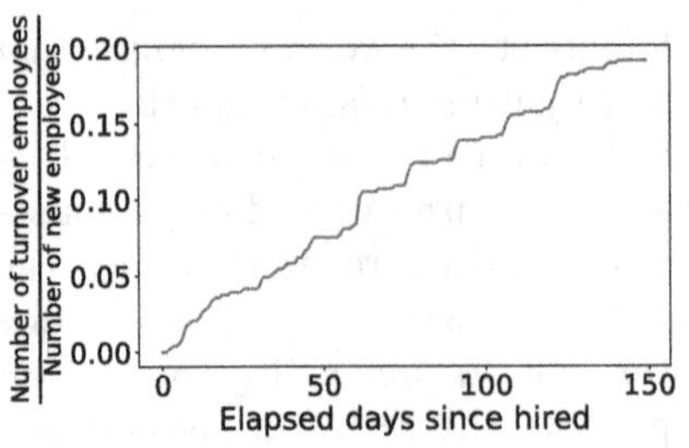

(a) Dynamics of turnover rate.

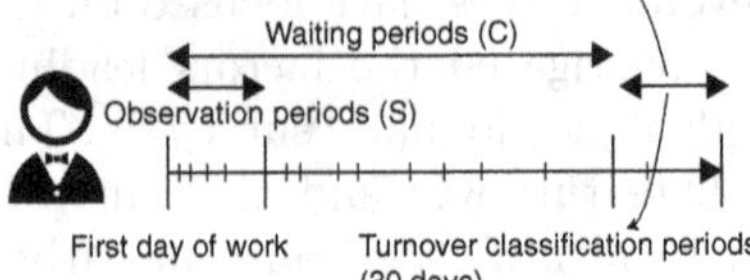

(b) Diagram of early turnover prediction.

Fig. 1. Relationship between elapsed days since employment and turnover rate (a), and diagram of early turnover prediction (b).

Table 1. Data in numbers.

S (days)	$C = 90$		$C = 150$	
	New employees	Turnover employees	New employees	Turnover employees
7	545	68	521	100
28	533	56	509	88

employee, and any other employee (who attended at least once within 30 days after C) as a *non-early turnover employee.* We defined S days as the observation period, and we performed the prediction using S days of attendance and employee attributes. To prevent artifacts, we included only the new employees whose first attendance day plus $C + 30$ days did not exceed the final day of the data (March 21, 2018). Additionally, it was unnecessary to predict employees who left within S days, so we analyzed only the new employees who attended at least once within C days after S days. To test the classification performance in various situations, we set $C = [90, 150]$ and $S = [7, 28]$. Table 1 shows the numbers of target new employees and early turnover employees.

We use the employee's attendance records and attributes for prediction. The attributes include age, gender, hiring channel (previously employed as part-time employee, hourly contractual employee, monthly contractual employee, or newly employed), previous working periods if the employee worked as a temporary employee, role in a restaurant (restaurant manager, vice restaurant manager, cooking manager, or general employee), and employment month. We determined the attributes based on discussions with some store managers. The attendance records include starting time, finishing time, working hours, working type (attendance, non-attendance day, absenteeism, public holidays, paid holiday, transfer holiday, or bereavement leave), and day of the week. Gender, hiring channel, role in a restaurant, employed month, working type, and days of the week are categorical variables, and the other attributes are continuous variables. The starting time, finishing time, and working hours are 0 when an employee is absent from

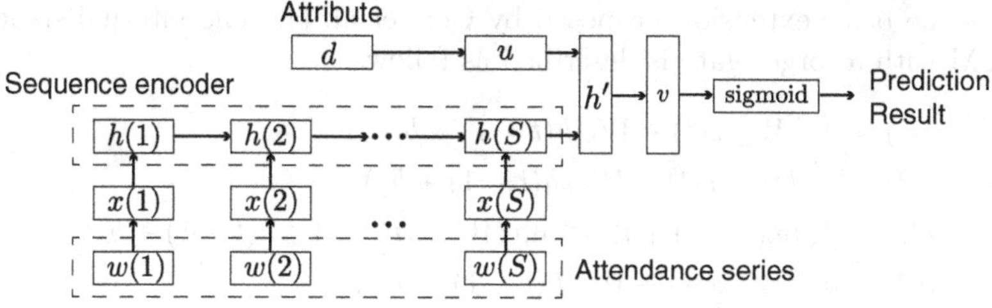

Fig. 2. Overview of the proposed model.

work. Note that detailed statistics of each feature have not been reported for privacy concerns.

4 Proposed Model

Figure 2 shows an overview of the proposed model. It outputs the rate of early turnover from an employee's attendance records and attributes.

We define an employee's *attendance series* as $\mathbf{w} = (w(1), ..., w(t), ..., w(S))$. t is the number of elapsed days since the employee's first attendance at work. $w(t)$ is a vector of the employee's attendance on day t. We obtain the hidden attendance representation ($x(t)$) by inputting $w(t)$ to a single-layer perceptron: $x(t) = \text{ReLU}\left(W_w w(t) + b_w\right)$. We define the employee's attributes as d and, similarly, we input d to a single-layer perceptron to obtain the hidden representation, $u = \text{ReLU}\left(W_d d + b_d\right)$.

We encode the attendance series by using two deep learning models with a recursive structure, as described below.

Recurrent Neural Network. An RNN is a deep learning model that can encode temporal correlations by using a recursive structure [10], and it is defined as follows:

$$h(t) = \tanh\left(W_x' x(t) + W_h' h(t-1) + b_h\right). \tag{1}$$

$h(t)$ is the output of the sequence encoder at t, which is produced by considering the previous output $h(t-1)$. Because of this recursive structure, an RNN can capture temporal correlations, but it is known that RNN cannot maintain long-term correlation.

Long Short-Term Memory. To maintain long-term correlation, long short-term memory (LSTM)—an extension of the RNN—was proposed [8]. LSTM has many variations in internal structure. We used LSTM with a forget gate, which is the basic extension proposed by Gers et al. [4]. The internal structure of LSTM with a forget gate is described as follows:

$$\begin{aligned} i(t) &= \sigma_h \left(W_{xi}x(t) + W_{hi}h(t-1) + b_i\right), \\ f(t) &= \sigma_h \left(W_{xf}x(t) + W_{hf}h(t-1) + b_f\right), \\ c(t) &= f(t)c(t-1) + i(t)\tanh\left(W_{xc}x(t) + W_{hc}h(t-1) + b_c\right), \\ o(t) &= \sigma_h \left(W_{xo}x(t) + W_{ho}h(t-1) + b_o\right), \\ h(t) &= o(t)\tanh\left(c(t)\right). \end{aligned} \tag{2}$$

$i(t)$ is the input gate, which determines the amount of input information to maintain. $f(t)$ is the forget gate, which determines the amount of information to forget the previous cell state ($c(t-1)$). $o(t)$ is the output gate, which determines the amount of information of the current cell state that is applied to the final output of the LSTM layer. $h(t)$ is the final output of the LSTM layer in each t. σ_h is the recurrent activation function, which is a sigmoid function.

The proposed model performs the prediction by combining the final output of the sequence encoder $h(S)$, encoded by the RNN or LSTM, and the hidden representation of attributes u. We concatenate u and $h(S)$ to obtain the vector $h' = [h(S), u]$, which maintains the employee's attendance series and attributes. Through one more hidden layer, we obtain $v = \text{ReLU}\left(W_{h'}h' + b_{h'}\right)$, which is the input to the output layer.

To classify early and non-early turnover employees, we assign the labels $\hat{y} = 1$ and $\hat{y} = 0$ to early and non-early turnover employees, respectively. To obtain the probability of being an early turnover employee, we used the sigmoid function as the output layer ($y = \text{sigmoid}(W_y v + b_y)$).

For a small C, the labels are highly imbalanced. To address this imbalance, we used weighted binary cross-entropy as the loss function:

$$loss(y, \hat{y}) = -\left((1-w)\hat{y}\log y + w(1-\hat{y})\log\left(1-y\right)\right). \tag{3}$$

w is the proportion of early turnover employees in the learning data. We minimize this loss with back-propagation through time [11].

5 Experiment

To evaluate the effectiveness of the proposed model, we compare the classification performance of the proposed model with the performance of the baseline models that are often used for employee turnover prediction.

5.1 Evaluation Criteria

To evaluate the classification performance of each model, we selected the following three indices: area under the receiver operating characteristic curve (ROC), area under the precision-recall curve (PR), and F-measure. They range from 0 to 1, with values close to 1 indicating good classification performance.

Table 2. Classification performance of each model.

S (days)	Model	$C = 90$ (days)			$C = 150$ (days)		
		ROC	PR	F-measure	ROC	PR	F-measure
7	LR	0.716	0.291	0.345	0.694	0.357	0.386
	SVM	0.700	0.277	0.000	0.691	0.321	0.000
	RF	0.680	0.259	0.050	0.658	0.296	0.188
	MN	0.691	0.295	0.340	0.688	0.327	0.410
	RNN	0.638	0.319	0.235	0.645	0.337	0.379
	LSTM	0.632	0.278	0.273	0.636	0.354	0.349
	RNN+AT	**0.719**	0.334	**0.352**	**0.706**	**0.375**	0.397
	LSTM+AT	0.691	**0.372**	0.305	0.698	0.371	**0.420**
28	LR	0.739	0.342	0.296	0.614	0.320	0.309
	SVM	0.694	0.290	0.029	0.665	0.336	0.040
	RF	0.633	0.306	0.114	0.647	0.277	0.126
	MN	0.684	0.289	0.300	0.666	0.355	0.354
	RNN	0.587	0.311	0.262	0.661	0.336	0.249
	LSTM	0.611	0.322	0.300	0.582	0.314	0.235
	RNN+AT	0.705	0.265	0.284	**0.722**	0.333	**0.375**
	LSTM+AT	**0.765**	**0.357**	**0.333**	0.706	**0.365**	0.350

Because ROC and PR are calculated by varying the classification threshold, they are independent of a threshold. When the value of C is small, the labels are highly imbalanced; in this case, PR is a more suitable evaluation criterion than ROC [2]. The F-measure assesses the classification performance with a certain threshold.

We performed $k = 10$ fold cross-validation to evaluate the performance on test data. We treated 20% of the training data as validation data and determined hyperparameters by using the validation data; we show the result using the test data when each criterion against the validation data is maximized.

5.2 Baseline Models

We used three machine learning models (logistic regression (LR), support vector machine (SVM), and random forest (RF)) and one deep learning model (3-layer neural network (MN)) as baseline models. We name the RNN-based and LSTM-based proposed models as RNN+AT and LSTM+AT, respectively. We perform prediction using RNN or LSTM only from the attendance series, not the employee attributes ($h' = h(S)$); we call these models RNN and LSTM, respectively. We applied min–max normalization to continuous variables before inputting them to each model.

We implemented each machine learning model using scikit-learn, which is a Python machine learning package[1]. To address label imbalance, we set the class_weight parameter as balanced in each model. To determine hyperparameters, we conducted a grid search using the validation data.

We implemented deep learning models using Keras, which is a deep learning API for Python[2]. The number of units in each layer was $[10, 20]$. We applied a dropout before inputting data to each layer, with a dropout rate of $[0.2, 0.3]$. We used Adam to train the model. The batch size was $[32, 16]$. We stopped learning when the validation loss stopped decreasing. We optimized the hyperparameters by grid search. Other hyperparameters used were the default values of Keras.

5.3 Results

Table 2 shows the mean value of each measure. We found that the ROC, PR, and F-measure of RNN+AT or LSTM+AT showed the highest values compared with the baseline models. We also found that the difference between the classification performances of RNN+AT and LSTM+AT varies depending on S and C.

We confirmed the effectiveness of the proposed model from the fact that RNN+AT or LSTM+AT showed the highest performance. However, we found that the performance difference from using the sequence encoding method was small. The length of the input series is small (28 at most), so it appears that the model did not need to maintain long-term correlations for prediction.

6 Feature Analysis

In this section, we analyze how each feature affects the early turnover prediction of new employees. We observe the change in the mean loss (Eq. 3) against test data when the model predicts the early turnover excluding each feature in turn. We define the baseline loss as the mean test loss obtained by the model to predict the early turnover from all features. We compare this baseline loss with the mean test loss by the model excluding specific features. If the mean test loss is greater than the baseline loss, it indicates that the excluded feature has a positive effect on the classification. If the mean test loss is less than the baseline loss, the excluded feature has a negative effect. We conducted this feature analysis by using the LSTM+AT, using $k = 10$ fold cross-validation and determining hyperparameters in the same way as the previous section. We conducted a paired-samples t-test to determine the differences from the baseline loss.

Figure 3 shows the mean and standard deviation of the mean test loss. Each blue dot and horizontal bar indicates the mean and standard deviation, respectively. The vertical dotted line is an eye guide and mean of the baseline loss. There was a significant difference in the gender and hiring channel for $S = 7$ and $C = 90$ ($p < 0.05$). There was also a significant difference in the gender for $S = 7$ and $C = 150$ ($p < 0.05$). There were no significant differences in the others.

[1] https://scikit-learn.org/stable/.
[2] https://keras.io/.

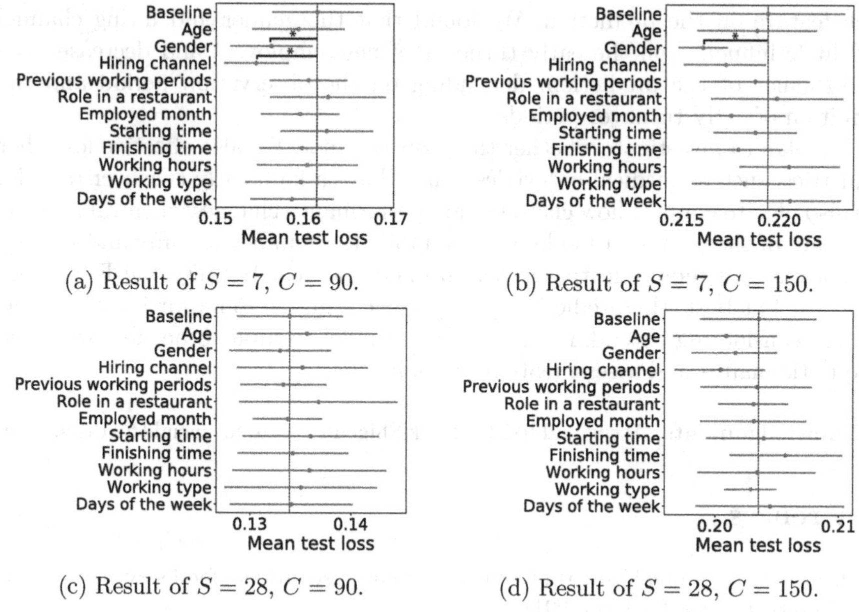

(a) Result of $S = 7$, $C = 90$.

(b) Result of $S = 7$, $C = 150$.

(c) Result of $S = 28$, $C = 90$.

(d) Result of $S = 28$, $C = 150$.

Fig. 3. Mean and standard deviation of the mean test loss ($* : p < .05$).

The mean test losses for gender and hiring channel were lower than the baseline loss for almost every combination of S and C, even when there was no significant difference. Therefore, we concluded that gender and hiring channel had little influence on the early turnover of new employees and decreased the performance of the prediction, to some extent, depending on the S and C. The reason for the negative influence of gender is that there were no gender-specific reasons for early turnover (such as childbirth). The reason for the negative influence of hiring channel is that early turnover is caused by the duty and heavy responsibility as an employee, which cannot be perceived by working only as a temporary employee (for example, increased working hours or restaurant management). In contrast, we did not observe any features that had a significant positive effect on prediction. It appears that there was a high correlation between each excluded feature and the remaining features.

7 Conclusion and Future Work

In this work, we have proposed a new model that predicts the early turnover of new restaurant employees by using the employee's attendance records and attributes. We evaluated the effectiveness of the proposed model on anonymized data from a restaurant chain in Japan. We confirmed that the proposed model exhibits a higher classification performance than baseline models that are often used for employee turnover prediction. Furthermore, we analyzed the effect of each feature on the prediction. We found that the gender and hiring channel had little influence on the early turnover of new employees and decreased the performance of the prediction, depending on the observation period and the definition of early turnover periods.

We plan to investigate whether the proposed model is also effective for other industries, such as medical industries, which have a high early turnover rate [5]. We also plan to confirm how classification performance changes when varying the structure of the proposed model (for example, the number of units and layers). This leads to a more effective hyperparameter search. Because the F-measure is 0.35–0.4 at best, the prediction performance still has room for improvement. We are considering extending the proposed model to input other features such as questionnaires and assessments from managers.

Acknowledgements. We would like to thank Shigehiro Kato for insightful discussion.

References

1. Cantrell, N., Sarabakhsh, M.: Correlates of non-institutional food service turnover. Hospitality Rev. **9**(2), 6 (1991)
2. Davis, J., Goadrich, M.: The relationship between precision-recall and ROC curves. In: ICML, pp. 233–240 (2006)
3. Ebrahimi Kahou, S., Michalski, V., Konda, K., Memisevic, R., Pal, C.: Recurrent neural networks for emotion recognition in video. In: ICMI, pp. 467–474 (2015)
4. Gers, F.A., Schmidhuber, J.A., Cummins, F.A.: Learning to forget: continual prediction with LSTM. In: ICANN, pp. 850–855 (1999)
5. Geun, H.G., Redman, R.W., McCullagh, M.C.: Predictors of turnover among Asian foreign-educated nurses in their 1st year of US employment. JONA **48**(10), 519–525 (2018)
6. Graves, A., Mohamed, A.R., Hinton, G.: Speech recognition with deep recurrent neural networks. In: ICASSP 2013, pp. 6645–6649 (2013)
7. Gray, A.M., Phillips, V.: Turnover, age and length of service: a comparison of nurses and other staff in the national health service. JAN **19**(4), 819–827 (1994)
8. Hochreiter, S., Schmidhuber, J.: Long short-term memory. Neural Comput. **9**(8), 1735–1780 (1997)
9. McFillen, J.M., Riegel, C.D., Enz, C.A.: Why restaurant managers quit (and how to keep them). Cornell Hotel Restaurant Adm. Q. **27**(3), 36–43 (1986)
10. Medsker, L., Jain, L.C.: Recurrent Neural Networks: Design and Applications. CRC Press, Boca Raton (1999)
11. Rumelhart, D.E., Hinton, G.E., Williams, R.J.: Learning representations by back-propagating errors. Nature **323**(6088), 526–533 (1986)
12. Strohmeier, S., Piazza, F.: Domain diven data mining in human resource management: a review of current research. Expert Syst. Appl. **40**(7), 2410–2420 (2013)
13. The Japan Ministry of Health, Labour and Welfare: Job introduction situation according to an occupation (2019). https://www.mhlw.go.jp/content/11602000/G38-3101.pdf
14. Thoms, P., Wolper, P., Scott, K.S., Jones, D.: The relationship between immediate turnover and employee theft in the restaurant industry. JBP **15**(4), 561–577 (2001)

An Efficient Premiumness and Utility-Based Itemset Placement Scheme for Retail Stores

Parul Chaudhary[1], Anirban Mondal[2(✉)], and Polepalli Krishna Reddy[3(✉)]

[1] Shiv Nadar University, Greater Noida, India
pc230@snu.edu.in
[2] Ashoka University, Sonipat, India
anirban.mondal@ashoka.edu.in
[3] International Institute of Information Technology, Hyderabad, India
pkreddy@iiit.ac.in

Abstract. In retail stores, the placement of items on the shelf space significantly impacts the sales of items. In particular, the probability of sales of a given item is typically considerably higher when it is placed in a *premium* (i.e., highly visible/easily accessible) slot as opposed to a non-premium slot. In this paper, we address the problem of maximizing the revenue for the retailer by determining the placement of the itemsets in different types of slots with varied *premiumness* such that each item is placed *at least once* in any of the slots. We first propose the notion of *premiumness of slots* in a given retail store. Then we discuss a framework for *efficiently* identifying itemsets from a transactional database and placing these itemsets by mapping itemsets with different revenue to slots with varied premiumness for maximizing retailer revenue. Our performance evaluation on both synthetic and real datasets demonstrate that the proposed scheme indeed improves the retailer revenue by up to 45% w.r.t. a recent existing scheme.

Keywords: Utility mining · Retail management · Pattern mining · Data mining

1 Introduction

In retail stores, the placement of items on the shelf space significantly impacts sales of items and consequently, sales revenue as well [1–5]. However, strategic placement of items in a manual manner for maximizing the sales revenue is not practically feasible in case of medium-to-large retail stores, which may typically occupy more than a million square feet of retail floor space. Examples of such medium-to-large retail stores include Macy's Department Store at Herald Square (New York City, US), Shinsegae Centumcity Department Store (Busan, South Korea) [6] and Dubai Mall.

Consider a large retail store with multiple aisles, where each aisle contains items that are stocked in the slots of the shelves. Notably, all of these slots are not created equal i.e., they differ in terms of *premiumness*. The premiumness of a given slot represents the extent to which the slot is visible and/or physically accessible to the shoppers. It is a well-established fact in the retail industry that the probability of sales

S. Hartmann et al. (Eds.): DEXA 2019, LNCS 11706, pp. 287–303, 2019.
https://doi.org/10.1007/978-3-030-27615-7_22

of a given item *i* increases with increase in the premiumness of the slot, in which *i* is placed [7]. Hence, this work considers that the probability of sales of a given item typically varies depending upon the *premiumness* of the slot in which the item is placed. Examples of slots with high premiumness include the *"impulse buy"* slots at the checkout counters of retail stores and slots that are near to the eye or shoulder level of users. Slots with low premiumness are typically located very high or very low in the retail store shelves.

In general, we can have N types of slots based on premiumness. Moreover, there would be ***multiple blocks of each of the slot types*** across the different aisles in a given large retail store, and each block would contain several slots of a given slot type. The research issue here is to devise a methodology for strategically placing the items in the slots of varied premiumness such that the revenue of the retailer is maximized.

We address the problem of maximizing the revenue for the retailer by determining the placement of the itemsets in the different types of slots with varied premiumness such that each item is placed *at least once* in any of the slots [24]. We need to ensure that every item is represented *at least once* in the slots of a retail store so that the retailer does not lose customers due to lack of the convenience of "one-stop" shopping.

A possible solution for the retailer could be to sort all of the items in descending order of their respective prices and then place the highest-priced items in the most premium (i.e., highly visible) slots and the lowest-priced items in the least premium slots. Similarly, mid-priced items could be placed in the slot types, whose premiumness is in the mid-range (i.e., moderately visible). However, this solution does not consider the fact that customers typically buy sets of items (i.e., *itemsets*) together; hence, it fails to exploit the *associations* between the items, thereby compromising retailer revenue. Alternatively, the retailer could fill the highly-premium slots with the most *frequent itemsets* [9, 10]. However, frequent itemsets do not consider item prices (utility values), which can vary considerably across items (e.g., a Rolex watch versus a can of Coke). Since revenue depends upon both frequency of sales and item prices, this solution may not maximize revenue as some of the frequent itemsets could have low revenue.

Utility mining approaches have been proposed for extracting the patterns by considering the respective utility values of items and itemsets. In this regard, utility can be defined in various ways e.g., in terms of revenue, interestingness and user convenience, depending upon the application. Thus, utility mining approaches focus on determining high-utility itemsets from transactional databases. Existing utility mining approaches focus on the following aspects: determining minimal high-utility itemsets [11], creating representations of high-utility itemsets [12], reducing candidate itemset generation overheads by using the utility-list [16] and the UP-Tree [18], and identifying the upper-bounds and heuristics for pruning the search space [14, 15].

Notably, *none* of the existing utility mining and pattern mining approaches have considered the problem of *itemset placement in retail slots of varied premiumness.*

In this paper, we use a utility mining approach for maximizing the revenue for the retailer by determining the placement of the itemsets in the different types of slots with varied premiumness. We designate this approach as the **Premiumness and Revenue-based Itemset Placement (PRIP)** scheme. We consider that if a set of items (i.e., an itemset) is frequently purchased together, it indicates that a typical customer is enticed to buy all of these items *together* instead of buying each of the items (in that itemset) individually. In other words, the customers, who purchased this itemset, would possibly not have purchased it if one or more of the items had been missing from the itemset. We focus on identifying high-revenue itemsets from the past history of the user purchase transactions and placing them in slots with high premiumness. Consequently, the probability of sales of the high-revenue itemsets would increase, thereby contributing to higher revenue for the retailer. The key contributions of this work are three-fold:

1. We propose the notion of *premiumness of slots* in a given retail store and introduce the problem of placement of itemsets in slots of varied premiumness for facilitating improved revenue for the retailer.
2. We propose a framework for *efficiently* identifying itemsets from a transactional database and placing these itemsets by mapping itemsets with different revenue to slots with varied premiumness for maximizing retailer revenue.
3. We conducted an extensive performance evaluation using both synthetic and real datasets to demonstrate that the proposed scheme is indeed effective in improving the total retailer revenue by up to 45% w.r.t. a recent existing scheme.

We use revenue as an example of a utility measure throughout this paper; hence, we use the terms *revenue* and *utility* interchangeably. The remainder of this paper is organized as follows. Section 2 discusses related works and background information. Section 3 formulates the problem. Section 4 presents the proposed PRIP scheme. Section 5 reports the performance evaluation. Finally, we conclude in Sect. 6.

2 Related Work and Background

This section discusses existing works and background information about the kUI index.

2.1 Related Work

Association rule mining approaches [8–10] use the downward closure property [8] for finding frequent itemsets above a support threshold [9], but they do not consider item utility. Hence, utility mining approaches [11–19] have been proposed for finding high-utility patterns, but they do not satisfy the downward closure property.

The notion of MinHUIs (minimal high-utility itemsets) and a representation of high-utility itemsets [21] have been proposed in [13]. MinHUIs allude to the smallest itemsets that generate high utility. The HUG-Miner and GHUI-Miner algorithms [12] mine concise representations of high-utility itemsets and prune the number of candidate itemsets for extracting the high-utility itemsets. The EFIM algorithm [14] determines high-utility itemsets by using two upper-bounds, namely *sub-tree utility* and *local utility*, for pruning the search space. The EFIM-Closed algorithm [15] finds closed high-utility itemsets by leveraging pruning strategies and upper-bounds for utility.

The Utility Pattern Growth (UP-Growth) algorithm [18] uses the Utility Pattern Tree (UP-Tree) to maintain information about high-utility itemsets for generating candidate itemsets. The HUI-Miner algorithm [16] uses the utility-list for storing information (including utility) about the itemsets for avoiding expensive candidate itemset generation. The CHUI-Miner algorithm [17] for mining closed high-utility itemsets also computes itemset utility values without generating candidates. The LHUI-Miner algorithm [25] mines local high-utility itemsets by considering that itemset utilities vary over time. The DMHUPS algorithm [26] uses a data structure, designated as IUData List, to discover multiple high-utility patterns simultaneously. The work in [19] prunes away low-utility itemsets and considers business goals as well when evaluating utility values.

Approaches for improving the placement of itemsets in retail stores have been discussed in our previous works in [22, 23]. The work in [22] focused on placing itemsets in the slots of the retail stores when the physical sizes of the items are variable e.g., large-sized golfing equipment versus a small can of Coke. The work in [23] aimed at diversifying the placement of itemsets in retail stores for improving the retailer's long-term revenue. Moreover, the work in [27] uses a mixed-integer programming model to study retail assortment planning and retail shelf space allocation for maximizing retailer revenue. However, these works do not consider the aspect of slot premiumness.

Notably, *none* of the existing utility mining and pattern mining approaches consider the problem of itemset placement in retail slots of varied *premiumness*. This limits their applicability to building practical systems for itemset placement in retail stores for maximizing the revenue of the retailer.

2.2 Background

This section discusses the background about kUI index. We discuss the kUI index [23], which our proposed scheme PRIP exploits for extracting high-revenue itemsets of varied sizes. kUI is a multi-level index, where each level concerns a given itemset size. At the kth level, the kUI index stores the top-λ high-revenue itemsets of itemset size *k*. Each level corresponds to a hash bucket. For indexing itemsets of N different sizes, the index has N hash buckets i.e., one hash bucket per itemset size. Hence, given a query for finding the top-λ high-revenue itemsets of a given size *k*, one can traverse quickly to the kth hash bucket instead of having to traverse through all the hash buckets corresponding to $k = \{1, 2, \ldots, k-1\}$.

Now, for each level *k* in the kUI index, the corresponding hash bucket contains a pointer to a linked list of the top-η itemsets of size *k*. (Here, η > λ.) The entries of the linked list are of the form (*itemset*, σ, ρ, NR), where *itemset* refers to the given itemset under consideration, σ refers to the frequency of sales of the given itemset and ρ refers to the price of the given itemset. Here, NR (net revenue) is the product of σ (frequency of sales) and ρ (price) of items in *itemset*. The entries in the linked list are sorted in descending order of NR for quick retrieval of the top-λ itemsets of a given size *k*.

Figure 1 depicts an illustrative example of the kUI index. Observe how the itemsets (e.g., {A}, {I}) of size 1 correspond to level 1 of the index, the itemsets of size 2 (e.g., {A, P}, {A, L}) correspond to level 2 of the index and so on. Notice how the itemsets are sorted in descending order of NR.

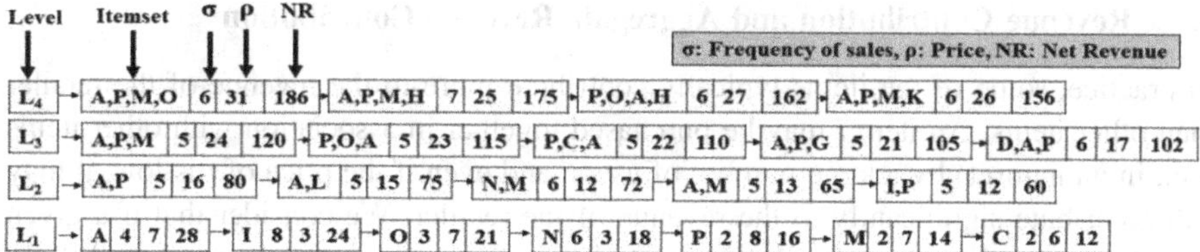

Fig. 1. Illustrative example of the kUI Index

3 Context of the Problem

Consider a set D of user purchase transactions on a finite set ϒ of m items$\{i_1, i_2, i_3, \ldots, i_m\}$, where each transaction comprises a set of items from set ϒ. Each item i_j of set ϒ is associated with a price ρ_j and a frequency of sales σ_j. We assume that each item occurs in any given transaction *only once*; hence, σ_j is essentially equivalent to the number of transactions in which a given item has appeared. Furthermore, an itemset comprises a set of items from set ϒ. We assume that each item of set ϒ is physically of the same size i.e., each item consumes an equal amount of space e.g., on the retail store's shelves.

We assume that all slots are of equal size, but they vary in terms of *premiumness*. Recall the notion of premiumness of the slots as discussed in Sect. 1. Slots in a retail store have varied premiumness i.e., certain slots have a higher probability of sales for items placed in them. We envisage N different types of slots with varied premiumness.

Problem Statement: Consider a set D of user purchase transactions on a finite set ϒ of m items. Each item is associated with a price ρ_j and a frequency of sales σ_j. Consider N types of slots with varied premiumness. This paper addresses the problem of maximizing the total revenue for the retailer by determining the placement of the itemsets in the different types of slots with varied premiumness such that each item is placed *at least once* in any of the slots to facilitate one-stop shopping (as motivated in Sect. 1).

4 Proposed Scheme

We first explain the notions of *revenue contribution* and *aggregate revenue contribution*, which constitute the basis of our proposed scheme. Next, we discuss the scheme.

4.1 Revenue Contribution and Aggregate Revenue Contribution

In practice, some of the items typically contribute more to the revenue of the retailer than other items. An item *i* may be purchased together in association with other items (i.e., in an itemset) by a large number of users, and even if the price of *i* is low, it may still contribute significantly to the revenue of the retailer. We consider that if a given item *i* was missing from the slots, many of the users may not have purchased those itemsets i.e., the presence of item *i* has encouraged the sales of those itemsets.

Based on this motivation, we introduce the notion of the **aggregate revenue contribution (ARC)** of a given item across all of the user purchase transactions. Given a set of user purchase transactions, ARC essentially quantifies the extent to which a given item contributes (in terms of revenue) to the overall revenue of the retailer. For purposes of computing ARC for a given item, we consider that each of the items in a given transaction contributed *equally* towards the user's purchase decision. This is because in practice, we would typically have no knowledge of the direction of causation of the sales i.e., given a transaction {A, B, C}, we do not know whether the customers purchased A because of B and C or if she purchased C because of A and B.

Purchase value (PV) of a given transaction T is the amount of money (revenue) that the user paid for buying all of the items in T. We compute the ***revenue contribution (RC)*** for each item in a given transaction T by dividing the purchase value (PV_T) of T by the number of items in T. Given a transaction T containing n_T items, we define the revenue contribution (RC_T) of each item in T as $RC_T = PV_T/n_T$. Now, for each item, we sum up its values of RC across all of the user purchase transactions to

PV = Purchase Value
RC = Revenue Contribution
ARC = Aggregated Revenue Contribution

Transaction	PV	RC
A, E	60	60/2= 30
A, C, G, I	63	63/4= 15.75
A, C, G	36	36/3= 12
B, C	24	24/2= 12
D, F, G	28	28/3= 9.33

Item	ARC
A	30+15.75+12= 57.75
C	15.75+12+12= 39.75
G	15.75+12+9.33= 37.08
E	30
B	12
D	9.33
F	9.33

Fig. 2. Illustrative example for the computation of RC and ARC

obtain the ***aggregated revenue contribution (ARC)*** for each item. Given q transactions, we compute the value of ARC for the kth item i_k as $ARC_k = \sum_{j=1}^{q} (RC_j)$ Observe that if the jth transaction does not contain the item i_k, the value of RC_j would be zero.

Figure 2 depicts five transactions involving items A to I and the corresponding purchase values of each of these transactions. Here, PV of transaction {A, C, G, I} is 63 and RC of each of the items in the itemset is 63/4 i.e., 15.75. Observe how the value of RC for item A differs across the itemsets. Finally, we sum up all of the RC values of item A to obtain the ARC of item A as (30 + 15.75 + 12 = 57.75) i.e., 57.75.

4.2 Premiumness and Revenue-Based Itemset Placement (PRIP)

It is well-known in the retail industry that strategic placement of items in the retail store slots has significant impact on sales [7]. Slots in a retail store typically vary in terms of *premiumness* and items placed in slots of higher premiumness have a higher probability of sales [7]. If we arbitrarily (or randomly) assign and place the items in the slots, it may fail to maximize the revenue of the retailer. For example, a high-revenue item may get placed in a slot with low premiumness, thereby compromising retailer revenue. Thus, there is an opportunity to improve the revenue of the retailer by considering user purchase patterns in conjunction with a methodology for placing items in slots based on the premiumness of the slots and the revenue generation potential of items.

We propose a framework for *efficiently* identifying itemsets from a transactional database and placing these itemsets by mapping itemsets with different revenue to slots with varied premiumness for maximizing retailer revenue. For placement of items in the retail slots, we extract potential 1-itemsets based on the values of ARC. We make the observation that the notion of ARC is only applicable to individual items and as such, it does not apply to multiple items in tandem. Hence, we extract itemsets with sizes greater than 1 based on the net revenue of those itemsets.

Our proposed scheme works as follows. First, for placing the 1-itemsets, we place the items with higher ARC values in the slots with higher premiumness, and we place the items with lower values of ARC in the slots with lower premiumness and so on. Then, for placing itemsets with sizes greater than 1 in the remaining available slots, we extract itemsets of varied sizes (starting from itemset size of 2 onwards) from the kUI index [23] (discussed in Sect. 2.2) to allocate itemsets of varied sizes to the slots of varied premiumness based on the net revenue of the itemsets. Observe that placing itemsets of varied sizes in the slots can potentially provide improved revenue for the retailer because it would possibly address the needs of different user segments.

The proposed approach primarily consists of two phases. In the first phase, individual items are assigned to slots with varied premiumness based on the ARC of the items and the premiumness of the slots. In the second phase, itemsets of size greater than 1 are mapped to slots based on itemset revenue and (slot) premiumness. Notably, itemsets (of different sizes) are extracted from the kUI index. As discussed earlier in Sect. 2.2, recall that the kUI index contains itemsets of different sizes as well as the frequency of sales, price and the net revenue of each itemset. Moreover, the itemsets in the kUI index are kept sorted in descending order of net revenue. Algorithm 1 depicts our proposed PRIP scheme. It takes as input a database D of user purchase transactions, slots of N types and the number n_i of slots pertaining to each of the slot types, and the kUI index.

In Lines 1–4, we perform the necessary initializations. In Lines 5–7, observe how the database D of user purchase transactions is scanned to compute ARC for each item (see Sect. 4.1). In Lines 8–9, notice how we determine the number of 1-itemsets to be allocated to each slot type based on the relative percentages of slots corresponding to each slot type. In Line 10, the items are sorted in descending order of ARC.

In Lines 11–16, starting from the slot type with the highest premiumness, we progressively fill up the corresponding highest-premiumness slots with the individual items (i.e., the 1-itemsets) starting with the item having the highest ARC. By traversing this list of items sorted in descending order of ARC, once we have filled up the required number of high-ARC items (as determined from the formula in Line 7) in the highest-premiumness slots, we move on to the next high-premiumness slot and repeat the same process as above. This process is essentially continued until all of the 1-itemsets have been assigned to slots. Notably, after using the formula in Line 7, some of the low-ARC items may still remain unassigned to any slot due to the *"floor"* function used in the formula. Hence, in Line 17, for handling this corner case, we assign all of these remaining low-ARC items to the lowest-premiumness slot. Now all of the 1-itemsets have been assigned to slots i.e., each item is represented in the slots *at least once*.

In Lines 18–19, we update the number of slots available for each slot type to reflect that the 1-itemsets have already been placed in the respective slots. In Lines 20–34, observe how we use the kUI index to populate the remaining slots corresponding to each of the slot types. Starting from the highest-premiumness slot type, we keep filling up the highest-premiumness slots as follows. We extract the top-revenue 2-itemset from the kUI index and place it in the highest-premiumness slot. Then we extract the top-revenue 3-itemset from the kUI index and place it in the highest-premiumness slot and so on. Upon reaching the topmost level of the kUI index, we do a round-robin and circle back to the next top-revenue 2-itemset, and then the next 3-itemset and so on until all of the highest-premiumness slots have been filled up. Once all the highest-premiumness slots have been exhausted, we move on to the next highest-premiumness slot and so on. This process is repeated until all of the slots have been filled up by items/itemsets.

Notably, in this work, we extract the itemsets in the afore-mentioned manner. However, it is also possible to fill up the slots with combinations such as (2 + 2 + 2), (3 + 3) or (2 + 4) itemsets and so on. We shall investigate such strategies in our future work.

Algorithm 1: Premiumness and Revenue-based Itemset Placement (PRIP)

Inputs: (a) A Transactional Database D on set I= $\{i_1,\ldots..i_m\}$ of m distinct items where each transaction T_i is of the form <T_i, S_i ,P_i>, where S_i is set of items such that S_i is subset of I and P_i is the purchase value of T_i

(b) Slots of N types and n_i of slots pertaining to each of the slot type

(c) kUI Index

Output: Placement of the items/itemsets in the slots with varied premiumness

Begin

1. Initialize the aggregated revenue contribution ARC[*i*] of each item *i* to zero
2. Initialize slot type s_type to 1 and no. of items already placed ip to zero
3. Initialize an empty slot type array ST[i]
4. Initialize current available slot array CAS[i] and assign number of slot instances n_i to the array

/*Scan D to compute ARC of each item*/

5. **for each** transaction T_i in D {
6.   **for each** item x belong to S_i {
7.   ARC[x]= ARC[x]+ P_i/ $|S_i|$; } }

/*1-itemsets allocated to each slot type */

8. **for** (i = 1 to N) {
9.   Y[i] = $[(n_i/\sum_{i=0}^{n} n_i)*m)]$; }
10. Sort the items in descending order of ARC[k] in list k-items into list Item[ ]
11. **for** each item k in Item[ ] {
12.   **if** (Y[s_type]!=0) {
13.   assign item k to slot type s_type
14.   ip++ ; Y[s_type]— ; }
15.   **if** (Y[s_type] = = 0) { s_type++; }
16.   **if** (Y[s_type] = = N) { break; } }
17. Place the remaining items (m – ip) items into slot type s_type

/*Compute remaining CAS after 1-itemset placement*/

18. **for** (i = 1 to N) {
19.   CAS[i] = CAS[i] - Y[i]; }

/*Scan kUI index to fill remaining slots*/

20. Scan the kUI index into a 2D array kUI[L, num[L]] ;
21. Set all elements of array h[lv] to 0
22. **for** each slot type (i=1 to N) {
23.   **for** each level lv (lv = 2 to maxL){
24.   h[lv]++;
25.   **if** (CAS[i] < lv) {
26.   lv--; h[lv]++;
27.   Select the itemset h[lv]
28.   Place itemset h[lv] in slot type i
29.   CAS[i] = CAS[i] – lv;
30.   break }
31.   **else** {
32.   Select the itemset h[lv] from lv
33.   Place itemset h[lv] in slot type i
34.   CAS[i] = CAS[i] – lv; }} **End**

4.3 Illustrative Example of Proposed Scheme

Figure 3 depicts an illustrative example for our proposed PRIP scheme.

Figure 3 indicates transactions with their respective values of RC and ARC. For the sake of clarity, this example considers three types of slots i.e., N = 3 (high-premiumness, mid-premiumness and low-premiumness slot types) and the total number of slots corresponding to these slot types are 12, 20 and 35 respectively. Observe how the items are

Items: A to P, N= type of Slots,
PV = Purchase Value, RC= Revenue Contribution
ARC= Aggregated Revenue Contribution

Transactions	PV	RC
A, D, J, N	6*(7+1+4+3)= 48	48/4= 22.5
A, C, G, I, N	3*(7+6+5+3+3)= 72	72/5= 15.75
A, C, G	2*(7+6+5)= 36	36/3= 12
B, C,J	4*(2+6+4)= 48	48/3= 16
D, F, G, O	4*(1+1+5+7)= 56	56/4= 14
B, C, I, F	3*(2+6+3+1)= 36	36/4= 9
E,H, I	3*(3+4+3)= 30	30/3= 10
C, F, H, I	2*(6+1+4+3)= 28	28/4= 7
E, H, I	2*(3+4+3)= 20	20/3= 6.66
A, B, C, G, H	1*(7+2+6+5+4)= 24	24/5= 4.8
G, H, I, O	1*(5+4+3+7)= 19	19/4= 4.75
J, K, P, N,E	2*(4+2+9+3+3)= 42	42/5= 8.4
P, M, L	2*(9+8+6)= 46	46/3= 15.33
O, N	4*(7+3)= 40	40/2= 20
M, K, L	3*(8+2+6)= 48	48/3= 16

Item	Total ARC	
C	15.75+12+16+9+7+4.8= 64.55	High revenue items
N	22.5+15.75+20= 58.25	
A	22.5+15.75+12+4.8= 56.55	
I	15.75+9+10+7+6.66+4.75= 53.16	
G	15.75+12+14+4.8+4.75= 51.3	
O	14+4.75+20= 38.75	Mid revenue items
J	22.5+16+8.4= 36.5	
D	22.5+14= 36.5	
F	14+9+7+4= 34	
H	10+7+6.66+4.8+4.75= 33.21	
E	10+6.66+8.4= 25.06	Low revenue items
L	15.33+ 9.67= 25	
M	15.33+9.17= 24.5	
B	16+ 3.6+ 4.8 = 24.4	
K	8.4+ 16= 24.4	
P	8.4+15.33=23.73	

Considers three types of slots i.e., N=3 (High priority, Mid priority and Low priority)
N_1 (Number of high-priority slots) : 12; N_2 (Number of mid-priority slots) : 20; N_3 (Number of low-priority slots) : 35

L_4 → A,P,M,O 186 → A,P,M,H 175 → P,O,A,H 162 → A,P,M,K 156 → A,P,M,C 150 → A,P,M,G 145 → A,P,M,J 140

L_3 → A,P,M 120 → P,O,A 115 → P,C,A 110 → A,P,G 105 → D,A,P 102 → A,P,J 100 → E,P,A 95

L_2 → A,P 80 → A,L 75 → N,M 66 → A,M 65 → I,P 60 → A,O 56 → I,L 54

L_1 → C 64.5 → N 58.2 → A 56.5 → I 53.1 → G 51.3 → O 38.7 → J 36.5 → D 36.5 → F 34 → H 33.2 → E 25.1 → L 25 → M 24.5 → B 24.4 → K 24.4 → P 23.7

High-Priority Slots Placements — All available slots (AAS) = 12

1-itemset placement: C 64.5, N 58.2, A 56.5, I 53.1, G 51.3 — Current available slots (CAS) = 12 – 5 = 7

top-1 (2-itemset) L_2: A,P 80 — CAS = 7-2 = 5

top-1 (3-itemset) L_3: A,P,M 120 — CAS = 5-3 = 2

top-2 (2-itemset) L_2: A,L 75 — CAS = 2-2 = 0

CAS < 4- itemset, Select top-2 itemset from L_2

Select top-2 itemset from L_2

High-Priority Slots Itemsets: { {C}, {N}, {A}, {I}, {G}, {A,P}, {A,P,M}, {A,L}}

Mid-Priority Slots Placements — All available slots (AAS) = 20

1-itemset placement: O 38.7, J 36.5, D 36.5, F 34, H 33.2 — Current available slots (CAS) = 20 – 5 = 15

top-2 (3-itemset) L_3: P,O,A 115 — CAS = 15-3 = 12

top-2 (4-itemset) L_4: A,P,M,H 175 — CAS = 12-4 = 8

top-3 (2-itemset) L_2: N,M 66 — CAS = 8-2 = 6

top-3 (3-itemset) L_3: P,C,A 110 — CAS = 6-3 = 3

top-4 (3-itemset) L_3: A,P,G 105 — CAS = 3-3 = 0

CAS < 4- itemset, Select top-3 itemset from L_3

Mid-Priority Slots Itemsets: { {O}, {J}, {D}, {F}, {H}, {P,O,A}, {A,P,M,H}, {N,M}, {P,C,A}, {A,P,G}}

Low-Priority Slots Placements — All available slots (AAS) = 35

1-itemset placement: E 25.1, L 25, M 24.5, B 24.4, K 24.4, P 23.7 — Current available slots (CAS) = 35–6 = 29

top-4 (2-itemset) L_2: A,M 65 — CAS = 29-2 = 27

top-4 (4-itemset) L_4: P,O,A,H 162 — CAS = 27-4 = 23

top-5 (2-itemset) L_2: I,P 60 — CAS = 23-2 = 21

top-5 (3-itemset) L_3: D, A, P 102 — CAS = 21-3 = 19

top-5 (4-itemset) L_4: A,P,M,K 156 — CAS = 19-4 = 15

top-6 (2-itemset) L_2: A,O 56 — CAS = 15-2 = 13

top-6 (3-itemset) L_3: A, P, J 100 — CAS = 13-3 = 10

top-6 (4-itemset) L_4: A,P,M,C 150 — CAS = 10-4 = 6

top-7 (2-itemset) L_2: I,L 54 — CAS = 6-2 = 4

top-7 (4-itemset) L_3: A,P,M,G 145 — CAS = 4-4 = 0

Low-Priority Slots Itemsets: { {E}, {L}, {M}, {B}, {K}, {P}, {A,M}, {P,O,A,H}, {I,P}, {D,A,P}, {A,P,M,K}, {A,O}, {A,P,J}, {A,P,M,C}, {I,L}, {A,P,M,G}}

Fig. 3. Illustrative example of PRIP scheme

sorted in descending order of ARC. Using the formula in Line 8 of Algorithm 1, items {C, N, A, I, G} are selected as high revenue items, {O, J, D, F, H} are selected as mid-revenue items and the remaining {E, L, M, B, K, P} are the low-revenue items.

High-revenue items are mapped to high-premiumness slots, mid-revenue items are mapped to mid-premiumness slots and low-revenue items are mapped to low-premiumness slots. After placement of 1-itemsets, the remaining slots of high-premiumness = 12 − 5 i.e., 7, mid-premiumness = 20 − 5 i.e., 15 and low-premiumness = 35 − 6 i.e., 29.

To fill the remaining slot instances in each slot type, we extract itemsets from the kUI index. Starting from the highest-premiumness slot, we keep filling up the highest-premiumness slots as follows. We extract top-1 itemset from level 2 of the index i.e., itemset {A, P} and place it. Then we compute the number of remaining slots as (7–2) i.e., 5. Then, we extract the top-1 itemset from level 3 i.e., {A, P, M} and place it. Next, we update the remaining available slots as (5–3) i.e., 2. To fill the remaining 2 slots, itemsets from the next level of the kUI index cannot be extracted because remaining slots are less than itemset size at the current level i.e., 2 < level 4-itemset. Hence, we extract the top-2 itemset from level 2 of the kUI index i.e., {A, L} and place the itemset in the remaining slots. Now, there are no high-premiumness slots left.

Next, the mid-premiumness slot are filled in a similar manner. We extract top-2 itemset from level 2 of the index i.e., itemset {A, L}, but the itemset is already placed in the high-premiumness slot. Hence, we move to the next level i.e., L = 3 and extract top-2 itemset from level 3 i.e., {P, O, A}. Next, we extract the top-2 itemset from level 4 i.e., {A, P, M, H} and we have now reached the highest level of the kUI index and still, there are 8 remaining available slots. Now, we extract the top-3 itemset from level 2 i.e., {N, M} and top-3 itemset from level 3 i.e., {P, C, A} and place it. Next, we update remaining available slots as (6–3) i.e., 3. Now, the remaining slots are less than the itemset size of the next level i.e. 3 < level 4-itemset. Hence, we extract the top-4 itemset from level 3 i.e., {A, P, G} and place it. Now, there are no slots left in the mid-premiumness slot type. Similarly, the low-premiumness slots are filled by extracting the next top-revenue itemset from each level until we exhaust all of the available slots.

5 Performance Evaluation

This section reports the performance evaluation. We performed our implementation and experiments on a 64-bit Core i5 processor running Windows 10 with 8 GB memory.

We conducted the experiments using a synthetic dataset as well as a real dataset. The real dataset is *Retail* dataset, which we obtained from the SPMF open-source data mining library [20]. It is an anonymous retail market basket data from an anonymous Belgian retail store. The dataset has 16,470 items and the number of transactions in the dataset is 88,162. We also generated a synthetic dataset, designated as T10I6N15K|D| 2,000K, using the IBM data generator. The parameters of the synthetic dataset are as follows: T (the average size of the transactions) is 10; I (the average size of potential maximal itemsets) is 6; N (the number of distinct items) is 15K; |D| (the total number of transactions) is 2,000K. The dataset has 15,000 items. For the sake of brevity, we shall henceforth refer to this data as the **synthetic dataset**.

The Retail dataset and the synthetic dataset do not provide utility (price) values. Hence, for each of these datasets, we generated the price of the items in the range [0.01, 1.0] as follows. We divided the price range into six approximately equal-width categories i.e., [0.01–0.16], [0.17–0.33], [0.34–0.50], [0.51–0.67], [0.68–0.84] and [0.85–1]. For generating the price for each item, we randomly select one of these categories and generate a random number within the price range of that category.

For our experiments, we divided the transactions dataset into two parts, namely *training set* (containing 70% of the transactions) and *test set* (containing the remaining 30% of the transactions). For our PRIP scheme, we placed the itemsets based on the kUI index built from the training set. We evaluate the performance on the test set.

Our performance metrics are execution time (ET), the number of patterns generated (N_P) and Total revenue (TR). ET is the execution time for the determination of the placement of itemsets in the slots for the training set. N_P is the number of patterns (itemsets) that a given scheme needs to examine for the itemset placement for the training set. TR, which is the total retailer revenue for the test set, is computed as follows. We iterate through each transaction *t* in the test set and add to TR only the prices of the items (in *t*), which had already been placed in the retail slots during the training phase.

To quantify the notion of premiumness, we assign probability of sales to the different slot types in the range of [0.01, 1]. Given N slot types, we divide the [0.01, 1] range into N sub-ranges and assign these sub-ranges to the slot types to quantify the respective probability of sales for items/itemsets placed in these slot types. Consider N = 3. Hence, the corresponding probabilities for the three different slot types in increasing order of premiumness would be in the following sub-ranges: [0.01–0.34], [0.35–0.68] and [0.69–1.0]. These ranges are *approximately* equally distributed.

Table 1. Parameters of performance evaluation

Parameter	Default	Variations
Total number of slots (10^4) (T_S)	10	2, 4, 6, 8
Number of slot type (S)	6	3, 9, 12, 15
Revenue threshold (α)	30%	
Zipf factor (Z_{NS})	0.7	0.1, 0.3, 0.5, 0.9

The number λ of itemsets at each level of the kUI index was 5000, and the kUI index had 10 levels. Table 1 summarizes the parameters of the performance study. In Table 1, the Zipf factor (Z_{NS}) represents skew in the number of slot instances across slot types. Z_{NS} is defined in the range [0.1, 0.9], where 0.1 indicates a relatively uniform distribution, while 0.9 indicates a highly skewed distribution. When Z_{NS} = 0.1, there would be approximately an equal number of slots corresponding to each slot type. In contrast, when Z_{NS} = 0.9, there would be a disproportionately higher number of low-premiumness slots and a relatively fewer number of high-premiumness slots.

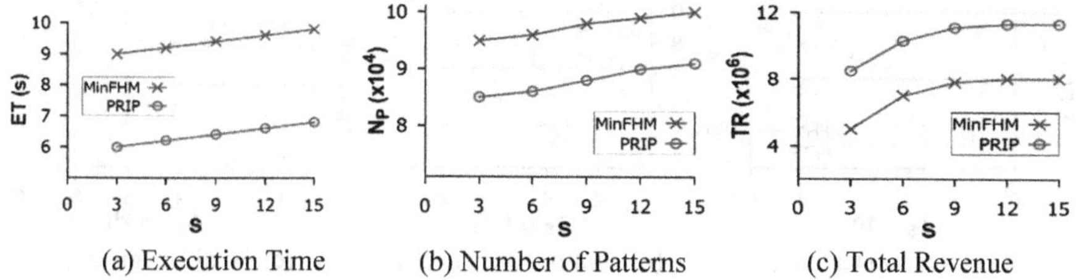

(a) Execution Time (b) Number of Patterns (c) Total Revenue

Fig. 4. Effect of variations in the number of slot types (Synthetic dataset)

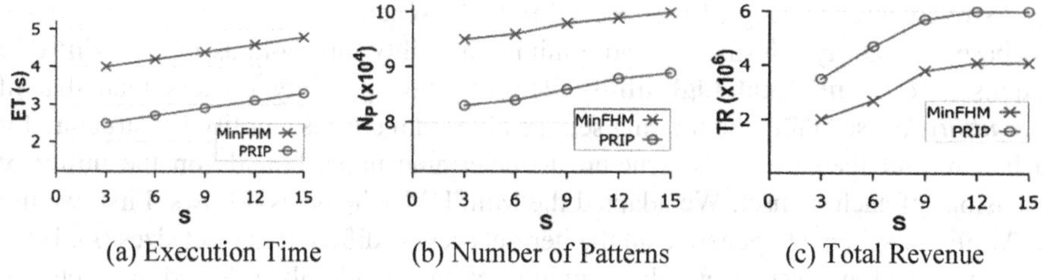

(a) Execution Time (b) Number of Patterns (c) Total Revenue

Fig. 5. Effect of variations in the number of slot types (Retail dataset)

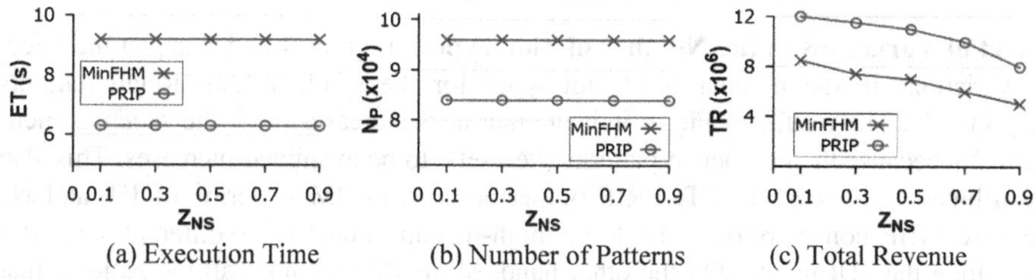

(a) Execution Time (b) Number of Patterns (c) Total Revenue

Fig. 6. Effect of variations in skew across slot types (Synthetic dataset)

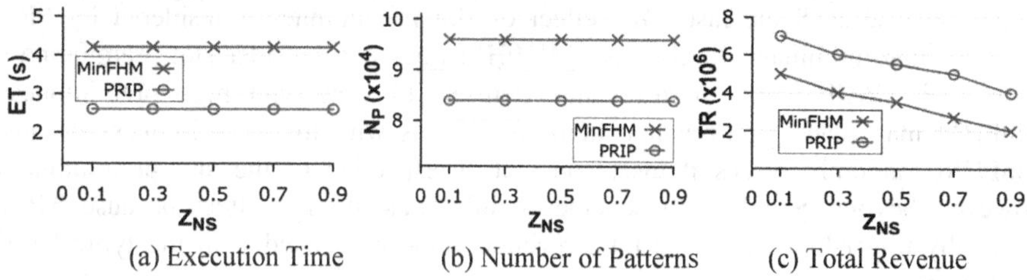

(a) Execution Time (b) Number of Patterns (c) Total Revenue

Fig. 7. Effect of variations in skew across slot types (Retail dataset)

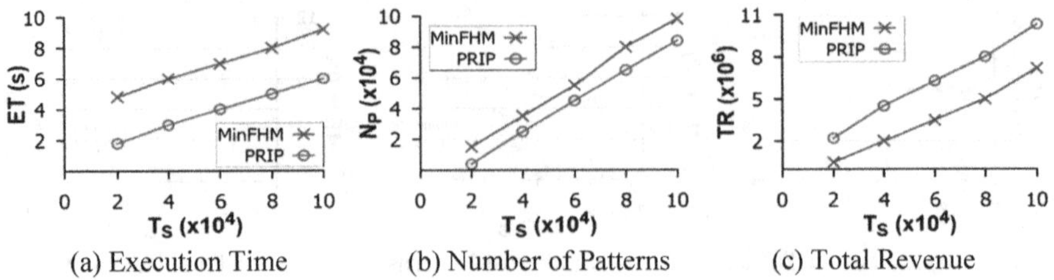

(a) Execution Time (b) Number of Patterns (c) Total Revenue

Fig. 8. Effect of variations in the total number of slots (Synthetic dataset)

As reference, we adapted the recent MinFHM scheme [13]. Given a transactional database with utility information and a minimum utility threshold as input, **MinFHM** outputs a set of **minimal high-utility itemsets** having utility no less than that of ***min_utility***. By scanning the training set, the algorithm creates a utility-list structure for each item and then uses this structure to determine upper-bounds on the utility of extensions of each itemset. We adapted the MinFHM scheme as follows. First, we use the MinFHM scheme to generate all the itemsets across different itemset sizes (k). From these extracted itemsets from the training set, we randomly selected itemsets for placement at the different types of available slots until we exhausted all of the available slots. We also ensured that each item is represented at least once on any type of slot. We shall henceforth refer to this scheme as **MinFHM**.

Effect of Variations in the Number of Slot Types: Figures 4 and 5 depict the effect of variations in the number S of slot types for the synthetic and Retail datasets respectively. The results in Fig. 4 indicate that as S increases, both the schemes incur more N_P because the number of patterns (itemsets) to be examined increases. This also contributes to increase in ET. PRIP outperforms MinFHM in terms of ET and N_P because PRIP considers only the top-λ high-revenue itemsets of different sizes (by exploiting the kUI index). On the other hand, MinFHM examines all the patterns that exceed a pre-specified minimum utility threshold. Hence, the number of patterns to be examined in case of MinFHM is greater than that of PRIP.

The results in Fig. 4(c) indicate that PRIP provides higher TR than MinFHM. This occurs because as S increases, the effect of slot premiumness considered by PRIP becomes more dominant. In contrast, MinFHM does not consider slot premiumness. Thus, while PRIP places high-revenue itemsets to slots with high premiumness, MinFHM may place high-revenue itemsets in slots with low premiumness because MinFHM randomly places itemsets on slots irrespective of the slot premiumness. However, beyond S = 9, both schemes exhibit a saturation effect because TR is essentially upper-limited by the total revenue value associated with the typical user purchase transactions.

The results in Fig. 5 follow similar trends as those of Fig. 4; the difference in the actual values of the performance metrics arises due to the respective dataset sizes.

Effect of Variations in Skew Across Slot Types: Figures 6 and 7 depict the effect of variations of skew in the number of slots across the slot types for the synthetic and Retail datasets. As discussed previously, such skew is modelled by Z_{NS}. The results in Fig. 6 indicate that ET and N_P remain comparable for both schemes as Z_{NS} increases.

This occurs because the total number of slots to be filled remain the same, irrespective of the value of Z_{NS}. Hence, the time required for extraction and placement of itemsets remains comparable as Z_{NS} changes. PRIP outperforms MinFHM in terms of ET and N_P due to the reasons explained for Fig. 4. The results in Fig. 6(c) indicate that both schemes show decrease in the value of TR with increase in Z_{NS}. This occurs because as the value of Z_{NS} increases, the skew increases. Hence, the number of lower premiumness slots become more w.r.t. the higher-premium slots. Due to a decreased number of slots with higher premiumness, TR decreases.

The results in Fig. 7 follow similar trends as those of Fig. 6; the difference in the actual values of the performance metrics arises due to the respective dataset sizes.

Effect of Variations in the Total Number of Slots: Figure 8 depicts the effect of variations in the total number T_S of slots for the synthetic dataset. The results in Figs. 8 (a) and (b) indicate that ET and N_P increase for both schemes with increase in T_S. As T_S increases, more slots need to be filled, thereby necessitating more patterns to be examined. PRIP outperforms MinFHM in terms of ET and N_P due to the reasons explained earlier for Fig. 4. The results in Fig. 8(c) indicate that TR increases for both the schemes as T_S increases. As T_S increases, more itemsets are used to fill up an increased number of slots, thereby increasing TR. PRIP provides higher TR than that of MinFHM due to the reasons explained earlier for Figs. 4(c). We also performed this experiment with the Retail dataset. The results exhibited similar trends. Due to space constraint, we do not present those results in this paper.

6 Conclusion

Given that retail slots typically vary in terms of premiumness, we have proposed a scheme for mapping and placing itemsets to slots based on itemset revenue and slot premiumness with the objective of maximizing the revenue for the retailer. Based on the knowledge extracted from users' purchase patterns, we have proposed a framework and a scheme for determining the placement of the itemsets in different types of slots with varied premiumness such that each item is placed at least once in any of the slots. Our framework is capable of *efficiently* identifying itemsets from a transactional database and placing these itemsets by mapping itemsets with different revenue to slots with varied premiumness for maximizing retailer revenue. Our performance evaluation using both synthetic and real datasets indicates that the proposed scheme is indeed effective in improving the total retailer revenue by up to 45% w.r.t. a recent existing scheme. Given that pattern mining is an important and active research area, the proposed framework is versatile and generic in the sense that different kinds of existing pattern mining approaches can also be investigated for further improving the revenue of the retailer.

References

1. Hansen, P., Heinsbroek, H.: Product selection and space allocation in supermarkets. Eur. J. Oper. Res. **3**(6), 474–484 (1979)
2. Yang, M.H., Chen, W.C.: A study on shelf space allocation and management. Int. J. Prod. Econ. **60**, 309–317 (1999)
3. Yang, M.H.: An efficient algorithm to allocate shelf space. Eur. J. Oper. Res. **131**(1), 107–118 (2001)
4. Chen, M.C., Lin, C.P.: A data mining approach to product assortment and shelf space allocation. Expert Syst. Appl. **32**(4), 976–986 (2007)
5. Chen, Y.L., Chen, J.M., Tung, C.W.: A data mining approach for retail knowledge discovery with consideration of the effect of shelf-space adjacency on sales. Decis. Support Syst. **42**(3), 1503–1520 (2006)
6. World's largest retail store. https://www.thebalance.com/largest-retail-stores-2892923
7. Hart, C.: The retail accordion and assortment strategies: an exploratory study. Int. Rev. Retail Distrib. Consum. Res. **9**(2), 111–126 (1999)
8. Agrawal, R., Srikant, R.: Fast algorithms for mining association rules. Proc. VLDB **1215**, 487–499 (1994)
9. Han, J., Pei, J., Yin, Y.: Mining frequent patterns without candidate generation. Proc. ACM SIGMOD **29**, 1–12 (2000)
10. Pasquier, N., Bastide, Y., Taouil, R., Lakhal, L.: Discovering frequent closed itemsets for association rules. In: Beeri, C., Buneman, P. (eds.) ICDT 1999. LNCS, vol. 1540, pp. 398–416. Springer, Heidelberg (1999). https://doi.org/10.1007/3-540-49257-7_25
11. Liu, Y., Liao, W.K., Choudhary, A.: A fast high utility itemsets mining algorithm. In: Proceedings of 1st International Workshop on Utility-Based Data Mining, pp. 90–99 (2005)
12. Fournier-Viger, P., Wu, C.-W., Tseng, V.S.: Novel concise representations of high utility itemsets using generator patterns. In: Luo, X., Yu, J.X., Li, Z. (eds.) ADMA 2014. LNCS (LNAI), vol. 8933, pp. 30–43. Springer, Cham (2014). https://doi.org/10.1007/978-3-319-14717-8_3
13. Fournier-Viger, P., Lin, J.C.-W., Wu, C.-W., Tseng, V.S., Faghihi, U.: Mining minimal high-utility itemsets. In: Hartmann, S., Ma, H. (eds.) DEXA 2016. LNCS, vol. 9827, pp. 88–101. Springer, Cham (2016). https://doi.org/10.1007/978-3-319-44403-1_6
14. Zida, S., Fournier-Viger, P., Lin, J.C.-W., Wu, C.-W., Tseng, V.S.: EFIM: a highly efficient algorithm for high-utility itemset mining. In: Sidorov, G., Galicia-Haro, S.N. (eds.) MICAI 2015. LNCS (LNAI), vol. 9413, pp. 530–546. Springer, Cham (2015). https://doi.org/10.1007/978-3-319-27060-9_44
15. Fournier-Viger, P., Zida, S., Lin, J.C.-W., Wu, C.-W., Tseng, V.S.: EFIM-closed: fast and memory efficient discovery of closed high-utility itemsets. MICAI 2016. LNCS (LNAI), vol. 9729, pp. 199–213. Springer, Cham (2016). https://doi.org/10.1007/978-3-319-41920-6_15
16. Liu, M., Qu, J.: Mining high utility itemsets without candidate generation. In: Proceedings of CIKM, pp. 55–64 (2012)
17. Tseng, V.S., Wu, C.W., Fournier-Viger, P., Philip, S.Y.: Efficient algorithms for mining the concise and lossless representation of high utility itemsets. In: IEEE TKDE, pp. 726–739 (2015)
18. Tseng, V.S., Wu, C.W., Shie, B.E., Yu, P.S.: UP-Growth: an efficient algorithm for high utility itemset mining. In: Proceedings of ACM SIGKDD, pp. 253–262 (2010)
19. Chan, R., Yang, Q., Shen, Y.D.: Mining high utility itemsets. In: Proceedings of ICDM, pp. 19–26 (2003)
20. http://www.philippe-fournier-viger.com/spmf/dataset

21. Fournier-Viger, P., Wu, C.-W., Zida, S., Tseng, V.S.: FHM: faster high-utility itemset mining using estimated utility co-occurrence pruning. In: Andreasen, T., Christiansen, H., Cubero, J.-C., Raś, Zbigniew W. (eds.) ISMIS 2014. LNCS (LNAI), vol. 8502, pp. 83–92. Springer, Cham (2014). https://doi.org/10.1007/978-3-319-08326-1_9
22. Chaudhary, P., Mondal, A., Reddy, P.K.: A flexible and efficient indexing scheme for placement of top-utility itemsets for different slot sizes. In: Reddy, P.K., Sureka, A., Chakravarthy, S., Bhalla, S. (eds.) BDA 2017. LNCS, vol. 10721, pp. 257–277. Springer, Cham (2017). https://doi.org/10.1007/978-3-319-72413-3_18
23. Chaudhary, P., Mondal, A., Reddy, P.K.: A diversification-aware itemset placement framework for long-term sustainability of retail businesses. In: Hartmann, S., Ma, H., Hameurlain, A., Pernul, G., Wagner, R.R. (eds.) DEXA 2018. LNCS, vol. 11029, pp. 103–118. Springer, Cham (2018). https://doi.org/10.1007/978-3-319-98809-2_7
24. Corsten, D., Gruen, T.: Desperately seeking shelf availability: an examination of the extent, the causes, and the efforts to address retail out-of-stocks. Int. J. Retail Distrib. Manag. **31**, 605–617 (2003)
25. Fournier-Viger, P., Zhang, Y., Lin, J.Chun-Wei, Fujita, H., Koh, Y.S.: Mining local high utility itemsets. In: Hartmann, S., Ma, H., Hameurlain, A., Pernul, G., Wagner, Roland R. (eds.) DEXA 2018. LNCS, vol. 11030, pp. 450–460. Springer, Cham (2018). https://doi.org/10.1007/978-3-319-98812-2_41
26. Jaysawal, B.P., Huang, J.W.: DMHUPS: discovering multiple high utility patterns simultaneously. Int. J. Knowl. Inf. Syst. **59**(2), 337–359 (2019)
27. Flamand, T., Ghoniem, A., Haouari, M., Maddah, B.: Integrated assortment planning and store-wide shelf space allocation: an optimization-based approach. Int. J. Manag. Sci.: Omega **81**, 134–149 (2018)

Data Lakes: Trends and Perspectives

Franck Ravat[1] and Yan Zhao[1,2(✉)]

[1] Institut de Recherche en Informatique de Toulouse, IRIT-CNRS (UMR 5505), Université Toulouse 1 Capitole, Toulouse, France
{Franck.Ravat,Yan.Zhao}@irit.fr

[2] Centre Hospitalier Universitaire (CHU) de Toulouse, Toulouse, France

Abstract. As a relatively new concept, data lake has neither a standard definition nor an acknowledged architecture. Thus, we study the existing work and propose a complete definition and a generic and extensible architecture of data lake. What's more, we introduce three future research axes in connection with our health-care Information Technology (IT) activities. They are related to (i) metadata management that consists of intra- and inter-metadata, (ii) a unified ecosystem for companies' data warehouses and data lakes and (iii) data lake governance.

Keywords: Data lake · Architecture · Metadata

1 Introduction

In the big data era, a great volume of structured, semi-structured and unstructured data are created much faster than before by smart-phones, social media, connected objects, and other data creators. These data have a great value for companies' Decision Support System (DSS) whose cornerstone is built upon data. Nevertheless, handling heterogeneous and voluminous data is especially challenging for DSS. Nowadays, Data Warehouse (DW) is a commonly used solution in DSS. Data have been extracted, transformed and loaded (ETL processes) according to predefined schemas. DW is popular thanks to its fast response, consistent performance and cross functional analysis. However, according to [4,5], DWs are not adapted for the big data analytics for the following reasons: (i) only predefined requirements can be answered. (ii) some information is lost through ETL processes. And (iii) the cost of a DW can grow exponentially because of the requirements of better performance, the growth of data volume and the complexity of database.

To face the challenges of big data and the deficiencies of DW, Dixon [4] put forward the concept data lake (DL): "If a data warehouse may be a store of bottled water - cleansed and packaged and structured for easy consumption - the data lake is a large body of water in a more natural state." This explication sketches the outline of DL, but it can not be considered as a formal definition. DL is a relatively new concept. Even though there are some so-called DL solutions in the market, there is not a standard definition nor an acknowledged architecture.

S. Hartmann et al. (Eds.): DEXA 2019, LNCS 11706, pp. 304–313, 2019.
https://doi.org/10.1007/978-3-030-27615-7_23

The goal of this prospective/survey paper is twofold. Firstly, we summarize the state of the art work and present a more complete vision of DL concept and a generic architecture. Secondly, we present future research axes by identifying major issues that appeared in our health-care IT activities. The remainder of the paper is organized as follows: Sect. 2 introduces the DL concept, we analyze different definitions and propose our own definition; Sect. 3 discusses DL architectures and introduces a generic and extensible architecture; Sect. 4 describes future research axes that includes metadata management, position of a DL in an information system and data lake governance.

2 Data Lake Concept

2.1 State of the Art

Data lake, as a relatively new concept, is defined in both scientific community and industrial world [3,5,7,15,18,21,25,31]. All the existing definitions respect the idea that a DL is a repository storing raw data in their native format. Yet, different definitions have different emphases. Regarding input, [5] introduces that the input of a DL is the data within an enterprise. Regarding process, [21] emphasizes that there is no process during the ingestion phase and [3,7,21,25] introduce that data will be processed upon usage. Regarding architecture, [5] presents that DLs are based on an architecture with low cost technologies. Regarding governance, metadata management is emphasized in [7,31]. And regarding users, [18] presents that data scientists and statisticians are DL users.

2.2 Dara Lake Definition

Existing definitions have evolved over time from experience feedback. Nevertheless, as mentioned in the previous paragraph, these different definitions are vague, they are not integrated with each other or even contradictory. To be as complete as possible, we propose a definition that includes input, process, output and governance of data lakes.

In the context of big data analytics, user requirements are not clearly defined at the time of the initial design and the implementation of a DL. A data lake is a big data analytics solution that ingests heterogeneously structured raw data from various sources (local or external to the organization) and stores these raw data in their native format, allows to process data according to different requirements and provides accesses of available data to different users (data scientists, data analysts, BI professionals etc.) for statistical analysis, Business Intelligence (BI), Machine Learning (ML) etc., and governs data to insure the data quality, data security and data life-cycle.

3 Data Lake Architecture

To the best of our knowledge, there does not exist an acknowledged DL architecture in literature. Firstly, we present different existing architectures and then propose a generic and extensible architecture.

3.1 State of the Art

Data lake functional architecture has evolved from mono-zone to multi-zone, and it is always presented with technical solutions.

The first vision of DL architecture is a flat architecture with a mono-zone that stores all the raw data in their native format. This architecture, closely tied to the HADOOP environment, enables load heterogeneous and voluminous data with low cost. Nevertheless, it does not allow users to process data and does not record any user operations.

A second vision of DL architecture contains five data ponds [10]. A *Raw data pond* that stores the just ingested data and the data that do not fit in other ponds. *Analog, application and textual data ponds* stores classified data from raw data pond by their characteristics. And *achival data pond* stores the data that are no longer used. This architecture classifies different types of data and achieves useless data, which make data finding faster and data analytics easier. However, the division of different ponds, especially the archival pond can not ensure the availability of all the raw data, contradicts the general recognition of DL which is to ingest all the raw data and process them upon usage.

To overcome these drawbacks, a third vision of architecture with multi-zones is proposed with a more diverse technological environment in the academic and industrial world. The author of [22] presents Amazon Web Services (AWS) DL architecture with four zones: ingestion, storage, processing and govern & secure. Raw data are loaded in the ingestion zone. The ingested raw data are stored in the storage zone. When data are needed, they are processed in the processing zone. The objective of Govern & secure zone is to control data security, data quality, metadata management and data life-cycle. The author of [19] separates the data processing zone into batch-processing and real time processing zones. He also adds a processed data zone to store all the cleansed data. Zaloni's DL architecture [14] separates the processing and storage zones into refined data zone, trusted data zone and discovery sandbox zone. The refined zone allows to integrate and structure data. Trusted data zone stores all the cleansed data. Data for exploratory analysis moves to the discovery sandbox.

As mentioned, a lot of DL architectures are supported with technical solutions. They are not independent of the inherent technical environment. Consequently, none of the existing architectures draws a clear distinction between functionality-related and technology-related components. What's more, the concept of multi-zone architecture is interesting and deserves further investigations. We believe that some zones are essential, while others are optional or can be regrouped. Concerning the essential zones, based on our DL definition, a data lake should be able to ingest raw data, process data upon usage, store processed data, provide access for different uses and govern data.

3.2 Data Lake Functional Architecture

Unlike several proposals, we want to distinguish functional architecture from technical architecture. Because a functional architecture concerns the usage perspective and it can be implemented by different technical solutions. By adopting to the existing DL architectures and avoiding their shortcomings, we propose a functional DL architecture (see Fig. 1), which contains four essential zones, and each zone, except the govern zone, has a treatment area (dotted rectangle) and a data storage area that stores the result of processes (gray rectangle):

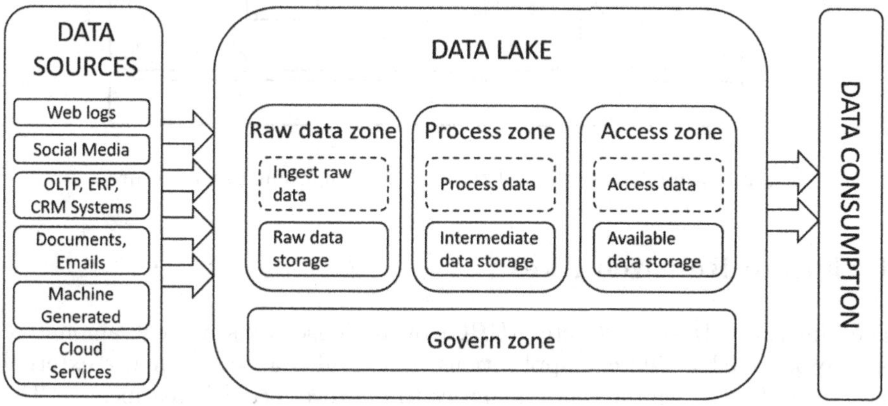

Fig. 1. Data lake functional architecture.

- *Raw data zone*: all types of data are ingested without processing and stored in their native format. The ingestion can be batch, real-time or hybrid. This zone allows users to find the original version of data for their analytics to facilitate subsequent treatments. The stored raw data format can be different from the source format.
- *Process zone*: in this zone, users can transform data according to their requirements and store all the intermediate data. The data processing includes batch and/or real-time processing. This zone allows users to process data (selection, projection, join, aggregation, etc.) for their data analytics.
- *Access zone*: the access zone stores all the available data for data analytics and provides the access of data. This zone allows self-service data consumption for different analytics (reporting, statistical analysis, business intelligence analysis, machine learning algorithms).
- *Governance zone*: data governance is applied on all the other zones. It is in charge of insuring data security, data quality, data life-cycle, data access and metadata management.

To exemplify our architecture, we propose an example of implementation (Fig. 2). Raw datasets (RD1, RD2) are ingested in data lake and stored in the raw data zone in their native format. Data are processed in the process zone and all the intermediate datasets (PD1, PD2, PD3, PD4) are stored in this area too. All the available data (AD1, AD2, AD3) are stored in the access zone for data consumption.

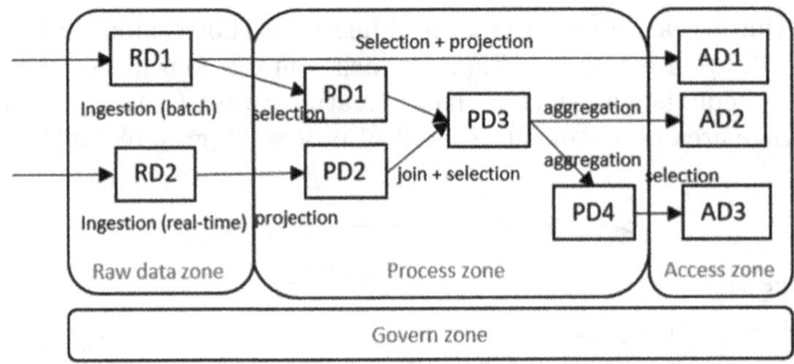

Fig. 2. An implementation of the data lake functional architecture.

4 Future Research Axes

The University Hospital Center (UHC) of Toulouse owns a great amount of data produced by different applications, it can also access to many external data. In order to facilitate data analytics to improve medical treatments, UHC of Toulouse lunched a project of DL to combine data from different individual sources. In this context, we encounter some problems: How to integrate a DL in the existing DSS? How to ensure the quality of data analytics by tracing back to the various transformations of data since the ingestion? Based on the questions that we are facing, we propose some research axes.

4.1 Integration of a Data Lake in an Information System

In a data lake, different users can access and process data for data exploration or statistical analysis for the purposes of decision making. Thus, DLs should be considered as one part of the DSS in enterprises' Information Systems (IS). Nowadays, the commonly used DSS solution is DW. According to the authors of [5,15,18], DLs and DWs are both created for extracting value of data to support decision makings but they also have differences. DWs are data repositories which store cleansed data based on predetermined schema. DLs ingest all types of raw data in their native format with low cost technologies to provide more flexibility and scalability. Regarding the similarities and differences, some questions like how do a DL and DWs work together, will a DL replace DWs need to be answered.

Many papers compared DLs and DWs but only a few papers introduced the impact of a DL for a data management ecosystem. Some authors present DL as a advanced version of DW [14,31]. The author of [5] introduces data lake cloud which is an elastic data storing and computing platform, DWs are constructed based on the data in the data lake. The author of [15,18] introduced that a DL can be fed by DWs and a DL can also be the source of DWs.

We think DLs should coexist with DWs because they have different objectives and users, a DL cannot simply replace a DW. To the best of our knowledge, a coexisting ecosystem has not been studied and implemented. To propose a such ecosystem, different research problems are induced. The first problem relates to the functional architecture definition which must determine precisely the information flow between DWs and a DL. If DWs feed a DL, the questions to solve are: where are the extracted data (from a particular DW or DM (Data Mart))? Do the data get into the ingestion zone or the process zone of a DL (because they have been processed in the DW)? Is it the same type of ingestion without data transformation as the ingestion from other data sources? If a DL is the source of a DW, the questions are similar on the sources (data in process zone or access zone), the target (a particular DW or DM) and the transfer process (ETL, ELT or only EL). Once these problems are solved, we must answer the problem on refreshing and updating data. We need to therefore answer the following questions: when is the data transformation done (real time, near real time, batch)? What is the type of refreshment (never, complete, incremental)? Finally, the third issue concerns the technical architecture which ensures the power and reliability of data flows between a DW and a DL).

4.2 Metadata

The main idea of data lake is to ingest raw data without process and process data upon usage. Therefore, data lakes keep all the information and have a good flexibility. Nevertheless, data lakes, which contain a lot of datasets without explicit models or descriptions can easily become invisible, incomprehensible and inaccessible. So that it is mandatory to set up a metadata management system for DL. In fact, the importance of metadata has been emphasized in many papers [1,7,31]. The first research problem that needs to be solved is the content of the metadata.

Data lake metadata, inspirited by DW metadata classification [16,26], are mainly classified in two ways. The first classification has three categories: *technical metadata* for data type, format and data structure, *operational metadata* for process history and *business metadata* for the descriptions of business objective. This classification focus on each single dataset, but not on the relationships between different datasets. Nevertheless, for a DL, datasets relationships are important to help users to find relevant datasets and verify data lineage. The second classification has two categories: *inter-metadata* for the relationships between data and *intra-metadata* for specifying each single dataset [17]. Inter-metadata is classified into dataset containment, provenance, logical cluster and content similarity by the author of [9]. Intra-metadata is classified into data characteristics, definitional, navigational, activity, lineage, rating and assessment [2,6,30]. The second classification is evolved, but it can still be improved. Some sub-categories are not adapted, for instance, the rating sub-category is not adaptive, because a DL can be accessed by different users who have different objective [30]. Moreover, the classification can be extended by some sub-categories, for instance, data sensitivity needs to be verified. Due to the specificity of DL with different zones including both storage and transformation processes, it is important to include intra- and inter-metadata. Therefore, we propose an metadata classification which contains inter- and intra-metadata and adapted subcategories:

- For *inter-metadata* [28], we propose to integrate *Dataset containment* which means a dataset is contained in another dataset. *Partial overlap* which means that some attributes with corresponding data in different datasets overlap. *Provenance* which means that one dataset is the source of another dataset. *Logical clusters* which means that some datasets are from the same domain (different versions, duplication etc.). And *Content similarity* which means that different datasets share the same attributes.
- For *intra-metadata* [28], we retain data characteristics, definitional, navigational and lineage metadata proposed in [2] and add the access, quality and security metadata.
  - *Data characteristics*: attributes describing a dataset, such as identification name, size and creation date.
  - *Definitional metadata*: datasets' meanings. Structured and unstructured datasets can be described semantically with a textual description or a set of keywords (vocabularies). Definitional metadata help users to understand datasets and make their data exploitation easier.
  - *Navigational metadata*: location information like file paths and database connection URLs.
  - *Lineage metadata*: information concerns data life-cycle. For example, data source, data process history.
  - *Quality metadata*: data consistency and completeness [26] to ensure dataset's reliability.
  - *Security metadata*: data sensitivity and access level. Some datasets may contain sensitive information that can only be access by certain users. Security metadata can support the verification of access.

The second research problem is to define an appropriate solution for metadata management. To the best our knowledge, there isn't a general metadata management system that works on heterogeneous data for the whole data life-cycle in DLs. We only have partial solutions in the literature. Some works concentrate on the detection of relationships between different datasets [1,9,27]. Some other work focus on the extraction of metadata for unstructured data (mostly textual data) [27,29].

To propose a appropriate solution, different research problems are induced. The first one relates to the way that metadata are stored: what is the conceptual schema of metadata? Which attributes should be recorded? How should we store the metadata (distributed RDBMS, NOSQL DBMS, RDF Stores etc.)? The second problem relates to the data feeding process: how to extract metadata from structured data as well as semi or unstructured data. What is the text analysis engine to detect automatically the keywords from unstructured data?

4.3 Data Lake Governance

A DL ingests and stores various types of data and can be accessed by different users. Without best practices in place to manage it, many issues that concern accessing, querying, and analyzing data can appear [23]. Given this context, data lake governance is required. In the state of the art work we find some partials solutions. A "Just-enough Governance" for DLs is proposed by [8] with data quality policy, data on-boarding policy, metadata management and compliance & audit policy. [11] indicated that the data governance needs to ensure data quality and availability throughout the full data life-cycle. [24] presented data governance with data lineage, data quality, security and data life-cycle management. However, these papers don't integrate the data lake governance through a complete vision.

We firstly propose a definition: the data governance of DL enables policies, standards and practices to be applied to manage data from heterogeneous sources and associated process (transformation and analysis) to ensure an efficient, secure usage, and a reliable quality of the analysis results. We propose to classify the DL data governance into data assets and IT assets [12,32]. Data assets refers to the value or potential value of data, while IT assets refers to technologies. Secondly, we identify some research axes based on the classification:

- Concerning the IT assets, the future researches must integrate the four following points: *data lake principles* concerns the position of the data lake in an information system. *Data lake functional architecture* concerns the evolution of different zones adapted to the enterprises' needs. *Data lake technical infrastructure* concerns the decisions about the technological solutions. *Data lake investment and prioritization* concerns the decisions about how much to invest for a data lake and the distribution for different zones.
- Concerning the data assets, the future research must be related to:
  - *Metadata management* (cf. previous section).
  - *Data quality management*: it influences the reliability of analytics result. In a data lake, data quality should at least be evaluated upon usage. New models must be proposed to make sure that data are valid in different data lake zones to ensure the data quality. Automatic validation [13,20] in the raw data zone, users' comments of analytics in the access zone can be one of the solutions.
  - *Data life-cycle management (DLM)*: it's necessary to define specific workflows to model the life-cycle of all the data that stored in different zones of a data lake and the relationships between different data.
  - *Security/Privacy*: a data lake can be accessed by different users with different privileges. Data sensitivity has to be authenticated, users access has to be managed [24]. New models and systems must be defined to be adapted to the data lake governance.

5 Conclusions

In this prospective/survey paper, we propose a complete definition of data lake and an extensible functional architecture based on 4 zones. Our definition has the advantages of being more complete than the literature and includes both input and output, different functions as well as users of data lakes. In our data lake architecture, each zone is defined more formally than the literature and is composed of a process layer and a data storage layer.

We also introduce future research axes. In the metadata context, we identify two issues: (i) identifying and modelling intra- and inter-metadata, (ii) implementing these metadata in an adequate data management system. In the information system context, the future work concentrates on the definition of a unified ecosystem in which the enterprise data warehouses and data lakes coexist. Finally, in the data lake governance context, we propose to work in the fields of data assets and IT assets.

References

1. Alserafi, A., Abelló, A., Romero, O., Calders, T.: Towards information profiling: data lake content metadata management. In: 2016 IEEE 16th International Conference on Data Mining Workshops (ICDMW), pp. 178–185. IEEE (2016)
2. Bilalli, B., Abelló, A., Aluja-Banet, T., Wrembel, R.: Towards intelligent data analysis: the metadata challenge. In: Proceedings of the International Conference on Internet of Things and Big Data, Rome, Italy, pp. 331–338 (2016)
3. Campbell, C.: Top five differences between data lakes and data warehouse, January 2015. https://www.blue-granite.com/blog/bid/402596/top-five-differences-between-data-lakes-and-data-warehouses
4. Dixon, J.: Pentaho, Hadoop, and data lakes, October 2010. https://jamesdixon.wordpress.com/2010/10/14/pentaho-hadoop-and-data-lakes/
5. Fang, H.: Managing data lakes in big data era: what's a data lake and why has it became popular in data management ecosystem. In: 2015 IEEE International Conference on Cyber Technology in Automation, Control, and Intelligent Systems (CYBER), pp. 820–824. IEEE (2015)
6. Foshay, N., Mukherjee, A., Taylor, A.: Does data warehouse end-user metadata add value? Commun. ACM **50**(11), 70–77 (2007)
7. Hai, R., Geisler, S., Quix, C.: Constance: an intelligent data lake system. In: Proceedings of the 2016 International Conference on Management of Data, pp. 2097–2100. ACM (2016)
8. Haines, R.: What is just enough governance for the data lake?, February 2015. https://infocus.dellemc.com/rachel---haines/just--enough--governance--data--lake/
9. Halevy, A.Y., et al.: Managing google's data lake: an overview of the goods system. IEEE Data Eng. Bull. **39**(3), 5–14 (2016)
10. Inmon, B.: Data Lake Architecture: Designing the Data Lake and avoiding the garbage dump. Technics publications (2016)
11. Kaluba, K.: Data lake governance - do you need it?, March 2018. https://blogs.sas.com/content/datamanagement/2018/03/27/data-lake-governance/
12. Khatri, V., Brown, C.V.: Designing data governance. Commun. ACM **53**(1), 148 (2010). https://doi.org/10.1145/1629175.1629210. http://portal.acm.org/citation.cfm?doid=1629175.1629210
13. Kwon, O., Lee, N., Shin, B.: Data quality management, data usage experience and acquisition intention of big data analytics. Int. J. Inf. Manag. **34**(3), 387–394 (2014)

14. LaPlante, A., Sharma, B.: Architecting Data Lakes. O'Reilly Media, Sebastopol (2014)
15. Llave, M.R.: Data lakes in business intelligence: reporting from the trenches. Procedia Comput. Sci. **138**, 516–524 (2018)
16. Lopez Pino, J.L.: Metadata in business intelligence, January 2014. https://www.slideshare.net/jlpino/metadata-in-business-intelligence
17. Maccioni, A., Torlone, R.: Crossing the finish line faster when paddling the data lake with kayak. Proc. VLDB Endow. **10**(12), 1853–1856 (2017)
18. Madera, C., Laurent, A.: The next information architecture evolution: the data lake wave. In: Proceedings of the 8th International Conference on Management of Digital EcoSystems, pp. 174–180. ACM (2016)
19. Menon, P.: Demystifying data lake architecture, July 2017. https://medium.com/@rpradeepmenon/demystifying-data-lake-architecture-30cf4ac8aa07
20. Merino, J., Caballero, I., Rivas, B., Serrano, M., Piattini, M.: A data quality in use model for big data. Future Gener. Comput. Syst. **63**, 123–130 (2016)
21. Miloslavskaya, N., Tolstoy, A.: Big data, fast data and data lake concepts. Procedia Comput. Sci. **88**, 300–305 (2016)
22. Nadipalli, R.: Effective Business Intelligence with QuickSight. Packt Publishing Ltd., Birmingham (2017)
23. O'Leary, D.E.: Embedding AI and crowdsourcing in the big data lake. IEEE Intell. Syst. **29**(5), 70–73 (2014)
24. Patel, P., Greg, W., Diaz, A.: Data lake governance best practices, April 2017. https://dzone.com/articles/data-lake-governance-best-practices
25. Piatetsky-Shapiro, G.: Data lake vs data warehouse: key differences, September 2015. https://www.kdnuggets.com/2015/09/data-lake-vs-data-warehouse-key-differences.html
26. Ponniah, P.: Data Warehousing Fundamentals: a Comprehensive Guide for IT Professionals. Wiley, Hoboken (2004)
27. Quix, C., Hai, R., Vatov, I.: Metadata extraction and management in data lakes with gemms. Complex Syst. Inf. Model. Q. **9**, 67–83 (2016)
28. Ravat, F., Zhao, Y.: Metadata management for data lakes. In: East European Conference on Advances in Databases and Information Systems. Springer (2019)
29. Sawadogo, P., Kibata, T., Darmont, J.: Metadata management for textual documents in data lakes. In: 21st International Conference on Enterprise Information Systems (ICEIS 2019) (2019)
30. Varga, J., Romero, O., Pedersen, T.B., Thomsen, C.: Towards next generation BI systems: the analytical metadata challenge. In: Bellatreche, L., Mohania, M.K. (eds.) DaWaK 2014. LNCS, vol. 8646, pp. 89–101. Springer, Cham (2014). https://doi.org/10.1007/978-3-319-10160-6_9
31. Walker, C., Alrehamy, H.: Personal data lake with data gravity pull. In: 2015 IEEE Fifth International Conference on Big Data and Cloud Computing, pp. 160–167. IEEE (2015)
32. Weill, P., Ross, J.W.: IT Governance: How Top Performers Manage IT Decision Rights for Superior Results. Harvard Business Press, Boston (2004)

An Efficient Greedy Algorithm for Sequence Recommendation

Idir Benouaret[1(✉)], Sihem Amer-Yahia[1], and Senjuti Basu Roy[2]

[1] CNRS, Univ. Grenoble Alpes, Grenoble, France
{idir.benouaret,sihem.amer-yahia}@univ-grenoble-alpes.fr
[2] New Jersey Institute of Technology, Newark, NJ, USA
senjuti.basuroy@njit.edu

Abstract. Recommending a sequence of items that maximizes some objective function arises in many real-world applications. In this paper, we consider a utility function over sequences of items where sequential dependencies between items are modeled using a directed graph. We propose EdGe, an efficient greedy algorithm for this problem and we demonstrate its effectiveness on both synthetic and real datasets. We show that EdGe achieves comparable recommendation precision to the state-of-the-art related work OMEGA, and in considerably less time. This work opens several new directions that we discuss at the end of the paper.

Keywords: Sequence recommendation · Submodular maximization · Algorithms

1 Introduction

Recommender Systems are models and algorithms that provide suggestions for items that are most likely to be of interest to a particular user [11]. Traditional recommendation systems are usually classified in two main categories. Collaborative Filtering approaches [13], which consists of learning users' preferences according to similar users and Content-based approaches, which consists of matching users' preferences with item features [8]. In their simplest form, recommender systems predict the most suitable items based on the user's preferences, and often return them as a ranked list. However, there are many applications where the order in which items should be consumed, and hence recommended, plays an important role in the satisfaction of the user. In that case, the set of recommended items should not be considered as a set of alternatives. Rather, they should form a sequence of items to be consumed in a "good" order. Typical examples include the recommendation of a sequence of music tracks [3], learning courses [7,18], or points of interest (POIs) in a city [16].

While various existing efforts address the problem of recommending packages of items, e.g., [1,17], they do not account for sequential dependencies and ordered preferences that might exist between items. In this paper, we are interested in developing a framework to recommend to a user u a sequence of at most

S. Hartmann et al. (Eds.): DEXA 2019, LNCS 11706, pp. 314–326, 2019.
https://doi.org/10.1007/978-3-030-27615-7_24

k items that maximizes a given utility function f. The function f should capture both the utility of selecting k items individually, and the utility of providing them in a given order $\sigma_1 \rightarrow \sigma_2, ..., \rightarrow \sigma_k$. For instance, a sequence containing a museum visit followed by sauna visit could have a higher utility than the inverse sequence. When dealing with sequences of items, the search space becomes exponential in the length of the recommended sequences. Submodular set functions constitute a category of widely used utility functions in recommender systems [5]. Such functions have been used to solve a submodular optimization problem that finds a subset of k items whose aggregated utility is maximized. The greedy algorithm that selects at each step the next item which maximizes a submodular monotone function has been shown to enjoy a $(1-e^{-1})$-approximation guarantee [6]. Recently, Tschiatschek *et al.* [15] considered the sequence selection problem using an expressive class of utility functions over sequences, which generalizes the class of submodular functions and captures ordered preferences among items. They encoded these ordered preferences in a directed acyclic graph (DAG) over items, where a directed edge between two items models that, in addition to selecting the two items, there is an additional utility when selecting the "tail" item before selecting the "head" item that preserves their order. To the best of our knowledge, [15] is the only work that considers this submodular sequence recommendation problem, we will thus follow their problem formulation. The authors proposed OMEGA, an algorithm that exploits the DAG property of the underlying utility graph. For each set of items, its ordering can be computed by first finding a topological ordering of the graph and then sorting the set of items according to that. At each subsequent step, an edge is selected to maximize the given utility function according to the sequence of items induced by the selected edge at each step. However, the full sorting at each step coupled with computing the topological ordering of the graph leads to a high computational runtime. For example, selecting a sequence of 10 over 1000 items takes about 4 min in our settings.

In this paper, we propose EdGe, an Edge-based Greedy sequence recommendation algorithm that extends the classical greedy algorithm proposed in [6], but instead of selecting at each step the item (node) with a maximum gain, it selects the next valid edge with a maximum *gain according to the edges that are selected so far.* EdGe takes as input a directed graph of items and a submodular monotone function, and returns a sequence of length at most k. EdGe greatly improves the complexity of OMEGA [15]. The worst-case computational complexity of EdGe is $\mathcal{O}(m \cdot k)$, where m is the number of edges in the graph and k is the length of the recommended sequence.

Our experiments on both synthetic and real datasets verify the performance of EdGe in terms of response time and recommendation precision.

2 Problem Formulation

We now formally define the problem of recommending a sequence of items following the formulation in [15]. Let $\mathcal{V} = \{v_1, v_2, ..., v_n\}$ be the set of n

items that are available for recommendation. Our goal is to provide a user u with a sequence recommendation of length at most k, from the set of possible items, where the order in which items are recommended is important. Let $\Sigma = \{(\sigma_1, \sigma_2, ..., \sigma_k) | k \leq n, \sigma_1, \sigma_2, ..., \sigma_k \in \mathcal{V}, \forall i, j \in [k] : i \neq j \Rightarrow \sigma_i \neq \sigma_j\}$ be the set of all possible sequences of items without repetitions, that can be selected from the set of items $\mathcal{V}$. We denote by $|\sigma|$ the length of the sequence $\sigma \in \Sigma$, and for two sequences $\sigma, \pi \in \Sigma$, let $\sigma \oplus \pi$ be their concatenation.

Ordered preferences are modeled using a set of edges $\mathcal{E}$, which represents that there is a utility when selecting items in a certain order, as it is illustrated in Fig. 1.

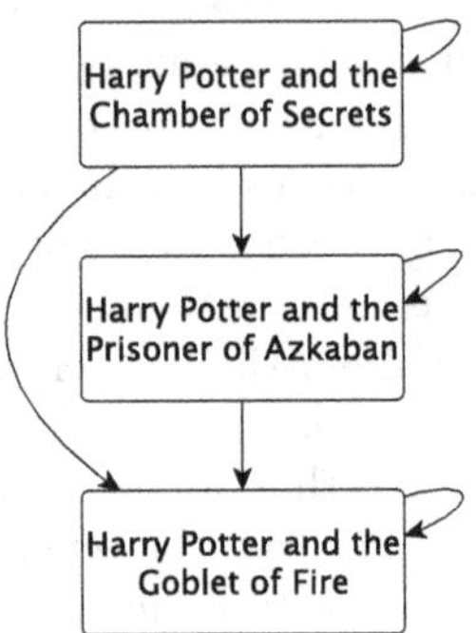

Fig. 1. An example of sequential dependencies between three *Harry potter* movies represented by a directed graph. Intuitively the order of watching these movies affects the satisfaction of the user. For example, watching *Harry Potter and the Chamber of Secrets* before watching *Harry Potter and the Prisoner of Azkaban* has a higher utility than watching them in the opposite order. Self-loops represent the utility of recommending an item individually.

More formally, an edge e_{ij} models the utility of selecting item v_j after item v_i. A self-loop edge e_{ii} models the utility of selecting item v_i individually. Hence, our model takes as input a directed graph $\mathcal{G} = (\mathcal{V}, \mathcal{E})$ whose nodes correspond to the set of items and whose edges correspond to the utilities of selecting items in a certain order. We define the utility function f for a sequence σ:

$$f(\sigma) = h(edges(\sigma)) \tag{1}$$

where $h(.)$ is a set-valued function and the function $edges(\sigma)$ maps the sequence of items σ to a set of induced edges according to the graph $\mathcal{G}$. More formally $edges(\sigma) = \cup_{j \in [|\sigma|]} \{(\sigma_i, \sigma_j) \mid (\sigma_i, \sigma_j) \in \mathcal{E},\ i \leq j\}$.

We assume that $h(E)$ is a non-negative monotone submodular set function over the edges $\mathcal{E}$, i.e, $h : 2^{\mathcal{E}} \rightarrow \mathbb{R}^{+*}$. Intuitively, submodularity describes the set of functions that satisfy a diminishing returns property. This property guarantees that the *marginal gain* of an edge $e \in \mathcal{E}$ is greater in the context of some set of edges A compared to a larger set of edges $B \supseteq A$. formally $h(A \cup \{e\} - h(A)) \geq h(B \cup \{e\}) - h(B)$ for all sets A, B s.t. $A \subseteq B \subseteq \mathcal{E} \backslash \{e\}$. Particularly, in our experiments (Sect. 4), we use a probabilistic coverage function. Intuitively, the utility of a set of edges E is large if the nodes $nodes(E)$ are well covered.

$$h(E) = \sum_{j \in nodes(E)} [1 - \prod_{(i,j) \in E} (1 - w_{i,j})] \tag{2}$$

where $w_{i,j}$ is the utility associated with edge (i, j) and $nodes(E)$ is the set of items that are induced by the set of edges E.

Our recommendation task is to select a sequence σ of at most k unique items that will maximize f. This can be formalized as the following maximization problem:

$$max_{\sigma \in \Sigma, |\sigma| \leq k} f(\sigma) \tag{3}$$

The search space for optimal solutions is exponentially larger for sequences in terms of k. In particular, there are $\binom{|\mathcal{V}|}{k}$ subsets of items of size k but $k!\binom{|\mathcal{V}|}{k}$ sequences of items of length k.

3 The EdGe Algorithm

In this paper, we develop EdGe, an Edge-based Greedy sequence recommendation algorithm, which takes as input a directed graph and outputs a sequence of at most k items. EdGe essentially extends the classical greedy algorithm proposed in [6], but instead of selecting at each step the item (node) with a maximum gain, it selects the next valid edge with a maximum *gain according to the edges that are selected so far*. More formally, we select at each step the edge e with the maximal obtained gain, i.e. the edge e maximizing $h(edges(\sigma) \cup e)$, according to the sequence of items σ that is selected so far.

EdGe (pseudo-code in Algorithm 1), takes as input a directed graph $\mathcal{G} = (\mathcal{V}, \mathcal{E})$, a maximum length k of the sequence to be recommended, a submodular monotone function $h(.)$ over the set of edges $\mathcal{E}$. EdGe outputs a sequence of items of length at most k. EdGe starts with the empty sequence σ (line 1). At each step, it computes the set of valid edges C (line 3). An edge e is said to be valid if it preserves the order of the recommended sequence so far, i.e., its endpoint node is not already in the sequence σ, which preserves the order of the corresponding sequence according to the topological ordering of the graph $\mathcal{G}$, and if adding the items induced by this edge to the actual sequence does not violate the maximum cardinality k. If the set C is not empty, EdGe selects the edge e_{ij} with the maximum gain (line 6). Note that the computation of $h()$ is incremental. It is not recomputed from scratch at every iteration. If the selected edge is a self-loop, EdGe appends a single node v_j to the sequence σ, if the start point of the selected edge is already in the sequence σ, it also appends the single node v_j (line 7, 8). Finally, if the selected edge e_{ij} induces two different items i and j and item i does not appear in the sequence σ, it appends node v_i

followed by node v_j to the sequence σ (line 10). The algorithm stops when it is not possible to select any further edge to grow the sequence σ (line 4).

Our algorithm does not need full sorting at each step compared to OMEGA [15]. It does a pseudo-sort while computing the set of valid edges and grows the set of selected edges, i.e., it discards all edges whose endpoints nodes are already in the selected sequence so far, to ensure that the recommended sequences will always be sorted according to the topological ordering of the graph.

Algorithm 1. EDGE-BASED GREEDY SEQUENCE ALGORITHM (EDGE)

Input: a directed graph $\mathcal{G} = (\mathcal{V}, \mathcal{E})$, a maximum length of the sequence k, a submodular monotone function $h(.)$ over the set of edges $\mathcal{E}$

Output: Recommended sequence with cardinality at most k

1 $\sigma \leftarrow \emptyset$
2 **while** $|\sigma| \leq k$ **do**
3     $C \leftarrow \{e_{ij} \in \mathcal{E}\ , |nodes(edges(\sigma) \cup e_{ij})| \leq k\ \ \&\ \ v_j \notin \sigma\}$
4   **if** $C = \emptyset$ **then**
5   break
6     $e_{ij} \leftarrow argmax_{e \in C}\ \ h(edges(\sigma) \cup e)$
7   **if** $(i = j) \vee (v_i \in \sigma)$ **then**
8         $\sigma \leftarrow \sigma \oplus v_j$
9   **else**
10         $\sigma \leftarrow \sigma \oplus (v_i \oplus v_j)$
11 **return** σ

3.1 Comparison with OMEGA Algorithm

In previous work on recommending item sequences [15], an algorithm (OMEGA) was proposed. OMEGA iteratively and greedily extends a set of edges E with the corresponding sequence of items σ. The algorithm is based on the assumption that the graph $\mathcal{G}$ is a DAG (directed acyclic graph). At each step, OMEGA selects an edge and reorders the sequence of items induced by the edges selected so far according to the topological ordering of the DAG graph $\mathcal{G}$. OMEGA has a runtime complexity of $\mathcal{O}(m + n + k\Delta m(k \cdot log(k)))$. Where m is the number of edges in the graph, n the number of nodes and $\Delta = min(\Delta_{in}, \Delta_{out})$, where Δ_{in}, Δ_{out} are the indegree and outdegree of the graph. The first term $(n + m)$ is the runtime complexity for computing a topological sort of the graph. In the second term, $k\Delta$ results from the fact that for a sequence of length k, the algorithm can select at most $k\Delta$ edges, and the factor $k \cdot log(k)$ is the complexity for sorting the sequence according to the topological ordering induced by the graph.

The bottleneck of the OMEGA algorithm is that at each step where an edge is selected to greedily extend the set of edges and the corresponding set of items, the algorithm requires a full reordering of the corresponding items that are induced by all the selected edges so far according to the topological ordering of the input graph.

Our EDGE algorithm builds on the same approach of greedily selecting and extending a set of edges and the corresponding sequence of items. But without the requirement of reordering the sequence at each step. Instead, our approach considers only those candidate edges that are not violating the ordering of the greedily selected sequence. At each iteration, we heuristically discard all edges having their endpoints nodes in the selected sequence so far. This process insures that the final recommended sequence will always be sorted according to the topological ordering of the graph. Furthermore, our EDGE algorithm has the advantage that it does not require the graph $\mathcal{G}$ to be a DAG, as it does not compute an ordering of the graph. This may be beneficial if we consider application scenarios where it is possible to recommend the same item multiple times in the sequence.

3.2 Runtime Complexity

The computational complexity of EDGE is $\mathcal{O}(m \cdot k)$, where m is the number of edges in the graph and k is the maximum length of the recommended sequence. Indeed, selecting the set of valid edges C and retrieving the edge with the maximum gain requires at worse to loop through all the m edges in the graph. This has to be done at most k times since k is the maximum length of the recommended sequence. This complexity is better than OMEGA's [15] whose computational complexity is $\mathcal{O}(m + n + k\Delta m(k \cdot log(k)))$.

4 Experiments

We run experiments on synthetic and real datasets to study the performance of EDGE in terms of response time, achieved value of the objective function and recommendation precision. Our implementation is in Python 3.7.0 and is running on a 2.7 GHz Intel Core i7 machine with a 16 GB main memory, running OS X 10.13.6.

Table 1. Experimental parameters (default values in bold)

Description	Parameter	Values
Number of nodes	n	10, 50, **1000**, 5000, 10000
Maximum outdegree	Δ_{out}	1, 2, 3, 4, 5, **6**, 7, 8, 9, 10
Sequence length	k	1, 5, **10**, 25, 50, 75, 100

4.1 Synthetic Datasets

For this experiment, we follow the same setting as in [15]. We create the input graph $\mathcal{G}$ as follows: let $\mathcal{A}$ be the adjacency matrix of $\mathcal{G}$, i.e. $A_{ij} = 1$ if there exists an edge from item i to item j. For every $i \in [n]$ we select a subset of size

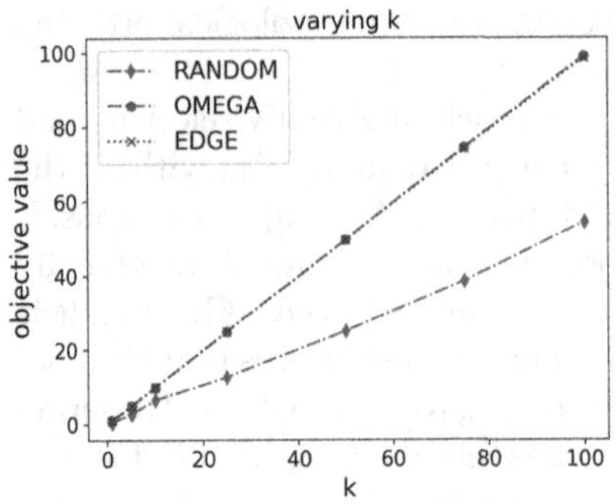

Fig. 2. Varying sequence length k

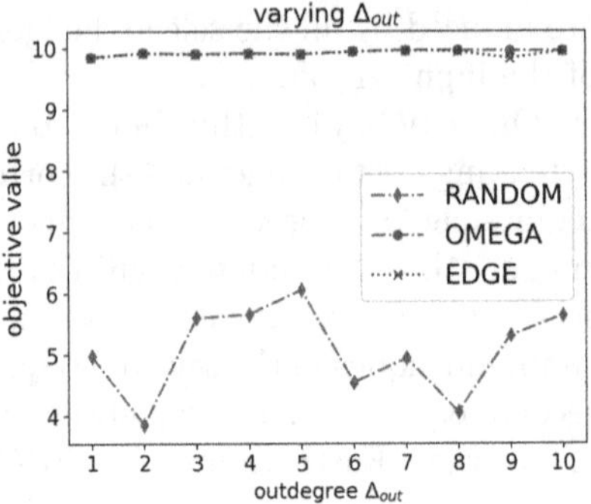

Fig. 3. Varying outdegree Δ_{out}

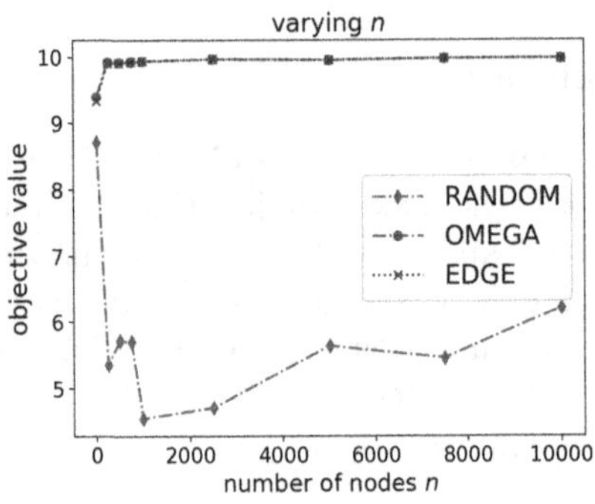

Fig. 4. Varying number of nodes n

$min(\Delta_{out}, n-i)$ uniformly at random from $[i+1, n]$ and set the corresponding entries of the matrix A to 1. This construction process of the graph $\mathcal{G}$ ensures a desired maximum outdegree Δ_{out}. It also guarantees that the input graph is a DAG which lets us compare EdGe with OMEGA [15]. We assign every edge of $\mathcal{G}$ a utility value that is independently drawn from a uniform distribution in $\mathcal{U}(0,1)$. We then use those utilities to define the submodular function $h(.)$ as follows:

$$h(E) = \sum_{j \in nodes(E)} [1 - \prod_{(i,j) \in E} (1 - w_{i,j})] \tag{4}$$

where $w_{i,j}$ is the utility associated with edge (i,j) and $nodes(E)$ is the set of items that are induced by the set of edges E.

In the following, we report the values of the objective function for solutions computed by EdGe and by OMEGA. We also use as a baseline a random selection of k items. In each setting, we compare the objective values for different possible combinations of the parameters: k (sequence length), n (number of nodes) and Δ_{out} (maximum outdegree). Table 1 summarizes those parameters. In each run, we vary one parameter and set the others to their default values which are written in bold.

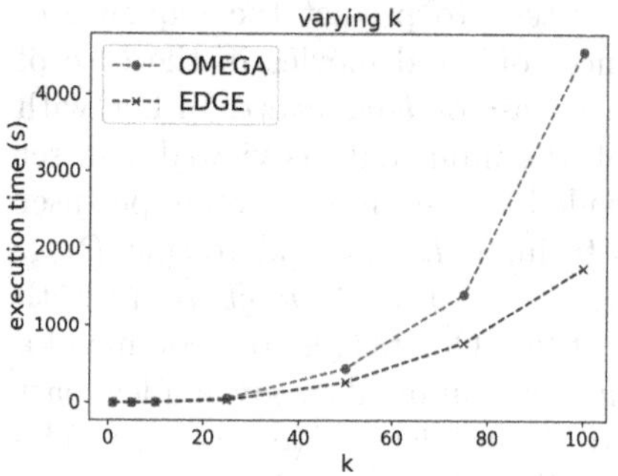

Fig. 5. Varying sequence length k

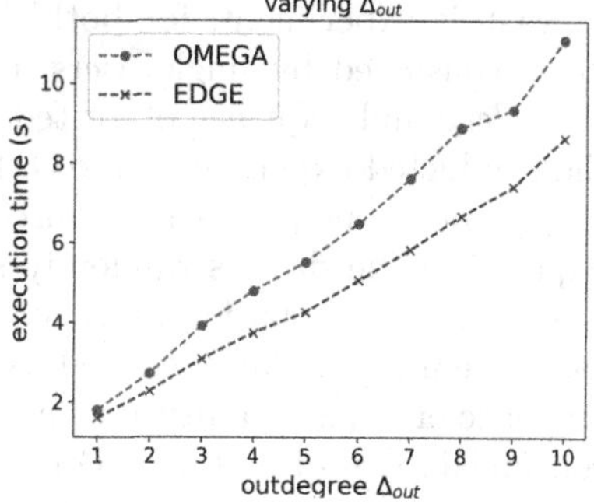

Fig. 6. Varying outdegree Δ_{out}

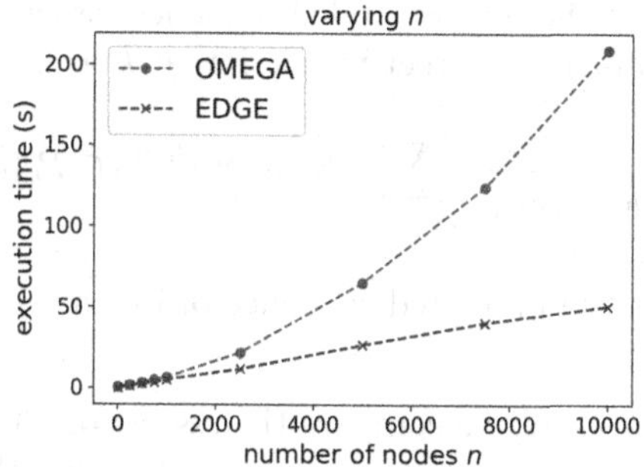

Fig. 7. Varying number of nodes n

Results are shown in Figs. 2, 3 and 4. For varying the length sequence k, the maximum outdegree Δ_{out} and the number of nodes n, we can observe that EDGE performs very closely to OMEGA according to the achieved value of the objective function f, while being much faster. Results on runtime according to each setting are reported in Figs. 5, 6 and 7. We notice that EDGE runs approximatively 3 times faster than OMEGA when $k = 100$, and 4 times faster when $n = 10,000$.

4.2 Real Datasets

Datasets and Setup. We use two real datasets in our experiments: *Movielens 1M*[1] and a dataset extracted from *Foursquare* [19]. The *Movielens 1M* dataset contains 1,000,209 ratings of 6,040 users for 3,706 movies. Every rating takes a value in $\{1, ..., 5\}$ and has a timestamp associated with it. For the *Foursquare* dataset, in order to make our setup realistic, we choose POIs from a single city: "New York". We excluded POIs that were visited by fewer than 20 users, and excluded users that visited less than 20 POIs. We end up with a dataset having 840 users and 4,750 POIs. Each POI is associated to a category, including restaurants, bars, parks, etc.

[1] http://grouplens.org/datasets/movielens/1m/.

In this experiment, for both datasets, we seek to predict the sequence of items consumed by single users. i.e., sequences of rated movies in the case of *Movielens* and sequence of visited POIs in the case of *Foursquare*. In line with the conducted experiments for OMEGA [15], the input data is viewed as a set of sequences, one per user. We used the provided timestamps to create per-user sequences. The data is randomly split into training $\mathcal{D}_{Train}$ and testing $\mathcal{D}_{Test}$ where $|\mathcal{D}_{Test}| = 500$ for *Movielens* and $|\mathcal{D}_{Test}| = 200$ for *Foursquare*. In both cases, we use randomly selected users for our tests. Our task is to recommend a sequence of items to a test user not present in the training data given a few past consumed items of that user. Formally, for a test user u let $\sigma = (\sigma_1, \sigma_2, ..., \sigma_m)$ be the sequence of items consumed by the user u. Then, given the first half of these items, i.e. $\sigma^{previous} = (\sigma_1, \sigma_2, ..., \sigma_l)$, where $l = \frac{m}{2}$, our goal is to make predictions about which other items the user consumed later, i.e. $\sigma^{left} = (\sigma_{l+1}, \sigma_{l+2}, ..., \sigma_m)$.

Given the length k of the recommended sequence for each user in the training set, *Prec@k* is the number of correct prediction at k averaged over the test set:

$$Prec@k = \frac{1}{k.|\mathcal{D}_{Test}|} \cdot \sum_{\sigma \in \mathcal{D}_{Test}} |nodes(\sigma^{left}) \cap P_k(\sigma^{previous})| \quad (5)$$

where, $P_k(\sigma^{previous})$ is the predicted sequence of items.

Baseline. For both datasets, we compare the results of EDGE to OMEGA [15]. For a fair comparison, we conduct the experiments using the same input graph model that is shown in Fig. 8. In this graph, solid nodes represent the items in $\sigma^{previous} = (\sigma_1, \sigma_2, ..., \sigma_l)$, empty nodes represent the candidate items. We model the dependencies between the last z items in $\sigma^{previous}$ and the items that can be selected, where $z = \{1, 2, 5\}$ (for example, $z = 2$ means that the 2 last items in $\sigma^{previous}$ are considered). Note that, the input graph is test instance dependent, i.e., we construct one graph per test user.

The value associated with an edge linking item i to item j noted as $p_{j|i}$ is estimated as the conditional probability that a user consumes j given that she has consumed i before. These conditional probabilities are estimated using the training data. The value associated with every self-loop edge (i, i) corresponds to the empirical frequency p_i of item i (which is also conveniently noted as $p_{i|i}$). These conditional probabilities are then used for estimating the utility function:

$$h(E) = \sum_{j \in nodes(E)} \left[1 - \prod_{(i,j) \in E} (1 - p_{j|i}) \right] \quad (6)$$

Results. Recommendation precision for EDGE and OMEGA is reported in Table 2 for *MovieLens* and Table 3 for *Foursquare*. We vary k, the length of the predicted sequences and z the length of the user's history considered for constructing the input graph. For both experiments, we observe that EDGE

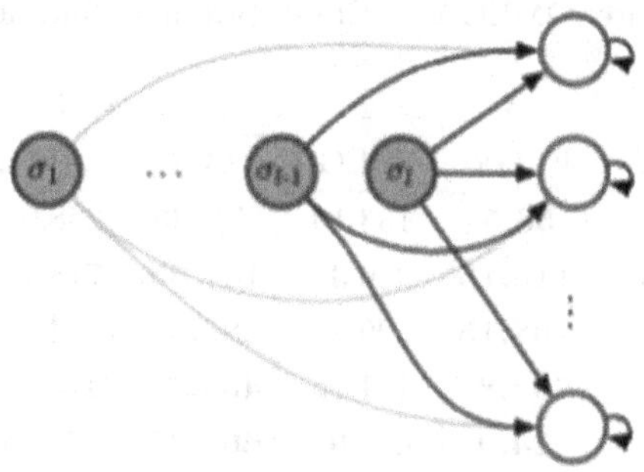

Fig. 8. Input graph [15]. Solid nodes represent the items in $\sigma^{previous} = (\sigma_1, \sigma_2, ..., \sigma_l)$, empty nodes represent the items that can be recommended.

Table 2. MovieLens: OMEGA vs EDGE for Precision (*Prec@k*)

	z = 1		z = 2		z = 5	
	OMEGA	EDGE	OMEGA	EDGE	OMEGA	EDGE
k = 1	0.346	**0.346**	0.37	**0.37**	0.38	**0.38**
k = 2	0.331	**0.331**	0.359	**0.349**	0.374	**0.345**
k = 3	0.3166	**0.3166**	0.346	**0.3406**	0.3699	**0.3330**
k = 4	0.3045	**0.3045**	0.3385	**0.327**	0.3665	**0.324**
k = 5	0.2916	**0.2916**	0.3304	**0.316**	0.3712	**0.314**

Table 3. Foursquare: OMEGA vs EDGE for Precision (*Prec@k*)

	z = 1		z = 2		z = 5	
	OMEGA	EDGE	OMEGA	EDGE	OMEGA	EDGE
k = 1	0.426	**0.426**	0.513	**0.513**	0.70	**0.70**
k = 2	0.42	**0.416**	0.518	**0.49**	0.683	**0.678**
k = 3	0.404	**0.398**	0.496	**0.474**	0.688	**0.667**
k = 4	0.382	**0.38**	0.481	**0.472**	0.665	**0.658**
k = 5	0.372	**0.37**	0.478	**0.47**	0.641	**0.637**

Table 4. MovileLens: OMEGA vs EDGE (average runtime in milliseconds)

	z = 1		z = 2		z = 5	
	OMEGA	EDGE	OMEGA	EDGE	OMEGA	EDGE
k = 1	72.3	**55.26**	106.16	**79.5**	207.04	**151.9**
k = 2	170.6	**116.64**	263	**171.74**	541.32	**334.54**
k = 3	306.32	**192.44**	514.04	**284.62**	1087.2	**569.24**
k = 4	484.02	**304.62**	836.88	**442.14**	1894.6	**861.14**
k = 5	729.3	**410.7**	1240.04	**620.88**	3030.26	**1253.44**

Table 5. Foursquare: OMEGA vs EDGE (average runtime in milliseconds)

	z = 1		z = 2		z = 5	
	OMEGA	EDGE	OMEGA	EDGE	OMEGA	EDGE
k = 1	90.9	**56.5**	133.15	**79.65**	258.95	**147.35**
k = 2	226.05	**115.15**	356.3	**167.95**	718.1	**328.15**
k = 3	417.3	**198.15**	699.35	**286.35**	1454.35	**579.1**
k = 4	677	**294.8**	1121.75	**439.2**	2474.4	**907.1**
k = 5	977.25	**424.9**	1653.9	**661.45**	3778.75	**1285.55**

performs very closely to OMEGA and equally for some settings. We also notice that the precision increases with increasing z, indicating that considering a longer history of user interactions is beneficial. We also observe that EDGE is much faster than OMEGA, especially when increasing the length of the recommended sequences. Results on the average runtime per user are reported in Table 4 for *MovieLens* and in Table 5 for *Foursquare*.

5 Related Work

In [12], the authors propose an interactive itinerary planning for ranking travel packages. However, in this work the order of visiting POIs is mainly due to the POI visit and transit times induced by the recommended itinerary. The utility of selecting POIs in a certain order is neglected and the system does not take into account ordered preferences. Some existing works address the problem of package recommendations [1,17]. However, in such works, authors assign a utility for sets of items but not for sequences, i.e., sequential dependencies are neglected. Another research area which deals with the problem of selecting sequences is sequential pattern mining. This problem was introduced by Agrawal and Srikant [14]. Sequential pattern mining is the task of finding all frequent subsequences in a sequence database. Numerous algorithms have been designed to discover sequential patterns in sequence databases, such as FPGrowth [2], Spade [21] and Prefixspan [9]. However, these works do not take into account user preferences and output the same subsequences for every user which leads to a lack of personalization.

Another related research area is *next basket recommendation* [10,20] which has gained much attention especially in e-commerce scenarios. Two different approaches were used to tackle this problem, the well-known collaborative filtering (CF) models such as Matrix factorization (MF) [4] which capture user's general interests but have a lack to consider sequential behaviors, and Markov chain models which extract sequential features from the history on users' interactions and then predict next purchase, but have lack to consider general preferences of the user. Rendle *et al* [10] combined the well-known collaborative filtering (CF) using Matrix factorization (MF) [4] and Markov chain models which extract sequential features from the history on users' interactions to account for both users' interests and sequential features in order to generate ranking lists. However, these works are different from the one we proposed in this paper. They don't formalize the optimization problem of recommending a sequence of items by modeling the sequential dependencies with a graph data model.

Another related work is the recommendation with prerequisites on items, which was studied in [7]. Authors addressed the problem of recommending the best set of k items when there is an inherent ordering between items expressed as prerequisites. Various prerequisite structures were studied and several heuristic approximation algorithms were developed to solve the problem. However, the problem is formulated as a set recommendation problem rather than a sequential recommendation problem.

Our formalization of sequence recommendation builds on the one proposed recently in [15]. We show that our algorithm, EDGE produces recommendations that are as good and has much better runtime complexity and response time.

6 Conclusion

We set out to solve the sequence recommendation problem. We proposed a submodular maximization formulation that is based on marginal edge utility, and developed an efficient greedy algorithm to find the best sequence to recommend. We validated our results on synthetic and real datasets and showed that EDGE, our Edge-based Greedy sequence recommendation algorithm, returns recommendation with high accuracy in little time. This work opens several directions. Our immediate improvement is to study the addition of constraints to the objective function including constraints on the type of items to consider and on the types of transitions between items. Additionally, we would like to examine how to simultaneously recommend sequences to multiple users. We also plan to test our solutions on other real datasets acquired from industry partners.

References

1. Amer-Yahia, S., Bonchi, F., Castillo, C., Feuerstein, E., Mendez-Diaz, I., Zabala, P.: Composite retrieval of diverse and complementary bundles. IEEE Trans. Knowl. Data Eng. **26**(11), 2662–2675 (2014)
2. Han, J., Pei, J., Yin, Y., Mao, R.: Mining frequent patterns without candidate generation: a frequent-pattern tree approach. Data Min. Knowl. Discov. **8**(1), 53–87 (2004)
3. Hariri, N., Mobasher, B., Burke, R.: Context-aware music recommendation based on latent topic sequential patterns. In: Proceedings of the Sixth ACM Conference on Recommender Systems, RecSys 2012, pp. 131–138. ACM, New York, NY, USA (2012). https://doi.org/10.1145/2365952.2365979, https://doi.org/10.1145/2365952.2365979
4. Koren, Y., Bell, R., Volinsky, C.: Matrix factorization techniques for recommender systems. Computer **8**, 30–37 (2009)

5. Krause, A., Golovin, D.: Submodular Function Maximization, pp. 71–104. Cambridge University Press, Cambridge (2014). https://doi.org/10.1017/CBO9781139177801.004
6. Nemhauser, G.L., Wolsey, L.A., Fisher, M.L.: An analysis of approximations for maximizing submodular set functions-I. Math. Program. **14**(1), 265–294 (1978)
7. Parameswaran, A., Venetis, P., Garcia-Molina, H.: Recommendation systems with complex constraints: a course recommendation perspective. ACM Trans. Inf. Syst. (TOIS) **29**(4), 20 (2011)
8. Pazzani, M.J., Billsus, D.: Content-based recommendation systems. In: Brusilovsky, P., Kobsa, A., Nejdl, W. (eds.) The Adaptive Web. LNCS, vol. 4321, pp. 325–341. Springer, Heidelberg (2007). https://doi.org/10.1007/978-3-540-72079-9_10
9. Pei, J., et al.: Mining sequential patterns by pattern-growth: the PrefixSpan approach. IEEE Trans. Knowl. Data Eng. **16**(11), 1424–1440 (2004)
10. Rendle, S., Freudenthaler, C., Schmidt-Thieme, L.: Factorizing personalized Markov chains for next-basket recommendation. In: Proceedings of the 19th International Conference on World Wide Web, pp. 811–820. ACM (2010)
11. Ricci, F., Rokach, L., Shapira, B.: Recommender systems: introduction and challenges. In: Ricci, F., Rokach, L., Shapira, B. (eds.) Recommender Systems Handbook, pp. 1–34. Springer, Boston, MA (2015). https://doi.org/10.1007/978-1-4899-7637-6_1
12. Roy, S.B., Das, G., Amer-Yahia, S., Yu, C.: Interactive itinerary planning. In: 2011 IEEE 27th International Conference on Data Engineering (ICDE), pp. 15–26. IEEE (2011)
13. Afoudi, Y., Lazaar, M., Al Achhab, M.: Collaborative filtering recommender system. In: Ezziyyani, M. (ed.) AI2SD 2018. AISC, vol. 915, pp. 332–345. Springer, Cham (2019). https://doi.org/10.1007/978-3-030-11928-7_30
14. Srikant, R., Agrawal, R.: Mining sequential patterns: generalizations and performance improvements. In: Apers, P., Bouzeghoub, M., Gardarin, G. (eds.) EDBT 1996. LNCS, vol. 1057, pp. 1–17. Springer, Heidelberg (1996). https://doi.org/10.1007/BFb0014140
15. Tschiatschek, S., Singla, A., Krause, A.: Selecting sequences of items via submodular maximization. In: AAAI, pp. 2667–2673 (2017)
16. Wörndl, W., Hefele, A., Herzog, D.: Recommending a sequence of interesting places for tourist trips. Inf. Technol. Tour. **17**(1), 31–54 (2017)
17. Xie, M., Lakshmanan, L.V., Wood, P.T.: Breaking out of the box of recommendations: from items to packages. In: Proceedings of the Fourth ACM Conference on Recommender Systems., pp. 151–158. ACM (2010)
18. Xu, J., Xing, T., Van Der Schaar, M.: Personalized course sequence recommendations. IEEE Trans. Signal Process. **64**(20), 5340–5352 (2016)
19. Yang, D., Zhang, D., Zheng, V.W., Yu, Z.: Modeling user activity preference by leveraging user spatial temporal characteristics in LBSNs. IEEE Trans. Syst. Man Cybern. Syst. **45**(1), 129–142 (2015)
20. Yu, F., Liu, Q., Wu, S., Wang, L., Tan, T.: A dynamic recurrent model for next basket recommendation. In: Proceedings of the 39th International ACM SIGIR conference on Research and Development in Information Retrieval, pp. 729–732. ACM (2016)
21. Zaki, M.J.: SPADE: an efficient algorithm for mining frequent sequences. Mach. Learn. **42**(1–2), 31–60 (2001)

Discovering Diverse Popular Paths Using Transactional Modeling and Pattern Mining

P. Revanth Rathan[1], P. Krishna Reddy[1](✉), and Anirban Mondal[2]

[1] IIIT Hyderabad, Hyderabad, India
revanth.parvathaneni@research.iiit.ac.in, pkreddy@iiit.ac.in
[2] Ashoka University, Sonipat, India
anirban.mondal@ashoka.edu.in

Abstract. While the problems of finding the shortest path and k-shortest paths have been extensively researched, the research community has been shifting its focus towards discovering and identifying paths based on user preferences. Since users naturally follow some of the paths more than other paths, the popularity of a given path often reflects such user preferences. Moreover, users typically prefer diverse paths over similar paths for gaining flexibility in path selection. Given a set of user traversals in a road network and a set of paths between a given source and destination pair, we propose a scheme based on transactional modeling and pattern mining for performing top-k ranking of these paths based on both path popularity and path diversity. Our performance evaluation with a *real dataset* demonstrates the effectiveness of the proposed scheme.

Keywords: Popular paths · Transactional modeling · Pattern mining

1 Introduction

The problems of finding the shortest path [6,7,12] and k-shortest paths [10,11,15] have been extensively researched. Path finding and discovery has significant applications in several important and diverse domains such as city planning, transportation and vehicular navigation, disaster management and tourism.

Interestingly, the recent focus of the research community as well as that of most commercial route planning and navigation systems has been shifting towards discovering and identifying paths based on user preferences. Such user preferences may include roads with relatively high thoroughfare (i.e., better for safety), smoother roads with better infrastructure (e.g., less potholes), roads with more facilities or points of interest nearby, lighted roads as opposed to dark and unsafe streets, roads with relatively lower crime rates, roads with better scenic beauty and so on. In practice, given that users naturally follow some of these paths significantly more as compared to other paths, the **popularity of a given path** often reflects such user preferences. *In this work, we use the notion of*

S. Hartmann et al. (Eds.): DEXA 2019, LNCS 11706, pp. 327–337, 2019.
https://doi.org/10.1007/978-3-030-27615-7_25

popularity as the overarching theme for reflecting the path preferences of users. Besides traversal time/cost of a route, users also consider the *popularity* of a route as an important factor. Furthermore, users often prefer to obtain **diverse paths** [4,5] for gaining flexibility in path selection e.g., when their preferred routes become blocked (unusable) possibly due to a wide gamut of reasons such as flooding, traffic congestion and road maintenance works.

We model a given road network as a weighted graph $G(V, E)$, where V is the set of nodes of the graph. Here, each node represents an intersection of the road segments. Thus, $V = \{v_1, v_2, ..., v_n\}$, where v_i represents the ith node. E is the set of edges of the graph, where each edge represents a given road segment. The edges are of the form $< v_i,\ v_j >$ such that $v_i,\ v_j \in V$. A user traversal is a spatial trajectory, which is represented by a sequence of edges in G.

Given a set of user traversals in a road network and a set SP of paths between a given source and destination pair, we address the problem of performing ranking of the paths in SP based on both path popularity and path diversity. We refer to this problem as the **Diverse Popular Paths (DPP) problem**.

For a road network G and a set of user traversals, we consider a path p_i between a given source s and a destination t to be more popular than other paths between s and t, if more users follow p_i as compared to the other paths. We develop a model for computing the popularity score of a given path based on the number of user traversals by introducing the notion of the *popularity contribution of a user traversal to a given path.* We also develop the methodology for computing the popularity scores of all the input paths between s and t by processing the set of user traversals. Determining the popularity of the path p_i from the road network G using the proposed model poses scalability issues as the number of edges in p_i increases. We propose an efficient algorithm for computing the popularity score of a given path using the knowledge of frequent patterns. Based on the popularity score and a pre-defined diversity threshold, we output a ranked list of popular and diverse paths from the set of input paths.

Our main contributions are three-fold:

1. We introduce the notion of *popular paths* and propose a model for computing the popularity scores of paths.
2. We propose a framework for modeling user traversals in a road network as transactions and computing the popularity score of any path based on the itemsets extracted from the transactions using pattern mining techniques.
3. We conducted a performance evaluation with a *real dataset* to demonstrate the effectiveness of the proposed scheme.

The remainder of the paper is organized as follows. In Sect. 2, we discuss related work and background. In Sect. 3, we present the modeling of popularity score of a path. In Sect. 4, we discuss the proposed scheme. In Sect. 5, we report the performance evaluation. Finally, we conclude in Sect. 6.

2 Related Work and Background

This section discusses related works and background about path diversity.

The problems of finding the shortest path and the top-k shortest paths have been addressed in [6,7,12] and [10,11,15] respectively. The work in [5] proposes exact and approximate algorithms to discover the top-k diverse shortest paths based on the concept of *limited overlap* by using Yen's algorithm [15].

Research efforts are being made to find popular paths based on the spatial trajectories of users [2,3,14]. The work in [3] uses adaptive Markov model for computing the popularity of a given sequence of edges in order to compute the most popular route. The work in [2] improves the above approach for finding the top-k personalized routes from spatial trajectories. The proposal in [14] constructs a route inference model and a routable graph construction from uncertain trajectories for determining the top-k popular paths, which comprise only trajectory points. The work in [13] uses a route score function that strikes a balance between user preference degrees and the length of the route as a measure of the popularity. The work in [8] reviews trajectory data mining techniques. Furthermore, frequent pattern (FP) mining [1] extracts the interesting information about the associations among the items in patterns in a transactional database based on a user-specified *support (frequency)* threshold.

Our proposed approach differs from existing popular path extraction approaches in two ways. First, it defines the popularity of the path based on the *effective number of user traversals*, which cover the path. Second, unlike existing approaches (e.g., the proposal in [14]), we extract top-k popular paths between a given source and destination pair.

About Diversity Among Paths: Now we shall explain the notion of *diversity among paths*. In this work, we model the notion of diversity based on the overlap among paths. We consider a given path as a set of edges. Each edge (v_i, v_j) in path p_i has a positive weight, which represents the cost of traversing from v_i to v_j, e.g., travel time or distance. The weight $l(p_i)$ of a given path p_i is computed as the sum of the weights of all of the edges present in p_i. The notion of overlap defined in [4,5] is as follows. Given two paths p_i and p_j in a network G, **overlap** OV of p_i w.r.t. p_j is the ratio of weight of common edges in p_j to the weight of all the edges in p_j. The notion of diversity used in this work is inversely proportional to the notion of overlap between two paths. The equation for **diversity value** between two paths p_i and p_j based on overlap is as follows:

$$DV(p_i,\ p_j) = 1 - \frac{\sum_{e \in p_i \cap p_j} l(e)}{\sum_{e \in p_j} l(e)} \quad (1)$$

Here, $l(e)$ represents the weight of the edge e and $0 \leq DV(p_i,\ p_j) \leq 1$. The maximum value of DV is achieved when intersection of p_i and p_j is the null set, while the minimum value of 0 is achieved when p_i equals p_j.

3 Model of Computing Popularity Score

In this section, we develop a new model for computing the popularity score of a given path. We defer the discussion on the proposed approach for the efficient computation of top-k diverse popular paths from a set of paths to Sect. 4.

Now we shall introduce the notion of *popularity score* of a given path. We shall henceforth designate the popularity score of a given path p_i with n edges as $\omega(p_i)$. Intuitively, the value of $\omega(p_i)$ depends upon the number of user traversals that traverse p_i. In practice, all user traversals need not necessarily traverse all of the edges in the path. We can divide the whole traversals of all users into a set of traversals, which cover only one edge, only two edges and so on up to the set of traversals that cover all of the edges of p_i. These partially covered traversals also contribute to the popularity of the path. To incorporate the effect of such partially covered traversals, we shall now introduce the notion of the *popularity contribution* $PC(q_i, p_i)$ of a user traversal q_i w.r.t. a given path p_i. Intuitively, the popularity contribution PC indicates the extent to which a user traversal contributes to the popularity of a given path.

Consider a path p_i between a source s and a destination t with n edges, and Q_{p_i} user traversals, which cover *at least* one edge of p_i. The **popularity score** $\omega(p_i)$ of p_i equals the sum of the individual popularity contributions of Q_{p_i} user traversals. We define $\omega(p_i)$ as follows:

$$\omega(p_i) = \sum_{i=1}^{Q_{p_i}} PC(q_i, p_i) \tag{2}$$

Here, $0 \leq PC(q_i, p_i) \leq 1$. When all edges of p_i are covered by q_i, $PC(q_i, p_i) = 1$. Conversely, when none of the edges of p_i is traversed by q_i, $PC(q_i, p_i) = 0$.

Equation 2 captures PC of all kinds of user traversals. The issue is to compute the value of $\omega(p_i)$ by considering all kinds of user traversals. First, for simplicity, we present a method for estimating $\omega(p_i)$ by considering only those user traversals, which cover all of the edges of the path. (This hypothetical case may not hold good in practice.) Second, we discuss a method for estimating $\omega(p_i)$ by considering the user traversals, which pertain to the traversal of a fixed number of edges in the path. Third, we present a method for the generalized case arising in real-world scenarios by considering all kinds of user traversals. Now let us discuss these three cases in detail.

Case 1 - User traversals which cover all edges of p_i: Given a path p_i and a user traversal q_i, if q_i traverses all edges of p_i, $PC(q_i, p_i) = 1$; otherwise, it is 0. Using Eq. 2, for Q_{p_i} user traversals, which cover all edges of p_i, $\omega(p_i) = Q_{p_i}$. Given two paths p_1 and p_2 between source s and destination t with the number of user traversals, which cover all edges of p_1 and p_2, being Q_{p_1} and Q_{p_2} respectively, $\omega(p_1) > \omega(p_2)$ iff $Q_{p_1} > Q_{p_2}$. This matches our intuitive understanding of the notion of popularity score, which depends on the number of user traversals of the path.

Case 2 - User traversals which cover exactly k edges of p_i: In practice, any given user may use only one or more edges of the path and then take a detour into other edges/paths. Case 1 does not capture this real-world scenario. In Case 1, if q_i covers all n edges of p_i, $PC(q_i, p_i)$ is 1. Now, if q_i covers only k edges of p_i, $PC(q_i, p_i) = W(k)$, where W is a function with k as the parameter. W can be any monotonically increasing function w.r.t. k. Thus, $W(k) \propto k \; and \; 0 \leq W(k) \leq 1$. If q_i does not cover any edge of p_i, $PC(q_i, p_i) = W(0) = 0$. (The k edges in p_i of a user traversal need not be continuous edges.)

Different functions can be selected for W. Intuitively, given two traversals, a traversal, which covers more edges of p_i contributes more to $\omega(p_i)$. Hence, W should be selected such that given user traversals q_1 and q_2, which cover k_1 and k_2 edges of p_i such that $k_1 < k_2$, $W(k_1)$ should be less than $W(k_2)$. Using Eq. 2, given Q_{p_i} user traversals, which cover only k edges in p_i, $\omega(p_i)$ equals ($\omega(p_i) = \sum_{i=1}^{Q_{p_i}} PC(q_i, p_i) = Q_{p_i} * W(k)$). Given two paths p_1 and p_2 between source s and destination t with the number of user traversals, which cover k edges of p_1 and p_2, being Q_{p_1} and Q_{p_2}, $\omega(p_1) > \omega(p_2)$ iff $\left(Q_{p_1} * W(k)\right) > \left(Q_{p_2} * W(k)\right)$.

Case 3 - All user traversals which cover p_i: We can divide Q_{p_i} user traversals into user traversals, which cover only one edge, only two edges and up to all n edges of the path. Let C_k represent the set of all combinations of edges in p_i of size k, while $Q_{p_i}(C_k)$ represents the number of user traversals, which traverse *only* C_k. Thus, the value of Q_{p_i} is $\sum_{k=1}^{n} Q_{p_i}(C_k)$. Each of these user traversals contribute to the popularity score of the path. Thus, using Eq. 2, we can compute $\omega(p_i)$ for the generalized real-world case as follows:

$$\omega(p_i) = \sum_{i=1}^{Q_{p_i}} PC(q_i, p_i) = \sum_{k=1}^{n} Q_{p_i}(C_k) * W(k) \tag{3}$$

Intuitively, given two paths, we consider that a path, which is covered by more user traversals with a larger combination of edges, is more *popular* than the path, which contains a relatively lower number of user traversals. Observe how this intuition of popularity is reflected in Eq. 3.

4 Diverse Popular Paths Query and Proposed Scheme

The *diverse popular path* (DPP) query is as follows. Given a set of user traversals in a road network G, source s, destination t and the user-specified diversity threshold value θ and a set SP of paths between s and t, the DPP query determines the ranked list L of diverse popular paths in SP (in descending order of popularity score) such that diversity threshold criterion is satisfied. (Recall the computation of popularity score in Eq. 3.) Here, any ties in path popularity scores are resolved arbitrarily. Furthermore, recall the computation of the diversity (DV) between any two given paths in Eq. 1 of Sect. 2. We deem the **diversity threshold criterion** to be satisfied for any two paths p_i and p_j if $DV(p_i, p_j) \geq \theta$.

4.1 Basic Idea

To process a given DPP query, we need to compute the popularity scores of paths in set SP of paths (using Eq. 3). However, this requires the frequencies of potential combinations of edges of user traversals, which traversed at least one edge of p_i. Hence, given a path p_i comprising n edges, we need to examine the frequencies of $2^n - 1$ combinations of these edges. Notably, for every path p_i, when a DPP query comes in, it would be prohibitively expensive to generate all combinations of edges and compute their frequencies in an *online* manner because each path may have a different set of edges. However, there is an opportunity here for *offline* pre-processing and extraction of the knowledge of frequency of edge combinations; we designate such knowledge as *patterns*. Thus, when a DPP query comes in, we can use the extracted patterns towards *efficiently* processing the DPP query *online*.

Our DPP query processing scheme is as follows. First, we convert the user traversals into a transactional dataset D. Next, we extract the knowledge of patterns from D by using any existing pattern mining algorithm. Then we compute the popularity score of each path of SP using Eq. 3, extract the set PP of popular paths and order the paths based on popularity score. Finally, we extract top-k DPP from PP such that the diversity threshold criterion is satisfied.

4.2 Proposed Scheme

The proposed scheme has three phases: (1) modeling of user traversal transactions ($UTTs$) (2) extraction of combinations of edges and their frequencies from $UTTs$ (3) extraction of diverse popular paths. Now we shall discuss each phase.

1. Modeling of user traversal transactions ($UTTs$): Here, the input is a set of user traversals (UTs). We convert each UT into one or more $UTTs$ between two nodes in a road network. We assume that a UT contains the details of user arrival time and departure time at each node. We discuss two options to form $UTTs$ from UTs. First, we specify a threshold time window t_w. Each UT is traversed from the starting node. All parts of a given UT occurring within t_w are considered to be one UTT. Second, whenever a user backtracks or visits one of the already visited nodes, the edges visited so far are considered as one UTT; the traversal process is then continued from the preceding node. For both cases, the same process is repeated till the last node to identify further $UTTs$.

2. Extraction of combinations of edges and their frequencies from $UTTs$: By considering $UTTs$ as transactions and edges as items, **the database of edge combinations ($DBEC$)** with support $\geq$ user-specified *support* threshold is extracted by the existing frequent pattern mining algorithms.

3. Extraction of diverse popular paths: Algorithm 1 depicts the computation of top-k diverse popular paths (DPP). We compute the popularity score $\omega(p_i)$ of a given path p_i in SP using Eq. 3. We compute $\omega(p_i)$ by extracting (from $DBEC$) the knowledge of the respective frequencies of UTTs of all edge

Algorithm 1. $DPP(SP, \theta, DBEC, W, k)$

Input: SP: set of paths; θ: diversity threshold; $DBEC$: Database of edge combinations; W: weight function; k: required number of paths;
Output: DPP: set of diverse popular paths;
Variables: PP: list of $<p_i, \omega(p_i)>$;

1: Initialize PP to null
2: **for** each $p_i \in SP$ **do** {
3: ps = Stores the popularity score computed using Equation 3
4: Insert $<p_i, ps>$ to PP }
5: Sort PP w.r.t. *ps* in the descending order
6: Intialize DPP to null
7: Remove p_i having the highest *ps* value from PP and insert in DPP
8: **while** PP $\neq$ null **do** {
9: Remove p_i having the highest *ps* value from PP and insert in DPP
10: for all $p_k \in DPP$ { if $(DV(p_i, p_k) \geq \theta)$ **then** add p_i to DPP }
11: **if** (size of (DPP) == k) **then break** }

combinations, which are formed with the edges of p_i. After computing the popularity scores for all paths in SP, we sort the paths in descending order based on their respective popularity scores into a list PP (see Line 5).

Now, from PP, we identify the list DPP, which contains the paths that have a minimum diversity with other paths in the list. Based on the order of popularity, the following procedure is followed until DPP contains the required number of the (top) k paths or until we have exhausted iterating over all of the paths in PP. A path p_i added to list DPP if $DV(p_i, p_j) \geq \theta$, where $p_j \in DPP$.

5 Performance Evaluation

We conducted the performance evaluation on an Intel i7 processor with 8 GB RAM running Ubuntu Linux. We used OpenStreetMap[1] for Beijing and did the implementation using Osmnx[2] framework. Experiments are conducted using Microsoft's *Geolife* real dataset [16]. This GPS trajectory dataset was collected in the Geolife project by 182 users during a period of over 5 years. The dataset contains 17,621 trajectories with a total distance of 1,292,951 km and a total duration of 50,176 h. These trajectories were recorded by different GPS loggers/phones with varied sampling rates. In this work, we considered 14,175 traversals in Beijing. First, we mapped each user trajectory to the road network using a map-matching tool Graphhopper[3]. Then we used nominatim[4] to obtain the corresponding sequence of edge IDs; we consider this sequence as a path.

[1] https://www.openstreetmap.org.
[2] https://osmnx.readthedocs.io/en/stable/.
[3] https://graphhopper.com/api/1/docs/map-matching/.
[4] https://nominatim.openstreetmap.org/.

We used the backtracking method for dividing user traversals into multiple user traversal transactions ($UTTs$). We deployed the FP-Growth algorithm [9] to obtain $DBEC$ from $UTTs$ with support greater than 300. To select the candidate set of paths between s and t, we introduce the notion of distance threshold δ. Let sl be the shortest path length between s and t. We extract all paths, whose length is less than $(sl + \delta * sl)$.

We set the default value of the distance threshold δ to 0.2 and its variations are 0.1, 0.3, 0.4 and 0.5. We conducted our experiments with three representative queries, where each query is a source and destination pair. To generate these queries, we randomly selected 10 queries each from areas of low, medium and high spatial densities such that the distance between each of the source and destination pairs is less than 1 km. Then, for each type of region (i.e., regions having low, medium and high spatial densities), we randomly selected one query from among these 10 queries. We shall henceforth refer to the queries from regions of high, medium and low spatial densities as $Q1$, $Q2$ and $Q3$ respectively.

Our performance metrics include average length (AL) of a set of paths, average popularity score (APS) of a set of paths, number of popular paths (NPP) and execution time (ET).

As reference, we used the implementation of our model for computing the popularity scores of paths (see Sect. 3) without using frequent pattern information. Given a transactional database with user traversal transactions, we traverse all of the transactions online to obtain the frequencies of each and every edge combination of each path. In this approach, we construct $DBEC$ for a specific path through online traversal. We shall henceforth refer to this reference scheme as the *online approach*. We refer to our proposed scheme as **DPP**.

Given a source and destination pair, for Q1, Q2, and Q3, we compare our proposed DPP scheme with the shortest paths and diverse shortest path approaches. We also compare the performance of DPP with the reference approach.

1. Performance of AL and APS: For $Q1$, $Q2$, and $Q3$, Fig. 1a depicts the results of AL of top-k shortest paths (SP), popular paths (PP), diverse shortest paths (DSP) and diverse popular paths (DPP). For three queries, it can be observed that AL of SP is less than AL of PP. Also, AL of DSP is less than

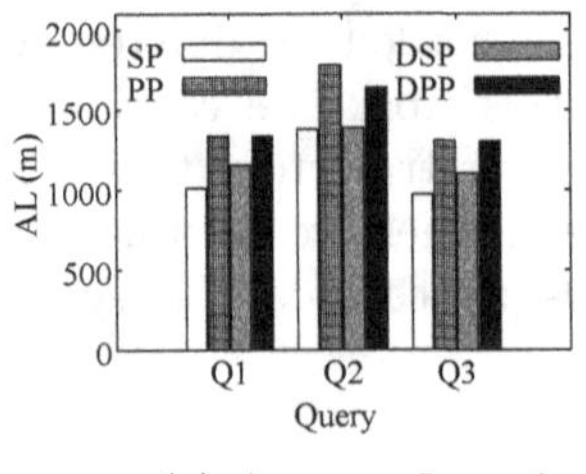

(a) Average Length

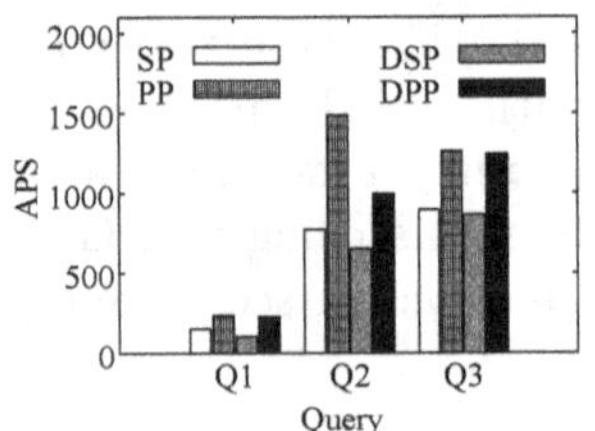

(b) Average Popularity Score

Fig. 1. Results on queries in regions of varying spatial densities

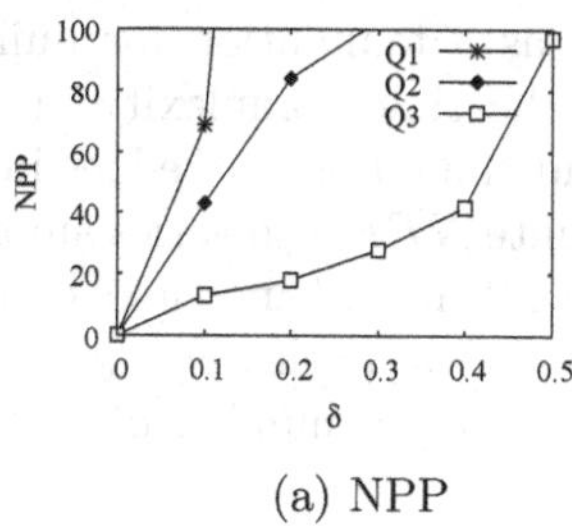

(a) NPP

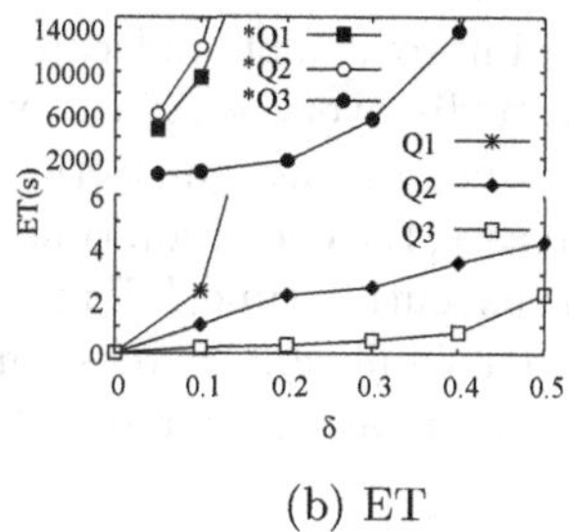

(b) ET

Fig. 2. Effect of variations in the distance threshold

AL of DPP. For $Q1$, $Q2$, and $Q3$, Fig. 1b shows the results of APS of top-k SP, PP, DSP and DPP. For three queries, it can be observed that APS of SP is less than APS of PP. Also, APS of DSP is less than APS of DPP.

It can be observed from Fig. 1 that the popular path is longer than the shortest path, and the diverse popular path is longer than the diverse shortest path. However, the additional distance is not significant. Hence, the results show that it is possible to obtain popular paths for a given source and destination by covering a small additional distance. The results also show that, independent of spatial density, it is possible to obtain popular paths for $Q1$, $Q2$, and $Q3$ as there are a reasonable number of user traversals in the order of thousands for each query. The popularity score is independent of spatial density w.r.t. roads because the proposed notion of popularity considers user traversals.

2. Effect of varying the distance threshold (δ): Figure 2 depicts the results of varying the value of δ. The experiment was conducted to determine the number of popular paths as we vary δ. The results indicate that the number of paths increases with δ; hence, the number NPP of popular paths also increases. Observe that for $Q1$ (which is a query in a dense region), 34 popular paths could be obtained even by adding a small δ (say 5%). Similarly, for $Q2$ and $Q3$, it is possible to obtain 22 and 6 popular paths respectively with a small δ of 5%. *The results are very encouraging because they indicate that users need not traverse any significant additional distance to obtain popular paths.* We believe that traversing a small additional albeit distance is a small price to pay for realizing the several potential benefits provided by popular paths.

Figure 2b depicts the execution of $Q1$, $Q2$, and $Q3$ of DPP w.r.t. the online approach. In the graph, the performance of the proposed offline approach is indicated by $Q1$, $Q2$ and $Q3$, while the performance of the online approach is indicated by $*Q1$, $*Q2$, and $*Q3$. Observe that the proposed approach improves the performance significantly over the online approach as we extract frequencies of each combination of edges in an offline manner using pattern mining techniques and we also employ a hashmap to *efficiently* search the patterns. On the other hand, in the online approach, the database of UTTs need to be scanned to extract frequency of combinations of each path in an online manner, which is computationally prohibitively intensive. As a result, DPP exhibits significant performance improvement and can be used as a foundation for building near real-time path discovery services. Notably, the time complexity of the online approach depends on the transaction size and the number of edges in the path.

As δ increases, the execution time to compute NPP also varies among $Q1$, $Q2$ and $Q3$. The execution time ET to extract NPP for $Q1$ increases exponentially with δ due to a significant increase in the number of paths. For $Q2$ and $Q3$, ET increases gradually as compared to $Q1$ due to a lower number of NPPs.

6 Conclusion

We have addressed the problem of ranking the input paths between a given source and destination pair based on popularity and diversity of paths. We have proposed a model to compute the popularity score of a given path by combining the effect of the number of user traversals and the number of edges of the path covered by each user traversal. We have also presented an efficient approach for computing the popularity score of a given path by modeling user traversals as transactions and deploying pattern mining techniques. Our performance study with a real dataset shows the effectiveness of the proposed scheme. We believe that the proposed approach would be beneficial in designing popular path discovery services as a complement to services for finding only the shortest paths.

References

1. Aggarwal, C.C., Han, J. (eds.): Frequent Pattern Mining. Springer, Cham (2014). https://doi.org/10.1007/978-3-319-07821-2
2. Chang, K.P., Wei, L.Y., Yeh, M.Y., Peng, W.C.: Discovering personalized routes from trajectories. In: Proceedings of ACM SIGSPATIAL, pp. 33–40 (2011)
3. Chen, Z., Shen, H.T., Zhou, X.: Discovering popular routes from trajectories. In: Proceedings of ICDE, pp. 900–911 (2011)
4. Chondrogiannis, T., Bouros, P., Gamper, J., Leser, U.: Alternative routing: k-shortest paths with limited overlap. In: Proceedings of ACM SIGSPATIAL (2015)
5. Chondrogiannis, T., Bouros, P., Gamper, J., Leser, U.: Exact and approximate algorithms for finding k-shortest paths with limited overlap. In: Proceedings of EDBT 2017, pp. 414–425 (2017)
6. Chondrogiannis, T., Gamper, J.: ParDiSP: a partition-based framework for distance and shortest path queries on road networks. In: Proceedings of MDM, vol. 1, pp. 242–251 (2016)
7. Dijkstra, E.W.: A note on two problems in connexion with graphs. Numer. Math. **1**(1), 269–271 (1959)
8. Feng, Z., Zhu, Y.: A survey on trajectory data mining: techniques and applications. IEEE Access **4**, 2056–2067 (2016)
9. Han, J., Pei, J., Yin, Y.: Mining frequent patterns without candidate generation. ACM SIGMOD Rec. **29**, 1–12 (2000)
10. Hershberger, J., Maxel, M., Suri, S.: Finding the k shortest simple paths: a new algorithm and its implementation. ACM Trans. Algorithms **3**(4), 45 (2007)
11. Martins, E.Q., Pascoal, M.M.: A new implementation of Yen's ranking loopless paths algorithm. Q. J. Belg. Fr. Ital. Oper. Res. Soc. **1**(2), 121–133 (2003)

12. Sommer, C.: Shortest-path queries in static networks. ACM Comput. Surv. (CSUR) **46**(4), 45 (2014)
13. Wei, L.Y., Chang, K.P., Peng, W.C.: Discovering pattern-aware routes from trajectories. Distrib. Parallel Databases **33**(2), 201–226 (2015)
14. Wei, L.Y., Zheng, Y., Peng, W.C.: Constructing popular routes from uncertain trajectories. In: Proceedings of ACM SIGKDD, pp. 195–203 (2012)
15. Yen, J.Y.: Finding the k shortest loopless paths in a network. Manage. Sci. **17**(11), 712–716 (1971)
16. Zheng, Y., Xie, X., Ma, W.Y.: Geolife: a collaborative social networking service among user, location and trajectory. IEEE Data Eng. Bull. **33**(2), 32–39 (2010)

Data Mining and Warehousing

Representative Sample Extraction from Web Data Streams

Michael Scriney[1(✉)], Congcong Xing[1], Andrew McCarren[2], and Mark Roantree[1]

[1] VistaMilk Research Centre, School of Computing, Dublin City University, Dublin, Ireland
{michael.scriney,mark.roantree}@dcu.ie, congcongxing2@mail.dcu.ie
[2] Insight Centre for Data Analytics, School of Computing, Dublin City University, Dublin, Ireland
Andrew.McCarren@dcu.ie

Abstract. Smart or digital city infrastructures facilitate both decision support and strategic planning with applications such as government services, healthcare, transport and traffic management. Generally, each service generates multiple data streams using different data models and structures. Thus, any form of analysis requires some form of extract-transform-load process normally associated with data warehousing to ensure proper cleaning and integration of heterogeneous datasets. In addition, data produced by these systems may be generated at a rate which cannot be captured completely using standard computing resources. In this paper, we present an ETL system for transport data coupled with a smart data acquisition methodology to extract a subset of data suitable for analysis.

1 Introduction

Data from the digital city can be generated from a wide variety of sources across many services such as housing, healthcare, transport, the environment etc. Similar to traditional web data, this data is available on-line, typically in either XML, JSON or CSV format, meaning it must be processed in some form before it can be properly used. These processing tasks include acquisition and interpretation, transformation, integration, and analysis or machine learning of data from sensors, devices, vehicles etc. There have been a number of approaches to building smart city applications [5] and to cluster or integrate query graphs on smart city data [10,12]. Decision makers who use smart city applications require OLAP type services to generate the datasets from a warehouse upon which they make their decision. OnLine Analytical Processing (OLAP) queries offer the richest form of data extraction with dimensional data providing powerful dimensional queries.

This research is supported by Science Foundation Ireland (SFI) and the Department of Agriculture, Food and Marine on behalf of the Government of Ireland under Grant Number 16/RC/3835.

S. Hartmann et al. (Eds.): DEXA 2019, LNCS 11706, pp. 341–351, 2019.
https://doi.org/10.1007/978-3-030-27615-7_26

The challenge for smart city researchers lies in incorporating web, sensor and streaming data to these datasets usually in RDF format [2]. Sources are often new, not always available, are not suited to integration, and use heterogeneous data models.

Motivation. DublinBus [4] provides a range of online services related to the running of their bus network for the city of Dublin. One of these services provides a real-time location for all buses on the network. Occasionally, times are provided for buses that do not exist. While these can be detected using an analysis of the real-time data streams, there is no indication as to why buses may appear and disappear from the system. In order to facilitate a deeper analysis, it requires a methodology to acquire and transform data into a usable format, and at sufficiently regular intervals so as not to lose any real-time information. This poses two problems. Firstly, the system requires considerable resources to ensure that all of the data, for the entire network of bus journeys is extracted. Secondly, sufficiently flexible OLAP functionality is required to manipulate datasets for the types of analyses required.

Contribution. In this research, we present a framework for manipulated transport data streams so as to create datasets for machine learning algorithms. Our data model represents one subset of a larger smart city application where our bus transport data can be integrated with other smart city datasets using selected attributes from time and geo dimensions. Our contribution is to tackle the problem of big data by using an algorithm to extract smaller representative samples. We also provide an extended OLAP which uses keyword extensions to present data so as to meet specific requirements for machine learning algorithms. A longer version of this paper can be found at [13].

Paper Structure. The paper is structured as follows: in Sect. 2, we provide a discussion on state of the art; in Sect. 3, we provide a description of our Extract-Transform-Load architecture which manages source transport data into our multidimensional model; in Sect. 4, we present an algorithm to optimise the acquisition of data; in Sect. 5, we present our evaluation and discussion; and finally in Sect. 6, we conclude the paper.

2 Related Research

In [11], the authors present a system to construct a data warehouse from user requirements. The system is given a domain ontology which is used to derive facts which are subsequently presented to the user. Once the user selects their desired fact, the data is extracted and presented to them. Similar to our approach, the authors attempt to identify facts without user interaction. However, the authors' approach requires a domain ontology to discover these facts while ours is static. In addition, our system provides several OLAP extensions to provide datasets in an analysis-ready format. In [1], the authors present an automatic ETL system. The authors use an ontology to provide semantic data integration, where that ontology is derived from a data warehouse schema and a lexicon. Similar to our approach, the authors then use clustering to determine the similarity between data sources while we use similarity matrices in order to determine the best possible subset of data to be extracted. The authors in [9] present an ETL approach which uses domain-specific modelling. Their own language (DSL) is used to model the different steps in the ETL process. A domain expert using this language designs the ETL framework which is subsequently deployed. Domain modelling is used to describe the data sources and these sources are then linked using an ontology. We also use an ETL process to extract data, however we also have a pre-processing step before data acquisition to determine a suitable subset of the data which can be acquired. In [8], the authors use RDF and OWL ontologies to create integrated data marts. Each data source in the mart is provided with an ontology which is used to construct an RDF version of the data. From this, an OWL definition of the mart uses RDF queries to extract the required data from each source. In contrast, our approach uses the original source format to extract dimensional data. Our system provides integration dimensions for integrated data marts. In addition, we provide OLAP extensions to convert the data when being queried from the data warehouse. In [3] the authors present a big data architecture for smart city data. The system is composed of a series of layers, starting from the data source to a presentation layer providing analysis applications. Similar to our approach the authors use transport as a use case and a custom ETL workflow to deliver a k-means analysis, while our approach presents a generic model for transport data coupled with a series of transport specific OLAP extensions. Finally, the authors in [7] present a smart city platform which allows a user to search and integrate smart city data to suit their end requirements. A user selects the data source(s) they wish to integrate and the data undergoes an ETL process to convert all source data into RDF triplets. The authors use RDF as their common data model while our approach uses a data model of our own design. Furthermore, our system strives to provide data which is suitable for analysis.

The state of the art in ETL and data warehousing shows that some form of common data model either on a data source layer or globally is required to provide holistic integration. We adopt this approach through the use of our data model. However, in limiting ourselves to a particular domain (i.e. transport) we are able to provide a set of domain specific operations to facilitate OLAP analysis.

3 Data Acquisition and Transformation

In this section, we begin with a description of the Extract-Transform-Load (ETL) process to acquire and transform raw data. We then describe our multi-dimensional (cube) data model, and provide a description of the data acquisition and transformation process.

3.1 System Architecture

In Fig. 1, we show the system workflow as a series of layers, each containing data in different formats, where layers are connected by data transformation processes. The *Source Data* layer is generally a web service providing an API to facilitate JSON or XML data extraction. In this instance, the DublinBus Real Time Passenger Information (RTPI) API generates XML elements based on bus routes or stops. The Data Acquisition process is effectively a purpose built wrapper to the *Data Staging Layer*, which manages JSON, XML or relational data. The Data Transformation processor is a generic process converting one of 3 data types to the system data model (the predefined smart city data mart defined in Sect. 3.2). Our customised OLAP (OnLine Analytical Processing) API described in Sect. 4, generates the datasets for analyses.

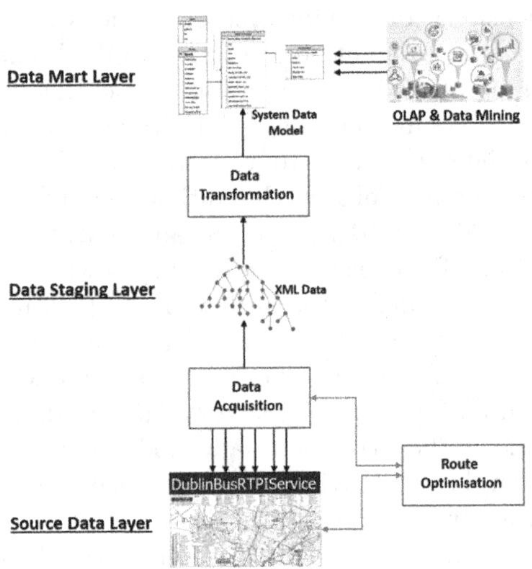

Fig. 1. System architecture

The cube structure for bus transport data is presented in Fig. 2, in the form of a star schema, centred around the `FtRealTimeData` fact. It comprises 5 dimensions: `Stop`, `RouteStops`, `Route`, `Time` and `Date`. The Stop, Date and Time dimensions are used for integration with other Cubes through the attributes *latitude*, *longitude*, *Date* and *Time*. These are underlined within Fig. 2.

3.2 Transport Data Model

It is important to note, while we use Dublin Bus for our case study, the model and algorithms presented are sufficiently generic for use in analysing any transport network provided the network can be modelled as a graph; where nodes indicate points of interest (e.g. bus stops, train stations) connected by edges. A

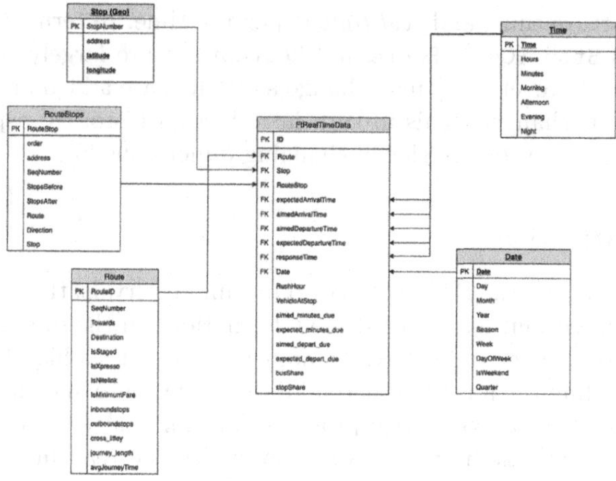

Fig. 2. Transport data schema

route indicates a specific path through the network. This coupled with the availability of a real-time data stream providing arrival times to points of interest could be used by our model for analysis. The `Stop` dimension describes every bus stop across the network with dimensional values for: stop identifier (StopID); a textual address for the stop (address); and the co-ordinates of the stop (latitude, longitude) effectively making this the `Geo` dimension. The `Route` dimension contains *static* information for bus routes such as the start and end points of the route and a number of flags indicating the type of route (e.g. `IsXpresso` indicating if the bus is a commuter route). The `RouteStops` dimension models the stops for each route. This is required in order to examine the route for a specific bus. Information on each stop and their related route lines (stops of one route, stop numbers, stop orders, destinations, directions, etc.) are stored in this dimension. The `RouteStops` dimension models the many to one relationship held between a stop and a route. It comprises: the stop number within that route `order`; a full textual address of the stop `address`; the number of stops preceding the current stop on the route `StopsBefore`; the number of stops after the current stop on the route `StopsAfter`; the route number `Route`; the direction of the route `Direction` (this can either be 'Inbound' or 'Outbound'); and the stop number itself `Stop`. The `Time` dimension is a `role-playing dimension` [6], providing a *many-to-1* relationship to the fact table `FtRealTimeData`. This is used where we wish to analyse across any of the time values extracted from real time information. It stores time in a 24-h format with each dimensional instance populated with additional flags such as `Morning`, which can be used in different types of analyses. The `Date` dimension is similar to the Time dimension, capturing data such as `Day`, `Week` and `DayOfWeek`. Finally, the `FtRealTimeData` fact table stores for each instance of real-time data: if it is rush hour; if the bus is currently at the stop; the schedule time the bus is due at the stop; the *real-time* due time; the scheduled departure time; and *real-time* departure time. In terms of persistence the dimensions `Stop (Geo)`, `Route` and `RouteStops` are largely static and are harvested once. However, they may change as Routes are added and removed or a specific route is changed. This in turn introduces a historical aspect to these dimensions, meaning they are slowly changing dimensions [6].

3.3 Data Extraction

Data is extracted from the RTPI service [4] in 2 min intervals: this is our limit for data acquisition without losing real-time information. Of the functions provided by the service, our system uses: `GetRoutes`, which provides a high level overview of a route, including the start and end address of the route and the name of the route; `GetStopDataByRoute`, which provides data for each stop, on a particular route.; and `GetRealTimeStopData`, which provides the realtime component of the system. It takes a stop number as input and returns a list of routes which are due at the stop. It provides 28 attributes for each bus due, most of which are repeated metadata values. The data used for our work are: the bus number due at the stop, a flag indicating whether or not the bus is in congested traffic, a flag indicating whether or not the bus is at the stop, the time of the server response and the scheduled and actual arrival times for the bus.

3.4 Data Transformation

The data is obtained from the service in XML format. For some attributes, the data is taken directly from the raw source and placed into the data warehouse. However, a number of attributes are provided by data enrichment. They are as follows: `Route.cross_liffey` indicates if the route travels across the city, this is manually annotated for each route. `Route.inboundstops` & `Route.outboundstops` are the number of stops on the route. `Route.journey_length` approximates the length of the route based on the distances between all stops on the route. `Route.avgJourneyTime` is populated from historical data denoting the travel time in minutes of the route. Finally, the attributes `busShare` and `stopShare` in the fact table denote how many buses are due at a particular stop within 5 min and the total number of routes which share this stop respectively.

4 Extracting Representative Samples

There are close to 4,000 bus stops in the Dublin Bus network, each updated every minute. In order to extract the entire network, this requires between 12,000 and 20,000 queries and requires up to 30 min on a typical quad-core workstation. If we limit each extraction to 2 min, this allows for up to 600 bus stops to be extracted. If we were to limit extraction to one minute intervals we would effectively halve the size of our dataset, limiting the number of complete routes that can be obtained, providing us with less data for analysis. Therefore, a strategy is required to maximise the extracted dataset. A small number of key steps are needed to obtain the *best* subset of the data. By *best*, we meant to maximise the amount of data acquired while minimising the number of queries (the set of bus stops). In order to maximise the dataset for the minimum queries, a route similarity matrix is constructed, which compares routes based on their number of *shared* stops. When constructing the similarity matrix, each route can be considered as a distinct set of stops which a bus must visit, in a specific order. The routes were then compared to each other by comparing the number of stops they share and the total number of stops for both routes. This produces the matrix shown in Table 1. We can use this matrix to determine how similar two routes are. For example, the routes `120` and `116` share 60% of their stops. Using this matrix, the algorithm then selects routes with the highest degree of interaction. The Data Acquisition command is specified in Definition 1 using a sql-like syntax. It informs the system how to acquire the required dataset based on routes or specific bus stops.

Table 1. Route similarity matrix

Route	1	11	116	118	120	122	123	13
1	1	0.45	0.43	0.65	0.54	0.46	0.48	0.34
11		1	0.57	0.76	0.63	0.56	0.58	0.44
116			1	0.85	0.6	0.52	0.55	0.38

In Definition 1, the acquisition expression uses 3 sub-expressions: INSERT INTO which must specify the name of the fact (always `FtRealTimeData` in our case); the optional WHERE clause is used to provide a set of routes or a set of stops; and the optional LIMIT clause contains either a time limit expressed in seconds or a bus stop limit. If the user does not specify a limit, the default limit, the number of stops which can be obtained in 2 min, will be used. Additionally, if no route or stop list is presented the system will choose two stops with the highest number of routes as default. Example 1 provides a sample query which would be used to populate the fact table for the routes 145, 42 and other routes which have the highest degree of interaction. Example 2 allows a user to select specific stops they wish to obtain. The system examines routes which pass through these stops, and then, using the route matrix selects the additional stops to be obtained.

Definition 1. *Data Acquisition Command*

```
INSERT INTO <FACT>
WHERE [ROUTE IN <ROUTELIST>] || [STOP IN <STOPLIST>]
LIMIT [TIME_LIMIT] || [STOP_LIMIT]
```

Example 1. Data acquisition for route

```
INSERT INTO FtRealTimeData
WHERE ROUTE IN (145,42)
```

Example 2. Data acquisition by Stops

```
INSERT INTO FtRealTimeData
WHERE STOP IN (4331, 7475)
```

The minimum value for the stop limit is 5, as that is the smallest number of stops which make up a particular route. The system starts by examining the list of potential routes which can be obtained. As each route can be considered as a set of stops, the system seeks to obtain the maximum number of stops by examining commonalities across routes using the route similarity matrices. When a route is selected, it is added to a list and the similarity matrix is consulted to add the next route with the highest similarity to those already selected. This process continues until either the upper limit of stops is reached, or that there are no more routes which can be added without passing the upper limit. For our experiments, the query used to gather the data used was based on the default values, effectively making our query `INSERT INTO FtRealTimeData`. Due to the nature of the bus network, it is inevitable that we will also obtain information on incomplete routes. In total, there are 10 full routes and 105 routes in total (95 route fragments).

5 Evaluation

5.1 OLAP Extensions

OLAP functionality (such as Slice, Dice and Pivot) can be used on the fact table to extract and view data for further analysis. However, due to the complexity of the data, these commands may not generate datasets suited to the required analyses to extract insights from the dataset. Because of this, we introduce three OLAP extensions which are used to produce analysis-ready datasets. These are: `series`, `lag` and `interval`. **Series:** If an analyst wishes to examine a particular route and direction a query, such as the one shown in Example 3 would be used.

Example 3. Sample SQL for route and direction

```
SELECT *
FROM FtRealTimeData join RouteStops
where Route='145' and Direction ='I'
ORDER BY Date, responseTime;
```

Output for query shown in Example 3

responseTime	Stop	aim_min_due
10:00 AM	7475	5
10:00 AM	4133	10
10:02 AM	7475	3

Series

responseTime	7475	4133
10:00AM	5	10
10:02AM	3	8

Lag

responseTime	7475	4133
2	-1	-1

Interval

responseTime	7475 - 4133
10:00AM	5
10:02AM	5

Fig. 3. OLAP extension examples

This would produce a list of all data grabbed from the real-time system based on the time they were scraped. For an individual bus, there are several measurements per individual stop on the route each taken at different times. Such data would need to be pivoted in order to be useful. To accomplish this we provide an OLAP extension called 'series' which produces a time-series representation of the data. This is non-trivial for the dataset, as it requires a list of all stops in order for a route be constructed. This command returns millions of records, a small sample of the output of the `Series` command can be seen in the `Series` table in Fig. 3. In this example we can see the data has been pivoted in order of stops on the route, with the stops occupying the columns, each row represents all data gathered at a specific time, and each cell occupies the measure `aimed_minutes_due`. As the series works based on a set of ordered stops the extension *requires* a `route` and `direction` as input. **Lag:** A common method used in time-series analysis is *lagging*. This is used to quantify a features rate of change with respect to time. In conjunction with the `series` keyword this can be used to produce a dataset showing the rate of change by time for all measurements. The time taken to execute a series query for 2,291,621 rows was 239 s. A sample output can be seen in the `Lag` table in Fig. 3. The time taken to execute a lag query for a series of size $133124 * 73$ was 3 s. **Interval:** The series keyword displays the progression of a bus over time, however an analyst may wish to examine the relationships between stops. This functionality is provided by the keyword **interval**. It can be considered similarly to lagging data in standard time series analysis. This keyword is used in conjunction with `series` to produce a dataset listing the intervals between stops. A sample of the output can be seen in the `Interval` table of Fig. 3. The time taken to execute an interval query for a series of size $133124 * 73$ was 51 s.

5.2 Data Collection and Consistency

The purpose of our approach is to collect a subset of big data which is suitable for analysis. To determine the feasibility of our approach we must satisfy two conditions: that our data collection process is *consistent*, and that collected data is suitable for analysis. **Data Collection:** We collected data for 30 days and collected 31,589,416 instances of real-time data (fact table records). Table 2 details the number of instances scraped per day. Looking at the `Count` column we can see that Wednesday and Sunday have the lowest number of instances (a limited number of routes run on Sunday). Wednesday has the lowest number of records at 3762178. This is because the real time system was unavailable one Wednesday. On average we obtain 1000000 records per day. Using this figure, we estimate, had the system been available that we would have obtained 47000000 records, bringing the figure for Wednesday in line with other weekdays. The minimum records count for Tuesday at 1777 can be explained by the fact that the system was first initialised on Tuesday.

Data Consistency: We have shown that we can collect data in a consistent manner, however the next step is to show that the data itself is *consistent*.

Table 2. Total data acquisition by weekday

Weekday	Count	Average	Max	Min	StDev
Monday	5104382	1701461	1783311	1589465	81958
Tuesday	4398690	1099673	1566792	1777	637228
Wednesday	3762178	1254059	1399659	990699	186570
Thursday	5252280	1050456	1501763	35223	558590
Friday	4577739	1144435	1539966	394970	464937
Saturday	4531959	1510653	1530066	1474605	25515
Sunday	3962188	1320729	1393328	1242274	61806

Missing data is expected as the system may go down, or provide inconsistencies. However, in aggregate, all data for an individual bus should be highly correlated. With this as our metric, we extracted all data for 4 full bus routes as a time series using the `series` keyword. A matrix for all correlations for individual buses on a route were then calculated for each bus; the results are shown in Table 3. `RouteID` is the identifier of the bus route, `Records` is the count of records obtained for the route. `mean`, `std`, `min` and `max` are the average, standard deviation, minimum and maximum correlation found respectively and finally the values `25%`, `50%` and `70%` denote the correlation for their percentiles. The difference in records is due to the differing lengths of the routes. As we can see, the data is highly correlated with all routes having >0.8 at the 50th percentile and >0.9 at the 75th. As is common with real world datasets, missing data is the cause for the minimum values. The missing data can manifest either due to the fact that we are obtaining data at 2 min intervals, or can be due to outages and anomalies within the real time system itself (the focus of our future work).

Table 3. Correlations for bus routes

RouteID	Records	Mean	Std	Min	25%	50%	75%	Max
1	2793896	0.77	0.37	−0.95	0.7	0.96	0.99	0.99
2	2136743	0.74	0.44	−0.91	0.73	0.97	0.99	0.99
3	129474	0.72	0.33	−0.447	0.54	0.88	0.98	0.99
4	114442	0.64	0.42	−0.675	0.34	0.85	0.98	0.99

6 Conclusions

As more smart city services come online, more information becomes available to make strategic decisions for our cities. The velocity and volume of this data may prove too large to manage with limited computing resources. In this work, we present an ETL system for modelling a real-time smart city service and a process which can be used to extract a usable dataset from a service which provides high-volume data at speeds which cannot be captured completely without an investment in computing resources. Our evaluation shows that the data is being captured at a consistent rate and that the datasets themselves are `consistent` and suitable for analysis. Our future work is to conduct a deeper analysis into detecting anomalies across the bus network where delays are caused by an event not captured in traditional urban data streams.

References

1. Bergamaschi, S., Guerra, F., Orsini, M., Sartori, C., Vincini, M.: A semantic approach to ETL technologies. Data Knowl. Eng. **70**, 717–731 (2011)
2. Cappellari, P., De Virgilio, R., Maccioni, A., Roantree, M.: A path-oriented RDF index for keyword search query processing. In: Hameurlain, A., Liddle, S.W., Schewe, K.-D., Zhou, X. (eds.) DEXA 2011. LNCS, vol. 6861, pp. 366–380. Springer, Heidelberg (2011). https://doi.org/10.1007/978-3-642-23091-2_31
3. Costa, C., Santos, M.Y.: Basis: a big data architecture for smart cities. In: SAI Computing Conference (SAI). IEEE (2016)
4. DublinBus RTPI Service WSDL site (2019). http://rtpi.dublinbus.ie/DublinBusRTPIService.asmx
5. Hernández-Muñoz, J.M., et al.: Smart cities at the forefront of the future Internet. In: Domingue, J., et al. (eds.) FIA 2011. LNCS, vol. 6656, pp. 447–462. Springer, Heidelberg (2011). https://doi.org/10.1007/978-3-642-20898-0_32
6. Kimball, R., Ross, M.: The Data Warehouse Toolkit, 2nd edn. Wiley, Hoboken (2002)
7. Nesi, P., Po, L., Viqueira, J.R.R., Trillo-Lado, R.: An integrated smart city platform. In: Szymański, J., Velegrakis, Y. (eds.) IKC 2017. LNCS, vol. 10546, pp. 171–176. Springer, Cham (2018). https://doi.org/10.1007/978-3-319-74497-1_17
8. Niinimaki, M., Niemi, T.: An ETL process for OLAP using RDF/OWL ontologies. J. Data Semant. XIII **5530**, 97 (2010)
9. Petrović, M., Vučković, M., Turajlić, N., Babarogić, S., Aničić, N., Marjanović, Z.: Automating ETL processes using the domain-specific modeling approach. Inf. Syst. e-Business Manage. **15**, 425–460 (2017)
10. Roantree, M., Liu, J.: A heuristic approach to selecting views for materialization. Softw. Pract. Exp. **44**(10), 1157 (2013)
11. Romero, O., Abelló, A.: A framework for multidimensional design of data warehouses from ontologies. Data Knowl. Eng. **69**, 1138–1157 (2010)
12. Scriney, M., O'Connor, M.F., Roantree, M.: Integrating online data for smart city data marts. In: Calì, A., Wood, P., Martin, N., Poulovassilis, A. (eds.) BICOD 2017. LNCS, vol. 10365, pp. 23–35. Springer, Cham (2017). https://doi.org/10.1007/978-3-319-60795-5_3
13. Scriney, M., Xing, C., McCarren, A., Roantree, M.: Using a similarity matrix to extract sample web data streams. Dublin City University Online Repository, Article 23435, pp. 1–15 (2019)

LogLInc: LoG Queries of Linked Open Data Investigator for Cube Design

Selma Khouri[1(✉)], Dihia Lanasri[1], Roaya Saidoune[1], Kamila Boudoukha[1], and Ladjel Bellatreche[2]

[1] Ecole nationale Supérieure d'Informatique (ESI), Algiers, Algeria
{s_khouri,ad_lanasri,r_saidoune,k_boudoukha}@esi.dz
[2] LIAS/ISAE-ENSMA, Futuroscope, Chasseneuil-du-Poitou, France
bellatreche@ensma.fr

Abstract. By avoiding the 'data not invented here' syndrome (NIH) (Data not invented here (NIH) syndrome is a mindset that consists in focusing solely on using data created inside the walls of a business (https://urlz.fr/9Yo9)), companies realized the benefit of including external sources in their data cube. In this context, Linked Open Data (LOD) is a promising external source that may contain valuable data and query-logs materializing the exploration of data by end users. Paradoxically, the dataset of this external source is structured whereas logs are "ugly", and in the case, they are turned into rich structured data, they will contribute to building valuable data cubes. In this paper, we claim that the NIH syndrome must be also considered for query-logs. As a consequence, we propose an approach that investigates the particularity of SPARQL query logs performed on the LOD and augmented by the LOD to discover multidimensional patterns when leveraging and enriching a data cube. To show the effectiveness of our approach, different scenarios are proposed and evaluated using DBpedia.

Keywords: Multidimensional modeling · Data leverage · LOD query-logs · LOD dataset

1 Introduction

A key challenge of the decision-making process for companies over these last years is to analyse different types of data sources that are increasingly heterogeneous, combining structured and unstructured data and that are often external to the company. Teasing insight from the massive universe of surrounding new data sources is the second era of data warehousing and data cubes, as claimed by Ralph Kimball [13]. With the proliferation of new external sources (social media, text, scientific data, etc.), a new process called *Data Leverage*[1], has been established for finding ways to use existing and external data assets for making

[1] https://www.ibm.com/downloads/cas/MQBM7GOW.

S. Hartmann et al. (Eds.): DEXA 2019, LNCS 11706, pp. 352–367, 2019.
https://doi.org/10.1007/978-3-030-27615-7_27

data-driven decisions. *Data enrichment* process is then conducted to augment data by combining internal and external sources.

In the context of data cube design, the Linked Open Data (LOD), supported by the growth of the Semantic Web community, is an example of representative external source alimenting the data cube of the company. The LOD is a set of design principles for sharing machine-readable data on the web for public administrations, business and citizens usage[2], using Semantic Web standards like RDF definition language and SPARQL query language. Both languages follow a graph-based representation. In this context, LOD datasets are seen as a typical data lake of open semantic data that can augment internal data of different operational systems; which is an emerging architecture that many organizations have to manage[3]. Recently, different studies considered this issue of enriching data cubes with LOD datasets [1,2,4,9]. On the one hand, the availability of open data lakes easily accessed by all companies decreases the value of this source if a substantial analysis is not conducted. On the other hand, these studies mainly focus on *Data enrichment* process and assume that the designer has a prior knowledge of the relevant concepts to extract from LOD datasets for enriching the cube. Looking for significant data from very large LOD datasets is a complex task conducted during *Data leverage* process. In our study, we propose an approach that fills this gap and that helps to discover relevant information to *leverage* data cubes from LOD spaces. The direct impact of this inclusion is the multidimensional (MD) modelling of the data cube. Thinking dimensionally (i.e. dividing the world into dimensions and facts) is cited among the best practices for Big Data analytics and new DW applications [13].

The LOD environment includes two main components: LOD datasets supplied by data providers and logs of SPARQL queries provided by data consumers. Interestingly, query-logs are gathered and published through initiatives like USEWOD[4] or the Linked SPARQL Queries Dataset (LSQ)[5]. One main characteristic of the LOD query-logs concerns their provenance, i.e., they are issued from various users with different profiles and skills (simple user, expert users, analyst users, etc.). Therefore, the pool of these logs may contain a variety of queries including naive queries as well as complex and federated queries over multiple datasets[6]. By deeply exploring existing studies incorporating LOD in data cube design, we figure out that they mainly use one component of LOD environment (the dataset) and *ignore SPARQL query-logs.* In this study, we present our approach called *LogLInc* that investigates the log queries and their LOD through three main scenarios. The traces of users reflected by their queries are first analyzed to identify relevant information. The dataset is reached to validate and to enrich the discovered patterns. In this way, we believe that our approach is complementary to existing ones, where logs are used as a summary-like view

[2] https://www.w3.org/DesignIssues/LinkedData.html.
[3] https://urlz.fr/9Yoc.
[4] http://usewod.org/.
[5] http://aksw.github.io/LSQ/.
[6] http://wp.sigmod.org/?p=2277.

and exploited for an initial investigation to identify relevant MD concepts. Note that matching MD patterns of the existing cube with the obtained ones is an important issue related to schema matching, which is out of the scope of this paper. Each scenario is formalized and evaluated through a set of experiments using DBpedia[7] which is one of the largest open base. Optimization techniques have been used for dealing with the important amount of graph-data generated for analyzing the logs. This paper is organized as follows: Sect. 2 discusses the related work. Section 3 provides the background. Section 4 details our approach and formalizes the proposed scenarios. Section 5 presents the experiments we conducted. Section 6 concludes the paper.

2 Related Work

In this section, we organize the main important works related to our proposal as follows:

1. Data Cubes & LOD. Different studies proposed to combine Data cubes and LOD, their propositions range a wide spectrum of issues, we can cite studies that brings OLAP analysis to LOD datasets like [7,8] which proposes vocabularies for publishing data cubes on the web or [21] who proposed ASPG to automatically generate OLAP queries from LOD. Other studies bring LOD datasets at different stages of the data cube life-cycle, mainly for: the definition of MD schemas (star or snowflake) from LOD datasets [2,9] or creation of a unified cube integrating internal DW & LOD dataset [17], definition of ETL processes for LOD integration [4,14] and OLAP querying of LOD datasets [10]. These studies only exploit LOD datasets and they assume prior knowledge on the relevant fragments (or at least the entries) of LOD to exploit. In our previous study [12], we explored LOD query-logs for data cube design. In this paper, we propose to exploit the synergy between query-logs and LOD datasets through different scenarios and to analyze their complementarity. From this perspective, our study can be seen as an approach that reconsiders mixed-driven approach from LOD vision (considering external dataset and external users' needs), according to the three established approaches for data cube design: demand-driven based on users' requirements, source-driven based on data sources and mixed approaches.

2. LOGs for Data Cubes. In the context of data cubes, different forms of logs have been used for resolving different issues like optimization [5] and recommendation [3]. The used logs are of different forms like XML query logs [22], SQL query logs [18], etc. Our study follows the lead and investigates an emerging type of logs (LOD query-logs). Existing studies exploit either operational or analytical logs [3]. In our study, the first scenario investigates the nature of LOD query-logs to identify the rate of analytical aggregate queries that are formulated by users.

[7] http://wiki.dbpedia.org/.

3. LOGs for LOD Analysis. Query logs of LOD are analyzed by some studies for different issues like visualization [16] or personalized ranking of RDF Data [15]. Other studies provide statistics related to graph structure of queries [6]. To the best of our knowledge, no study has been devoted to explore LOD query-logs for MD discovering.

3 Background

LOD dataset is published as a set of triples or statements, following the Resource Description Framework (RDF)[8] which is a W3C standard for knowledge representation and interchange. Each statement in RDF takes the form <s, p, o, g> such that in the graph label *g*, subject *s* has the predicate (i.e. property) *p*, and the value of that property is the object *o*. Valid statements take the following form: s ∈ (I ∪ B), p ∈ (I), o ∈ (I ∪ B ∪ L) such as I, B, L are disjoint sets of IRIs, blank nodes (non-distinguished variables), and literals. Sparql is the standard query language defined for the LOD. A SPARQL query example is illustrated in Fig. 1, it is formalized as a 3-tuple <Query Type, Graph Pattern, Solution modifier> [6]:

Fig. 1. Example of a SPARQL query

Query Type determines the output of the query. The different types are: Select, Ask, Describe and Construct.

Graph Pattern is the main part of the SPARQL query which contains patterns for matching a defined subgraph in the queried RDF graph. A basic graph pattern (GP) is a set of triple patterns that may contain query variables at the subject, predicate, or object position. Triple patterns are defined based on RDF language, as follows: Let V = ?x, ?y, ?z,... be a set of variables, disjoint from the sets I, B, and L. A triple pattern is an element of (I ∪ B ∪ V) × (I ∪ V) × (I ∪ B ∪ L ∪ V). A Sparql pattern is an expression that can be generated using triple patterns, property path expressions[9] or operators (And, Union, Opt and Filter and Graph). SPARQL[10] Patterns can be represented as graphs as illustrated in Fig. 2 (graph G1 presents the GP of query illustrated in Fig. 1).

Solution-modifier allows aggregation, grouping, sorting, duplicate removal, and returning a range of the multiset of mappings returned by the pattern.

[8] https://www.w3.org/RDF/.

[9] Property path expressions are negligible in our corpus, our study do not focus on these expressions.

[10] Details about the operations can be found in W3C recommandation https://urlz.fr/9Yqk.

4 Our Proposed Approach

Our approach discusses three main scenarios (illustrated in Fig. 3) for discovering MD patterns from LOD logs and dataset. The first scenario aims to identify the analytical aggregate queries among the whole query logs in order to analyse whether MD patterns can be directly identified from each one. The experiments showed that the analysed LOD query logs contains a very small amount of valid aggregate queries (cf. Sect. 5). This motivates us to consider a second scenario that analyses the graph patterns of all valid queries and consolidates the candidate patterns that share common resources. The third scenario extends the second one by adding a step consisting in extending the MD patterns identified from the logs by new concepts from the dataset. All scenarios share common steps: (i) a substantial cleaning from syntactic and semantic errors, (ii) the analysis of the queries/dataset using a graph-based approach, (iii) the annotation of identified concepts with MD patterns, based on the fact/dimension dichotomy using the universal concepts of: Fact, dimension, base dimension (i.e hierarchy level, according to the terminology used in [19]), measure, fact attribute, dimension attribute. Because there is no standard MD algebra, our work is based on study [18] that identified four mandatory constraints to be guaranteed to represent a combination of MD operators that we adapted to our context. These constraints are: (C1) Multidimensional compliance: the query must be a valid query template. A cleaning procedure is required at this step to identify correct queries. (C2) Star-schema: facts are related to levels by to-one relationships. (C3) Uniqueness: Every two different data instances retrieved by the query must be placed in different points of the MD space (either facts or dimensions). (C4) Orthogonality: The set of concepts that produce the MD space must be orthogonal. This constraint is used to determine that two different levels of one single

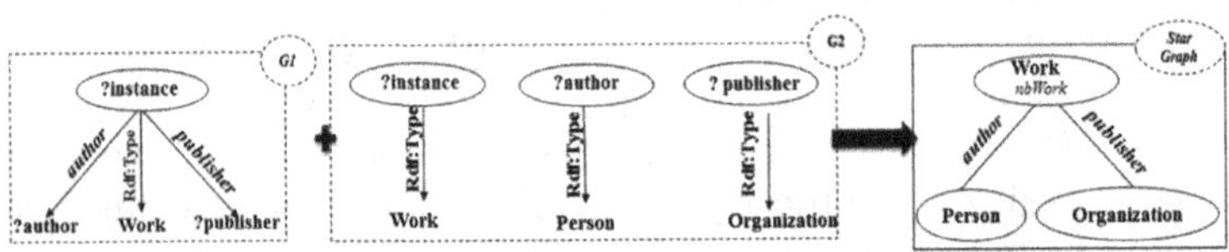

Fig. 2. Queries graph patterns

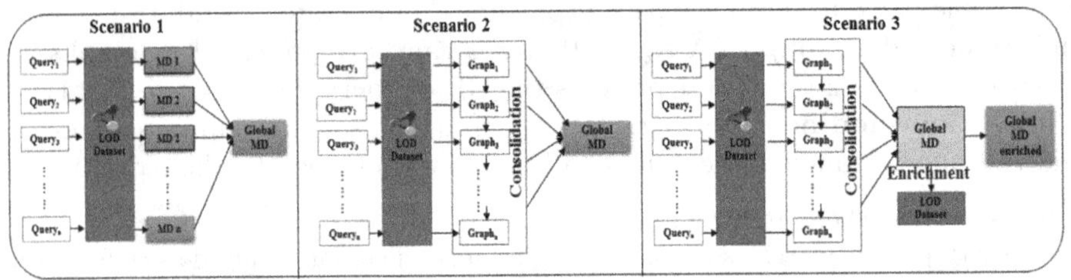

Fig. 3. Proposed scenarios for our approach

hierarchy cannot be related to the same fact. Algorithm 1 formalizes the common functions shared by all scenarios, which are detailed in scenario 1. Algorithm 2 formalizes the first scenario, and Algorithm 3 formalizes scenarios 2 and 3.

4.1 Scenario 1

This scenario consists in identifying aggregate queries , for which the identification of MD patterns is straightforward. The steps of the process are detailed in what follows: the first two steps and the two last ones are shared by all scenarios, while the graph construction step is specific to the first scenario.

Algorithm 1. Functions common to all scenarios

```
                         /*******Query-Logs Cleaning*******/
function LogCleaning(Query-logs files)          ▷ Output ValidFile: file of valid queries
    for all log file LF do                      ▷ Parsing the queries and eliminate extra strings
        CleanLog(LF)
    end for
    Remove_duplicate_Queries(ValidFile)
    for all resulting file RF do                ▷ Extract Select queries and check syntax and semantic
        ValidFile = ExtractValidQueries(RF)
    end for
    Filter Queries with aggregate functions      ▷ Only for Scenario 1
end function
                         /*******Graph Patterns Analysis*******/
function GP_Analysis(Queries of ValidFile)      ▷ Output QueryGraph: Graph of the query
    for all Query Q in ValidFile do Extract the graph patterns GP (in triple algebra)
        for all Node Subject Ns in GP of type Instance or variable do
            SubjectClass = rdfType(Ns)
            for all Property P of node Ns do
                Identify Type of P (Datatype or Object Property)
                ObjectClass= getRange(P)
                Add triple <Ns,rdfType,SubjectClass>
                Add triple <O,rdfType,ObjectClass>
            end for
        end for
        Construct the complete graph of the query
    end for
end function
          /*******MD Annotation: Labelling the Graphs with MD patterns*******/
function MD_Annotation(Graph)
        ▷ Input: <Vertex S, nodes O, Edges, InfoVertex, InfosNumVertex, InfoNode>; Output: MD patterns
    for all Graph G of vertex S do
        Label Vertex as Facts
        Label first level Nodes as Dimensions
        Label other Nodes as Base dimensions
        Label InfosVertex as Fact attributes
        Label InfosNumVertex as Measures
        Label InfoNode as Dimension attributes
    end for
    /**MD Annotation: Validation**/
    if Graphs G and G' share common Vertex(Fact f) and Node(Dimension d) then
        Choose Graph G of Vertex f
        Consolidate G' in G
    end if
    if Graph G has common nodes N1 (dimension) and N2 (base dimension) then
        Consolidate Path of N1 in Path of N2 (discard path of N1)
    end if
    Build MD schema according to identified concepts and graphs annotated.
end function
```

Algorithm 2. MD patterns exploration for scenario 1

```
Input: LF: Query-logs files
Output: MD Graph: a multidimentional Annotated Graph in the form of a star schema

File file:= LogCleaning(LF)                     ▷ use LogCleaning function defined in Algorithm1
for all query Q in file do
    Initialize Graph(G)          ▷ Construct graphs <Vertex S, nodes O, Edges, InfoVertex, InfoNodes>
    Graph QG:= GP_Analysis(Q)    ▷ QG: the Query Graph using GP_Analysis function presented in Algorithm1
    for all Var in Select Clause and agg_fct(Var) do              ▷ Var is a candidate attribute measure
        Identify path in QG formed by triple <S, P, Var>    ▷ Find resource class S of Var, candidate fact
        Vertex= Resource (S)  ▷ Resource is the function that returns the RDfType of S if Var is a variable or an instance
        if !G.contains(Vertex) then
            G.Vertex=Vertex              ▷ Fact are put as a vertex, dimensions are their related nodes
            G.Add_InfoNodes(Vertex, <Aggregate_function,var, rdftype(var))>)
        end if
        for all path in QG formed by triple <S, P, O> where (P is an ObjectProperty and P!=rdfType) do  ▷ resources O are added as candidate dimensions
            Node = Resource(O); G.Add_Node(Node); G.Add_Edge(Vertex,Node)
        end for
    end for
    for all Var in Group By clause do                      ▷ Var is a candidate dimension (or attribute)
        Identify path in QG formed by triple <S, P, Var> where (P is an ObjectProperty and P!=rdfType)  ▷ Find resource class S of Var, candidate fact
        Vertex= Resource(S)
        Node= Resource(Var)
        if !G.contains(Vertex) then
            G.Add_Vertex(Vertex); G.Add_Node(Node); G.Add_Edge(Vertex,Node)
        end if
        Identify path in QG formed by triple <Vertex, P, O> where (P is a datatype Property)
        G.Add_InfoNodes(Vertex, O)
    end for
    MD_Annotation(G)                          ▷ Basing on MD_Annotation function defined in Algorithm1
end for
```

Query-Logs Cleaning. Figure 4 illustrates one query of the query-logs. The query-logs cleaning includes the following steps (*Algorithm* 1 - *Lines 2–11*):

(i) *Parse Query-logs*: the query logs are presented as raw sets where URIs are encoded in UTF-8 format, a parsing using HTML parser is required to obtain a string query in a understandable format.

(ii) *Extract queries*: consists in extracting the query string starting just after 'query=' to 'HTTP/' containing the body of the query. Non-relevant information (IP address, date-time, etc) are discarded.

(iii) *Keep SELECT queries*: we keep just SELECT queries or SELECT sub-queries in CONSTRUCT queries, reflecting the user requirements in term of data analysis. (our process doesn't consider ASK and Describe queries)

(vi) *Check the query syntactically*: the correctness of each query is verified in order to meet constraint C1. The queries are verified using a regular expressions in order to check their syntactic correctness. Rules are defined in order to enrich incomplete queries (eg. incomplete brackets, operators, etc.). The semantic correctness of the query is verified if it can be executed on the dataset using a Sparql endpoint[11].

(v) *Eliminate duplicate queries*: aims to eliminate similar (at 100%) queries.

(iv) *Filter aggregate queries*: this step is executed only for scenario 1. It identifies aggregate queries using the template provided in [11]: Given variables var_i,

[11] A Sparql endpoint is an HTTP-based query service that executes SPARQL queries over the linked dataset. eg. http://dbpedia.org/sparql.

Algorithm 3. MD patterns exploration for scenario 2 & 3

```
Input: LF: Query-logs files
Output: MD Graph: a multidimentional Annotated Graph in the form of a star schema
File file:= LogCleaning(LF)                          ▷ use LogCleaning function defined in Algorithm1
Initialize listGraphs        ▷ Construct graphs <Vertex S, nodes O, Edges, InfoVertex, InfoNodes>
for all Query Q in file do
    Graph QG:= GP_Analysis(Q)                                             ▷ QG: the Query Graph
    Add(QG) to listGraphs
end for
/**Consolidation of graphs sharing common vertexes**/
Initialize graphsMap: Map <String, Graph> ▷ Create Map <key, graph> where key is the common vertexes of
the graphs
for all Graph G in listGraphs do
    if !exists(<G.getSubject, G>) in graphsMap then
        Put (<G.getSubject, G>) in graphsMap
    else
        Merge (graphsMap.get(G.getSubject), G)
    end if
end for
/**Start merging graphs**/
for all <subject, graph> in graphsMap do
    for all node N in graph different form subject do
        if g thenraphsMap contains <subj, graph'> where subj.equals(N)
            Merge (graph, graph')
        end if
    end for
end for
/**Form a graph of structure<Vertex S, nodes O, InfoNodes, InfoNodesNum, Info Vertex, Edges>**/
for all Path P<S,P,O> in graph G of graphsMap do
    if P is a datatype Property then
        Delete edge P; G.Add_InfoNodes(S, O);                             ▷ S can be a vertex or a node
    end if
    if P=rdfType then
        Replace S by its resource O then Delete P
    end if
end for
for all graph G of graphsMap do
    MD_Annotation(G)                               ▷ Basing on MD_Annotation function defined in Algorithm1
end for
    /*******This block is specific to Scenario 3: exploring the dataset to enrich the MD graph identified from
                                                  query- Logs*******/
for all Node/Vertex S in graph G of graphsMap do
    Identify O in the dataset such that <S,P,O>
    if !G.contains(O) then
        if P.objectProperty then
            G.Add_Node(O); G.Add_Edge(P)
        else
            G.Add_InfoNodes(S,O)
        end if
    end if
    Explore (rdfs:subClassOf) paths to enrich nodes by inherited properties in the same way as above.
end for
```

aggregate functions agg_fct (such that agg_fct $\in$ {*min, max, count, sum, average*}) and optional *Group By*; an aggregate query is defined in this form:

$\texttt{Select ?var}_1\texttt{, ?var}_2\texttt{, ..., ?var}_n\texttt{, AggFct}_1\texttt{(?var}_x\texttt{1), ..., AggFct}_m\texttt{(?var}_x m\texttt{)}$
$\texttt{Where\{...\} Group By ?var}_1\texttt{, ?var}_2\texttt{, ..., ?var}_n$

For example, the query of Fig. 1 illustrates an aggregate query.

Graph Patterns Analysis. This is the core step of our solution and consists in exploring and analyzing the graph patterns (GP) of queries retained (*Algorithm* 1 - *Lines 12–25*). For discovering MD patterns, each GP in WHERE clause have to be explored first, by extracting the set of triple patterns algebra in the form <Subject(S), Predicate(P), Object(O)> (using Jena[12] API). The GP can contain variables (?var) and instances, their related concepts and properties have to be identified. Two cases are possible: if an (rdf:Type) property is already defined in the query for a given variable or instance, we consider it as is. Otherwise, a triple pattern of the form <?var rdf:Type ?class> is added to the query, for every unknown variable or instance; to be executed in the SPARQL endpoint in order to get the related concepts from the dataset. The type of predicates P is also identified using Jena API (owl:DatatypeProperty or owl:ObjectProperty). A graph G considering all these triples can be constructed for each query as illustrated in Fig. 2 (graph G2).

For instance, the process generates these statements from the aggregate query illustrated in Fig. 1; GP= {(?instance rdf:type <http://dbpedia.org/ontology/Work>), (?instance <http://dbpedia.org/ontology/author> ?author), (?instance <http://dbpedia.org/ontology/publisher> ?publisher), (?publisher rdf:type ?sub1), (?author rdf:type ?sub2)}

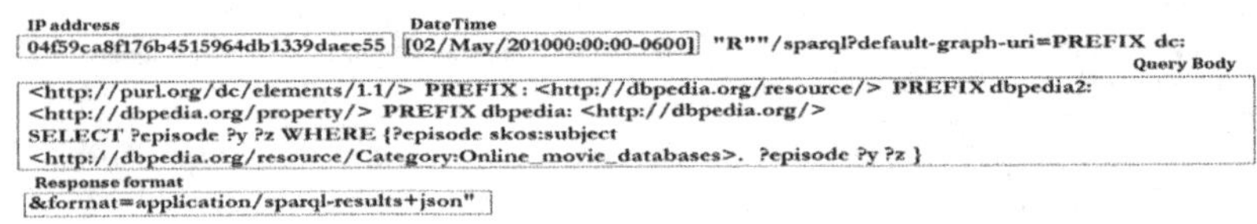

Fig. 4. One entry in the query-log

Graph Construction. This step is specific to Scenario 1. The algorithm (*Algorithm* 2) identifies the MD patterns based on the template of aggregate queries presented previously. The following process is executed for each query:

(i) Explore Aggregate functions (*Algorithm* 2 - *Lines 5–15*): according to the aggregate query template, variables that are associated with aggregate functions are considered as measures. Consequently, the variable (Var) associated to the aggregate function is an attribute and its domain concept have to be identified. For doing this, the process considers the path *Path* from S to Var (identified by the triple <S, P, Var>) in the graph G generated during the previous step, and looks for the resource S (Type_S) (identified by the path rdf:type in the previous step). The resource of Var (Type_Var) is also identified in the same way. A new graph G_{MD} is constructed having (Type_S) as a vertex, considered as a fact. (Type_Var), its data-type and the aggregation function applied are added as additional information to the vertex *S*. An exception is made for variables of type (Date) which cannot be considered as measures. The other paths from S to O (identified by the triples <S, P', O>) where P' is of type ObjectProperty (resp. Datatype Property) are also identified, a new node (O) is added to graph G_{MD}; they represent candidate dimensions (resp. attributes of the fact S).

[12] https://jena.apache.org/documentation/ontology/.

To carry on with our example 4.1, the execution of the process with the new GP generates these statements GP'= {(?instance rdf:type <http://dbpedia.org /ontology/Work>), (?instance <http://dbpedia.org/ontology/author> ?au thor), (?instance <http://dbpedia.org/ontology/publisher> ?publisher), (?publisher rdf:type DBpedia:organization), (?author rdf:type DBpedia:person)} From this, the process considers *nbWorks* as a measure , and *Work* as a Fact.

(ii) Explore Group by clause (Algorithm 2 - *Lines 16–25)*: if the query contains a group by clause (*Group By* (Var)), the same process is executed where the paths P (formed by the triples <S, P, Var>) are identified, resources (Type_S and Type_Var) are also identified. The resource (Type_Var) is considered as a candidate dimension, and the resource (Type_S) linked to (Type_Var) by P of type object property is considered as the related fact. Graph G_{MD} is enriched as follows: if (Type_S) is already created as a vertex, (Type_Var) is added as a node; else, the two nodes (Type_S) and (Type_Var), and the edge (from (Type_S) to (Type_Var)) are created. Continuing with our example, *Organization* and *Person* are considered as dimensions linked to *Work* fact.

(iii) Explore the rest of graph G: Finally, if the vertex or the nodes are related to other resources (using a data-type property) in G, these attributes are added as information in the corresponding nodes in G_{MD}. The obtained graphs G_{MD} are formalized as follows: < <Vertex S, InfoVertex, InfoVertexNum>, <Nodes O, InfoNodesNum, InfoNodes>, Edges>, such that each graph G_{MD} contains one vertex S and its information (InfoVertexNum for numerical attributes, InfoVertex for categorical ones), a set of nodes that may contain information (InfoNodesNum for numerical attributes, InfoNodes for categorical attributes) and a set of edges. For the sake of simplicity, the algorithms are formalized using only InfoNodes (numerical and categorical) without distinction, but the implementation distinguishes these types. Figure 2 illustrates the obtained graphs from the aggregate query in the example (Fig. 1).

MD Annotation. Before identifying MD patterns, some validation steps are required. The algorithm (*Algorithm* 1 - *Lines 26–34*) ignores graphs G_{MD} containing less than three nodes because they are not rich enough to reflect an MD pattern. The set of vertexes without numerical attributes have to be checked by the designer. The annotation process associates to each node its MD label assuring the form of Star schema (constraint C2) : *Fact* label to Vertex, measure label to InfosVertexNum, *Fact attribute* label to InfoVertex. The nodes that are directly related to the vertex are labeled as *dimensions*, the other related nodes of the hierarchy are labeled *base dimensions*. The InfoNodes of dimensions and base dimensions are labeled as *dimension attributes.* The resuted graph from the query is: {(nbWork annotatedAs measure),(Work annotatedAs fact),(Organization annotatedAs dimension), (Person annotatedAs dimension),(........)}

Validation. This step checks special cases (*Algorithm* 1-*Lines 35–43*): (i) if a node is identified as a fact in graph G1 and as a dimension in graph G2, constraint (C3) is not respected. This case should be validated by the designer, either by retaining the fact or by retaining the dimensions and the base classes.

In our experimentation, we gave priority for facts. (ii) In the same graph G, if a node is identified as a dimension in path P1 and as a base dimension class in path P2, constraint (C4) is not respected. In this case, and because P2 includes P1, path P1 is rejected and path P2 is retained.

4.2 Scenario 2

As we will detail in section experiments (*cf. Sect.* 5), scenario 1 did not provide a high amount of aggregate queries. Thus, we decided to explore another scenario that considers the *interactions* between the queries provided by users, for discovering MD patterns. The second scenario analyses all the valid queries, constructs the graphs incrementally and consolidates graphs sharing the same vertex for all the queries. MD patterns are explored based on the graphs obtained. As explained previously, the first two steps presented for the first scenario are the same for this scenario (cleaning and graph patterns analysis). The new steps are the following:

Graph Consolidation. The previous step provided as a result the triple algebra of each graph pattern and identified the related resources (concepts) of the variables and instances used. This step considers the triple patterns that are joined in the queries (basing on join operator). According to how joins connect triple patterns, study [20] identified four types of graphs that can be constructed from triple patterns, among them the *Star* structure which has multiple outgoing links but no incoming links, which corresponds to constraint (C2). Consequently, our task for this step is to discover star graphs from the queries, which is achieved as follows:

(i) For each triple pattern <subject, predicate, object> in the queries, a star graph (which is actually a DAG) is defined having the subject of the triple as a vertex (*Algorithm* 3 - *Lines 7–15*). Note that we will use the terms (subject, predicate and object), but we refer to the resources of these terms identified in the previous step. The edges linking the vertex to other nodes are constructed by analyzing the predicate of the triple pattern. Two scenarios are possible: the predicate is either a data-type role (i.e. attribute), an object property role (i.e. relationship), or a blank node. If the predicate is a real node, the type of resource is analyzed (relationship or attribute). If the predicate is a data type property (numerical or categorical), it is added as additional information in the vertex. If the predicate has type ObjectProperty, the multiplicity of the predicate is then analyzed to check if it is a to-one relationship (constraint (C2)). This is checked using owl#FunctionalProperty constraint: *ASK* {*predicate rdf:type owl:FunctionalProperty.*}. In this case, a node defining the object of the triple is added to the graph, then it is linked to the vertex by an edge (labeled by the predicate name).

(ii) The graphs are consolidated incrementally for the different triples of the queries (*Algorithm* 3 - *Lines 16–32*). The set of graphs sharing common vertex are consolidated incrementally to form one single graph. Note that the consolidation can create different paths from a single node in the obtained graph. The obtained graphs G are formalized as described in the previous scenario (*Algorithm* 3 - *Lines 33–35*): < <Vertex S, InfoVertexNum, InfoVertex>, <Nodes O, InfoNodes>, Edges>. In Fig. 5, we illustrate the obtained graphs from different queries after consolidation. The step (MD Annotation and Validation) are proceeded as explained in the first scenario (*Algorithm* 3 - *Lines 43–45*). From an implementation viewpoint, we used Jena TDB[13] to manage and query the graphs generated. An illustration of some star graphs obtained from DBpedia is available on: https://urlz.fr/9Yo1.

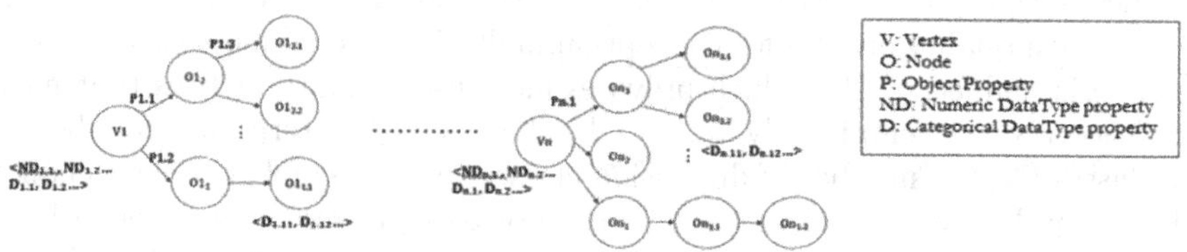

Fig. 5. Graphs of queries consolidated

4.3 Scenario 3

This scenario investigates more deeply the interactions between the query-logs and the LOD dataset (*Algorithm* 3 - *Lines 36–46*). This scenario executes the same steps of scenario 2 and adds a new step: the enrichment of the Global MD Schema with new properties and concepts explored from the dataset. The process is executed for vertexes and for nodes levels: for each vertex/node (S) in the consolidated graph, we look for its associated concept (O) in the LOD knowledge base, and we extract its related triples <S, P, O> by executing the query { *Select ?P ?O where S ?P ?O* }. For each obtained object (O): if (O) doesn't exist in the consolidated graph, it is added to the MD schema after exploring the type of the property P (either data-type or object property). These resources (O) will enrich the MD schema by new attributes or by new concepts labeled as dimensions (if (O) is related to a vertex), or as base dimensions (if (O) is related to a node). (ii) During the analysis, we also consider the property (rdfs:subClassOf) defining the inheritance relationship between parent and child concepts. When finding the nodes of the graph defined as child concepts, we enrich these nodes with the properties of the parent concept. Note that existing studies like [9] provide a more complex process to enrich an existing data cube from LOD dataset, and can be more useful. Our goal from this scenario is to simply illustrate the complementarity between LOD components (logs and datasets).

---

[13] https://jena.apache.org/documentation/tdb/.

5 Results

The experiments we conducted analyse the logs obtained from LSQ containing queries formulated against dbpedia-3.5.1 (log file of *3.193.672* queries). The algorithms are developed in Java and Scala. In order to manage the big volume of graphs created and consolidated from the log analysis (scenario 2, $\sim$ 40 Go of graphs managed in Jena TDB), we used the following optimization programming techniques: (i) loops' parallelism using threads (for independent steps), (ii) Scala functional programming for better parallelism and for its data structures adapted to big data volume. This experiment was performed on a machine OS Windows 10, 128 Go of RAM, Processor Intel(R) Xeon(R) CPU E5-2673 V4 @ 2.30 GHz.

Since our proposed approach generates MD patterns, we first analysed studies proposing quality criteria for conceptual multidimensional models. We follow the criteria defined by [19] which provides most used quality metrics that have been theorically & empirically validated. The selected criteria are: number of fact classes (NFC), number of dimension classes (NDC), number of base classes (NBC), Total number of classes (NC), ratio of base classes i.e. number of base classes per dimension class (RBC), number of fact attributes of the fact classes (NAFC), number of dimension attributes of the dimension classes (NADC), number of attributes of the base classes (NABC), number of hierarchy relationships (NH), maximum depth of the hierarchy relationships (DHP), ratio of attributes of the model i.e. number of fact attributes divided by the number of dimensions and its attributes (RSA). Note that there is no established thresholds defined for each criterion, since it depends on the specific context and model measured, but the authors state that the metrics can be used during the modeling phase in order to compare different design options. This motivates us to consider these metrics to compare the diffrent scenarios and choose the most appropriate model. After all the cleaning steps, 1.351.281 queries were valid. Considering scenario 1, only 83 distinct queries contain aggregate functions, 4.3% of queries presented valid aggregate queries, many of these queries were simply used for counting the number of lines of the query result. The process analyzed at the final step, 3.84% queries. This explains the low rate of MD patterns identified (see Table 1 line 1, a (.) delimits every three digits).

Table 1. Total result of the proposed scenarios

Results	NFC	NDC	NBC	NC	RBC	NAFC	NADC	NABC	NH	DHP	RSA
Scenario 1	5	3	0	8	0	8	3	0	0	0	2,66
Scenario 2	18.889	433.362	486.710	938.961	1,12	18.889	708	0	39	15	0,04
Scenario 3	18.889	433.420	486.738	939.047	1,12	5.305	3.900.966	1.460.214	39	15	0,001

Table 2. Results of scenario 2

Results	*Album*	*Book*	*Game*	*OlympicSport*	*University*	*School*	*Company*	*Artist*	Average
NFC	1.255	114	263	155	236	300	220	135	334,75
NDC	10.577	1.862	4.330	5.457	6.341	5.205	9.095	4.148	5.876,88
NBC	4.828	685	1.203	1.473	5.420	1.772	10.056	5.169	3.825,75
NC	16.660	2.661	5.796	7.085	11.997	7.277	19.371	9.452	10.037,37
RBC	0,46	0,37	0,28	0,27	0,85	0,34	1,11	1,25	0,61
NAFC	1.255	114	263	155	236	300	220	135	334,75
NADC	11	3	5	12	15	10	25	12	11,625
NH	26	25	16	6	39	4	2	11	16,13
DHP	13	14	10	5	12	15	4	7	10
RSA	0,12	0,06	0,06	0,03	0,04	0,06	0,02	0,03	0,05
T. Triples (From logs)	3.781	3.782	3.829	1.719	6.689	29.801	1.916	3.829	12.553
T. Triples (From Dataset)	1.390.773	5.089.434	1.329.008	560.608	1.642.707	791.128	2.605.764	3.834.305	2.155.465,875

This result convinced us to propose and evaluate scenario 2 that considers the interactions between all the valid queries, which actually provided better results (see Table 1 line Scenario 2). Because DBpedia is a generalist knowledge base that covers many domains, we filtered the results according to some topics given as inputs (before the consolidation step) The last column shows the average values for each criterion. We notice that the number of attributes of the base classes (NABC) is null and the number of attributes identified of all classes is low, this is explained by the low rate of attributes in the formulation of queries. The two last lines of the table refer to the total amount of triples retrieved from the graph patterns of the queries after the cleaning step (reflecting volume of data to consider before MD pattern exploration) *Vs* the volume to consider from only the dataset (without considering the logs). For this last line, we used DBpedia lookup[14] and a script that calculates the related concepts of each topic in terms of triples. The results show the complexity of dealing with a large volume of data when considering only the dataset (millions of triples), and illustrate the relevance of considering the log for data leverage process, to identify investigation entries in order to help the CDO or BI designer. To sum up, this scenario identified an important rate of star graphs that have been discovered from the behavioral information of users queries. Scenario 3 has been proposed to study the complementarity between logs and the dataset. As noticed in Table 1 (the table shows the total results of all-star graphs identified of the whole log, without filtering by topic), scenario 3 augments the results of MD patterns (especially for the number of attributes of fact, base and dimension classes). This is because the scenario considers the information provided by the dataset once the main entries are identified from the logs. These entries provide to designers investigation points of MD knowledge they can exploit during data enrichment of the cube, and allows them to decide what to keep according to the decisions makers requirements. These results convinced us about the relevance of scenario 3 for

[14] https://wiki.dbpedia.org/lookup.

considering the dataset as a complementary source of value. Furthermore, These MD patterns have been presented to a business intelligence (BI) expert who provided a semantic validation by regrouping similar concepts. Further investigation should be provided to automate this step. Examples of the MD patterns identified for topic "School" are provided in this link: https://urlz.fr/9Yo5. These results show the synergy between LOD components in providing added-value for an internal data cube. Thus, the objective of our study is achieved because the obtained results can help BI designers to select what they need before attacking the whole huge complex dataset thanks to the invesitagtion points provided.

6 Conclusion

Finding value in big amounts of available data is one of the main challenges brought by the big data era. Different actors join their efforts for *data leverage* process consisting in finding ways to augment data by combining internal and external data. Our study focuses on data cube leverage by using external LOD environment. Two main contributions are provided by our study: exploit the two components of LOD (LOD query logs and datasets) and analyzing the synergy between these components, in order to identify basic concepts that can generate MD knowledge for data leverage. Our approach proposes three scenarios for analyzing the LOD (logs and dataset). Dbpedia portal has been used to evaluate the process. our solution is not limited to LOD, but it can be applied to any Knowledge graph with it's query logs gathered from its asociated SPARQL endpoint.The main perspectives of this study are: a further investigation on logs provenance and users profiles for enriching the data cube, integration of discovered MD patterns with internal DW according to companies' business (data enrichement process) and working on the possibility to generalize our approach to other types of external sources like Social data and natural language which may be more complicated.

References

1. Abelló, A., et al.: Using semantic web technologies for exploratory OLAP: a survey. IEEE Trans. Knowl. Data Eng. **27**(2), 571–588 (2015)
2. Abelló Gamazo, A., Gallinucci, E., Golfarelli, M., Rizzi Bach, S., Romero Moral, Ó.: Towards exploratory OLAP on linked data. In: 2016 24th Italian Symposium on Advanced Database Systems, SEBD 2016, Italy, June 2016, pp. 86–93 (2016)
3. Aligon, J., Gallinucci, E., Golfarelli, M., Marcel, P., Rizzi, S.: A collaborative filtering approach for recommending olap sessions. DSS **69**, 20–30 (2015)
4. Baldacci, L., Golfarelli, M., Graziani, S., Rizzi, S.: QETL: an approach to on-demand etl from non-owned data sources. DKE **112**, 17–37 (2017)
5. Bonchi, F., et al.: Web log data warehousing and mining for intelligent web caching. Data Knowl. Eng. **39**(2), 165–189 (2001)
6. Bonifati, A., Martens, W., Timm, T.: An analytical study of large SPARQL query logs. Proc. VLDB Endowment **11**(2), 149–161 (2017)

7. Cyganiak, R., Reynolds, D., Tennison, J.: The RDF Data Cube Vocabulary. World Wide Web Consortium, Cambridge (2014)
8. Etcheverry, L., Vaisman, A.A.: QB4OLAP: a new vocabulary for OLAP cubes on the semantic web. In: Proceedings of COLD (2012)
9. Gallinucci, E., Golfarelli, M., Rizzi, S., Abelló, A., Romero, O.: Interactive multidimensional modeling of linked data for exploratory OLAP. IS **77**, 86–104 (2018)
10. Hilal, M.: A proposal for self-service OLAP endpoints for linked RDF datasets. In: Ciancarini, P., et al. (eds.) EKAW 2016. LNCS (LNAI), vol. 10180, pp. 245–250. Springer, Cham (2017). https://doi.org/10.1007/978-3-319-58694-6_38
11. Hung, E., Deng, Y., Subrahmanian, V.S.: RDF aggregate queries and views. In: International Conference on Data Engineering ICDE, pp. 717–728. IEEE (2005)
12. Khouri, S., Bellatreche, L.: LOD query-logs as an asset for multidimensional modeling. In: Benczúr, A., et al. (eds.) ADBIS 2018. CCIS, vol. 909, pp. 45–53. Springer, Cham (2018). https://doi.org/10.1007/978-3-030-00063-9_6
13. Kimball, R.: Newly emerging best practices for big data. Whitepaper, Kimball Group, September 2012
14. Komamizu, T., Amagasa, T., Kitagawa, H.: SPOOL: a SPARQL-based ETL framework for OLAP over linked data. In: IIWAS, p. 49. ACM (2015)
15. Marx, E., Zaveri, A., Moussallem, D., Rautenberg, S.: Dbtrends: exploring query logs for ranking RDF data. In: Semantic Systems, pp. 9–16. ACM (2016)
16. Mazumdar, S., et al.: SEMLEX-A framework for visually exploring semantic query log analysis. In: Semantic Web Conference-Poster and Demo Session (2011)
17. Ravat, F., Song, J.: Enabling OLAP analyses on the web of data. In: 2016 Eleventh International Conference on Digital Information Management (ICDIM), pp. 215–224. IEEE (2016)
18. Romero, O., Abelló, A.: Automatic validation of requirements to support multidimensional design. Data Knowl. Eng. **69**(9), 917–942 (2010)
19. Sabharwal, S., Nagpal, S., Aggarwal, G.: Empirical analysis of metrics for object oriented multidimensional model of data warehouse using unsupervised machine learning techniques. Int. J. Syst. Assur. Eng. Manag. **8**(2), 703–715 (2017)
20. Saleem, M., Ali, M.I., Hogan, A., Mehmood, Q., Ngomo, A.-C.N.: LSQ: the linked SPARQL queries dataset. In: Arenas, M., et al. (eds.) ISWC 2015. LNCS, vol. 9367, pp. 261–269. Springer, Cham (2015). https://doi.org/10.1007/978-3-319-25010-6_15
21. Wang, X., Staab, S., Tiropanis, T.: ASPG: generating OLAP queries for SPARQL benchmarking. In: Li, Y.-F., et al. (eds.) JIST 2016. LNCS, vol. 10055, pp. 171–185. Springer, Cham (2016). https://doi.org/10.1007/978-3-319-50112-3_13
22. Zhang, J., Ling, T.W., Bruckner, R.M., Tjoa, A.M.: Building XML data warehouse based on frequent patterns in user queries. In: Kambayashi, Y., Mohania, M., Wöß, W. (eds.) DaWaK 2003. LNCS, vol. 2737, pp. 99–108. Springer, Heidelberg (2003). https://doi.org/10.1007/978-3-540-45228-7_11

Handling the Information Backlog for Data Warehouse Development

Naveen Prakash[1] and Deepika Prakash[2(✉)]

[1] Department of Computer Science and Engineering, IIITD, New Delhi 110020, India
praknav@hotmail.com
[2] Department of Computer Engineering, NIIT University, Neemrana 301705, India
dpka.prakash@gmail.com

Abstract. In both, bus of data marts and agile data warehouse development, it is required to select the next product part to be developed. Whereas this issue is not addressed in the former, agile development approaches leave selection from a backlog to the product owner. We provide an approach to backlog selection that is rooted in the decision-making process. Our task is to first select a decision from a backlog of decisions and then select from the backlog of information relevant to it. Three yardsticks are considered for decision selection, business importance, decision structure and picking complete decisions in preference to partial decisions. The backlog of information for a selected decision may be for the Intelligence or Choice phases of the decision-making process.

Keywords: Iterative data warehouse development · Agile data warehouse development · Decision backlog · Semantic priority · Complete decisions · Syntactic priority

1 Introduction

Since data warehouse projects take long to deliver and are expensive [1], there is considerable pressure from stakeholders to show progress in the development task. This pressure led to the rejection, in practice, of the monolithic development approach of [2]. Reduced lead times is at the core of the "bus" approach of [3]. While promoting iterative and incremental development, this approach has two chief difficulties, (a) there is no guidance on which data mart is to be developed next, and (b) the lead time to delivery is still high. This is because Subject Oriented Requirements Engineering, SORE, [4–9] develops full requirements specifications for a subject.

Agile development has been adopted in DW development as well. Hughes [10] uses the notion of user stories of Scrum but did not address the issue of selection from a backlog. The agile approach in BEAM* [11] uses the concept of importance to enable selection. However, selection is an unguided, experience based process. In [12], user stories of Scrum were adapted for agile data warehousing. The approach organizes data warehousing in three levels, application, decision, and information. Some guidelines for selecting from the backlog at the application level are provided in [12]. However,

S. Hartmann et al. (Eds.): DEXA 2019, LNCS 11706, pp. 368–378, 2019.
https://doi.org/10.1007/978-3-030-27615-7_28

for the other two levels, issues in selection were identified but and the selection process itself was not elaborated. We propose to develop such a process in our paper here.

Data warehouses can be built to support complete decisions or, if decisions are complex, then data warehouses for supporting partial decisions can be built. Our partitioning approach is to identify backlog partitions of "complete" decisions and provide support for these. Completeness is defined at the end of Sect. 3. Data warehouse support for complete decisions is taken up earlier than for others.

We assume that selection from the backlog of decisions can be done on a syntactic or semantic basis. The former tells us the number of decisions that must be supported for a given decision to be supported. As far as we are aware, there is no literature about the lead time required to support decisions. Therefore, we make the simplifying assumption that all decisions take equal time. Our heuristic is that if the number of decision to be supported is larger for decision, A than for decision, B. then A shall require more time for delivery than B. Thus, a low value of syntactic priority promotes early delivery. Semantic priority is specified by the stakeholders of the data warehouse to-be. A number of methods for specifying it are available [13–18]. We adopt the strategy of AHP where relative priorities are assigned to candidates such that the sum of the priorities adds up to unity. Evidently, the larger the value of semantic priority, the more important it is from the business perspective.

It is possible that there is a conflict between syntactic and semantic priorities. We have three possible conflict resolution strategies, (a) Give preference to syntactic priority over semantic priority, (b) Semantic priority overrides syntactic priority, and (c) Perform selection as an interaction with the stakeholder and do not commit to any overriding. We superimpose the notion of completeness on these strategies. If the nature of the backlog is such that its decisions provide complete support in the decision making task, then we adopt the first strategy. If the backlog has decisions that provide partial support then we adopt the second strategy: important partial support is provided while we wait for full support. We follow the third option only when there is a conflict that requires human intervention.

Once a decision is selected, the next issue is to identify the information relevant to it. Following the decision model of Simon [19], we formulate procedures for supporting the Intelligence or Choice phases of the decision-making process.

The layout of the paper is as follows. In Sect. 2, we review the basic concepts underlying decision-making. In Sect. 3, we consider the different types of decision backlogs and some basic strategies to handle them. Sections 4 to 7 contain procedures for selection of decisions from the different kinds of backlogs. In Sect. 8, we present procedures for selecting from the information backlog.

2 The Decision Making Process

The decision making process of Simon [19] consists of three main activities, Intelligence, Design and Choice. In the phase of intelligence, the decision maker analyses the current situation with a view to discovering a problem requiring decision making. The decision-making problem is formulated during the design stage. During the Choice phase, the most appropriate alternative is selected.

A decision is a commitment to an action, course of action, or strategy. We will refer to these collectively as action unless required otherwise. This commitment implies that the action must be taken to completion. There are two kinds of decisions, simple decisions that cannot be decomposed and compound decisions that consist of other simpler decisions. It is also possible for decisions to be dependent on one another.

A decision problem is a design artifact built during the design phase. Two widely prevalent representation systems for the design problem are decision trees and influence diagrams. Whereas the former provides detailed decision logic and can be visualized as a flow chart, the influence diagram presents a global picture of the decision problem. This larger view point is interesting for us since it provides to us the concepts for selection from a backlog.

An influence diagram [20] is a directed acyclic graph with three kinds of nodes and two kinds of edges. Its nodes are as follows:

- Decision nodes represented as rectangles. These nodes represent the choices that are available. The probability of selecting a choice is the same as that of selecting any other one. Further, at least one of the choices shall be selected.
- Uncertainty nodes represented as ovals. Ovals represent situations that are beyond the control of the decision maker. The individual probability of any situation occurring can be only estimated within a certain range of possibilities. The term, uncertainty, conveys that the situation at the time the decision is taken and the situation when the effect of the decision is felt may be different. For example, product price after purchase may change by the time the product is delivered.
- Outcome nodes are rounded rectangles that represent decision outcome.

The influence diagram has two kinds of edges, namely

- Functional edges represented as solid arrows say that the node at the source of the arrow is relevant for assessing the value of the node at the arrow head.
- Information edges represented as dashed arrow indicate that knowledge of the predecessor node is available when the successor node is being considered.

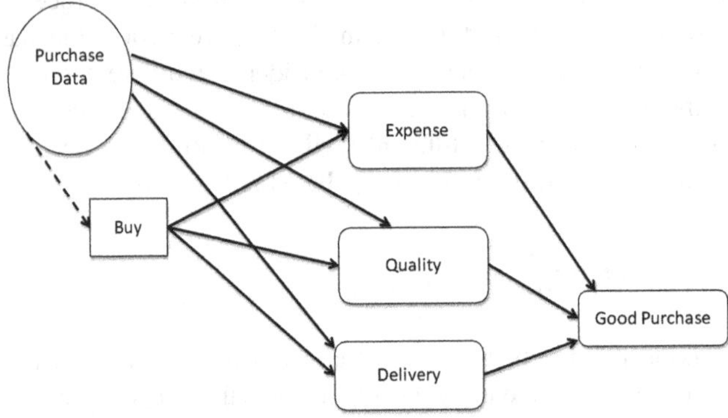

Fig. 1. The basic buy decision

Let us exemplify the foregoing with the decision, Buy stock. Our decision maker must decide which material is to be bought from which supplier. This is the Buy decision node of Fig. 1. The uncertainty, Purchase Data, in the figure includes, expected demand, current price, history of price movement, expected material delivery delay etc. The dashed arrow, i.e. the information edge says that Purchase Data for all products to be purchased is needed for the decision. The outcomes are Expense, Quality, and Delivery, all contributing to the main outcome, Good purchase. The objectives used to evaluate these are Minimize Cost, Maximize quality and Maximize punctual delivery.

3 Defining the Backlog of Decisions

One way of defining the backlog is to treat it as being the universe of all decisions of interest. If this universe is large, then assigning semantic priorities may present difficulties. Further, if the universe changes rapidly, then handling changing priorities may be a challenge. Therefore, we adopt the divide and conquer approach and create four partitions of the universal backlog.

The decision backlog is the aggregate of four kinds of backlogs, namely, Free, Specialized, Composite, and Mixed backlogs. The first of these contains complete decisions that are stand alone and are the **best candidates** for early data warehouse support. The second develops specialization hierarchies for complete decisions. Each node in the hierarchy is a special case of the root decision and is complete in its own right. However, the root of the hierarchy is only completely supported after all specialized decisions are supported. Given that completeness permeates the hierarchy, selection from the specialized hierarchy is given **second preference**. **Third preference** is given to selecting from the backlog of composite decisions because a composite decision is only supported after all its component decisions have been supported. The component decisions have only partial value and are interesting as part of the composite decision. Finally, the backlog of mixed decisions is the **last one** from which selection is made. For such decisions, first selection is made from its specialized decisions, then dependent and finally composite. This delivers special cases of high complexity early.

These four backlogs are formed as follows:

1. A decision that is atomic (cannot be decomposed into simpler ones), does not give rise to a specialization hierarchy, and is not dependent on any other decision, is a standalone decision, an individual decision. We refer to such decisions as “free” decisions and the backlog of such decisions is called a free_backlog.
2. The specialized_backlog consists of a set of atomic decisions. Each decision in this set **gives rise to a specialization hierarchy**, and is not dependent on any other decision.
3. The composite_backlog is a set of composite decisions that do not give rise to a specialization hierarchy, and are not dependent on any other decision.

4. Finally, the mixed_backlog is a set of decisions in which a decision is **composite** OR **gives rise to a specialization hierarchy** OR is **dependent** on any other decision.

Now, we can define a complete decision. A decision is complete if it is

(a) Not dependent on another decision or if dependent on decision D, then D has already been selected earlier, and
(b) Cannot be decomposed into others or if all its components have been selected earlier, and
(c) Cannot be specialized into others or if all the specialized decisions have been selected earlier.

4 Selecting from Free_Backlog

The stakeholder assigns semantic priority, sem to the decisions comprising the backlog. Since all decisions are atomic and are not roots of a specialization hierarchy they have the same structural priority. ND is the one with the highest semantic priority and post selection, it is removed from Free_backlog. Consider Free_backlog = {D_0, D_1, D_2, D_5} whose dependency graph in Fig. 2.

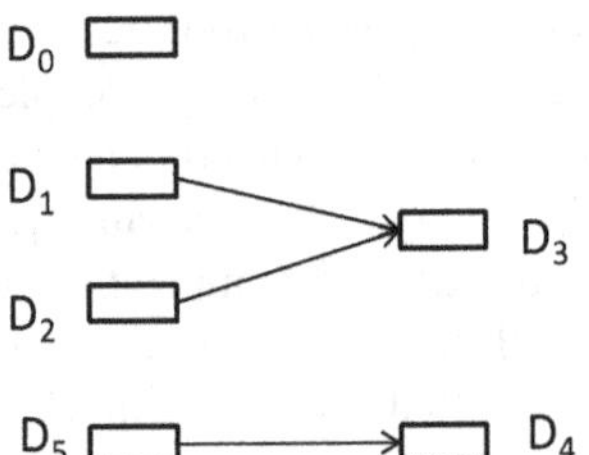

Fig. 2. Dependency between decisions

Let the semantic priorities be Sem(D_0) = 0.4, Sem(D_1) = 0.3, Sem(D_2) = 0.2 and Sem(D_5) = 0.1. The selected ND is D_0. It is removed from Free_backlog. The newly computed Free_backlog is {D_1, D_2, D_5}. The selected ND is D_1 and it is removed from Free_backlog giving us the free_backlog = {D_2, D_5}. The next selection is D_2 and its removal makes D_3 eligible for inclusion in Free_backlog that is now {D_3, D_5}. Since it has not been prioritized with respect to D_5, relative priorities are now assigned and the procedure continues.

5 Selecting from Specialized_Backlog

Consider a collection of decisions each of which gives rise to a specialization hierarchy. First we need to select the decision from the collection and then select from the hierarchy of the selected decision. We determine the specialized syntactic priority, Ssyn of each decision in the hierarchy as the number of its specialized decisions. We adopt the heuristic that a decision with a higher value of Ssyn will take more lead time to support than one with a lower value. This is because the number of decisions to be supported is larger in case of the former. The method of selection is as follows.

1. The decisions in the Specialized_backlog are assigned semantic priorities, Sem by assigning values that add up to unity. Present to the stakeholder, the specialized_backlog in increasing order of syntactic priority. Let the stakeholder select decision D. Mark the decision as selected.

   For the hierarchy rooted in D Repeat
2. Move to next lower level of the hierarchy. If no such level then go to 4.
3. Select the decision with the least syntactic priority, Syn. If several decisions have this value, then determine semantic priority of the decisions and
   i. The decision with the highest semantic value is the selected decision of the level.
   ii. If more than one decision has the least value of Sem then selection is made through stakeholder interaction.

   Until a leaf node is marked
4. The marked leaf node is the selection from the backlog and is removed from the hierarchy.
5. If there are marked nodes in the hierarchy then move to the next higher level and repeat from step 2

Table 1. Presenting the backlog

Decision name	Syntactic Priority, Ssyn	Semantic Priority, Sem
A	10	0.3
V	15	0.2
Z	17	0.5

To illustrate, consider a Specialized_backlog = {Z, V, A} each member of which is the root of a specialization hierarchy. The syntactic and semantic priorities are as shown in Table 1. The rows are ordered on the syntactic priority.

Let the selected ND be A. It is marked as selected (step 1). For the specialization hierarchy of A in Fig. 3, the next level is the set {B, C, D} (step 2). By step 3, Band C are potential candidates and using (ii), let the stakeholder select B which is marked. At the next level, we now have the backlog of {E, F}. These are leaves of the hierarchy and so both have Ssyn = 0. Their Sem values determine that F will be marked and by step (4), it is removed from the hierarchy after declaring it to be the selected decision from the backlog. Following step (5), we position ourselves at B again, and following step (2) E is marked, selected for the next iteration/sprint, and then removed. Now following step 5, we position ourselves at B again. This procedure continues till A is selected. At this stage, attention shifts to selecting from Table 1 again.

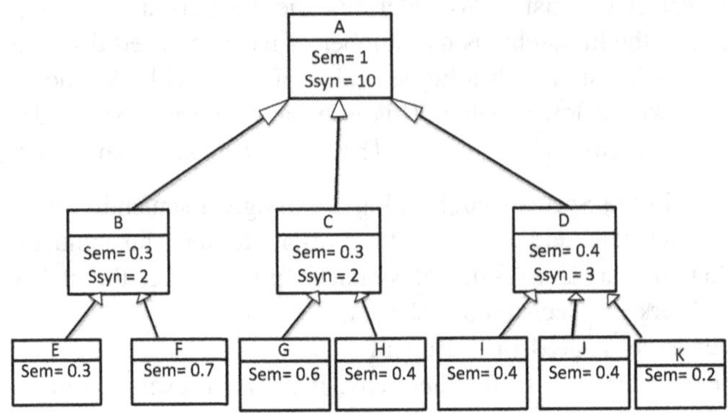

Fig. 3. Selecting from a specialization hierarchy

6 Selecting from Composite_Backlog

A component decision of a composition hierarchy is part of a larger decision. When support is provided to such a decision, then only a part of the entire decision, and not a full case of the decision, is supported. **Therefore, it makes better sense to select those component decisions that provide higher semantic benefit than those that can be delivered early**.

Table 2. Presenting the backlog

Decision name	Syntactic Priority Csyn	Semantic Priority (Sem)
A	10	0.5
V	15	0.2
Z	17	0.3

As before, we present in Table 2 the relative semantic priorities. The Csync attribute contains a count of the number of component decisions of a decision. Now, the method of selection of the decision to be taken up next is as follows:

1. Determine the decision with the highest value of Sem.
   While level is not the leaf level do
2. Move to next lower level of the composition hierarchy.
3. Select the decision with the highest value of Sem. If several such decisions exist then determine their Csyn and
   (a) The decision with the least value of Csyn is selected.

(b) If more than one decision has the least value of Csyn then selection is made through stakeholder interaction.

End

4. The leaf node is selected and then removed from the hierarchy.
5. Move to the next higher level, if it exists, and repeat from step 2

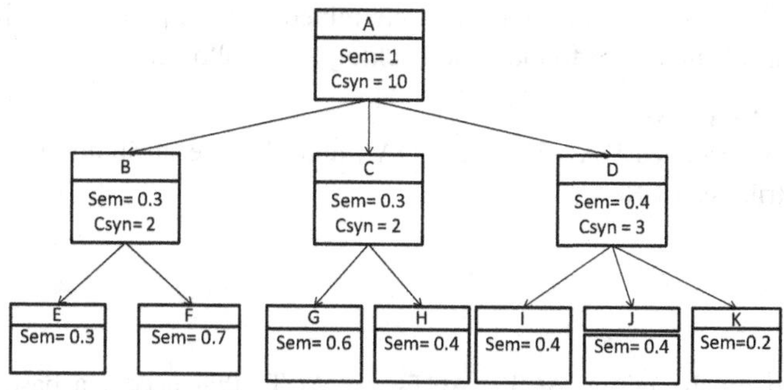

Fig. 4. A composition hierarchy

Applying step 1 to Table 2, we get A. Applying the rest of the procedure to the composition hierarchy of Fig. 4, we obtain the order I, J, K, D, F, E, B, G, H, C, A.

7 Selecting from Mixed_Backlog

Selection from the mixed backlog is done in the order as follows:

(a) For decisions that have csync = 0 and ssync $\neq$ 0, follow the procedure of Sect. 5.
(b) For decisions having cync $\neq$ 0 and ssync = 0 follow Sect. 6.
(c) For decisions with csyn $\neq$ 0 and ssync $\neq$ 0, first follow Sect. 5 and then Sect. 6.

For reasons of space, we do not illustrate selection from mixed_backlog here.

8 The Information Backlog

Having selected the decision, we now need to consider the information that must be stored in a data warehouse. The entire information backlog consists of the contribution made by Uncertainty and Outcome of the Influence Diagram, as well as that made by the action committed to by the decision. An outcome is evaluated by an objective function of the form y = F(x). Evidently, y and x respectively contribute to the information backlog. On the other hand, the contribution of an action is the record that it leaves behind. If the information backlog obtained is too large then we need to reduce it. At a global level, we do this based on the stage of the decision-making process, Intelligence or Choice, supported by the data warehouse. For the former, information generated by the actions is relevant since a collection of past actions tells us the current situation. For the latter, we need the contributions of Uncertainty and Outcome parts. This is because choice is enabled by the functional link between these.

For the **Intelligence phase** the data warehouse should contain the current situation of the enterprise. Relevant actions are those that create, read, update, or delete information. Let our decision be a commitment to action, ACT and let ACT leave behind a record of data, $R(A_1, A_2, \ldots, A_n)$. Now, R may be accessed by actions other than ACT, say, $ACT_1, ACT_2, \ldots, ACT_p$. The record of all ACT_i is of interest as well. Yet again, actions that access the new attributes are determined and the process continues till no new actions or attributes are found. The backlog is as follows:

- The record $R(A_1, A_2, \ldots, A_n)$
- All derived attributes $DA_1, DA_2, \ldots, DA_m$ and all base attributes $BA_1, BA_2, \ldots, BA_k$ of attributes of R

REPEAT

- The record of all actions $ACT_1, ACT_2, \ldots, ACT_p$ that access a base or derived attribute.
- Attributes that are derived from $ACT_1, ACT_2, \ldots, ACT_p$

UNTIL no new Action or new base/derived attributes

To illustrate consider the decision, Buy, once again. Its action, Order associated leaves the record Ordered_product(Order#, Supplier#, Product#, Quantity, Unit price). Now, Total_on_order is derived from quantity because Total_on_order = Total_on_order + Quantity. Similarly, Total_committed_amount = Total_committed_amount + Quantity × Unit price. This gives us the information backlog (a) Ordered_product(Supplier#, Product#, Quantity, Unit price) and (b) Total_on_order and Total_committed_amount.

Now, consider the **first iteration** for action, Cancel Order that reduces Total_on_order by Quantity_cancelled and also reduces Total_committed_amount by the amount of order cancelled. Its record is Cancelled_product(Cancellation#, Order#, Supplier#, Product#, Quantity_cancelled). At the end of this iteration, information backlog would be

(a) Ordered_product(Suplier#, Product#, Quantity, Unit price)
(b) Total_on_order and Total_committed_amount
(c) Cancelled_product(Cancellation#, Order#, Supplier#, Product#, Quantity_cancelled)

Assuming that this is manageable then subsequent iterations are done later.

Consider, now the **Choice phase**. Maximum value shall be provided if in the selected backlog, all outcomes of all alternatives could be evaluated. However, where the information backlog is too large, we could either pick a subset of the set of alternatives or the set of outcomes. Picking the former is highly restrictive since there are some un-evaluated alternatives and no comparison of the alternatives would be possible. Therefore, picking a subset of outcomes is our preference.

Let there be information backlog B for a decision, D, having the set of outcomes $O = \{O_1, O_2, \ldots, O_n\}$. It is possible to find subsets of B, $B_1, B_2, \ldots, B_n$ such that O_i is calculated from B_i. The decision maker gets value, if support is provided for achieving any one or more decision outcomes. The selection procedure is:

(a) Assign relative semantic priority to the outcomes of the decision.
REPEAT
(b) Determine the objective that evaluates outcome with highest priority. Base attributes of a parameter are determined recursively as above.
(c) Estimate the size of this information for all alternatives.
  i. If too small, then repeat from step (b) for the next highest priority outcome.
  ii. If too big then break into a number of sprints; each sprint handling a subset of the alternatives.
  iii. If right size then treat as backlog selection for sprint.
UNTIL no more outcomes

Consider the decision, Buy of Fig. 1 having three outcomes, Expense, Quality, and Delivery. During step (a) let the relative priorities be, 0.5, 0.2, 0.3 respectively. Following step (b), Expense is evaluated by the objective Minimize Cost and Cost = Unit_price * Quantity + Tax + shipment. In step (c), this is found too small for a sprint and the next outcome, Delivery is picked up and its information determined. Assuming that this backlog is enough, Quality is deferred for the next iteration.

9 Conclusion

The approach adopted here promotes stakeholder interaction during selection from the backlog. Selecting from the decision backlog identifies the decisional problem to be taken up. This leads to considering data warehouse information contents which are determined from the information backlog of the decision. We have factored, into our process, (a) business importance, (b) providing as much value as possible and (c) structural aspects of decisions.

References

1. Hayen, R., Rutashobya, C., Vetter, D.: An investigation of the factors affecting data warehousing success. Issues Inf. Syst. **VIII**(2), 547–553 (2007)
2. Inmon, W.H.: What is a Data Warehouse? W.H. Inmon, Prism, vol. 1, no. 1 (1995)
3. Kimball, R.: The Data Warehouse Toolkit. Wiley, Hoboken (1996)
4. Boehnlein, M., Ulbrich vom Ende, A.: Business process oriented development of data warehouse structures. In: Proceedings of Data Warehousing 2000, pp. 3–21. PhysicaVerlag HD (2000)
5. Bonifati, A., Cattaneo, F., Ceri, S., Fuggetta, A., Paraboschi, S.: Designing data marts for data warehouses. ACM Trans. Softw. Eng. Methodol. **10**(4), 452–483 (2001)
6. Prakash, N., Gosain, A.: An approach to engineering the requirements of data warehouses. Requir. Eng. J. **13**(1), 49–72 (2008)

7. Giorgini, P., Rizzi, S., Garzetti, M.: GRAnD: a goal-oriented approach to requirement analysis in data warehouses. Decis. Support Syst. **45**(1), 4–21 (2008)
8. Mazón, J.-N., Pardillo, J., Trujillo, J.: A model-driven goal-oriented requirement engineering approach for data warehouses. In: Hainaut, J.-L., et al. (eds.) ER 2007. LNCS, vol. 4802, pp. 255–264. Springer, Heidelberg (2007). https://doi.org/10.1007/978-3-540-76292-8_31
9. Prakash, D., Prakash, N.: A multi-factor approach for elicitation of Information requirements for data warehouses. Requir. Eng. J. https://doi.org/10.1007/s00766-017-0283-9
10. Hughes, R.: Agile Data Warehousing Project Management Business Intelligence Systems Using Scrum. Morgan Kaufman, Burlington (2013)
11. Corr, L., Stagnitto, J.: Agile Data Warehouse Design. Decision One Press, Leeds (2012)
12. Prakash, N., Prakash, D.: Model-driven user stories for agile data warehouse development. In: Proceedings of 19th IEEE Conference on Business Informatics, pp. 424–433 (2017)
13. Berander, P., Andrews, A.: Requirements prioritization. In: Aurum, A., Wohlin, C. (eds.) Engineering and Managing Software Requirements, pp. 69–94. Springer, Heidelberg (2005). https://doi.org/10.1007/3-540-28244-0_4
14. Firesmith, D.: Prioritizing requirements. J. Object Technol. **3**(8), 35–47 (2004)
15. Heusser, M.: 6 agile methods for backlog prioritization, downloaded on February 19, 2018 (2018). https://learn.techbeacon.com/units/6-agile-methods-backlog-prioritization
16. Maiti, R.R., Mitropoulos, F.J.: Prioritizing non-functional requirements in agile software engineering. In: Proceedings of the South East Conference, ACM SE 2017, pp. 212–214 (2017)
17. Mead, N.: Software Requirements Introduction, white paper Software Engineering Institute, Carnegie Mellon University (2006)
18. Thakurta, R.: Understanding requirements prioritization artifacts: a systematic mapping study. REJ **22**, 491–526 (2017)
19. Simon, H.A.: The New Science of Management Decisions. Prentice Hall, Englewood Cliffs (1977)
20. Marakas, G.M.: Decision Support Systems in the 21st Century, Eastern Economy Edition, PHI (2003)
21. Bhardwaj, H., Prakash, N.: Eliciting and structuring business indicators in data warehouse requirements engineering. J. Expert Syst. **33**(4), 405–413 (2016)

Ontario: Federated Query Processing Against a Semantic Data Lake

Kemele M. Endris[1,2(✉)], Philipp D. Rohde[2], Maria-Esther Vidal[1,2], and Sören Auer[1,2]

[1] L3S Research Center, Hannover, Germany
{endris,vidal,auer}@L3S.de

[2] TIB Leibniz Information Centre for Science and Technology, Hannover, Germany
{Philipp.Rohde,Maria.Vidal}@tib.eu

Abstract. Data lakes enable flexible knowledge discovery and reduce the overhead of materialized data integration. Albeit effective for data storage, query execution over data lakes may be expensive, being demanded novel techniques to generate plans able to exploit the main characteristics of data lakes. We devise Ontario, a federated query processing approach tailored for large-scale heterogeneous data. Ontario provides efficient and effective query processing over a federation of heterogeneous data sources in a data lake. Ontario resorts to source descriptions named RDF Molecule Templates, i.e., abstract descriptions of the properties of the entities in a unified schema and their implementation in a data lake. We empirically evaluate the effectiveness of the Ontario optimization techniques over state-of-the-art benchmarks. The observed results suggest that Ontario can effectively select plans composed of subqueries that can be efficiently executed against heterogeneous data sources in a data lake.

Keywords: Polystore · Federated engine · Semantic Data Lake

1 Introduction

In recent years, enormous amount of heterogeneous data became available through various platforms. The need for efficient management and query processing techniques for big and heterogeneous data has been gaining attention. Big data systems that integrate different data sources need to handle the variety and volume of data efficiently and effectively. The Semantic Web community has invested significant efforts in lifting existing tabular data into linked data and interlinking these datasets with existing LOD datasets [12], e.g., LODD (Linked Open Drug Data) [11], Bio2RDF [2], and Linked Government Data [8]. However, these transformations to linked datasets are costly and assumes a stable schema and data model. To provide scalable and flexible knowledge discovery, analysis, and reporting, *data lakes* have been proposed. Data lakes are composed of heterogeneous data sources in their original format. Management against data lakes reduce the costs of identifying, storing, cleansing, and integrating data significantly and promote flexibility in data analysis. However, they introduce

S. Hartmann et al. (Eds.): DEXA 2019, LNCS 11706, pp. 379–395, 2019.
https://doi.org/10.1007/978-3-030-27615-7_29

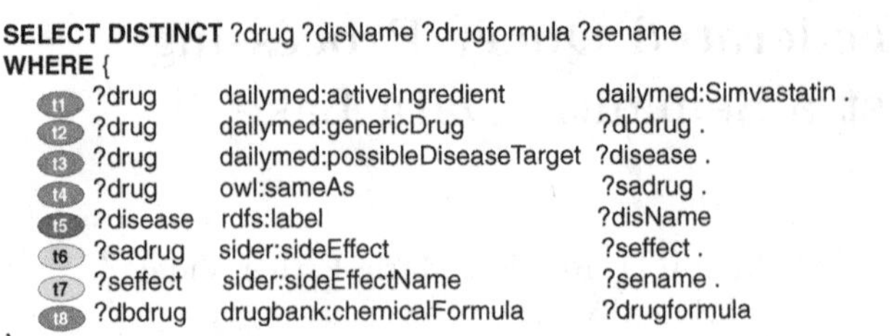

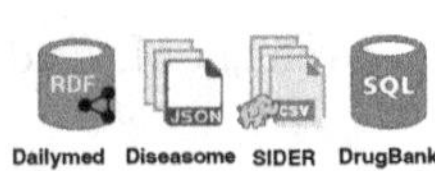

(a) SPARQL Query: Find targets, and side effects of drugs with active ingredient Simvastatin.

(b) Data Sources in a Data Lake

Fig. 1. Motivating Example. (a) A SPARQL query composed of four star-shaped subqueries accessing four data sources, Dailymed, Diseasome, SIDER, and DrugBank. (b) Data Sources: Dailymed (RDF in Virtuoso), Diseasome (Local JSON File), SIDER (TSV in HDFS), DrugBank (XML in MySQL)

complexity during query execution time. Contrary to existing SPARQL federated query engines, federated query processing over data lakes demands the integration and semantic description of data collected from heterogeneous data sources. Thus, selecting relevant data sources for a specific query, creating an efficient query execution plan considering the data source types, and combining partial results obtained from these sources are the main challenges in federated query processing on data lakes.

We tackle the problem of federated query processing over *semantic data lakes* and propose Ontario, a query engine able to efficiently interoperate among heterogeneous datasets. Ontario implements novel query processing methods, i.e., source selection, query decomposition, and query planning; they are capable of exploiting knowledge about the sources and the query to generate plans *customized* for the sources in a data lake. Ontario resorts to RDF Molecule Templates [4], i.e., abstract descriptions of the properties of entities in an RDF dataset in order to identify the star-shaped subqueries of an input query. Differently to state-of-the-art approaches, Ontario is able to classify star-shaped subqueries according to type of instantiations and joins. Additionally, star-shaped subqueries are characterized in terms of the data engines where they will be executed. Ontario exploits these meta-data to identify efficient query plans. We study the impact of Ontario query plans over the LSLOD benchmark [6]. Ontario is compared to existing federated query engines (FedX [13] and MULDER [4]), and to different configurations of the data lakes (only RDF datasets, or RDF and RDB datasets). Observed results suggest that the Ontario heuristics are able to guide the optimizer into the space of high quality query plans over heterogeneous data sources.

This paper is structured as follows. Section 2 motivates with an example, the problem of federated query processing tackled in this paper. Section 3 defines the problem statement of this work, as well as the proposed solution. We present the Ontario framework in Sect. 4, and the results of an empirical evaluation are presented in Sect. 5. Finally, we discuss the related work in Sect. 6 and an outlook on future directions in Sect. 7.

2 Motivating Example

In the biomedical domain, frequently complex questions need to be answered with multiple data sources and data models. Especially in this domain, flexible data management and integration techniques are required due to the variety of tools and formats data is collected, generated, and processed. To provide a unified view over these heterogeneous data sources, mapping rules are utilized to describe the required transformations from raw data into the unified schema. These mappings enable the translation of queries from the unified schema into queries against the sources using native access interfaces.

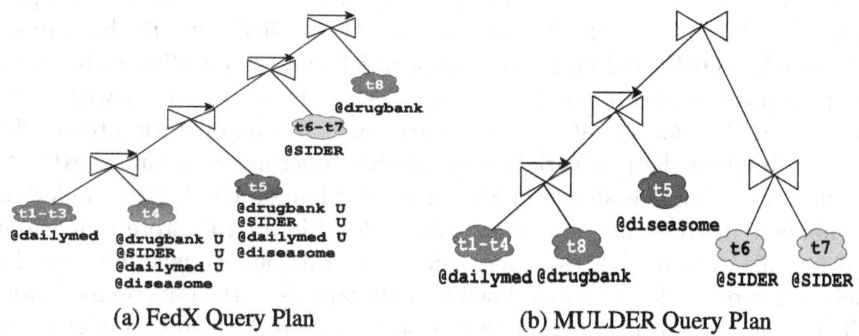

Fig. 2. Motivating Example. (a) FedX created a left-linear plan and used nested loop joins (arrows on top of join) (b) MULDER identifies a bushy-tree for star-shape groups.

We motivate our work by comparing the performance of federated SPARQL query engines over a federation of data sources that provide a SPARQL-based access interface. For instance, a SPARQL query in Fig. 1a requires to collect the name of possible drug targets, chemical formula, and side effects of drugs labeled by FDA that have the active substance Simvastatin. To answer this query, four (4) datasets (Fig. 1b) need to be accessed. Dailymed[1] publishes FDA label information about marketed drugs in the United States; Diseasome[2] makes available a network of disorders and disease genes; DrugBank[3] reports information about drugs and drug targets, and SIDER[4] presents information on drug side effects. Our running query comprises eight (8) triple patterns (identified with $t1$ to $t8$ in Fig. 1a). Dailymed can answer the triple patterns $t1$–$t4$, while triple pattern $t5$ can be answered by Diseasome. Further, SIDER can answer $t6$–$t7$, and $t8$ can be answered by DrugBank. The data access services of each datasets are implemented by different backends and provide different capabilities. For instance,

[1] https://dailymed.nlm.nih.gov.

[2] https://old.datahub.io/dataset/fu-berlin-diseasome.

[3] https://www.drugbank.ca.

[4] http://sideeffects.embl.de/.

the endpoint services for SIDER and Diseasome are Spark-based query processors that translate queries from SPARQL to SQL, where the raw data need to be loaded in memory to evaluate the query in these data sources. Similarly, the endpoint for DrugBank translates SPARQL to SQL and execute the translated query in MySQL, which provides efficient indexing and query optimization for relational data.

Federated query engines, FedX [13] and MULDER [4], provide a unified view over a set of data sources that respect SPARQL protocol. They rely on source descriptions to select relevant sources for a given query and for finding an efficient query execution plan. For instance, FedX contacts the data sources to decide where the triple patterns will be executed, while MULDER requires RDF Molecule Templates (RDF-MTs) to be collected in advance. FedX decomposes the query, in Fig. 1a, into five (5) sub-queries; $t1$–$t3$, $t6$–$t7$, and $t8$ that are sent to Dailymed, SIDER, and Drugbank, respectively, and $t4$ and $t5$ sent to all four (4) data sources, respectively. FedX creates a left linear tree plan with nested loop join operator, an operator that pushes down the join operation to the data sources by binding the join variables of the right operand with values extracted from the left operand, as shown in Fig. 2a. FedX planner assumes the underlying data model is in RDF and triples are materialized in a triple store that is optimized for this data model. However, since the data sources have different data models and capabilities, pushing down join operations to the data sources would result in a higher execution time, 20 min and incomplete results. On the other hand, MULDER decomposes the query into five (5) sub-queries; $t1$–$t4$ executed in Dailymed, $t8$ executed in Drugbank, Diseasome executes $t5$, while $t6$ and $t7$ executed in SIDER, respectively. MULDER creates a bushy-tree plan with nested hash join and GJoin [1] operators based on the selectivity of operands to decide the type of operator. MULDER, like FedX, assumes RDF as an underlying data model and uniform querying capabilities of the given data sources. Based on these assumption, MULDER selects a nested hash join operator for the first two joins, between sub-queries $t1$–$t4$, $t8$, and $t5$. Despite, creating an efficient bushy-tree plan that helps to parallelize the query execution, the selection of join operator ignores the data source capabilities and underlying data model, which results in higher execution time, 4.6 min. In this paper, we devise optimization techniques guided by heuristics that enable the creation of source-dependent query plans. First, the Ontario query optimizer resorts to data source descriptions in terms of RDF Molecule Templates to select the sources that will evaluate the query. Then, the query is decomposed into subqueries that can be executed in the selected sources. Finally, a plan that composes the subqueries is generated; physical operators are selected in order to minimize execution time and maximize answer completeness.

3 Problem Statement and Proposed Solution

Our formalization is based on RDF Molecule Templates, which represent an abstract description of entities that have the same semantic type.

Definition 1 (RDF Molecule Template). *An RDF Molecule Template is defined as a 5-tuple* $\sigma = \langle S, C, \gamma, IntraL, InterL \rangle$, *where:*

- *S - an interface to access dataset G;*
- *C - an RDF class such that the triple pattern (?s rdf:type C) is true in G;*
- *γ - is a set of pairs (p, T) such that p is a property with domain C and range T, and the triple patterns (?s p ?o), and (?s rdf:type C) are true in G;*
- *IntraL - is a set of pairs (p, C_j) such that p is an object property with domain C and range C_j, and the triple patterns (?s p ?o) and (?o rdf:type C_j) and (?s rdf:type C) are true in G;*
- *InterL - is a set of triples (p, C_k, SW) such that p is an object property with domain C and range C_k; SW is a Web service API that provides access to an RDF dataset K, the triple patterns (?s p ?o) and (?s rdf:type C) are true in G, and the triple pattern (?o rdf:type C_k) is true in K.*

Definition 2 (Semantic Data Lake). *A Semantic Data Lake (SDL) is a tuple* $SDL = \langle \psi, \mathbb{S}, M \rangle$ *where,* ψ *is a set of RDF Molecule Templates,* $\mathbb{S}$ *is a set of sources in raw formats (stored either in a file system or DBMS) in the Data Lake, M is a set of conjunctive rules that associate sources in* $\mathbb{S}$ *with RDF Molecule Templates in* ψ.

Definition 3 (Instantiation of an RDF-MT). *Instantiation of an RDF-MT,* $[\sigma]$, *is defined as a set of RDF molecules,* σ^*, *that are the instances of a class from data source(s) as described in the template.*

$$[\sigma] = \{\sigma^* | \forall p \in \sigma^*, p \subseteq \gamma, \text{ where } \gamma \subseteq \sigma\} \tag{1}$$

Definition 4 (Virtual Knowledge Graph). *Given a Semantic Data Lake* $SDL = \langle \psi = \{\sigma_1, \ldots, \sigma_k\}, \mathbb{S} = \{S_1, \ldots, S_n\}, M \rangle$, *a Virtual Knowledge Graph* (KG^*) *for SDL is a virtual RDF graph that corresponds to the union of all the RDF Molecules instantiations,* σ^*, *that are created by applying rules in M to the data sources in* $\mathbb{S}$:

$$KG^* = \bigcup_{i=1}^{n} \bigcup_{j=1}^{k} [\sigma_j]_{S_i} \tag{2}$$

In order to efficiently query the resulting *virtual Knowledge Graph*, SPARQL queries need to be rewritten into queries operating on the data sources. SPARQL language is based on matching graph patterns; a basic graph pattern (BGP) is a set of triple patterns and (optional) filter clauses.

Definition 5 (Basic Graph Pattern). [4] *Let I be the set of all IRIs, B be the set of blank nodes, L be the set of literals and* ε *be the set of variables. A SPARQL Basic Graph Pattern (BGP) expression is defined recursively as follows:*

1. *A triple pattern* $\tau \in (I \cup B \cup \varepsilon) \times (I \cup \varepsilon) \times (I \cup B \cup L \cup \varepsilon)$ *is a BGP;*
2. *The expression (P FILTER E) is a BGP, where P is a BGP and E is a SPARQL filter expression that evaluates to Boolean value;*
3. *The expression (P1 AND P2) is a BGP, where P1 and P2 are BGPs.*

A BGP in a SPARQL query contains at least one star-shaped subquery (SSQ), a non-empty set of triple patterns that share the same subject variable (constant).

Definition 6 (Star-shaped Subquery (SSQ)). [14] *A star-shaped subquery* `SSQ(S,?X)` *on a variable (constant)* `?X` *is defined as:*

1. `SSQ(S,?X)` *is a triple pattern* t={`?X p o`}, *and* `p` *and* `o` *are different to* `?X`.
2. `SSQ(S,?X)` *is the union of two stars,* `SSQ(S1,?X)` *and* `SSQ(S2,?X)`, *where triple patterns in* `S1` *and* `S2` *only share the variable (constant)* `?X`.

Definition 7 (Query Rewriting). *Let Q and $\beta(\mathrm{Q})$ be a SPARQL query and the set of Basic Graph Patterns (BGPs) in Q, respectively. Let $SDL = \langle \psi, \mathbb{S}, M \rangle$ be a Semantic Data Lake. A rewriting Q' of Q over sources in S corresponds to a SPARQL query, composed of BGPs in $\beta(Q')$ that meet the following conditions:*

- *$\beta(\mathrm{Q})$ has the same number of triple patterns as $\beta(Q')$, i.e., $\tau(Q) = \tau(Q')$*
- *there is a function μ: $\beta(\mathrm{Q}) \rightarrow \beta(Q')$ that maps BGPs in $\beta(\mathrm{Q})$ to its corresponding rewriting in the sources of* SDL. *$\mu\langle BGP_i \rangle = \{\langle BGP_{ij}, S\rangle | BGP_{ij} \subset BGP_i, S$ is a non-empty set and $S \subset \mathbb{S}\}$*

3.1 Problem Definition

Given a SPARQL query Q, a Semantic Data Lake $SDL = \langle \psi, \mathbb{S}, M \rangle$, and a Virtual Knowledge Graph KG^* of SDL. The problem of query rewriting in a federation of heterogeneous data sources in a Semantic Data Lake (SDL) is defined as follows. Given a set of BGPs in Q, find a query Q' that satisfies the following conditions:

- The evaluation of Q over heterogeneous data source in SDL is complete, i.e., the evaluation of Q in KG^* is equivalent to the evaluation of Q' in SDL

$$[[Q']]_{SDL} = [[Q]]_{KG^*} \tag{3}$$

- The cost of executing Q in SDL has a minimal execution cost, i.e., if $cost([[Q']]_{SDL})$ represents the execution time of Q' in SDL, then

$$[[Q]]_{SDL} = \underset{[[Q']]_{SDL}}{\operatorname{argmin}} \operatorname{cost}([[Q']]_{SDL}) \tag{4}$$

3.2 Proposed Solution

To solve the query rewriting problem, we propose Ontario, a federated query processing engine over heterogeneous data sources in a Semantic Data Lake. Ontario utilizes the SPARQL query language as a unified query language and, its decomposition and source selection technique is based on RDF Molecule Templates which keep the information about the type of the sources. Given a SPARQL query, Ontario creates a set of star-shaped subqueries that matches

RDF Molecule Templates in the Semantic Data Lake. Furthermore, Ontario is able to distinguish different types of star-shaped subqueries and decide which of them are more appropriate to be run in a given engine. The type of the star-shaped subqueries are further considered to decide the shape of the query plan tree, the more suitable join operators, and the location of the selections and projections in the plan. Ontario utilizes a greedy algorithm to find an efficient plan that minimized execution time and maximize answer completeness.

4 Ontario: Federated Query Processing over Semantic Data Lakes

In this section, we present the architecture of Ontario. First, we present an overview of basic components in a Semantic Data Lake architecture, then we describe the Ontario architecture with respect to these components. Figure 3 shows an overview of the Semantic Data Lake components. The top layer catalyzes, i.e., via the Lake Catalyst, a given SPARQL query into a set of star-shaped subqueries and selects matching RDF Molecule Templates. The Lake Catalyst passes the subqueries to the respective RDF-MT Catalysts that are responsible for specific RDF Molecule Templates. It dispatches the decomposed query and coordinates the global query planing and optimization. In addition, once the results are returned from lower layer, i.e., from Molecule Synthesizers, the Graph Synthesizer combines (synthesizes) the molecules and generate the final result. The second layer manages the instantiation of RDF-MTs, as defined in Definition 3. It catalyzes, via RDF-MT Catalysts, the star-shaped subqueries into a set of API calls to the Data Catalysts and synthesize results via Molecule Synthesizer. RDF-MT Catalysts are specialized components that deal with only

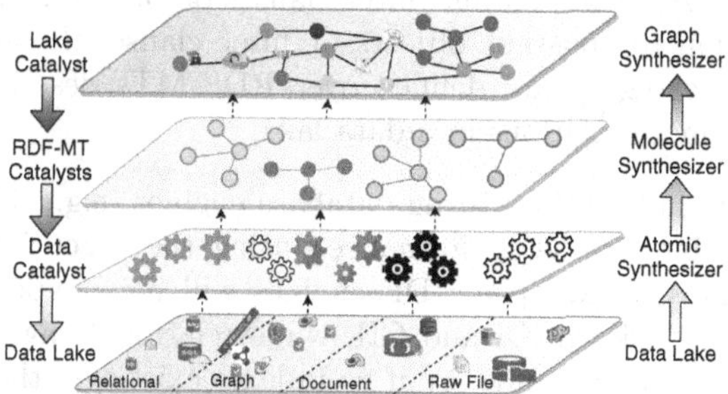

Fig. 3. Semantic Data Lake Basic Components: Lake Catalyst catalyzes a SPARQL query into a set of star-shaped subqueries and select RDF-MTs matching each subquery. RDF-MT Catalysts catalyze star-shaped subqueries into subqueries that can be executed in different Data Catalysts. Finally, Data Catalysts execute a subquery in a data source by translating a given SPARQL query to native query language of the data source. Conversely, the results returned from each these components need to be synthesized (by Atomic, Molecule, and Graph Synthesizers) and passed to the upper layer.

one specific RDF Molecule Template, and provide decomposition, planning and execution of a particular star-shaped subquery. The third layer provide access to a specific data source by translating queries from global querying mechanism, e.g., SPARQL, to the underlying native query mechanism of the data sources via Data Catalysts. Atomic Synthesizers, on the other hand, transform raw data to RDF on demand by applying mapping rules, e.g., defined by RML or R2RML mapping languages. They are specialized to a specific data model and system interface. Atomic Synthesizers perform transformation of the results from native data sources to RDF based on transformation rules. The bottom layer, Data Lake, provides an infrastructure to store raw data and access interface to a set of heterogeneous data sources. These data sources can be characterized with different properties, such as autonomy (sources can be autonomous), data format heterogeneity (provides different data formats), access interface heterogeneity (various query languages), semantic heterogeneity (different representation of same data points), volume (different sizes from small to large data sets), access restrictions, etc.

4.1 Categories of Star-Shaped Subqueries

Ontario is able to differentiate the following star-shaped subqueries:

1. CI: In this category, the star-shaped subqueries do not have a constant object (instantiation) or a filter clause on object variables in any of triple patterns.
2. CII: Star-shaped subqueries in this category do not have a constant object or a filter clause in any of triple patters. Further, these star-shaped subqueries are defined over RDF-MTs described in terms of joins of two or more relations in a data lake.
3. CIII: Star-shaped subqueries in this category are composed of triple patterns with constant objects or contains filter clauses on object variables.
4. CIV: In addition to constant objects or filter clause, the star-shaped subqueries in this category are defined over RDF-MTs described in terms of joins of two or more relations in a data lake.

As to be shown in Sect. 5, existing database engines (e.g., RDF or relational engines) may exhibit diverse performance during the execution of these star-shaped subqueries. For example, RDF engines will have expensive executions of star-shaped subqueries in CI and CII, while relational engines will perform badly on subqueries in CIII and CIV if no indexes exist over the instantiated or joined attributes. Contrary, since RDF engines always create indexes over the subject, predicates, and objects of the triples in an RDF graph [16], subqueries in CIII and CIV speed up in RDF engines.

Algorithm 1. Query Plan Generation: Φ - query decomposition, Q - SELECT query

```
procedure CreatePlan(Φ, Q)
    α ← []
    for SQ ∈ Φ do
        orderTriples(SQ)
        α.push(SQ)
    P ← Q.projs()                      ▷ Q.projs() - list of join and projected variables
    α ← OrderSSQs(α, P)
    while len(α) > 1 do
        SQ_i ← α.pop()
        δ ← []
        β ← [SQ_j for SQ_j ∈ α if shareVars(SQ_i, SQ_j)]
        β ← OrderSSQs(β, P)
        for SQ_j ∈ β do
            J ← join(SQ_i, SQ_j)
            α.remove(SQ_j)
            α.push(J)
            break
        if |β| = 0 then
            δ.push(SQ_i)
    if len(δ) > 0 then
        α ←join(α, δ)
    return α
```

4.2 Star-Shaped Subquery Based Heuristics and Query Planning in Ontario

Once the star-shaped subqueries and the sources where they will be executed are identified, the Ontario optimizer uses a set of heuristics to build query plans. These heuristics are guided by both the type of star-shaped subquery and the selected engine. They transform subqueries in a category into subqueries in another category.

Pushing Down Instantiations into a Star-Shaped Subquery. This rule is performed whenever a star-shaped subquery SSQ_i of type CI is executed over an RDF engine in a query Q. If SSQ_i is part of a join in Q, SSQ_i is selected as the inner subquery of the join; a nested loop join is chosen as the physical operator. Additionally, if variables in triple patterns of SSQ_i are part of a filter in Q, the filter is represented as an instantiation of SSQ_i. Thus, this rule transforms CI subqueries into CIII subqueries.

Breaking Up Joins in Star-Shaped Subqueries. This rule is performed whenever a star-shaped subquery SSQ_i of type CII is executed over an RDF engine in a query Q. In this case, SSQ_i is divided into as many subqueries as joins are defined in the corresponding RDF-MT and the attributes used in SSQ_i. These subqueries are connected by nested loop join operators that will be executed at Ontario level. Thus, this rule enables for transforming subqueries in CII into subqueries in CIV.

Pushing Up Instantiations into a Star-Shaped Subquery. This rule is performed when a star-shaped subquery SSQ_i of type CIII or CIV is executed over an RDB engine in a query Q and the instantiation is not over an indexed attribute. If SSQ_i is part of a join in Q, hash join (or gjoin [1]) is chosen as the physical operator. Further, the selection is represented as a filter which is performed at Ontario level. Thus, this rule enables for transforming subqueries in CIII and CIV into subqueries in CI and CII, respectively.

Combining Joins into a Star-Shaped Subquery. This rule is performed whenever two star-shaped subqueries SSQ_i and SSQ_j of type CI are executed over an RDB engine, and there is a join between them over an indexed attribute. SSQ_i and SSQ_j are merged into one star-shaped subquery $SSQ_{i,j}$, transforming CI subqueries into CII subqueries.

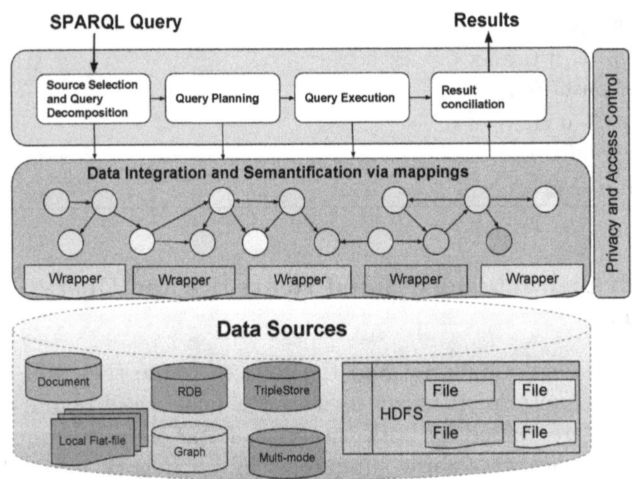

Fig. 4. The Ontario architecture

Given a SPARQL query, Ontario produces a decomposition composed of star-shaped subqueries (SSQs). Using RDF-MT descriptions, Ontario finds a matching RDF-MT for each SSQs. An RDF-MT matches an SSQ if it contains the same predicates as in SSQ. During the creation of SSQs, the heuristics presented in Subsect. 4.2 are used to decide if instantiations are pushed up or filters are pushed down. Also, they are utilized to decide if two SSQs required to be merged. The selected SSQs composed a decomposition, Φ, which represents the input for the optimizer sketched in Algorithm 1. The optimizer is guided by the heuristics in Subsect. 4.2. The planner first performs triple ordering within each SSQs, Line 3–5. The planner then orders SSQs in Φ based on their selectivity, categories, and the data sources, Line 7. Then, iteratively, it picks an SSQ_i (Line 9) and joinable SSQ_js (Line 11), i.e., that share same join variable and create join between them, Line 13–16. Joins are selected and SSQs are ordered according to the heuristics in Subsect. 4.2. Figure 4 illustrates the Ontario architecture.

Table 1. Characteristics of SSQs of the LSLOD benchmark Queries. LSLOD queries are described in terms of categories of star-shaped subqueries (SSQs). Categories are as follows: CI with no instantiations of properties; CII with no instantiations and joins at the RDF-MT definition; CIII with instantiations, and no joins at the RDF-MT definition; and CIV with instantiations with joins at the RDF-MT definition.

Query	Sub-query	Category	Source type	Source(s)
SQ_1	$SSQ_{1,1}$	CI	RDB	DrugBank
	$SSQ_{1,2}$	CI	RDF	Dailymed
SQ_2	$SSQ_{2,1}$	CI	RDF	KEGG
	$SSQ_{2,2}$	CII	RDF	KEGG
	$SSQ_{2,3}$	CIV	RDB	DrugBank
SQ_3	$SSQ_{3,1}$	CI	RDB	DrugBank
	$SSQ_{3,2}$	CI	RDF	ChEBI
	$SSQ_{3,3}$	CIV	RDF	KEGG
SQ_4	$SSQ_{4,1}$	CIV	RDB	DrugBank
	$SSQ_{4,2}$	CIII	RDF	KEGG
SQ_5	$SSQ_{5,1}$	CIV	RDB	DrugBank
	$SSQ_{5,2}$	CIV	RDF	KEGG ∪ ChEBI
	$SSQ_{5,3}$	CI	RDB	DrugBank
SQ_6	$SSQ_{6,1}$	CII	RDB	DrugBank
	$SSQ_{6,2}$	CI	RDF	Diseasome
SQ_7	$SSQ_{7,1}$	CIV	RDF	Dailymed
	$SSQ_{7,2}$	CI	RDB	SIDER
	$SSQ_{7,3}$	CII	RDB	SIDER
SQ_8	$SSQ_{8,1}$	CII	RDB	DrugBank
	$SSQ_{8,2}$	CI	RDF	Diseasome
SQ_9	$SSQ_{9,1}$	CIV	RDB	LinkedCT
	$SSQ_{9,2}$	CII	RDF	Dailymed
SQ_{10}	$SSQ_{10,1}$	CIV	RDB	LinkedCT
	$SSQ_{10,2}$	CI	RDB	DrugBank

5 Empirical Evaluation

We empirically study the behavior of Ontario; it is compared with the state-of-the-art RDF federated engines FedX [13] and MULDER [4]. We study the following research questions: **(RQ1)** What is the overhead of considering heterogeneity during federated query processing? **(RQ2)** Can RDF-MT based source descriptions be effectively applied for source selection, query decomposition, and optimization for non-RDF data sources? **(RQ3)** Are Ontario optimization techniques able to generate effective and efficient query plans for heterogeneous data sources? The experimental configuration is as follows:

Benchmark: LSLOD [6] is a benchmark composed of ten real-world datasets of the Linked Open Data (LOD) cloud from the life sciences domain. The RDF version of LSLOD datasets are transformed into RDB tables. Initially, all the RDF triples that correspond to an RDF-MT are included in one table, but functional and multivalue dependencies between the attributes of a table are utilized to produce normalized version of the table in 3NF. Thus, attributes containing multiple values for one subject are stored in a separate table. Tables and RDF graphs of each of the LSLOD datasets are uploaded in a dedicated Docker container. RDB tables are loaded into *MySQL 5.7.24* and indexes are created for the primary key of each table. We study the LSLOD simple queries [6] (Table 1).

Metrics: We report on the following metrics: **(a)** *Execution Time*: Elapsed time between the submission of a query to an engine and the delivery of the answers; it corresponds to absolute wall-clock system time reported by the Python `time.time()` function. Timeout is set to 300 s. **(b)** *Cardinality*: Number of answers returned by a query. **(c)** *Completeness*: Query result percentage with respect to the answers produced by the unified SPARQL endpoint created as the union of all the benchmark datasets. **(d)** *dief@t*: measures the continuous efficiency of an engine in the first t time units of query execution.

Implementation: Ontario[5] is implemented in Python 3.6. Two versions of Ontario are compared: **(i) RDF version** for queries over RDF graphs accessible via SPARQL endpoints; and **(ii) RDF+RDB version** for executing queries over RDF graphs accessible via SPARQL endpoints and RDB tables stored in MySQL.

Experiment I: Impact of Star-shaped Subquery Types. In this experiment, we analyze the impact of different star-shaped subqueries (SSQs) on the performance of the query engine. This analysis allows us to understand the behavior of the engines while evaluating subqueries and adopt in the planning and execution strategy. Figure 5 shows the performance of Ontario Semantic Data Lake query engine while performing semantification on-demand, compared to execution over materialized version of the same dataset in RDF for selected star-shaped subqueries in each category. The behavior of the engine on SSQs in categories CI ($SSQ_{1,1}$, $SSQ_{1,2}$, and $SSQ_{10,2}$) and CII (i.e., $SSQ_{6,1}$, and $SSQ_{8,1}$),

[5] https://github.com/SDM-TIB/Ontario.

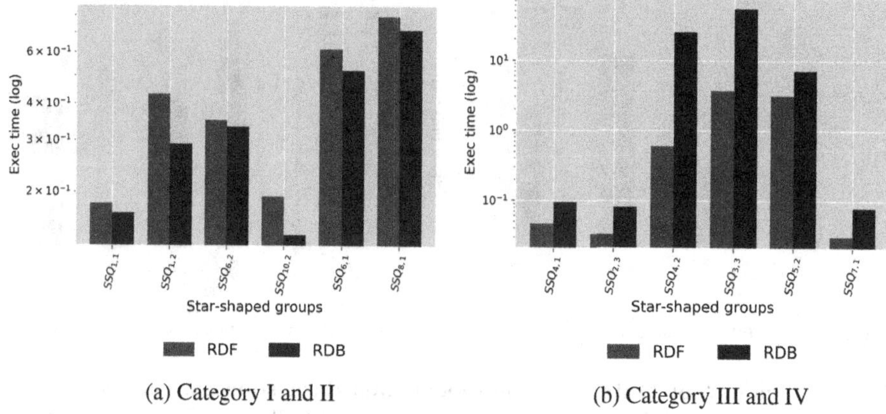

(a) Category I and II

(b) Category III and IV

Fig. 5. Star-Shaped Subqueries. Impact of star-shaped subquery categories is reported.

is presented in Fig. 5a. As can be observed, star-shaped subqueries in these categories are more expensive in RDF than in RDB. RDF engines have indexes over combination of subject, predicate, and object. When a triple pattern do not have instantiation either in subject or object part of any triple patterns, then the engine scan all available data for each predicate in the subquery. On the other hand, relational engine create indexes on primary keys (and optionally any other columns). In RDB, even if triple patterns do not have instantiations, they only scan a relation or a set of relation, unlike RDF triple stores that scan over all data. This leads to RDB engines perform better in these categories than RDF engines. Figure 5b presents SSQs in category CIII and CIV; they comprise triple patterns with object instantiations. The behaviour of the engines in this category shows that, RDF engine performs faster than RDB. This entails, RDB engine performs slower than RDF when the instatiations are not on the indexed predicates. Thus, we answer **RQ1** and suggest that considering different type of subqueries does not affect query execution time.

Experiment II: Impact of Considering Heterogeneity. All the data sources are in RDF, and we evaluate the overhead introduced by the heuristics and query planning techniques proposed in this paper and implemented in Ontario. We consider FedX and MULDER as the baselines of this evaluation and compare their performance with respect to Ontario over a federation of RDF data sources. Figure 6 presents the results of executing LSLOD queries; Fig. 6a shows the results of executing queries where MULDER and Ontario find the same plans while Fig. 6b presents six queries where Ontario generates different plans. Figure 6a suggests that Ontario and MULDER outperform FedX on queries $SQ4$, $SQ5$, $SQ9$, and $SQ10$. Even though, the plans for these queries are the same in Ontario and MULDER, Ontario pays the price of considering heterogeneous data sources and MULDER exhibits a slightly better performance

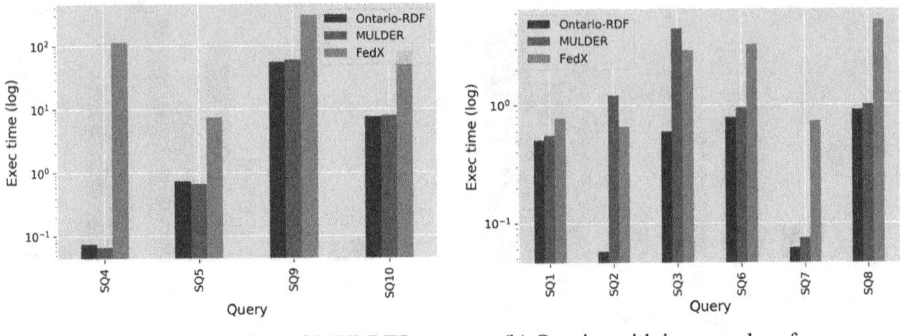

(a) Same plan by Ontario and MULDER (b) Queries with improved performance

Fig. 6. Efficiency of Ontario on homogeneous data sources. Ontario is compared with existing SPARQL federation engines, MULDER and FedX. (a) Ontario outperform FedX in all queries. (b) Ontario overcomes both FedX and MULDER by generating efficient plans and using optimization rules tailored for RDF sources.

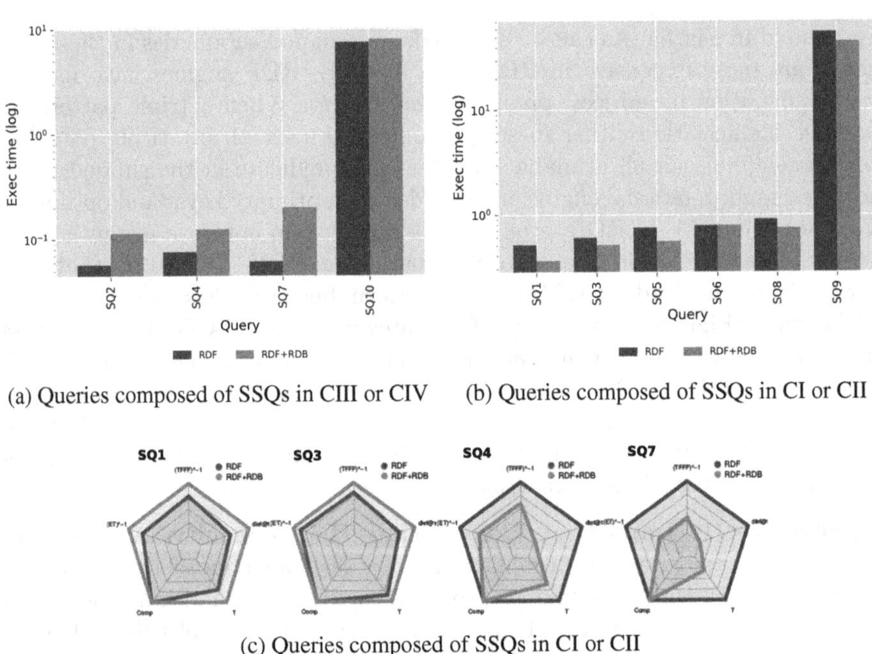

(a) Queries composed of SSQs in CIII or CIV (b) Queries composed of SSQs in CI or CII

(c) Queries composed of SSQs in CI or CII

Fig. 7. Performance of Ontario engine on heterogeneous sources. Executing queries composed of SSQs in Category III or IV are expensive in RDF+RDB data sources, whereas SSQs in Category I and II are expensive in RDF only data sources. $(TFFF)^{-1}$ - inverse time for first result, ET^{-1} - inverse execution time, Comp - Completeness, T - throughput, and $dief@t$ continuous efficiency in time t

for $SQ4$ and $SQ5$. On the other hand, Ontario outperforms both FedX and MULDER by generating efficient plans and using optimization rules tailored for RDF sources on the rest of the queries, as shown in Fig. 6b. In this way, we can answer **RQ1** and suggest that considering heterogeneity does not impact on query performance.

Experiment III: Impact of Heterogeneous Sources. The performance of Ontario over heterogeneous source, i.e., RDF and RDB, is evaluated and analyzed with respect to the SSQ categories; data sources are both in RDF and RDB. Figure 7 presents the results of executing LSLOD queries over two versions of Ontario: RDF only and RDF+RDB. For queries that are composed of SSQs in CIII and CIV, i.e., subqueries with object intantiations, the RDF version performs better than RDF+RDB, as shown in Fig. 7a. This is expected, as we have observed in the Experiment I, star-shaped subqueries with instatiations are cheaper in RDF than RDB; indices over RDF data empower RDF engines and enable a efficient execution of these queries. On the other hand, for queries that are composed of SSQs in CI and CII, the RDF+RDB version performs faster than the RDF version, as shown in Fig. 7b. These results illustrate the benefits of Ontario query processing and their effect on query processing; thus, **RQ2** and **RQ3** can be answered.

Experiment IV: Measuring the Continuous Efficiency. Figure 7c reports on the performance of Ontario in producing continuous answers. The continuous efficiency in time t, i.e., $dief@t$, Inverse of Time for the first tuple ($TFFF^{-}1$), Inverse of Total Execution ($TE^{-}1$), Number of answers produced (Comp), and Throughput (T), are presented in Fig. 7c using radar plots. The interpretation of these metrics in each axes is **'higher is better'**. For all queries, the completeness (Comp) of the queries is 100%, but the throughput varies as it correlates with the overall execution time. As clearly shown, the continuous efficiency of the RDF+RDB version is better in $SQ1$ and $SQ3$, while it is lower in $SQ4$ and $SQ7$ than the RDF version. These results are aligned with the previous experiments and answer **RQ2** and **RQ3**.

6 Related Work

Existing solutions to the problem of query processing over federated data sources rely on a unified interface that allows for executing queries on a federation of homogeneous data sources in a way that execution time is minimized while query completeness is maximized. Federated database engines employ the relational model to represent the unified view of the federation [5], and the query language SQL is utilized to express queries against the federation of databases. Multi-database systems have been proposed to overcome technical heterogeneity. They differ from distributed database systems in a higher degree of autonomy. Dealing with different kinds of data sources introduces new challenges to query processing. The capabilities of the DBMSs may be different, e.g. some systems may support complex queries including joins and aggregations while others do not.

Traditional Data Warehouses, unlike Data Lakes, are centralized data stores that ingest data from heterogeneous data sources after transforming them in a common static structure. Since this kind of data integration may lead eventually to information silos, more flexible data integration approaches have been introduced in recent years. To tackle the data integration problem of heterogeneous data, a few Data Lake systems have been proposed, mainly with focus on data ingestion and metadata extraction and management. For instance, GEMMS (Generic and Extensible Metadata Management System) [10] for Data Lakes extracts metadata from heterogeneous sources, stores it in an extensible metamodel, and enriches it with semantic annotations in order to provide basic querying support. A few other approaches like SeBiDA [9] and Personal Data Lakes [15] propose to keep data from various data sources in raw format in the Data Lake after serializing them in a common data format. PolyWeb [7] and BigDAWG [3] keep data sources in raw format, i.e., without serializing them in a common data format. In PolyWeb, SPARQL queries are translated to the native query language of the sources. PolyWeb indexes each data source predicates for query decomposition and creates left-deep plans. Albeit efficient, existing approaches are not able to exploit knowledge about the main features of the integrated data sources, and produce query plans *customized* for sources selected for collecting the data from a data lake.

7 Conclusions and Future Work

In this paper, we presented Ontario, a federated query processing engine over heterogeneous data sources in a Data Lake. Ontario relies on RDF Molecule Templates to describe heterogeneity of data sources; it is also able to decompose a SPARQL query into a set of star-shaped groups that can be efficiently executed. Ontario also identifies bushy-tree plans which are able to reduce execution time and increase answer completeness; Ontario optimizer is guided by a set of heuristics defined at the level of star-shaped groups and the data engines where they will be executed. We showed through our empirical analysis that, even though, data engines behave differently on diverse types of star-shaped groups, Ontario is able to create efficient and effective plans where physical operators are selected accurately. Thus, our work expands the series of techniques available for federated query processing, and we hope that our techniques will provide scalable solutions in real-world settings. In the future, we plan to investigate other engines and data models to expand the repertoire of heuristics over other data engines.

Acknowledgement. This work has been partially supported by the EU H2020 RIA funded project iASiS with grant agreement No 727658.

References

1. Acosta, M., Vidal, M.-E., Lampo, T., Castillo, J., Ruckhaus, E.: ANAPSID: an adaptive query processing engine for SPARQL endpoints. In: Aroyo, L., et al. (eds.) ISWC 2011. LNCS, vol. 7031, pp. 18–34. Springer, Heidelberg (2011). https://doi.org/10.1007/978-3-642-25073-6_2
2. Belleau, F., Nolin, M.-A., Tourigny, N., Rigault, P., Morissette, J.: Bio2RDF: towards a mashup to build bioinformatics knowledge systems. J. Biomed. Inf. **41**(5), 706–716 (2008)
3. Duggan, J., et al.: The BigDAWG polystore system. SIGMOD Rec. **44**(2), 11–16 (2015)
4. Endris, K.M., Galkin, M., Lytra, I., Mami, M.N., Vidal, M.-E., Auer, S.: Querying interlinked data by bridging RDF molecule templates. TLDKS **39**, 1–42 (2018)
5. Golshan, B., Halevy, A.Y., Mihaila, G.A., Tan, W.: Data integration: after the teenage years. In: 2017 Proceedings of the 36th ACM SIGMOD-SIGACT-SIGAI Symposium on Principles of Database Systems, PODS, pp. 101–106 (2017)
6. Hasnain, A., et al.: BioFed: federated query processing over life sciences linked open data. J. Biomed. Seman. **8**(1), 13:1–13:19 (2017)
7. Khan, Y., Zimmermann, A., Jha, A., Gadepally, V., D'Aquin, M., Sahay, R.: One size does not fit all: querying web polystores. IEEE Access **7**, 9598–9617 (2019)
8. Maali, F., Cyganiak, R., Peristeras, V.: A publishing pipeline for linked government data. In: Simperl, E., Cimiano, P., Polleres, A., Corcho, O., Presutti, V. (eds.) ESWC 2012. LNCS, vol. 7295, pp. 778–792. Springer, Heidelberg (2012). https://doi.org/10.1007/978-3-642-30284-8_59
9. Mami, M.N., Scerri, S., Auer, S., Vidal, M.-E.: Towards semantification of big data technology. In: Madria, S., Hara, T. (eds.) DaWaK 2016. LNCS, vol. 9829, pp. 376–390. Springer, Cham (2016). https://doi.org/10.1007/978-3-319-43946-4_25
10. Quix, C., Hai, R., Vatov, I.: GEMMS: a generic and extensible metadata management system for data lakes. In: 2016 28th International Conference on Advanced Information Systems Engineering CAiSE, pp. 129–136 (2016)
11. Samwald, M., et al.: Linked open drug data for pharmaceutical research and development. J. Cheminformatics **3**(1), 19 (2011)
12. Scharffe, F., et al.: Enabling linked data publication with the Datalift platform. In: AAAI 2012, 26th Conference on Artificial Intelligence, W10: Semantic Cities, Toronto, Canada, July 2012
13. Schwarte, A., Haase, P., Hose, K., Schenkel, R., Schmidt, M.: FedX: optimization techniques for federated query processing on linked data. In: Aroyo, L., et al. (eds.) ISWC 2011. LNCS, vol. 7031, pp. 601–616. Springer, Heidelberg (2011). https://doi.org/10.1007/978-3-642-25073-6_38
14. Vidal, M.-E., Ruckhaus, E., Lampo, T., Martínez, A., Sierra, J., Polleres, A.: Efficiently joining group patterns in SPARQL queries. In: Aroyo, L., et al. (eds.) ESWC 2010. LNCS, vol. 6088, pp. 228–242. Springer, Heidelberg (2010). https://doi.org/10.1007/978-3-642-13486-9_16
15. Walker, C., Alrehamy, H.: Personal data lake with data gravity pull. In: 2015 IEEE Fifth International Conference on Big Data and Cloud Computing, BDCLOUD 2015, pp. 160–167, Washington, DC, USA. IEEE Computer Society (2015)
16. Weiss, C., Karras, P., Bernstein, A.: Hexastore: sextuple indexing for semantic web data management. PVLDB **1**(1), 1008–1019 (2008)

A Model-Driven Framework for the Modeling and the Description of Data-as-a-Service to Assist Service Selection and Composition

Hiba Alili[1,2(✉)], Rim Drira[1], Khalid Belhajjame[2], Henda Hajjami Ben Ghezala[1], and Daniela Grigori[2]

[1] National School of Computer Sciences, University of Manouba, RIADI, 2010 Manouba, Tunisia
{hiba.alili,rim.drira,henda.benghezala}@ensi-uma.tn
[2] Paris-Dauphine University, PSL Research University, CNRS, [UMR 7243], LAMSADE, 75016 Paris, France
{hiba.alili,khalid.belhajjame,daniela.grigori}@dauphine.fr

Abstract. Data as a Service (DaaS) is seen as a promising cloud offering for wrangling the overload of information and making it available across cloud platforms anytime and anywhere. While there exist a large number of DaaS providers in the market, each one has a different way to describe its provided services as well as supplied datasets. The lack of a well-defined machine-readable model strongly hinders the automatic selection and composition of DaaSs. This paper presents MoDaaS, a model-driven framework for the modeling and the description of DaaS services. MoDaaS enables DaaS providers to describe their services capabilities and concerns according to a shared ontology, thereafter it enables them to automatically generate service views in order to assist the integration and data exchange between heterogeneous services.

Keywords: Cloud computing · Data-as-a-Service (DaaS) · Web services · Semantic annotation · Domain ontologies · Model driven engineering · Reuse and specialization

1 Introduction

Increasing number of companies and businesses of all sizes are storing some or all of their data in the cloud, in turn leading to the construction of large data marketplaces in cloud computing environments, such as Microsoft Azure Data Marketplace [5], Amazon Data sets [1], Oracle Data Cloud [6] and Enigma [4], as well as in Open Data Initiative. The data stored in these marketplaces is made available and accessible within the Data as a Service Model. Like all the 'as a Service' family, DaaS is built on the concept that the service, the data in this case, is provided to cloud users on demand regardless of their geographic location.

The use of a DaaS is bound to various concerns. Some of them are technical specific to the data while others concern the service properties such as the Quality

S. Hartmann et al. (Eds.): DEXA 2019, LNCS 11706, pp. 396–406, 2019.
https://doi.org/10.1007/978-3-030-27615-7_30

of Service (QoS) and the cost. The data quality also influences the decision making. All these concerns are critical for DaaS selection. QoS criteria has been extensively studied in favor of data quality. We further argue that the quality of data should be valued as much as QoS to improve the selection decisions. Actually, DaaS providers publish little information about their business models in their websites (HTML documentation). The non existence of a standard and a machine readable model for the description of DaaSs explains the fact that the selection of pertinent services have been mostly carried out manually by users. Furthermore, DaaS users need to deal with the semantic and structural heterogeneities that may exist between different DaaSs in the context of a service composition. While there exist a large number of DaaSs in the market, each DaaS provider has a different way to describe its provided services as well as supplied data sets. The resulting heterogeneity in the data exchanged between the involved services in a DaaS composition strongly hinders the execution of the composition.

In this context, we emphasize the need for a modeling framework to semi-automatically create DaaS models that define DaaSs' functionnal and non functionnal properties in a complete and machine readable description. In order to deal with the heterogeneity issue, this framework allows that DaaS providers annotate DaaSs' functional capabilities according to a domain ontology. Thereby, DaaS capabilities and their associated data concerns can be easily processed by service selection and composition tools. The key idea is to define a higher description level on top of the traditional service descriptions in WSDL, OWL-S, etc, among which DaaS providers define their DaaS capabilities. We highlight in the following the main contributions of this paper: (1) a meta-model that captures all the main concerns associated with DaaS services and the data it provide, (2) a modeling tool which provides an intuitive user interface allowing DaaS providers to define and model their services according to a domain ontology, and finally (3) a view generator which establish mappings between the data a DaaS provides and the domain ontology. The remainder of this paper is organized as follows. We discuss related work in Sect. 2. Section 3 presents MoDaaS, the whole modeling process giving a detailed description of each step. Then, we showcase the working of our solution using a case study in Sect. 4, and finally conclude the paper and discuss our future work in Section 5.

2 Related Work

Little effort have been spent on supporting the description and the modeling of DaaS services. [16] presents DEMODS, a general model to cover all basic information of DaaSs, such as the data service API is decoupled from the data asset description. DEMODS relies on external information models to describe the data assets, but it does not provide any special support for them, beyond links to the description documents. Mashroom [10] uses nested tables to model data services and provides a family of tools to transform HTML, XML, and JSON to the data model. In fact, various data sources are encapsulated as data services with nested tables as their unified data model both for internal processing and for external uses. Users can operate on the nested tables interactively, however it is not the case for data applications. Other frameworks have been developed for publishing information about Web services' capabilities but they neglect the publishing of data concerns. A generic framework for the evaluation and publishing of data concerns, associated with data exposed through DaaS, was proposed in [13]. The authors have analyzed most of DaaS concerns in [14] in details. We go further in this direction and we extend their model and implement it using the MDE capabilities. Furthermore, we implement a generative tool to generate the corresponding service views over a domain ontology.

Many works have proposed solution for automatic web service composition approaches such they regard DaaS as RDF views. Our work complements these efforts by providing an integrated framework to automatically define services' views given their descriptions. The Open Data as a Service (ODaaS) [12] approach uses multi-level modeling to construct open data applications. It consists of a set of domain models and meta-models and a library of 'injectors' to import data from heterogeneous source. The domain descriptions are the classification of concepts and their successive refinements by means of the multi-level modeling, that allows mapping data in some format into a semantically rich model. The data modeling is organized based on generic domain meta-models, making the system use domain-restrictive. In our case, DaaS providers can introduce new ontologies in the framework, thereafter new domain concepts are dynamically defined in the meta-model and can be used for the services annotation.

3 MoDaaS for the Modeling and the Description of DaaS

Our goal is to provide a complete solution that assists DaaS providers to describe their services' capabilities and properties.

3.1 Overall Architecture

Figure 1 shows MoDaaS architecture. It includes three main modeling tools: *DaaSMetaModel*, *MoDaaS Editor* and the *View Generator*.

First, DaaS providers describe their service capabilities and properties through the creation of a new DaaS model under MoDaaS Editor. The resulting DaaS model is saved as an XMI description consisting of several nodes. Each node may correspond to a DaaS provider, a DaaS operation, an I/O parameter or to a service property. Moreover, providers can check for the validity of their models during their creation.

Second, providers need to annotate their DaaS models through matching the different input and output parameters with the shared ontology concepts. This is an important step to deal with data heterogeneity. DaaS providers can also introduce new ontologies of their choice to the system (i.e., by specifying the ontology url/path). For this purpose, we implemented a parser to extract the concepts of the introduced ontology and to dynamically enrich the MoDaaS Editor. Thereafter new concepts are introduced in the editor within the I/O

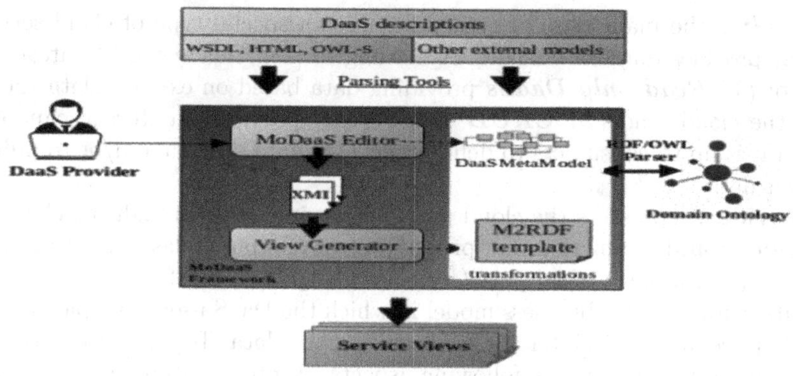

Fig. 1. Overall architecture of the MoDaaS framework

annotation property, enabling to define services through a conceptual representation of the domain of interest provided in terms of the introduced ontology, to which DaaSs are mapped.

Finally, the resulting XMI file can be leveraged to define service views over the shared ontology. In doing so, the M2RDF transformation pattern is executed to define the Resource Description Framework(RDF) view corresponding to the DaaS model.

3.2 Our Meta-Model for the Modeling of DaaS Concerns

MoDaaS is built on top of the meta-model DaaSMetaModel depicted in Fig. 2. It captures all the main DaaS concerns that need to be addressed in its description, and being represented as extensible and customizable classes in the model. In the following, we describe the different meta-model elements:

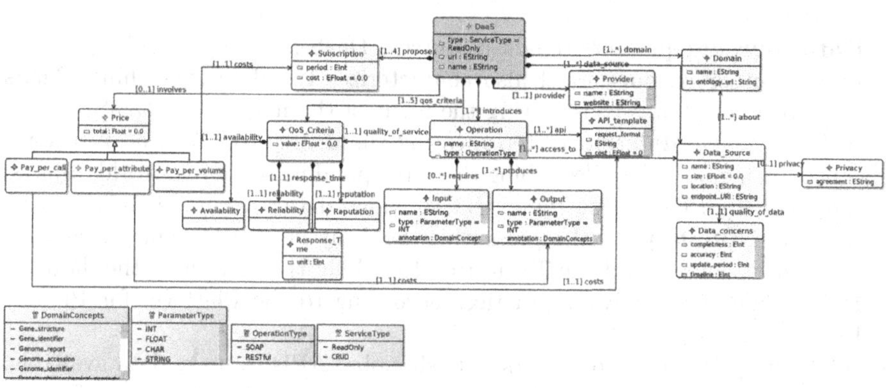

Fig. 2. DaaSMetaModel

- **DaaS:** is the main class of our model. It is a special type of cloud services that provides data on demand. According to [13], DaaSs can be categorized into: ***(1) Read-only DaaSs*** providing data based on existing data sources in the cloud, and ***(2) CRUD DaaSs*** only providing a storage capability and it is up to consumers to define their own data schema and/or to publish their data.
- **Provider:** represents the cloud provider serving data through the cloud.
- **Operation:** is a function that processes a set of inputs (class **Input** in Fig. 2) and results in a set of outputs (see class **Output** in Fig. 2).
- **Subscription:** is a business model in which the DaaS user must pay a recurring price at regular intervals for accessing the data. To quantitatively measure a service quality, the following aspects are often considered:
  - **Reputation:** is a measure of the service trustworthiness. It mainly depends on end user's experiences of using a DaaS.
  - **Reliability:** is the probability that a request is correctly responded within the maximum expected time frame indicated in the DaaS description.
  - **Availability:** is the percentage of time that a DaaS service is operating.
  - **Response time:** is the expected delay in seconds between the moment when the request is sent and the moment when results are received.
- **Price:** aims to represent the cost information when a DaaS web call is executed. Different payment/subscription models, describing how to charge consumers for using DaaS services, are proposed such as:
  - **Pay per call:** DaaS users are charged every time they call a DaaS API to retrieve data. The API usage fee is specified in the API description.
  - **Pay per attribute:** users are charged according to the type and the size of the requested data, such as only unit prices are described.
  - **Pay per volume:** users subscribe for data usage in a period (e.g., a week, a month, etc.) and only pay once in this period with or without limitations for how frequent they access data and how much data they retrieve.
- **Data source:** represents the data source a DaaS is accessing to.
- **Data concerns:** our model also characterizes the datasets behind DaaSs with some attributes constituting the focus of the majority of DaaSs' users, such as the ***Completeness*** representing the degree to which a given data collection includes data describing the corresponding set of real-world objects, the ***Accuracy*** measuring the extent to which data are correct, reliable and certified, **Update period** representing the delay between two major changes in the data source, and the **Timeline** describing the lifetime of the data.
- **Privacy:**describes privacy practices according to the Platform for Privacy Preferences project.
- **API template:** Data providers publish global APIs to make data available. A data request for a given DaaS service is expressed in the same format; given a set of values for the input parameters, the service execution generates values for the output parameters of the service.

- **Domain:** incorporating semantic information about DaaS services, besides the typical information on I/O parameters can help selecting the relevant DaaSs providing the data answering user queries without human assistance.

This is the core model which was extended incrementally by adding new and derived classes to manage all the DaaS and providers concerns. An important extension is the introduction of the "Domain" class, enabling the selection of a domain ontology. Each I/O parameter must be matched to its correspondence from the chosen ontology. Such information is stored in the attribute ***'annotation'***. The re-use factor can be further increased through the introduction of semantic information to DaaS models.

3.3 MoDaaS Editor and Semantic Annotation of DaaS

MoDaaS Editor provides an intuitive user interface that allows DaaS providers to define their services and the data they provide, through a simple graphical modeling tool. As shown in Fig. 3, MoDaaSEditor is composed of *Design Workspace* where users can define new DaaSs by designing and validating the corresponding DaaS model. The essential building blocks of a DaaS model, such as Operation, Input, Output, etc, can be dragged and dropped from a tools *palette* and a Configuration Tabs, each tab opens a view that displays the properties of the selected element in the design workspace.

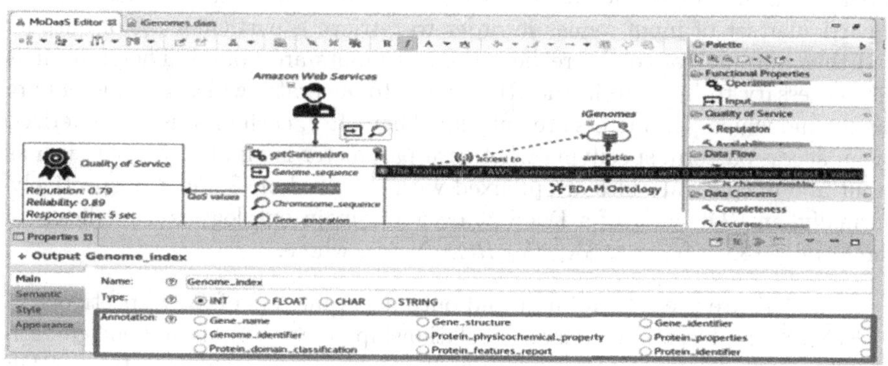

Fig. 3. DaaS modeling and annotation according to the EDAM ontology

The creation of a new DaaS model consists of defining the provider, I/O parameters, QoS values, etc. Such information must be specified in the configuration tab corresponding to each class. DaaS providers must also integrate semantic information about their services by modeling their capabilities according to a shared ontology in MoDaaS. As we said earlier, it is possible to introduce new ontologies in MoDaaS. The semantic annotation of services is performed manually by selecting the corresponding concept from the chosen ontology for each service parameter (cf. Fig. 3).

To implement the *DaaSMetaModel* meta-model, we conform to the Ecore meta-metamodel, and based on which, we built the *MoDaaS* editor.

3.4 Generation of Service Views

The main idea behind service views is to deal with the possible semantic and structural heterogeneity that may exist between the data formats and types returned by different DaaS services. As a second contribution, our framework provides a set of model-to-text transformations to be automatically executed in order to generate service views from input DaaS models. In the literature, most service selection and composition approaches [7,9,15], explicitly regard DaaS as Parametrized RDF Views (PRVs) over a mediated ontology to capture their semantics in a declarative way. Thus, we propose in this work a transformation template, we call ***M2RDF transformation pattern***, which given an input DaaS model, generates the corresponding RDF view over a domain ontology, initially shared in the MoDaaS framework. We note that to implement M2RDF, we used Acceleo which is an EMF (Eclipse Modeling Framework) model-to-text transformations tool.

Using the annotations of the form described in the previous section, M2RDF extracts the service view from the DaaS description file (xmi file) as follows:

I/O parameters are declaratively represented based on concepts and relations that are semantically defined in the domain ontology and selected by the DaaS provider during the service annotation. We recall that a data service requires a particular set of input values in order to retrieve a particular set of outputs and thus outputs cannot be retrieved unless inputs are bound. Therefore, it is also necessary to indicate in the RDF views to generate, which parameters are inputs and which parameters are outputs. Thereafter, each view is characterized by an access pattern, specifying whether a parameter is input or output such as input and output variables are prefixed with the symbols \$ and ?, respectively. Formally, an RDF view of a DaaS S_i over a domain ontology is a predicate of the form: $S_i(\$\overline{\mathrm{X_i}}, ?\overline{\mathrm{Y_i}}) =< \$\overline{\mathrm{X_i}}, ?\overline{\mathrm{Y_i}}, R(\overline{\mathrm{X_i}}, \overline{\mathrm{Y_i}}) >$, where:

- $\overline{X_i}$ and $\overline{Y_i}$ are the sets of input and output parameters of S_i, respectively.
- $\mathrm{R}(\overline{\mathrm{X_i}}, \overline{\mathrm{Y_i}})$ represents the semantic relationship between input and output variables using RDF triples of the form (subject,property,object). These RDF triples are extracted from the .owl file representing the domain ontology.

3.5 Assistance to DaaS Selection and Composition

As mentioned above, most service selection and composition approaches regard DaaSs as PRVs over a mediated ontology. This assumption requires the definition of service views beforehand, which has traditionally been carried out manually by experts when looking for services. This is a complex task that requires considerable effort, especially when considering a vast amount of services. Our goal is to automatically generate RDF views given DaaS descriptions in order to make them useful in the context of service selection and composition. Thereafter, answering a data query consists of identifying the services whose RDF views include all or a part of the conjuncts constructing the data query. Note that user queries must be reformulated over the domain ontology stored (internally) in the system repository. Answering a data query may also require investigating combinations of multiple and heterogeneous DaaSs. RDF views also help capturing all the possible and more certain interactions between the component services, resulting valid and executable compositions.

4 Case Studies

In this section, we present two different case studies in biology to demonstrate how our editor can be easily used to model and semantically annotate DaaS services.

As a first case study, we consider the DaaS *AWS-iGenomes* [2] from Amazon which represents a collection of reference sequences and annotation files for analyzed organisms. This information is made accessible using the service *getGenomeInfo*. Our aim is to model and semantically annotate it using MoDaaS. In our current prototype, we propose the well-known ontology *EDAM* according to which we want to annotate the iGenomes capabilities. EDAM is a comprehensive ontology of well-established, familiar concepts that are prevalent within bioinformatics and computational biology, including types of data and data identifiers, data formats, operations and topics [11]. We provide in the following a step-by-step explanation of the DaaS modeling process in MoDaaS:

Model Design. The first step was the creation of a DaaS model describing all the service capabilities and properties. We defined at the top of the model an instance of the DaaSprovider metaclass representing the provider AWS. We then created and defined an input "Genome_sequence" and three outputs, "Genome_index", "Chromosome_sequence" and "Gene_annotation" within the service operation. Each parameter has been matched to a concept from the chosen EDAM ontology in this case. Actually, the semantic annotation of an operation parameter consists of selecting its correspondence in *EDAM*, from the set of concepts proposed within the property *annotation* as shown in Fig. 3. This set can be automatically enriched each time a new ontology is introduced to MoDaaS. Finally, we specified the response time, the reputation and the availability values. Note that the validation of DaaS models can be performed during their creation. Once all errors are fixed, the provider can launch the generation of service views.

Service Views Generation. The generated RDF view is depicted in Listing 1.1. This is the outcome of the execution of the transformation pattern M2RDF on the DaaS model *"iGenomes.daas"*, created for the service *getGenomeInfo* (cf. Fig. 4). For the sake of the simplicity, we just represent the RDF triples representing the corresponding ontology concepts and not the semantic constraints. The RDF

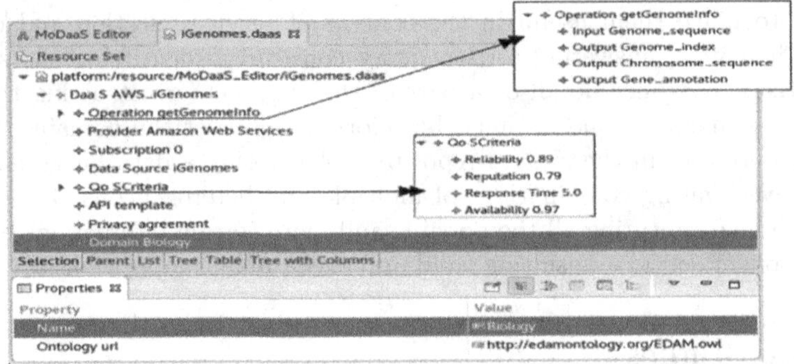

Fig. 4. EMF tree view and property sheet for the created DaaS Model (*.daas)

triples defined in Listing 1.1 were originally extracted from the EDAM ontology. Given the xmi file(iGenomes.daas), the M2RDF template extracts the RDF triple representing each service parameter in the mediated ontology.

```
getGenomeInfo($genome_sequence, ?genome_index, ?Chromosome_sequence,
    ?gene_annotation'):
$<http://edamontology.org/data_2909> <http://www.w3.org/2000/01/rdf-
    schema#label> ''Organism name"
?<http://edamontology.org/data_3210> <http://www.w3.org/2000/01/rdf-
    schema#label> ''Genome index"
?<http://edamontology.org/data_0919> <http://www.w3.org/2000/01/rdf-
    schema#label> ''Chromosome report"
?<http://edamontology.org/data_0916> <http://www.w3.org/2000/01/rdf-
    schema#label> ''Gene report"
```

Listing 1.1. Generated RDF View for the service *getGenomeInfo*

The same way, we generated the RDF view of the service *getProteinByName* (Listing 1.2) providing access to information stored in the DisProt database [3], a database of experimental evidences of disorder. This service returns the UniProtaccession, DisProtID, organism name, its taxonomy and the homologous entries, given a protein name.

```
getProteinByName($Protein_Name, ?DisProt_ID, ?UniProt_Accession, ?
    Organism, ?Taxonomy, ?Homologous_entries'):
$<http://edamontology.org/data_1009> <http://www.w3.org/2000/01/rdf-
    schema#label> ''Protein name"
?<http://edamontology.org/data_2723> <http://www.w3.org/2000/01/rdf-
    schema#label> ''Protein ID (DisProt)"
?<http://edamontology.org/data_3021> <http://www.w3.org/2000/01/rdf-
    schema#label> ''UniProt accession"
?<http://edamontology.org/data_2909> <http://www.w3.org/2000/01/rdf-
    schema#label> ''Organism name"
?<http://edamontology.org/data_1868> <http://www.w3.org/2000/01/rdf-
    schema#label> ''Taxon"
?<http://edamontology.org/data_3148> <http://www.w3.org/2000/01/rdf-
    schema#label> ''Gene family report"
```

Listing 1.2. Generated RDF view for the service *getProteinByName*

Assistance to Service Selection and Composition. The use of a mediated ontology helps detecting the same real world entities expressed differently by different DaaS providers. In this context, MoDaaS enables DaaS developers to match their services' capabilities to the ontology concepts. Thereafter, answering a data query consists of identifying the set of DaaSs whose RDF views contain the triples representing the requested data. Let consider a query Q asking for the organism name and the protein identifier in the DisProt database given an UniProt accession number; **Q($UniProt_accession, ?Organism_name, ?Protein_ID(DisProt)) = $data_3021, ?data_2909, ? data_2723 .**

We represent the RDF triples using their identifiers in the EDAM ontology in order to improve the readability. The RDF views of the services *getProteinByName* and *getUniProtEntryByAccession* contain some RDF triples constructing the query. Consequently, a composition tool can easily determine that these services can be used and combined to get the requested data. The first service can be used to output the protein identifier in the DisProt database and the organism name, given a protein name. No protein name is being specified in the query, the second service must be executed first to output the protein name given its accession, yielding an executable composition.

In practise, we used MoDaaS to generate service views, the input of our service selection and composition algorithms developed in a previous work [8]. Actually, our work was initially motivated by the need for a tool automating the definition of service views given their descriptions. Before the implementation of MoDaaS, we had to manually model data services as views in order to deal with the possible heterogeneity in data structure. This manual way is time-consuming and error-prone given the big number of considered data services.

5 Conclusion and Future Work

This paper presents a model-driven framework for the modeling and the description of DaaSs. Our goal was to provide flexible means of DaaS modeling and to allow their easy integration and automatic invocation by applications. We describe DaaSs as views over a domain ontology in order to deal with schemas heterogeneity. Specifically, the capabilities of each DaaS is mapped to its correspondence in the ontology which is used to define RDF views. The generated views can then be used for answering data queries reformulated over the same ontology. As a future work, we plan to implement a new user interface, enabling users to model a DaaS composition, with generative tools to estimate the corresponding QoS values and to automatically generate and execute the corresponding web calls. Actually, we are implementing the aggregation functions proposed in [7] for the estimation of QoS values, given a service composition. We will also be examining new techniques to generate data mapping workflows that map data from a specific type to another in order to handle the semantic and structural heterogeneities that may exist between different DaaSs in a composition.

References

1. Amazon data sets. http://aws.amazon.com/publicdatasets/
2. Aws igenomes. https://registry.opendata.aws/aws-igenomes/
3. Disprot. http://www.disprot.org/
4. Enigma. https://www.enigma.com/public-data
5. Microsoft data market. https://datamarket.azure.com/
6. Oracle data cloud. https://www.oracle.com/fr/data-cloud/
7. Alili, H., Belhajjame, K., Drira, R., Grigori, D., Ghézala, H.H.B.: Quality based data integration for enriching user data sources in service lakes. In: 2018 IEEE International Conference on Web Services, ICWS 2018, San Francisco, CA, USA, pp. 163–170, 2–7 July 2018
8. Alili, H., Belhajjame, K., Grigori, D., Drira, R., Ghezala, H.H.B.: On enriching user-centered data integration schemas in service lakes. In: Abramowicz, W. (ed.) BIS 2017. LNBIP, vol. 288, pp. 3–15. Springer, Cham (2017). https://doi.org/10.1007/978-3-319-59336-4_1
9. Barhamgi, M., Benslimane, D., Medjahed, B.: A query rewriting approach for web service composition. In: IEEE Transactions on Services Computing, pp. 206–222 (2010)
10. Han, Y., Wang, G., Ji, G., Zhang, P.: Situational data integration with data services and nested table. Serv. Oriented Comput. Appl. **7**, 129–150 (2013)
11. Ison, J.C., et al.: EDAM: an ontology of bioinformatics operations, types of data and identifiers, topics and formats. Bioinformatics **29**(10), 1325–1332 (2013)
12. Segura, Á.M., Cuadrado, J.S., de Lara, J.: ODaaS: towards the model-driven engineering of open data applications as data services. In: EDOC Workshops, pp. 335–339. IEEE Computer Society (2014)
13. Truong, H.L., Dustdar, S.: On evaluating and publishing data concerns for data as a service. In: 2010 5th IEEE Asia-Pacific Services Computing Conference, APSCC, Hangzhou, China, pp. 363–370, 6–10 December 2010
14. Truong, H.L., Dustdar, S.: On analyzing and specifying concerns for data as a service. In: 2009 IEEE Asia-Pacific Services Computing Conference (APSCC), pp. 87–94, December 2009
15. Vaculín, R., Chen, H., Neruda, R., Sycara, K.P.: Modeling and discovery of data providing services. In: 2008 IEEE International Conference on Web Services (ICWS 2008), Beijing, China, pp. 54–61, 23–26 September 2008
16. Vu, Q.H., Pham, T.V., Truong, H.L., Dustdar, S., Asal, R.: DEMODS: a description model for data-as-a-service. In: AINA, pp. 605–612. IEEE Computer Society (2012)

Named Entity Recognition in Local Intent Web Search Queries

Saloni Mittal and Manoj K. Agarwal(✉)

AI and Research, Microsoft India, Hyderabad 500032, India
{salmitta, agarwalm}@microsoft.com

Abstract. Semantic understanding of web queries is a challenging problem as web queries are short, noisy and usually do not observe the grammar of a written language. In this paper, we specifically study the user web search queries with local intent on Bing. Local intent queries deal with searching for local businesses and services in a location. Hence, local query parsing translates into the classical problem of Named Entity Recognition (NER) in NLP. State-of-the-art NER systems rely heavily on hand-crafted features and domain-specific knowledge to effectively learn from the small, supervised training corpora that is available. In this paper, we use deep learnt neural model that relies solely on features extracted from word embeddings learnt in an unsupervised way, using search logs. We propose a novel technique for generating domain specific embeddings and show that they significantly improve the performance of existing models for the NER task. Our model outperforms the existing CRF based parser currently used in production.

Keywords: Named Entity Recognition · Web queries · Embeddings · Deep learning

1 Introduction

Semantic parsing of user web query, i.e., to precisely identify the query intent, is an integral part of any query understanding system. It can be transformed into a semantic tagging task for every term in a query. In this paper, we investigate the queries with the specific intent of searching local businesses and services. For example, in the query, "Hotel Taj Krishna Hyderabad", the intent is to search for a specific business entity, represented by the phrase "Hotel Taj Krishna" located in "Hyderabad". Hence, our objective is to tag "Hotel Taj Krishna" as "*BusinessName*" and "Hyderabad" as "*Location*". Category tags "*BusinessName*" and "*Location*" are known as semantic tags.

Local intent queries such as "*restaurants near me*" or a more implicit natural language query like "*where can I have a cheese cake right now*", aim to search for local business entities in a location, ranging from hotels, restaurants to service providers like plumbers, hairdressers, etc.

In this paper, we cover 15 different category tags for query annotation for the queries seeking local information, covering a wide range of services and business types. Bing returns a single result if the name of the entity is specified in the query, or multiple entities if asked for a type of business, like hotels, restaurants.

S. Hartmann et al. (Eds.): DEXA 2019, LNCS 11706, pp. 407–417, 2019.
https://doi.org/10.1007/978-3-030-27615-7_31

Table 1. Query tagging examples

Tagged query 1	Tagged query 2
taj[businessName] **mahal**[businessName] **Agra**[location]	**hotels**[businessCatgory] **near** [separatorPreposition] **taj**[location] **mahal** [location]
Art[*businessName*] **of**[*businessName*] **living**[*businessName*] **centres**[*businessAttribute*] **in** [*separatorPreposition*] **hyderabad** [*location*]	**Art**[*businessCategory*] **gallaries**[*businessCategories*] **in** [*separatorPreposition*] **Mumbai**[*location*]
Art[*other*] **of**[*separatorPreposition*] **India**[*location*]	**madhubni**[*product*] **art**[*product*] **paintings**[*product*]

Semantic tagging of query terms helps provide more relevant and contextual results. Identifying named entities in queries is non-trivial as keeping a list of all locations, business names, persons is not feasible. Moreover, the surface form of an entity mention can be significantly different from its stored canonical form. For a web scale search engine, it is not feasible to have an exhaustive list covering all variations in the names (including partial lexicon matches) of the entities across the globe. Hence, for a significant fraction of entity mentions in the queries, there is no prior knowledge of the mentioned entity name. Further, in many cases, entities are tagged with the help of context in the query. For instance, in query "pink songs", the modifier word "songs" provides the context and helps disambiguate the entity name to a singer corresponding to mention "*pink*" (where as in query "*pink tops*", "pink" just refers to a color).

Another challenge in tagging the user queries is, query words can be mapped to conflicting categories. For example, *Milan* can be a location (a city) or an organization (a sports team), and *Paris* can be the name of a city or a person. This too makes it a context dependent task; tagging decision should be made by leveraging contextual information from the neighboring words. Traditionally, Named Entity Recognition (NER) in NLP has heavily relied on hand-engineered features [7] (e.g., suffix, prefix features and knowledge about capitalization patterns) and domain-specific knowledge resources (e.g., gazetteers, lexicon bins) to effectively learn from the small amount of supervised training corpora available.

Unfortunately, language and domain specific resources and features are not only costly to develop in new domains, but most of these features are not available for noisy web queries, making NER challenging. In Table 1, we show some of the examples with tagged queries, which highlight the challenges faced by NER systems for tagging terms in web search queries. For example, for queries "taj mahal agra" and "hotels near taj mahal", taj mahal is tagged both as *businessName* and as *location* for the two queries by the production model, which is a wrong annotation. However, word "art" must be tagged differently in each of the queries it is mentioned due to the context words around it (as correctly done by the current production model in Bing for the queries shown in Table 1). Therefore, tagging short and noisy text of web search queries, along with absence of context makes the task of NER for web search queries challenging.

Lample et al. [1] proposed to use bidirectional LSTM, along with a fully connected CRF layer at the top (cf. Sect. 3), for NER task. This network captures arbitrarily long context information around the target word (curbing the limitation of a fixed window size) and produces two fixed size vectors. In their architecture, Lample et al. use a CRF layer as the top layer, for the final entity tagging.

In this paper, we use the architecture presented in [1] for query tagging. However, we face the following additional challenges: (1) Web queries are short and noisy, as opposed to the corpora of well-structured sentences used to train the model in [1]; (2) Number of categories to be tagged are more in case of Bing local intent queries (in our set up, we have 15 categories, as opposed to 4 categories used in [1]).

To overcome these challenges, we propose a novel unsupervised method, to learn the word embeddings using Bing search logs. We exploit the model presented in [1], and propose an unsupervised method to learn features from word embeddings. These embeddings are used as input to the BiLSTM-CRF model proposed in [1]. Our experiments show that the query specific embeddings, learnt over Bing logs, outperform publicly published standards in word vectors like *Glove* [4, 6] on the task at hand.

1.1 Contributions

Specifically, our contributions in this paper are as follows:

- We use the model proposed in [1] for tagging user queries with the entity and intent types. We show that the model is generic in nature and can be adapted for entity tagging in the more complex web queries domain.
- We propose a novel method to generate word embeddings, using search data logs. The word embeddings are generic, i.e., they can be used for various tasks, other than the one addressed in this paper.
- In our experiments, we show that our model trained on using features from these embeddings outperforms the other widely used generic word embedding models such as *Glove* [4].
- Our experiments further show that our model outperforms the intent and entity tagging performance of the production model for local intent queries, for both natural language queries as well as keyword queries.

Rest of the paper is organized as follows: In Sect. 2, we give an overview of the related work. In Sect. 3, we briefly describe the model presented in [1]. In Sect. 4, we present different models used for generating word embeddings. In Sect. 5, we present our results, followed by conclusion in Sect. 6.

2 Related Work

Named Entity Recognition in search queries has been addressed using various statistical and probabilistic models like Conditional Random fields (CRFs) [9], Latent Dirichlet Allocation (LDA) [8] and other topic models. In [15], authors investigated a weakly supervised topic modelling approach using query log data and Latent Dirichlet Allocation. The model is constructed using partially labelled seed entities. The contexts of the entities are used as the documents and the classes of the entity are treated as the topics. In [13], Marius also proposed a weakly supervised method on the similar lines using a set of seeds for each class as input.

In another study [14], authors tackled the lack of context in short queries by capitalizing on information available in the same search session as extra contextual information. They trained CRF models [9], showing that using context from search sessions improves the performance significantly. More recently, authors in [11] proposed NERQ-2S I, a two-step named entity recognizer for open-domain search queries. The first step classifies each query term as token or part of a named entity based on a CRF. The second step takes advantage of these binary labels for categorizing query terms into a pre-defined set of 28 named entity classes using another trained CRF.

Demonstrating the efficacy of a probabilistic sequence classifier in a restricted domain, [7] applied a CRF model to travel-domain search queries. The model's high accuracy performance is attributed to numerous features derived from in-domain dictionaries and word clusters, and labor-intensive heuristic rules developed over several years of work on a rule-based pattern matching system.

In contrast, we have proposed a novel way of extracting domain-specific features in a completely unsupervised way from search sessions of the users, which requires no hand-feature engineering. We believe, it is the first method that has proposed a deep learnt approach to address NER for web search queries for the local intent queries.

3 Input Word Embeddings

In this section, we describe our methodology to generate input word embeddings. The input features to the model are vector representations of the individual words. These vector representations should capture both the semantics and syntactic structure of the word.

The distributional hypothesis of the language, set forth by [12] and [17], states that the meaning of a word can be inferred from the contexts in which it is used. Words that occur in the same context tend to have similar meaning. Many algorithms were developed over the years to make use of this property, latest being unsupervised embedding-based methods which represent each word as a vector such that similar words (words having similar distribution in the corpus) have similar vectors [2]. The key idea behind these unsupervised approaches is to create unlimited supervised training instances from raw text where the goal is to either predict the word from its context (CBoW model) or predict the context from the word (Skip-gram model) using a simple neural network model [2]. As a by-product of this training on large amounts of raw unannotated text, word embedding vectors are created. Our intuition is that named entities appear in regular contexts in large corpora. Therefore, we use pre-trained word embeddings learned from a large corpus that are sensitive to the distributional hypothesis.

Many languages have orthographic or morphological evidence that a word is an entity name (e.g. common suffix/prefix patterns in the names) [1]. We also want to generate the vector representations of words that are sensitive to spelling and can capture the form/structure of the word. We, therefore, use a model that derives representation of a word from the characters that compose it. Figure 1 shows the architecture used in our model, first proposed in [1, 5, 6]. To prevent the models from depending on one representation or the other too strongly, we use dropout method. We find that dropout mechanism is crucial for good performance.

3.1 Pre-trained Word Embeddings

We use pretrained embeddings instead of randomly initializing them. Significant improvement is reported using pretrained word embeddings [1]. Fine-tuning the embedding weights during training provides no difference on the model's performance; hence, the embedding layer weights are kept fixed for shorter training time. We experimented with two sets of publicly available embeddings, namely Stanford's Glove embeddings trained on Wikipedia [4] and Web text and Facebook's *FastText* trained on Common Crawl [3]. Among the two, Glove performed slightly better on the intended task. However, Glove word vectors pose the problem of out-of-vocabulary (OOV) named entities, where it is very unlikely for small businesses or remote locations/neighborhood names to have Wikipedia information about them. Glove vectors have only 40% vocabulary coverage for train, development and test sets vocabulary combined. Also, it is always desired that the similarity between word vectors learned by the algorithm captures the same aspects of similarity that are useful for performing the intended task.

Given these, we can safely hypothesize that word embeddings trained on in-domain text may perform better on our task. In next section, we describe our methodology to generate the in-domain embeddings.

3.2 Local Search Specific Word Embeddings

Web queries are short, do not observe the grammar of the written language, and users often overlook word order, and hence, cannot be potentially used to train an effective unsupervised embedding model. Therefore, for training local domain specific word embedding model, we used 20 million queries from Bing log. For all the queries for which local answers trigger, top five corresponding web snippets are fetched. These web snippets contain well-formed sentences that (1) contains most/all the tokens of the query, (2) semantically consistent with the query, and (3) provide enough contextual information. A corpus of these web snippets together is used to train a *word2vec* Skip-gram [2] model after appropriate pre-processing of text. We train many Skip-gram models with varied dimensions and window sizes to learn the embeddings. Among these, model with embedding dimension 200, minimum frequency cutoff of 4 (i.e., the word must occur at least 4 times in the training corpus) and a window size of 5 (i.e., two preceding and two subsequent words around the current word) for the skip-gram model, yielded the best results. We obtained word vectors for 1461489 unique words, including entity names. These new local specific word embeddings give the advantage of approximately 80% vocabulary coverage. These word embeddings become richer as more searches hit the search engine.

3.3 Character Based Embeddings

To learn embeddings for a word from its characters, a Bi-LSTM as given in Fig. 1 is used, as described in [1]. The character embedding corresponding to each character in a word is given in direct and reverse order to a forward and backward LSTM. The final states of the forward and backward LSTMs are then concatenated to form a character-level representation of the word. Further, this character embedding is concatenated with the pre-trained word embedding from a word look-up table to be used as final input to the model. Learning character level embeddings has the advantage of handling out-of-vocabulary problem for rare and unseen words.

4 Architecture

In the following section, we provide the description of the hybrid (Bi-LSTM + CRF) tagging architecture. The architecture is proposed in [1], and it is similar to the ones presented by Colobert et al. in [6], and Huang et al. in [5].

4.1 Bi-LSTM

The Long-Short-term Memory networks (LSTMs) is a family of RNNs [10]. For a given sentence $(x_1, x_2,.., x_n)$ of n words, each represented as a d-dimensional vector w_t; $1 \leq t \leq n$, an LSTM computes a representation l_t of the left context of the sentence for every word x_t. A second LSTM reads the sequence in reverse to generate a representation r_t of the right context. These are two distinct networks with different parameters, and the forward and backward LSTM pair is referred as a bidirectional LSTM (Bi-LSTM). Figure 1 represent architecture, presented in [1].

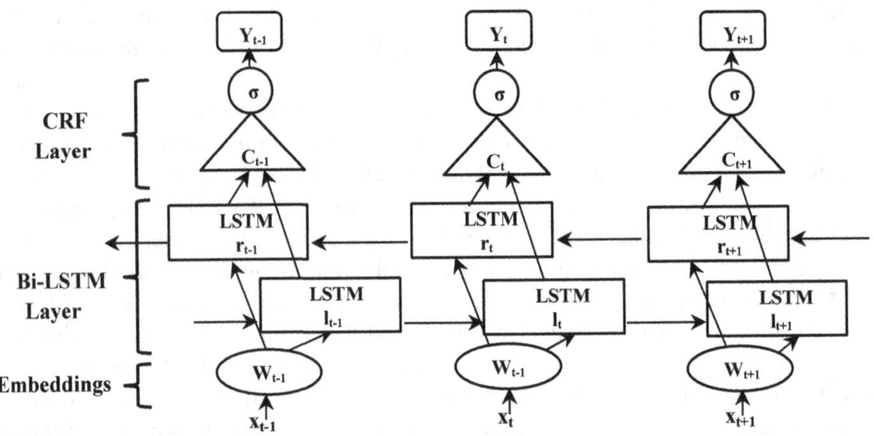

Fig. 1. Architecture proposed for training the NER model in [1]. We use a novel methodology to generate word-embeddings, which out-perform the widely used *Glove* and *FastText* word embeddings

The representation of a word is obtained by concatenating its right and left context vectors, $c_t = [l_t; r_t]$. This representation effectively takes into account an arbitrary length of context on both sides of a word and eliminates the problem of diminishing context for long sentences. The final concatenated output at each time step is decoded by a linear layer and a log-softmax layer into log-probabilities for each tag category, i.e., corresponding to each word in a *k*-sized vector at the output, where *k* is the number of tags (*k* is 15 in our case). Therefore, we generate a matrix of size of $n \times k$, where *n* is the size of input query. This matrix is fed to the CRF layer.

4.2 CRF Layer

Consider a query "Paris Hilton"; the first word can be tagged a *location* or a *person*, but if we choose it as *person* then second should be tagged as *person* with certainty. Therefore, independent classification decisions are highly sub-optimal when there are strong dependencies among input words and output labels. As reported in [1], for NER task, the "grammar" or the rules to characterizes interpretable sequences of tags imposes several hard constraints on the tag sequences (e.g., PERSON tag cannot be followed by LOCATION tag, etc.). Therefore, it would be impossible to model an interpretable tag sequence with independence assumptions. Therefore, instead of modelling tagging decisions independently, we model them jointly using a conditional random field. For an input query $(x_1, x_2, .., x_n)$.

Consider *P* to be the matrix of output scores by the bidirectional LSTM network. *P* is of size $n \times k$, where *k* is the number of distinct tags, and P_{ij} corresponds to the score of the jth tag for the ith word in the sequence. For a sequence of predictions

$$\boldsymbol{y} = (y_1, y, y_3 \ldots .y_n)$$

define its score to be

$$s(\boldsymbol{X}, \boldsymbol{y}) = \sum\nolimits_{i=0}^{n} A_{y_i, y_{i+1}} + \sum\nolimits_{i=1}^{n} P_{i, y_i} \tag{1}$$

where A is a matrix of transition scores such that A_{ij} represents the score of a transition from tag *i* to tag *j* ([1]).

A *SoftMax* over all possible tag sequences yields a probability for the sequence $\boldsymbol{y}$:

$$p(\boldsymbol{y}|\boldsymbol{X}) = \frac{e^{s(\boldsymbol{X}, \boldsymbol{y})}}{\sum_{\tilde{y} \in Y_X} e^{s(\boldsymbol{X}, \tilde{\boldsymbol{y}})}} \tag{2}$$

This becomes the objective function for combined Bi-LSTM + CRF model. During training, we maximize the log-probability of the correct tag sequence [1].

5 Experiments

In this section, we present the methods used to train the models and the impact of local domain specific embeddings on the overall task.

5.1 Training

We trained our network using the back-propagation algorithm updating our parameters on every training batch of size 20, using Adaptive Moment Estimation [1, 16] with a learning rate of 0.01. The LSTM-CRF model used a single layer for the forward and backward LSTMs. The hidden layer dimension for both forward and backward LSTM layers is set to 200 in the best performing model. For the character-level model, the hidden dimension of the forward and backward LSTMs is 70 each and it also uses a single layer. Tuning this dimension had no significant impact on the model performance. A dropout mask is applied to the final input word embedding layer with dropout rate set to 0.5. A significant improvement is observed in the model's performance after using dropout. We used the Xavier uniform initializer [18] to initialize the parameters of the LSTM, which draws the samples from a uniform distribution.

We tweaked several components of the model, keeping the basic architecture same, to understand their impact on the overall performance. We explored the impact that the character-level representations, pre-training our local domain specific word embeddings and dropout had on the LSTM-CRF model. To compare the performance of our word embeddings with Glove vectors, keeping the rest of the network configuration same, two models were trained with different input vectors.

5.2 Data

The model is trained on a set of 280,000 queries including both natural language queries (e.g., "where can I buy grocery near me") and keyword-based queries (e.g., "grocery stores near me"). 50% of these training samples are synthetically generated using specific grammar for each sequence of tags and a limited vocabulary, whereas the remaining queries were real user queries. While training, we validated our model against a validation set of real user queries, comprising 15000 queries, with early stopping parameter set as 5.

The model is evaluated on 3 different test sets, with varied characteristics, comprising real user queries on Bing.com, annotated manually by human judges, which are different from training data set; **Test Set 1**: 1400 keyword-based queries, **Test Set 2**: 19000 natural language queries and **Test Set 3**: 1200 queries on which the current production model regresses. Test Set 3 queries contain many unseen and misspelled business names that are not found in the lexicon list search.

5.3 Results

The evaluation metric used is complete query prediction accuracy percentage, i.e., the tagging is considered accurate only if all the n tags of a n-word length query are predicted correctly. Table 2 shows the results of all the model variants on the evaluation sets. Our best performing model has surpassed the existing CRF model on all 3 sets, suggesting that given enough data neural network learns relevant features for NER without hand feature engineering.

As we can see in Table 2, using our local domain specific embeddings (model id 3) gave an improvement of +9.33, +2.28 and +5.70 on the three sets respectively, over glove embedding vectors (model id 2). Clearly, the local vectors outperform glove on the intended task. Learning Character-level word embeddings further resulted in an additional improvement of about +2.3, +1.55 and +4.62 on all the three sets respectively (Table 2). Finally, our best performing model (model id 4) gave an improvement of +5.3, +2.0 and +22.32, signifying a huge improvement over existing production level model (based on CRF and hand-crafted rules).

Table 2. Results using different configurations. "***pretrain***" refers to models that include pretrained word embeddings, "***char***" refers to models that include character-based modeling of words, "***dropout***" refers to models that include dropout rate.

Model id	Model name	Variant	Test set 1	Test set 2	Test set 3
1	*Production*		68	73	30
2	*Bi-LSTM + CRF*	Glove pretrain + dropout	61.67	71.17	40
3	*Bi-LSTM + CRF*	Local pretrain + dropout	71	73.45	45.70
4	*Bi-LSTM + CRF*	Local pre train + char + dropout	73.3	75	50.32

To further estimate the model robustness, we performed the 10-fold cross validation of our model, for which the training data into 10 parts, with each part becoming the test data once. The variance and the mean accuracy of our model is plotted in Fig. 2. We see that the mean validation error falls to 8.93% after 25 iterations, i.e., the mean accuracy is over 91%. Further, the variation in the standard deviation of validation error rate across different models is small.

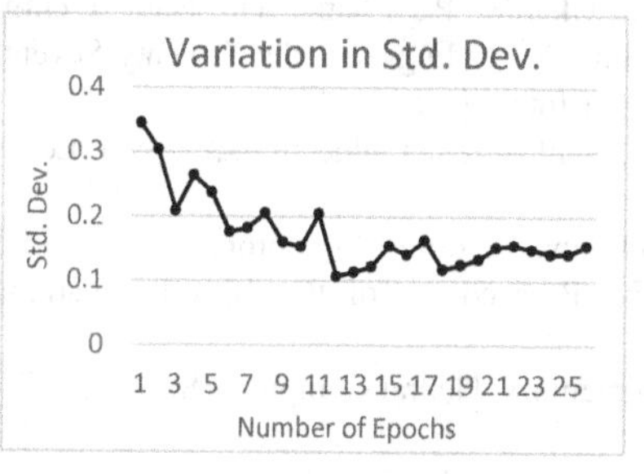

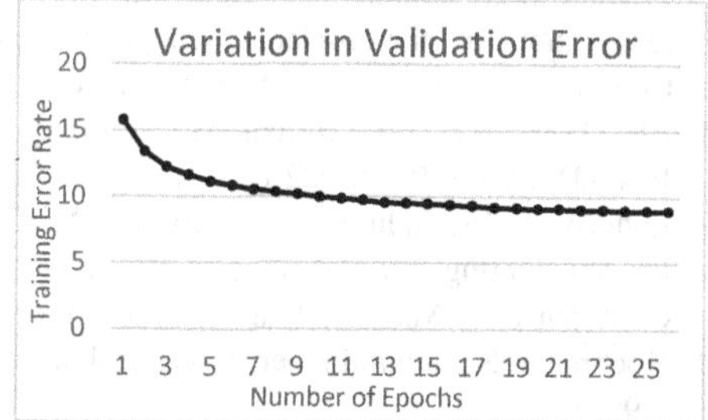

Fig. 2. Variation in Std. Dev and mean training error rate with the number of epochs for 10-fold cross validation of our model

6 Conclusion

Semantic parsing of web queries is extremely important as it reveals actionable signals to many downstream applications, including search ranking, etc. In this work, we train neural parsers and show the importance of creating in-domain embeddings to be used as features for NER models. We used the model initially proposed in [1, 5], for tagging noisy and short web queries. The number of classes or tags used for tagging the queries were also significantly higher compared to the classes used in [1].

Our experiments show that we achieved significant improvement over traditional parsers for Bing local queries. We showed that the word-embedding trained over Bing search logs significantly outperforms the existing production model as well as the models trained on off the shelf embeddings such as *FastText* [3] and *Glove* [4]. Our embeddings can be used for various semantic tagging tasks for web search queries over Bing. Our future work direction is to create a generic intent understanding model for search queries.

References

1. Lample, G., Ballesteros, M., Subramanian, S., Kawakami, K., Dyer, C.: Neural architectures for named entity recognition. In: Proceeding of NAACL-HLT (2016)
2. Mikolov, T., Chen, K., Corrado, G., Dean, J.: Efficient Estimation of Word Representation in Vector Space (2013). https://arxiv.org/abs/1301.3781
3. Joulin, A., Grave, E., Bojanowski, P., Mikolov, T.: Bag of tricks for efficient text classification. In: Proceedings of European Chapter of the Association for Computational Linguistics (EACL) (2017)
4. Pennington, J., Socher, R., Manning, C.D.: GloVe: global vectors for word representation. In: Proceeding of Empirical Methods in Natural Language Processing (EMNLP) (2014)
5. Huang, Z., Xu, W., Yu, K.: Bidirectional LSTM-CRF models for Sequence Tagging (2015). https://arxiv.org/abs/1508.01991
6. Collobert, R., Weston, J., Bottou, L., Karlen, M., Kavukcuoglu, K., Kuksa, P.: Natural language processing (almost) from scratch. J. Mach. Learn. Res. (JMLR) **12**, 2493–2537 (2011)
7. Cowan, B., Zethelius, S., Luk, B., Baras, T., Ukarde, P., Zhang, D.: Named entity recognition in travel-related search queries. In: Proceedings of the Twenty-Seventh Conference on Innovative Applications of Artificial Intelligence (2015)
8. Blei, D.M., Ng, A.Y., Jordan, M.I., Lafferty, J.: Latent dirichlet allocation. J. Mach. Learn. Res. (JMLR) **3**, 993–1022 (2003)
9. Lafferty, J., McCallum, A., Pereira, F.C.N.: Conditional random fields: probabilistic models for segmenting and labeling sequence data. In: Proceedings of the 18th International Conference on Machine Learning (2001)
10. Hochreiter, S., Schmidhuber, J.: Long short-term memory. Neural Comput. **9**(8), 1735–1780 (1997)
11. Eiselt, A., Figueroa. A.: A two-step named entity recognizer for open-domain search queries. In: Proceedings of the Fourth International Joint Conference on Natural Language Processing (IJCNLP 2013), pp. 829–833 (2013)
12. Harris, Z.: Distributional structure. Word **10**(23), 146–162 (1954)

13. Pasca, M.: Weakly-supervised discovery of named entities using web search queries. In: Proceedings of the Sixteenth ACM Conference on Conference on Information and Knowledge Management - CIKM 2007, pp. 683, New York (2007)
14. Du, J., Zhang, Z., Yan, J., Cui, Y., Chen, Z.: Using search session context for named entity recognition in query. In: Proceeding of the 33rd SIGIR (2010)
15. Guo, J., Xu, G., Cheng, X., Li, H.: 2009 named entity recognition in query. In: Proceedings of the 32nd SIGIR (2009)
16. Kingma, D.P., Ba, J.: Adam. A Method for Stochastic Optimization (2017). https://arxiv.org/abs/1412.6980
17. Firth, J.R.: A synopsis of linguistic theory 1930–1955. In: Studies in Linguistic Analysis, pp. 1–32. Oxford Philological Society, Selected Papers of 1952–1959, London (1968)
18. Glorot, X., Bengio, Y.: Understanding the difficulty in training deep feedforward neural networks. In: Proceedings of the 13th International Conference on Artificial Intelligence and Statistics (AISTATS), Sardinia, Italy (2010)

Database Processing-in-Memory: A Vision

Tiago R. Kepe[1,2(✉)], Eduardo C. Almeida[1(✉)], Marco A. Z. Alves[1(✉)], and Jorge A. Meira[3(✉)]

[1] Federal University of Paraná, Curitiba, Paran, Brazil
{trkepe,eduardo,mazalves}@inf.ufpr.br
[2] Federal Institute of Paraná, Curitiba, Paran, Brazil
[3] University of Luxembourg, Luxembourg City, Luxembourg
jorge.meira@uni.lu

Abstract. The recent trend of Processing-in-Memory (PIM) promises to tackle the memory and energy wall problems lurking in the data movement around the memory hierarchy, like in data analysis applications. In this paper, we present our vision on how database systems can embrace PIM in query processing. We share with the community an empirical analysis of the pros/cons of PIM in three main query operators to discuss our vision. We also present promising results of our ongoing work to build a PIM-aware query scheduler that improved query execution in almost 3× and reduced energy consumption in at least 25%. We complete our discussion with challenges and opportunities to foster research impulses in the co-design of Database-PIM.

Keywords: Processing-in-Memory · Query scheduler · Energy efficiency

1 Introduction

PIM is a hardware architecture with simple processing units attached to the memory chip to efficiently use the internal memory bandwidth. PIM was originally thought at the end of the 60's [14]. Over the years, PIM followed the progress of memory technology with processing components installed in magnetic disks to run particular database algorithms [6] that later evolved to Smart Disks [15] with embedded logical components. Intelligent RAM [20] also tried to add logical units inside the DRAM to support specific computations. Unfortunately, commercial products did not adopt those approaches due to limitations of the hardware technology and the continuous growth in CPU performance complied to the Moore's Law and Dennard scaling. In the 2010s flash disks also tried to add internal functional units to run database applications [5,7,22]. However, they missed a general programming interface and suffered from the low abstraction level when handling hardware errors.

With the increasing growth in CPU performance, the memory access became the bottleneck for many applications, a problem known as the "memory wall".

S. Hartmann et al. (Eds.): DEXA 2019, LNCS 11706, pp. 418–428, 2019.
https://doi.org/10.1007/978-3-030-27615-7_32

In the last years, the assumptions of Moore and Dennard came to an end due to hardware limitations and the emergence of multi-core, but the memory wall remains an open issue as well the current "energy wall". In data-centric systems, both walls are critical as huge amounts of data move around the memory hierarchy from disks, main memory, and caches to the CPU.

Why PIM Now and Again? Recently, PIM architectures came back to the spotlight due to the introduction of Through-Silicon Vias (TSVs) that made 3D integration technologies feasible: the integration of DRAM dies and logic cells in the same chip area. Furthermore, commercial GPUs already embed the emerging 3D-stacked memories, such as the Hybrid Memory Cube (HMC) [12] and the High Bandwidth Memory (HBM) [17]. These smart memories invert the traditional data processing approach: they move computation near to the data.

There is no surprise that PIM poses as an attractive approach to reduce the data movement in data-centric systems. Therefore, in our vision, it is time to discuss *how Database Management System (DBMS) can embrace PIM in query processing*. The major benefits to be explored in the Database-PIM co-design are the drastic reduction in energy consumption and the internal high memory bandwidth due to the high levels of data access parallelism and the on-chip processing. PIM has been used to accelerate isolated database operators: select [24] and join [19]. But, our recent work demonstrated [16,23] the trade-offs of PIM in query processing. The behavior of each query operator depends on the dataset characteristics and the system cache settings. This position paper aims to disclose the key requirements and insights for Database-PIM.

I. PIM-aware Query Processing. We make a case for designing a PIM-aware Query Processing engine that coordinates intra-query parallelism between CPU and PIM. We reinforce the idea of moving computation around instead of just deciding over the data movement, discussing: *When is it cheaper to process a record in main memory rather than moving around the memory hierarchy?*

II. PIM-aware Scheduler. We introduce a design of PIM-aware Scheduler [16] at operator granularity to interleave intra-query processing between CPU and PIM. We discuss the research question: *How the DBMS should coordinate intra-query execution between CPU and the memory processor to exploit the potential gains from each device?*

III. Findings and Results. We share our findings about the impact of PIM on traditional query processing. We discuss the energy-saving and drastic reduction in response time that goes more than 1 order of magnitude for certain query operators.

IV. Challenges and Opportunities. We share a list of challenges and opportunities for the co-design and integration of a Database-PIM.

2 Background and Related Work

PIM Architectures. Emerging 3D-stacked memories increase memory bandwidth by integrating one logic die with a stack of four or eight DDR-3 dies bonded by the TSV. Typically, the DDR-3 module consists of memory rows of

8-KB, which means that the memory controller must activate an entire 8-KB DRAM row to access any memory portion, even smaller portions than 8-KB, which limits the memory bandwidth and increases energy consumption. Therefore, the 3D-stacked memories split the 8-KB DRAM rows into small rows of 256-bytes. They organize the stack of DRAM dies vertically, where underlying memory banks of different dies are interconnected by the TSV to the logic die at the base, forming a vertical partition called as a "vault" (see Fig. 1). Current 3D-stacked memories have 32 interdependent vaults to access and process data in parallel leveraging the internal high bandwidth memory up to 320 GB/s. The logic layer of 3D-stacked memories supports arithmetic, logical and bitwise atomic instructions up to 16 bytes size.

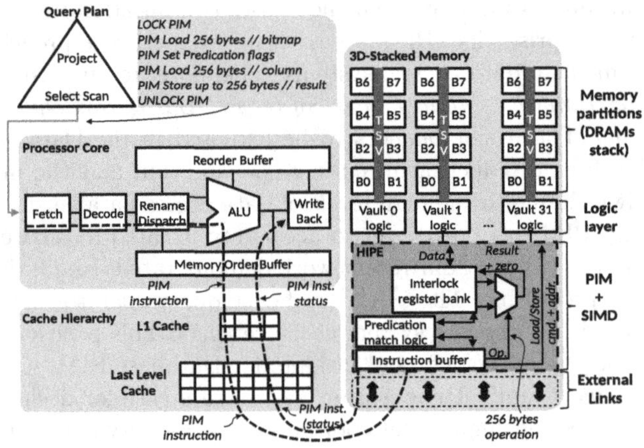

Fig. 1. A query datapath movement in a traditional von Neumann architecture plus a modern 3D-stacked memory with PIM and SIMD support.

To fully benefit from the 256-bytes vault's row buffer, a recent work [1] extended the logic layer to operate with vectorized instructions of 256-bytes, enabling *Single Instruction, Multiple Data* (SIMD) with PIM. Figure 1 shows the described PIM architecture with SIMD support to the right side and the traditional von Neumann architecture to the left: the processor core and a detached cache hierarchy. At the top of Fig. 1, the processor dispatches PIM instructions directly to the PIM device bypassing the cache hierarchy. At the end of the instructions, the PIM device only returns the instructions status to the CPU to continue the pipeline. This is a main advantage compared to current DBMS, such as Netezza and Exasol, that also filter data in hardware before passing to the CPU. They have to deal with packing qualifying tuples into condensed pages to avoid unnecessary bufferpool pollution, which is expensive and error prone. On the other hand, current memory protocols support all the idiosyncrasies of PIM instructions, such as cache coherence, Error-Correcting Code Memory (ECC) and Direct Memory Access (DMA). The execution flow works at instruction-granularity as the traditional CPU processing (e.g., AVX/SSE): programmers insert intrinsics PIM instructions into the code, like Intel Intrinsics, and the compiler flags them as special memory PIM instructions.

PIM as Query Accelerator. Recent works use PIM devices as isolated accelerators to boost query operators, such as selection [24], projection [23], and join [19]. However, this one-sided approach neglects the potential of CPU-PIM co-processing with caching and energy-saving benefits.

Kernel Scheduling on PIM-Assisted GPU. Related work in GPU architectures proposed scheduling techniques with PIM devices installed as GPU main memory. GPU applications are split into independent GPU-kernels and interleave the processing of each kernel between the GPU cores and the PIM device [11,21]. Although GPUs are highly parallel devices to boost processing power, data still need to be transferred around the memory hierarchy before moving to the GPU-PIM device.

Scheduling on Emerging Hardwares. Current intra-query scheduling focused on co-processing between GPU and CPU to improve execution time based on runtime learning model [3] and operator cost model [13]. In the Intel Xeon Phi co-processor [4] was also tested a similar scheduling idea. While these approaches tackle compute-intensive applications, they neglect the potential of PIM to run data-intensive applications.

Flash Disks. Recently, the attention moved to flash disks to accelerate [8] and save energy [18] of scan and join operators. However, there are two main problems in these works: (1) They rely on complex and database dedicated hardware that may reduce the general use for the hardware. Smart SSDs [7] use an embedded ARM processor into the SSD with a firmware for communication to evaluate the execution of database operators. Intelligent SSDs [5] add a reconfigurable stream processor to reach high processing performance with energy savings. (2) They are application-driven without a general interface to abstract hardware features.

3 Query Processing with PIM

The first step towards a PIM-aware query processing is understanding *the impact of PIM on traditional query processing.* We discuss the impact on analytic workloads through the execution time and energy consumption of the operators: projection, selection, and join when switching the processing unit from the CPU to PIM. Since current PIM hardware do not yet implement all the extensions depicted in Fig. 1, we performed the analysis through the SiNUCA cycle-accurate simulator [2] with the same parameters used by related work [24]. The energy estimations consider the DRAM parameters for state-of-the-art PIM devices [12]. The SiNUCA is extensively adopted by scientific articles in computer architecture [24], HPCC [2] and database [16,23].

We evaluate those operators using the 1 GB TPC-H database workload because their input data set fit into the caches (the best scenario for the CPU processing). It is therefore possible to investigate the data reuse behavior of such operators. The CPU implementations of the query operators use SIMD instructions of 64-bytes (i.e., Intel AVX-512) and the PIM versions use SIMD instructions of 256-bytes (see [16,23] for details). The results presented here guide us to exploit our Database-PIM co-design.

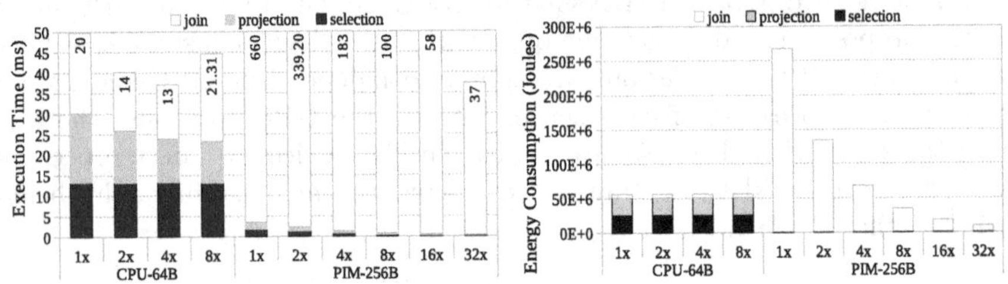

Fig. 2. The execution time and energy consumption breakdown of the TPC-H Query 03 on the operators: selection, projection and join in the CPU-64B and PIM-256B with different unrolling depths.

Projection: Figure 2 shows that the PIM-aware projection operator reduced the execution time by more than one order of magnitude, 60× for projection and 23× for projection-path. We also observe reductions in energy consumption around 36× for both projection primitives. **Selection**: The PIM-aware selection operator reduces the energy consumption by around 98% and reduces the execution time by 76× compared to the best CPU scenario.

The streaming behavior of the projection and selection operators results in low data reuse and less amount of off-chip data transfers when running into PIM. In a conventional CPU processing, the column under processing is streamed from main memory across the memory hierarchy to the CPU registers, then the CPU tests column chunks against the selection and projection predicates. However, all the column chunks are "dead" on arrival in the cache hierarchy, because the operators do not reuse any chunk during the execution. On the other hand, the counterpart PIM operators send the computation from CPU to the PIM, both operators access all the 32 vaults in parallel and evaluate the columns with 32× PIM+SIMD instructions of 256-bytes. These results endorse the feasibility of both operators for PIM. However, small enough tables may fit into the cache, and the PIM device might lead to performance loss. Here, the DBMS must take a careful decision between CPU and PIM.

Join: In our experiments, we implemented the tree main join algorithms, but the Hash Join and Sort-Merge Join algorithms generate random memory accesses that inhibit the internal data access parallelism and the high bandwidth. To give an intuition of the problems of join in PIM, we analyze the execution of the Nested Loop Join (NLJ). The NLJ-PIM unrolls the inner loop up to 32× to exploit the data access parallelism of the PIM device. Every inner loop iteration causes compulsory *load* and *store* instructions, i.e., the inner column is re-accessed every time in the inner loop. Such an effect is more evident in our tests because the inner column fits into the data caches that have lower memory latency than the DDR3 dies. Thus, the CPU processing outperforms the execution of the NLJ-PIM, as depicted in Fig. 2. The best CPU processing is 2.8× faster than PIM. Even the energy consumption, a major benefit of PIM, is 70% worst in the best PIM execution. However, the NLJ-PIM becomes appealing as long as the inner column does not fit into the LLC inhibiting data reuse. In this point, PIM shows an improvement of 1.38× against the CPU. Here, we observe opportunities to tackle random memory access of other join algorithms.

4 PIM-Aware Query Processing

Although the emerging PIM architectures stand as high performance memory technology removing part of the memory wall, the impact of PIM on query processing arises the issue: *Not every (instance of) database operator benefits from PIM.* The choice of the target architecture to process a query operator is not trivial. Indeed, we observed that query operators with high data reuse benefit from the caching mechanism and thus the CPU processing becomes appealing, such as for the NLJ. On the other hand, operators that perform data streaming (e.g., projections and selections) are best fit for PIM [16]. Other operators, such as aggregation, may have low or medium data reuse and partial streaming behavior, making it unclear where is the best fit device to process.

4.1 PIM-Aware Scheduler

Our investigation focus on how to interleave intra-query execution between the CPU and PIM. To the best of our knowledge, this is the first effort in that direction. Existing solutions usually direct sole data-intensive operators for PIM [19,23,24]. However, they neglect the potential of CPU processing boosted by caching mechanism for workloads with high temporal and spatial data locality. Thus, our insight is that *a database system requires a PIM-aware scheduler for intra-query processing.*

The DBMS scheduler is a critical performance component in query processing: it orchestrates the execution of physical primitives from the query plans. Thus, we design two scheduling strategies for a PIM-aware scheduler: static profile based scheduling and dynamic profile based scheduling. Inspired by related work, we discuss two potential strategies to receive as input the optimal plan generated by the query optimizer and coordinate the intra-query execution between PIM and the traditional CPU.

Static Scheduling. The static scheduling uses a classification model based on operator profiles to decide which architecture to process a given operator. Different from related work on operator scheduling for GPU, this strategy does not require a calibration step [3] nor heuristics [13]. Figure 3 depicts three execution plans from the optimal plan: two hardware-specific query plans, i.e., either CPU or PIM, and a hybrid query plan based on the static profile scheduling. In the hybrid query plan, the selection and projection operators run into the PIM hardware, while the join operator executes in the CPU. The hybrid query plan outperforms the hardware-specific query plans (Fig. 3d) in almost 3× exploiting the advantages of both architectures: the operators with high data reuse run in CPU, and the operators with data streaming run into PIM. Also, the hybrid plan reduces the energy consumption in at least 25%.

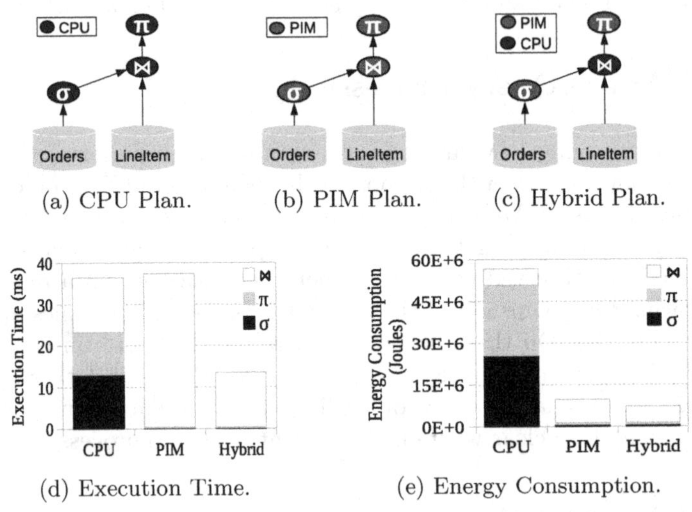

Fig. 3. Query execution plans: a sole CPU (**a**) and PIM (**b**) plans, a hybrid plan (**c**) with CPU/PIM. Respective execution time (**d**) and energy consumption (**e**).

Dynamic Scheduling. The dynamic scheduling is our ongoing work to coordinate, automatically, the intra-query execution between CPU and PIM. We envision the dynamic scheduler regarding two criteria: execution time and energy consumption.

We present a prototype of the dynamic scheduling in Fig. 4: (1) Initially, it assumes a target architecture based on the PIM-aware algorithms from each operator of the query plan; (2) The scheduler requests the observed operator profiles to the DB Profile that returns the stored profiles generated according to some criteria on database load monitoring (e.g., energy consumption); (3) The scheduler chooses the most prominent profile and (4) decides the target architecture and algorithm. (5) After the execution, the scheduler stores the observed workload information based on the chosen criteria to learning base.

At its heart, the learning model decides the target architecture for an incoming operator from a query plan. The decision model shall select the most prominent algorithm based on the defined criteria from a set of profile operators. A profile operator consists of an execution algorithm (e.g., NLJ), the target architecture (e.g., CPU, PIM), the measured criteria (e.g., energy consumption) and the features about the dataset (e.g., size, distribution, cardinality, and type). We bind the workload information W to any profile, such as data reuse and selectivity, and use in statistical methods to interpolate future profiles. The W is useful to measure the quality of the chosen profile. For instance, the PIM-aware scheduler measures the distance between the new W and the chosen profile (e.g., Dynamic time warping). We call this distance as "Energy ramp" when using the energy criteria. Also, the W might be used to insert, remove or update profiles.

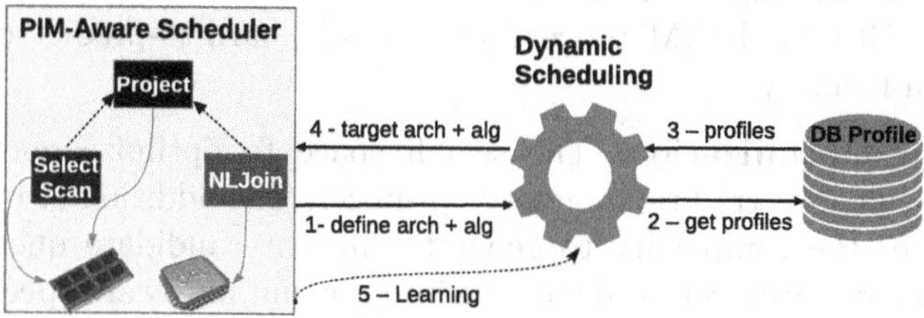

Fig. 4. Architecture overview of the PIM-aware Scheduler (dynamic scheduling).

Figure 5 exemplifies the energy consumption of expected and observed profiles for the NLJ. As the data sets do not fit into the caches (L1, L2, and LLC), then the data reuse distance increases, which leads to increasing data movement across cache levels and more energy consumption. When the dynamic scheduling detects an increasing energy ramp, a scheduling decision is made to switch the processing unit of the running operator based on the trend profile. In the example, we use the least squares method to set the energy trend profile. If the energy

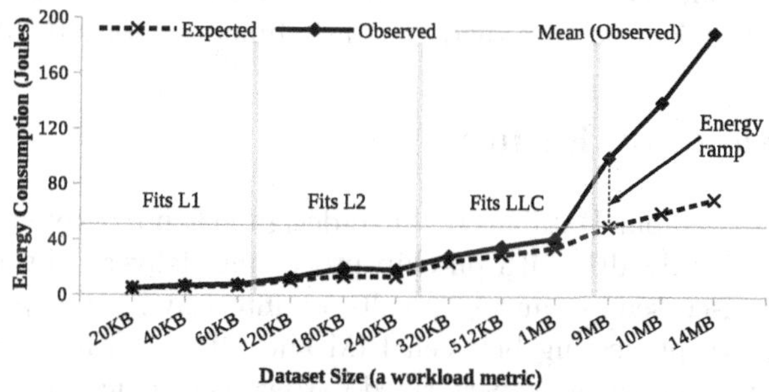

Fig. 5. The dynamic scheduler monitors the energy consumption (Joules) during the data movement (Bytes) of the NLJ for Query 03.

trend goes up and data do not fit in cache, then the operator is rescheduled to the PIM device. In our ongoing work, we also study the tradeoffs of restarting the running operator from scratch after rescheduling(e.g., cache invalidation).

5 Challenges and Opportunities

Hybrid query processing and optimization require a holistic view of a query optimizer to exploit heterogeneous co-processing. Next, we identify a list of challenges for the co-design of Database-PIM.

Simultaneous Co-processing: In heterogeneous processing environments, the optimizer needs to identify opportunities of co-processing to avoid idle devices or inefficient power-consumption (e.g., the power wall problem [25]). The operator-pipelining is also an important technique to be further investigated. Besides, the simultaneous CPU and PIM processing may add hard-to-predict concurrency into the main memory.

Query Plan Optimization: The search space to optimize query plans is already a problem for traditional query optimizer. The addition of hybrid query plans increases the complexity to generate efficient candidate query plans. A PIM-aware query optimizer needs to take into account hardware-specific features to choose the appropriate running device, such as limited processing power and reduced data movement for PIM, and fast processing with caching mechanisms for superscalar CPUs.

Transactions: Intrinsically, arithmetic and logical PIM update instructions are atomic [10]. This opens research opportunity for near-data transactions and Hybrid Transactional/Analytical Processing (HTAP). The current PIM ISA supports compare-and-swap instruction to evaluate values. Therefore, PIM update instructions can be synchronized in-memory without wasting cache-check time or extra memory bandwidth. The opportunity is to reduce the overhead of locking and latching, which correspond to 30% of the instructions in OLTP [9].

DBMS Adoption: We envision that DBMSes should invoke PIM instructions at the operator code base, similarly to the SSE and AVX approaches. We also consider code optimization to provide intrinsic functions for PIM ISA.

6 Conclusion and Future Work

This paper presented our Database-PIM co-design vision to exploit the unprecedented memory bandwidth with on-chip processing delivered by modern PIM hardware. We discussed promising results of interleaving the parallel execution of intra-query processing between PIM and CPU. With our static scheduler, the hybrid query plans outperformed hardware-specific plans in almost 3× and reduced energy consumption about 25%. Our ongoing work focuses on the dynamic scheduling strategy to create and update operator profiles and reschedule operators on-the-fly between the CPU and PIM devices. A future work is to investigate the hash join, sort-merge join, and aggregations, also to consider skewed data distribution. Finally, we share with the community a list of challenges and opportunities opened by our vision in the co-design of DBMS-PIM.

Acknowledgments. This work was partially supported by the Serrapilheira Institute (grant number Serra-1709-16621).

References

1. Alves, M.A.Z., Diener, M., Santos, P.C., Carro, L.: Large vector extensions inside the HMC. In: DATE (2016)
2. Alves, M.A.Z., Villavieja, C., Diener, M., Moreira, F.B., Navaux, P.O.A.: Sinuca: a validated micro-architecture simulator. In: HPCC/CSS/ICESS (2015)
3. Breß, S., Mohammad, S., Schallehn, E.: Self-tuning distribution of DB-operations on hybrid CPU/GPU platforms. In: 24th Grundlagen von Datenbanken (2012)
4. Cheng, X., He, B., et al.: Many-core needs fine-grained scheduling: a case study of query processing on intel xeon phi processors. J. Parallel Distrib. Comput. **120**, 395–404 (2018)
5. Cho, S., Park, C., Oh, H., Kim, S., Yi, Y., Ganger, G.R.: Active disk meets flash: a case for intelligent SSDs. In: ICS (2013)
6. DeWitt, D.J., Hawthorn, P.B.: A performance evaluation of data base machine architectures (invited paper). In: 7th VLDB (1981)
7. Do, J., Kee, Y., Patel, J.M., Park, C., Park, K., DeWitt, D.J.: Query processing on smart SSDs: opportunities and challenges. In: SIGMOD (2013)
8. Graefe, G., Harizopoulos, S., Kuno, H.A., Shah, M.A., et al.: Designing database operators for flash-enabled memory hierarchies. IEEE Data Eng. **33**, 21–27 (2010)
9. Harizopoulos, S., Abadi, D.J., Madden, S., Stonebraker, M.: OLTP through the looking glass, and what we found there. In: SIGMOD (2008)
10. HMC Consortium: Hybrid Memory Cube Specification 2.1 June 2015. http://www.hybridmemorycube.org/. hMC-30G-VSR PHY
11. Hsieh, K., Ebrahimi, E., Kim, G., Chatterjee, N., O'Connor, M., Vijaykumar, N., Mutlu, O., Keckler, S.W.: Transparent offloading and mapping (TOM): enabling programmer-transparent near-data processing in GPU systems. In: ISCA (2016)
12. Jeddeloh, J., Keeth, B.: Hybrid memory cube new DRAM architecture increases density and performance. In: VLSIT (2012)
13. Karnagel, T., Habich, D., Schlegel, B., Lehner, W.: Heterogeneity-aware operator placement in column-store DBMS. Datenbank-Spektrum (2014)
14. Kautz, W.H.: Cellular logic-in-memory arrays. IEEE Trans. Comput. **18**, 719–727 (1969)
15. Keeton, K., Patterson, D.A., Hellerstein, J.M.: A case for intelligent disks (IDISKs). In: SIGMOD Record (1998)
16. Kepe, T.R.: Dynamic database operator scheduling for processing-in-memory. In: PhD@VLDB (2018)
17. Kim, J., Kim, Y.: HBM: memory solution for bandwidth-hungry processors. In: 26th HCS (2014)
18. Kim, S., Oh, H., Park, C., Cho, S., Lee, S.: Fast, energy efficient scan inside flash memory. In: ADMS@VLDB (2011)
19. Mirzadeh, N., Kocberber, O., Falsafi, B., Grot, B.: Sort vs. hash join revisited for near-memory execution. In: ASBD@ISCA (2015)

20. Patterson, D.A., et al.: A case for intelligent RAM. IEEE Micro **17**, 34–44 (1997). https://dblp.uni-trier.de/rec/bibtex/journals/micro/PattersonACFKKT97
21. Pattnaik, A., Tang, X., Jog, A., Kayiran, O., et al.: Scheduling techniques for GPU architectures with processing-in-memory capabilities. In: PACT (2016)
22. Tiwari, D., Boboila, S., et al.: Active flash: towards energy-efficient, in-situ data analytics on extreme-scale machines. In: 11th USENIX/FAST (2013)
23. Tome, D.G., Kepe, T.R., Alves, M.A.Z., de Almeida, E.C.: Near-data filters: taking another brick from the memory wall. In: ADMS@VLDB (2018)
24. Tome, D.G., Santos, P.C., Carro, L., de Almeida, E.C., Alves, M.A.Z.: HIPE: HMC instruction predication extension applied on database processing. In: DATE (2018)
25. Wang, L., Skadron, K.: Implications of the power wall: dim cores and reconfigurable logic. IEEE Micro **33**, 40–48 (2013). https://dblp.uni-trier.de/rec/bibtex/journals/micro/WangS13

Context-Aware GANs for Image Generation from Multimodal Queries

Kenki Nakamura and Qiang Ma(✉)

Graduate School of Informatics, Kyoto University, Kyoto, Japan
kenki@db.soc.i.kyoto-u.ac.jp, qiang@i.kyoto-u.ac.jp

Abstract. In this paper, we propose a novel model of context-aware generative adversarial networks (GANs) to generate images from a multimodal query: a pair of condition text and context image. In our study, context is defined as the objects and concepts that appear in the image but not in the text. We construct two object trees expressing the objects and the corresponding hierarchical relationships described in the input condition text and context image, respectively. We compare these two object trees to extract the context. Then, based on the extracted context, we generate parameters for the generator in context-aware GANs. To guarantee that the generated image is related to the multimodal query, i.e., both the condition text and context image, we also construct a context discriminator in addition to the condition discriminator, similar to that of conditional GANs. The experimental results reveal that the prepared model generates images with higher resolutions, containing more contextual information than previous models.

Keywords: Context-aware GANs · Context · Multimodal query

1 Introduction

Image generation based on generative adversarial networks (GANs) [3] is widely studied [5,9,11,12]. In particular, Reed et al. proposed a model of conditional GANs for automatic image generation for a given (conditional) text query [12]. Conditional GANs make it possible to generate images from text and can generate copyright-free images for creating training datasets, sightseeing promotions [6–8], and so on.

However, a text-only query may miss important context information. For example, suppose Bob wants to share his trip to Paris with his friends. He wants to share his experiences visually by using photos with some comments. However, unfortunately, he missed the photos shot around the Eiffel Tower. In this case, if we simply apply the text-conditional GANs [12] to generate images, we may miss the context of "Bob's trip to Paris" and generate irrelevant images, although they may look like real ones. We need to consider the context of generating images to help Bob share his experience visually. Moreover, it is not easy to accurately

This work is partly supported by MIC SCOPE(172307001).

S. Hartmann et al. (Eds.): DEXA 2019, LNCS 11706, pp. 429–443, 2019.
https://doi.org/10.1007/978-3-030-27615-7_33

describe the details of the target object, or the components we want to generate, with a short text query. For example, the query "The man is sitting on the chair" does not contain information about what the man looks like and what kind of chair he is sitting on. A possible solution is that we can use multimodal input, i.e., text and images, to complement the input information, and then we can generate images according to the user's intentions.

In this paper, we propose a model named context-aware GANs to handle multimodal queries. A multimodal query is a pair of condition text and context image. In our study, the context denotes the objects and concepts expressed in the context images but not clarified in the text query. We construct two object trees expressing objects and their hierarchical relationships described in the condition text and context image, respectively. We then compare them to extract the context. After that, we generate the noise, which is the input to the generator of our context-aware GANs, based on the extracted context. The idea behind is that images sharing the same context should be generated from noises with the same distribution. Furthermore, we introduce an additional discriminator, named context discriminator, to check whether the generated images are related to the context or not.

To generate images with high resolutions, we train our model in a progressive manner, by adding new layers incrementally. We use the parameters learned in the previous steps as the initialization values of the current step. However, the original model of progressive GANs [5] is not trainable because of the complex structure of the context discriminator. As a solution, our model only activates the context discriminator at the low-resolution layers, and the subsequent layers are only trained for improving the resolutions.

The major contributions of this work can be summarized as follows.

- We propose a novel concept named "context" and a method of its extraction from multimodal queries. "Context" captures the relationship between multimodal inputs. It can complement and enhance the user's query for image generation (Sect. 3).
- We propose a model of context-aware GANs. By generating the parameters based on the context extracted from multimodal queries, and by introducing an additional context discriminator, our model can generate images without losing contextual information. Furthermore, we train the generator and discriminator in a progressive manner to increase the resolution of generated images (Sect. 3).
- The experimental results for a real dataset and a user study demonstrate that our context-aware GANs can generate high-resolution images related to query text and image (Sect. 4).

2 Related Work

Goodfellow et al. proposed the framework of GANs for image generation [3]. Radford et al. proposed a method to train GANs steadily and efficiently by using convolutional neural networks (CNNs) [11]. In case of image generation, the generator is trained to generate clear images looking like real ones. The discriminator is trained to identify its input image as a real image from the dataset or a fake one generated by the generator. In other words, this is a min-max game of two players (the generator and discriminator), which can be formulated as follows:

$$\min_G \max_D V(D,G) = \mathbb{E}_{x \sim p_{data}(x)}[\log D(x)] + \mathbb{E}_{z \sim p_z(z)}[\log(1 - D(G(z)))], \quad (1)$$

where G, D are the functions of the generator and discriminator, x is training data, and z is a random noise. The generator minimizes $\log(1 - D(G(z)))$ to cheat the discriminator, and the discriminator maximizes Formula 1 to improve its accuracy.

Reed et al. proposed the model of conditional GANs to generate images according to the conditional text [12]. In the original GANs, the discriminator distinguishes [real image, real text] from [fake image, arbitrary text]. In conditional GANs, the discriminator judges if the input is sampled from the dataset distribution and if it matches the text or not. The discriminator of conditional GANs distinguishes between [real image, real text], [real image, arbitrary text], and [fake image, arbitrary text]. Its loss function is defined as follows:

$$s_r = D(x, h) \ \{real\ image,\ right\ text\}, \quad (2)$$

$$s_w = D(x, \hat{h}) \ \{real\ image,\ wrong\ text\}, \quad (3)$$

$$s_f = D(\hat{x}, h) \ \{fake\ image,\ right\ text\}, \quad (4)$$

$$L_D = \log(s_r) + (\log(1 - s_w) + \log(1 - s_f))/2, \quad (5)$$

where the output of the discriminator with the input of [real image, real text] is s_r. The output from the input [real image, arbitrary text] and with [fake image, arbitrary text] are s_w and s_r, respectively. In our context-aware GANs, we introduce an additional context discriminator with a context loss function.

Vondrick et al. proposed a model to generate videos from image sequences [14]. Their model generated sequential images with the same background. Zhang et al. proposed the model of StackGANs to output high-resolution images by dividing the image generation process into two stages: outputting low-resolution images from text embeddings in the first stage and generating higher-resolution images from these images in the second stage [15].

Karras et al. proposed the method of generating high-resolution images by adding new layers during the training rather than training all parameters at the same time [5]. Their experimental results demonstrate that all the parameters can approach the appropriate values more efficiently. Training the generator and discriminator progressively makes it possible to capture the overall features of the image distribution in the first half of the training and then shift to learning fine-grained features.

However, to the best of our knowledge, there is no model of GANs that could handle multimodal queries, which are more reasonable in real applications. In our previous work [9], we proposed the concept of context-aware GANs and trained it based on similarity of generated images to the context image.

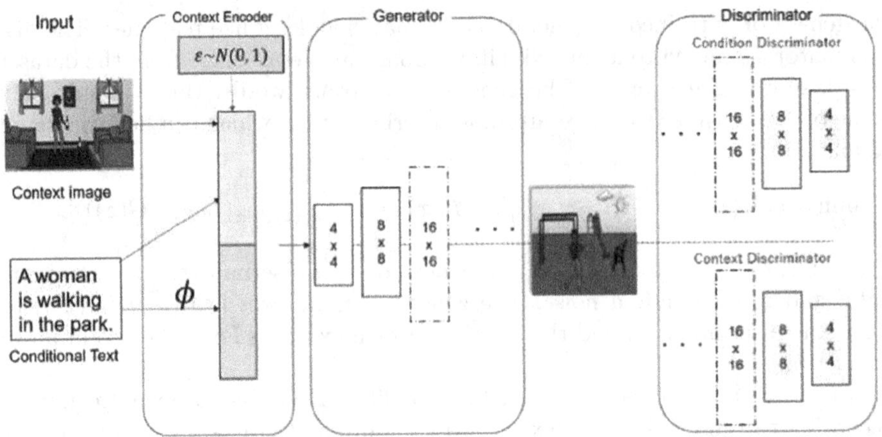

Fig. 1. Overview of context-aware GANs

The concept of context was not well-defined, and the previous context-aware GANs could not generate high-resolution images from both the conditional text and context image.

3 Context-Aware GANs

3.1 Overview

As shown in Fig. 1, our context-aware GANs consist of three components: (1) context encoder, (2) generator, and (3) discriminator. The input of a multimodal query consists of one conditional text and the corresponding context image. The output consists of images that satisfy the condition text and are related to the context image.

The context encoder analyzes the given multimodal query and generates the input vector for the generator. The generator generates images, and discriminators distinguish the real images from generated images. In a context-aware GAN, there are two discriminators. The first one is the condition discriminator, similar to that in conventional conditional GANs. The condition discriminator checks whether the image is (1) the real one or not, and (2) suitable for the conditional text or not. The second one, the context discriminator, checks whether the context is expressed in the image or not.

To generate images with high resolutions, we train the generator and discriminators in a progressive manner. That is, we start with training the networks to generate images with low resolutions (e.g., 4×4). Then, we incrementally add new layers to the generator and discriminator, and train the networks to generate higher-resolution images. The details of training GANs in a progressive manner can be found in [5].

3.2 Context Encoder

The context encoder (Fig. 2) represents the condition text and context image using object trees. The context is extracted by comparing these two object trees. Then, the context encoder draws a latent distribution from the context to sample the noises, which are the parameters to generate images in GANs. The context encoder also encodes the conditional text and concatenates it with the noise for use as the input vector to the generator.

Multimodal Encoding. To generate images from the given text, at first, we construct a multimodal space to capture the relationship between images and texts. That is, we embed text and its corresponding images to the same/close points to construct the multimodal space.

The input of constructing the multimodal space is a set of images and its corresponding text descriptions. At first, images are encoded by using an autoencoder [4]. Then, the text encoder is trained so that the text is projected to the position close to the corresponding image in the single multimodal space constructed by the autoencoder.

We basically use the model of long short-term memory (LSTM) to encode text. We convert each word to a vector by using GloVe [10]. In addition, to handle text with various lengths, especially for long text, we introduce the syntax analyzer [2] to convert texts into parser trees. As shown in Fig. 3, for example, if the text is "the book is on fire", the syntax analyzer will output parser trees, which can be represented by triple tuples like ["det", "the", "book"], ["nsubj", "book", "is"], ["cop", "is", "fire"], and ["advmod", "on", "fire"]. One simple way to represent the relationships (the first term in the triple tuple of a parser tree,

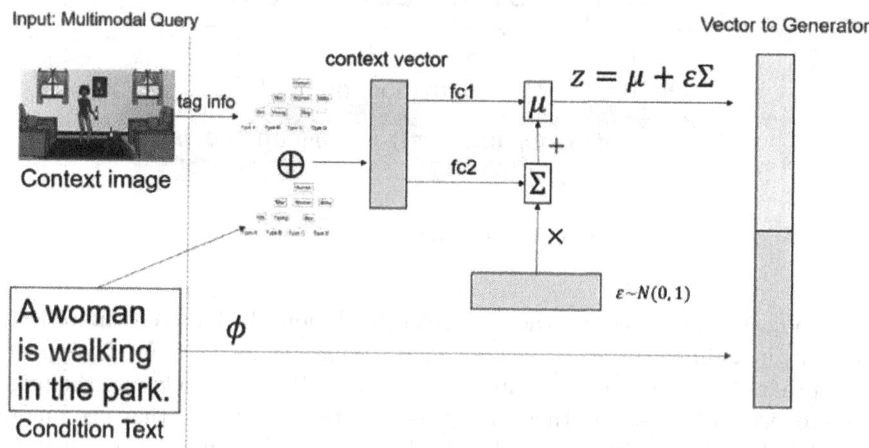

Fig. 2. Context encoder

e.g., "det", "nsubj"[1].) is using a one-hot vector [13]. However, this approach could not capture the difference between different relationships. As a solution, in this work, we apply the skip-gram approach to learn the distributed representation of relationships, as shown in Fig. 4.

In short, each triple tuple can be represented by a vector consisting of three distributed representations of relationships and two terms. Then, as shown in Fig. 4, as a result, each sentence can be translated into a sequence of vectors and used as the input to train the text encoder.

Context Extraction. As mentioned before, "context" denotes the objects and concepts expressed in the context image but not in the condition text. To extract the context, we construct two object trees from the condition text and context image, and compare them. For example, suppose the context image has the text descriptions "the young man in red" and "the old woman in blue". When the conditional text is "The man is sitting.", we want to clarify that "the man" indicates "the young man in red."

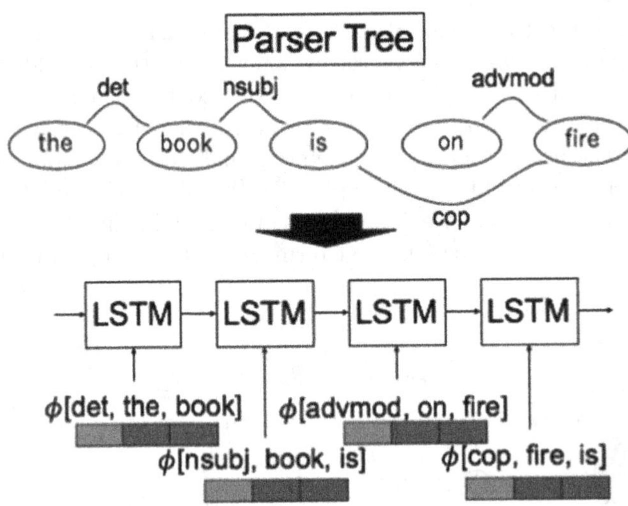

Fig. 3. Text encoder: an overview

An object tree represents the hierarchical relationship between the objects appearing in an image or text. Each node has a binary value denoting the appearance of that object (0: does not appear, 1: appears). The leaf nodes are the most concrete visual objects, and their ancestors are abstract objects that need more context to be detailed. The edges denote the relationships between two objects. Currently, we construct the object trees based on the object ontology of the VQA dataset [1].

[1] For details of these relationships, please refer to [2].

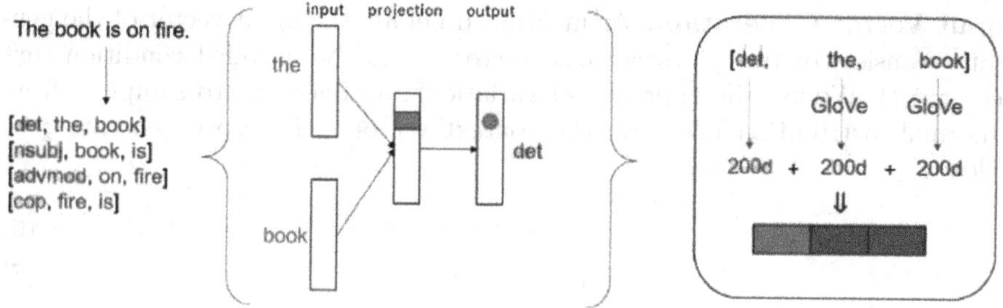

Fig. 4. Text encoder: learning distributed representations of relationships

At first, we construct the object trees o_t and o_i of condition text t and context images i, respectively. Currently, we use the tag information of context images to construct the object tree o_i. We detect the object names from the condition text and construct its object tree o_t. After that, we compare these two object trees to generate the context vector c.

- First, we compare the leaf nodes of o_t and o_i and find the candidate context nodes N with value 1 in o_i but 0 in o_t.
- Second, from N, we find the nodes M whose ascent node's value in o_t is 1.
- Third, we generate the context vector c with the same length as the leaf nodes. The value of the element in c is 1 if the corresponding leaf node is a member of M.

An example is shown in Fig. 5, in which the context vector $c = \{0, 0, 1, 0, 0\}$.

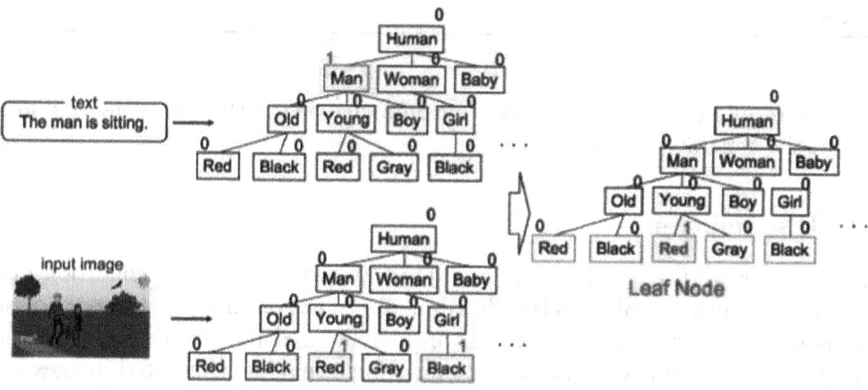

Fig. 5. Example of context extraction

Input Vector Generation. As mentioned before, the input vector of the generator consists of two parts: a noise vector z and the encoded condition text vector $\phi(t)$. We use the approach of variational autoencoder to sample z from a normal distribution based on the context vector c. Here, we generate z as follows:

$$z = \mu + \epsilon \Sigma, \tag{6}$$
$$\mu = fc1(c), \tag{7}$$
$$\Sigma = fc2(c), \tag{8}$$

where $\epsilon \sim N(0, 1)$; μ and Σ are the values given by two fully-connected layers ($fc1$ and $fc2$) with context vector c.

Algorithm 1. Training algorithm of the context discriminator

Input: minibatch images x_i , minibatch context images c_i, (image resolution: $2^i \times 2^i$)
Parameters: α: learning rate

for $i = 2$ to 6 **do**
  if $i = 2$ **then**
    Construct D_c by using CNNs (stride=2) with the sigmoid function. // D_c: Context Discriminator
  else
    Add new layers by using CNNs (stride=2) to the current D_c.
  end if
  for $n = 1$ to N **do**
    $L_{D_c} = H(c_i, x_i)$ // sigmoid cross entropy of c_i and x_i.
    $D_c \leftarrow D_c - \alpha \partial L_{D_c} / \partial D_c$
  end for
end for

As a result, the context encoder outputs the vector that consists of z and $\phi(t)$ to the generator.

3.3 Discriminator

In context-aware GANs, there are two discriminators: one is the condition discriminator, which is similar to the discriminator in the original conditional GANs and is used to distinguish real images from generated images; the other is the context discriminator, which is used to guarantee that the generated images are related to the context. The loss function of context L_{D_c} is defined as cross-entropy of the generated image and the contextual image.

As mentioned before, we train the generator and discriminators in a progressive manner for high-resolution images. This manner requires that we should compare the generated images of various resolutions with the contextual image of a fixed size. It makes the model hard to train. As a considerable solution, we resize the contextual image corresponding to the resolutions of generated images, and compare the images with the same size. In addition, the progressive manner helps us to generate high-resolution images by incrementally increasing the size. It means the content of generated image is decided at the early stages, and the later stages focus on increasing the resolutions. Therefore, it is possible that we only fire the context discriminator at the layers to generate low-resolution images and turn it off when the resolution becomes larger (128×128 in our study).

The method to train the context discriminator is shown in Algorithm 1. In lines 1–6, we add layers. In lines 7–10, we calculate the loss and update the model.

The training of context-aware GANs is done following the steps shown in Algorithm 2.

Algorithm 2. Training algorithm of context-aware GAN model

// G: generator, D: condition discriminator, D_c: context discriminator, I: fully-connected layer for the input context

Input: minibatch of images x_i (image size: $2^i \times 2^i$), minibatch of matching texts t, minibatch of mismatching texts $\widehat{t}$, minibatch of original context images c, minibatch of other context images c'

Parameters: α: learning rate, β, γ: the weights of context and Kullback-Leibler (KL) divergence.

Generate context vector v and v' from c and c', respectively.
for $i = 2$ to 8 **do**
    //training starts with the image resolution of $2^i \times 2^i$. for each resolution, the number of epochs is a hyperparameter.
    $\mu, \Sigma \leftarrow I(v)$ // calculate the mean μ and the variance Σ of the latent distribution based on the context c
    $z \leftarrow \mu + \epsilon\Sigma$ ($\epsilon \sim N(0,1)$) // generate noise z
    $\mu', \Sigma' \leftarrow I(v')$
    $z' \leftarrow \mu' + \epsilon\Sigma'$ // generate the fake noise z'
    $\widehat{x}_{ires} \leftarrow G(z,t)$ // generate image from t and z
    $\widehat{x}'_{ires} \leftarrow G(z',t)$ // generate image from t and z'
    $s_r \leftarrow D(x_i, t)$ // real image, right text
    $s_w \leftarrow D(x_i, \widehat{t})$ // real image, wrong text
    $s_f \leftarrow D(\widehat{x}_i, t)$ // fake image, right text
    $s'_f \leftarrow D(\widehat{x}'_i, t)$ // fake image (other context), right text
    $d \leftarrow D_c(\widehat{x}_i)$ // use D_c trained by Algorithm 1.
    $ce \leftarrow H(v, d)$ // sigmoid cross entropy of v and d.
    $L_I \leftarrow KL(\mu \| \Sigma)$ // loss function of I.
    $L_D \leftarrow s_r - (s_w + s_f)/2$ // loss function of condition discriminator
    $D \leftarrow D - \alpha\partial L_D/\partial D$ // update condition discriminator
    $L_G \leftarrow s_f + s'_f - \beta ce - \gamma L_I$ // loss function of generator
    $I \leftarrow I - \alpha\partial L_I/\partial I$ // update I
    $G \leftarrow G - \alpha\partial L_G/\partial G$ // update generator
end for

---

4 Experiments

To confirm that our proposed method can handle multimodal queries, we conducted comparative experiments. We compared three models as follows:

- C-GAN: the conditional GANs proposed by Reed et al. [12]. However, because the original conditional GANs could not handle multimodal queries, we only use the condition text part of the multimodal queries as the input to C-GANs. We also apply the progressive training mechanism to the original conditional GANs to train C-GAN.
- CA-GAN: the proposed context-aware GANs.
- SCA-GAN: a simple version of CA-GAN. In SCA-GAN, the noise parameter z is sampled from a normal distribution, and the input to the generator is a vector consisting of the noise z, the conditional text vector, and the context vector.

Table 1. Hyperparameters used in the experiments

Learning rate α	0.002
Weight of the context discriminator β	0.2
Weight of KL-divergence γ	1
Number of epochs of each resolution	600 epochs

Table 2. Reducing the batch size for higher image resolutions

Resolution	Batch size
4 × 4	256
8 × 8	256
16 × 16	128
32 × 32	64
64 × 64	32
128 × 128	16
256 × 256	8

We used the dataset of abstract scene images [1] to train the models. In this dataset, there are 20,000 images with metadata: each image has five explanatory sentences and tags with information about the objects appearing in the images. In this experiment, we generated 16 multimodal queries using 4 condition texts and 16 context images: each conditional text was complemented by four context images.

We trained these models using a machine with four GPUs of GTX 1080Ti. The hyperparameters are shown in Table 1. Moreover, to handle the trade-off of image resolutions and the available memory budget, we incrementally reduced the batch sizes when the resolution of generate images increased (see Table 2).

Because our model is trained incrementally by adding layers, this process is time-consuming. In our environment, which was mentioned above, training our model with max-resolution of 256 × 256 takes approximately 85 h.

4.1 User Study

We conducted a user study to evaluate the quality of the generated images from two aspects: their relevance to the given condition text, and their relatedness to the given context image. We invited seven graduate school students to check the

Fig. 6. Queries and generated images in the user study

generated images from these two aspects using a five-point scale: 5 and 1 denote the highest and lowest relevance or relatedness score, respectively.

For each multimodal query (a pair of condition text and context image), we generate three images and ask the evaluators to score them. The queries and corresponding outputs are shown in Fig. 6.

The average scores of our evaluators are shown in Table 3. SCA-GAN and CA-GAN achieved the best performance in relevance to condition text queries and relatedness to context images, respectively. Approximately 77% of the evaluators assigned higher scores of relevance to condition text (greater than or equal to 3) to SCA-GAN. Approximately 72% of the evaluators assigned higher scores (≥3) to CA-GAN for the aspect of relatedness to query images. Approximately 54% of the users agreed that CA-GAN was good at both relevance to query texts and relatedness to query images. These results demonstrate that our context-aware GANs can handle multimodal queries to generate images related to both text and image.

4.2 Accuracy of Discriminators

We compared the (conditional) discriminators of C-GAN, CA-GAN, and SCA-GAN. For this experiment, we sampled 100 images from our dataset as the positive results, and generated 100 images as the negative results. Then we used the discriminators of C-GAN, CA-GAN, and SCA-GAN to classify these images, respectively. The receiver operating characteristic (ROC) curves and area under the curve (AUC) scores are shown in Fig. 7.

Table 3. Experimental results of the user study

	C-GAN	SCA-GAN	CA-GAN
Average score of relevance to conditional text	2.85	**3.48**	3.01
Average score of relatedness to context images	2.47	3.09	**3.19**
Percentage of relevance score ≥3 to conditional text	0.50	**0.77**	0.70
Percentage of score of relatedness ≥3 to context images	0.46	0.69	**0.72**
Percentage of relevance score ≥3 and score of relatedness ≥3 to conditional text and images	0.23	0.50	**0.54**

As shown in Fig. 7, although our discriminator could not achieve the best performance, its AUC score is reasonably high. One of the reasons is that SCA-GAN and CA-GAN handle information from context images, and this makes the generated images less relevant to conditional text for C-GAN, which generates images only considering the conditional text.

4.3 Variation of Generated Images

To compare the images generated by SCA-GAN and CA-GAN, we constructed four multimodal queries: one conditional text and four different context images. We generated 28 images for each query. As shown in Fig. 8, the background information of outdoor/indoor is well-expressed in the generated images.

To confirm that the context affects the output images in our proposed method, the differences between the output images generated from different context images (C, $|C| = 4$) are calculated. In this study, we simply compared their pixels to calculate their difference as follows:

$$\mathbb{E}_{c,c' \sim P_{data}(c), c \neq c', c \in C, c' \in C} ||G(t,c) - G(t,c')||^2, \tag{9}$$

where c and c' denote two different context images in C, t is the conditional text, and $G()$ is the generator of our context-aware GANs.

The average values are shown in Table 4. This result shows that CA-GAN can generate more different images from different context images than SCA-GAN.

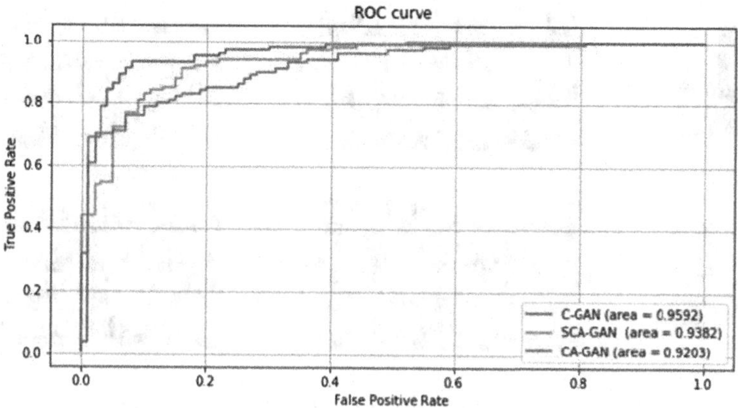

Fig. 7. ROC curves that show accuracy of discriminators

Table 4. Average difference scores of images generated from different context images with the same condition text

SCA-GAN	0.0012
CA-GAN	0.0041

Fig. 8. Output images of SCA-GAN and CA-GAN

5 Conclusion

In this paper, we proposed a novel model of context-aware GANs to generate images from multimodal queries. A multimodal query consists of a condition text and a context image. The condition text and context image can complement each other to present user's intentions in a more informative way. In our contextual GANs, we introduce a context decoder and a context discriminator to guarantee that the generated images are related to both the condition text and context image. Additionally, we train the generator and discriminator in a progressive manner to increase the resolution of images.

The experimental results on the public dataset demonstrate that the proposed model can generate images with high resolutions. The images generated by the proposed model are more relevant to the multimodal queries, and the variation of generated images is higher, compared with the baselines.

In future work, we will try to improve the context discriminator by utilizing the context encoder. Two major challenges are applying our model to other datasets and generating our context encoder.

References

1. Antol, S., Agrawal, A., Lu, J., Mitchell, M., Batra, D., Zitnick, C.L., Parikh, D.: VQA: visual question answering. ICCV **2015**, 2425–2433 (2015)
2. Chen, D., Manning, C.: A fast and accurate dependency parser using neural networks. EMNLP **2014**, 740–750 (2014)
3. Goodfellow, I., Pouget-Abadie, J., Mirza, M., Xu, B., Warde-Farley, D., Ozair, S., Courville, A., Bengio, Y.: Generative adversarial nets. In: NIPS 2014, pp. 2672–2680 (2014)
4. Hinton, G.E., Salakhutdinov, R.R.: Reducing the dimensionality of data with neural networks. Science **313**(5786), 504–507 (2006)
5. Karras, T., Aila, T., Laine, S., Lehtinen, J.: Progressive growing of GANs for improved quality, stability, and variation. CoRR abs/1710.10196 (2017). http://arxiv.org/abs/1710.10196
6. Ma, Q.: Utilization and analysis of user generated contents toward personalized and distributed sightseeing. Syst. Control Inf. **63**(1), 32–37 (2019)
7. Ma, Q.: Forefront of sightseeing informatics - technologies of collective intelligence for promotion of personalized and distributed sightseeing. Inf. Process. **58**(3), 220–226 (2017)
8. Zhuang, C.Y., Ma, Q., Liang, X.F., Yoshikawa, M.: Discovering obscure sightseeing spots by analysis of geo-tagged social images. ASONAM **2015**, 590–595 (2015)
9. Nakamura, K., Ma, Q.: Context-aware image generation by using generative adversarial networks. ISM **2017**, 516–523 (2017)
10. Pennington, J., Socher, R., Manning, C.D.: GloVe: global vectors for word representation. EMNLP **2014**, 1532–1543 (2014)
11. Radford, A., Metz, L., Chintala, S.: Unsupervised representation learning with deep convolutional generative adversarial networks. CoRR abs/1511.06434 (2015). http://arxiv.org/abs/1511.06434
12. Reed, S.E., Akata, Z., Yan, X., Logeswaran, L., Schiele, B., Lee, H.: Generative adversarial text to image synthesis. ICML **2016**, 1060–1069 (2016)
13. Teney, D., Liu, L., van den Hengel, A.: Graph-structured representations for visual question answering. CVPR **2017**, 3233–3241 (2017)
14. Vondrick, C., Pirsiavash, H., Torralba, A.: Generating videos with scene dynamics. NIPS **2016**, 613–621 (2016)
15. Zhang, H., Xu, T., Li, H.: StackGAN: text to photo-realistic image synthesis with stacked generative adversarial networks. ICCV **2017**, 5908–5916 (2017)

The experimental results on the [illegible] demonstrated that the proposed model [illegible] [illegible] [illegible] the multimodal query [illegible] of generated images [illegible] compared [illegible] the baselines.

[illegible]

References

[illegible]

Author Index

Printed in the United States
By Bookmasters